ISBN 978-1-334-00181-9
PIBN 10685928

# 1 MONTH OF
# FREE
# READING

at

## www.ForgottenBooks.com

By purchasing this book you are eligible for one month membership to ForgottenBooks.com, giving you unlimited access to our entire collection of over 700,000 titles via our web site and mobile apps.

To claim your free month visit:

www.forgottenbooks.com/free685928

(THE)

# JOURNAL OF BOTANY

## BRITISH AND FOREIGN.

Edited by

JAMES BRITTEN, F. L. S.,

BRITISH MUSEUM (NATURAL HISTORY), SOUTH KENSINGTON

VOL. XXVI.

·

ILLUSTRATED WITH PLATES.

LONDON:

WEST, NEWMAN & CO., 54, HATTON GARDEN.

1888.

LONDON :

WEST, NEWMAN AND CO., PRINTERS,

54, HATTON GARDEN, E.C.

# CONTRIBUTORS

TO THE PRESENT VOLUME.

C. C. Babington, M.A., F.R.S.
J. E. Bagnall, A.L.S.
L. H. Bailey.
E. J. Baillie, F.L.S.
J. G. Baker, F.R.S.
R. H. Beddome, F.L.S.
W. H. Beeby, A.L.S.
Arthur Bennett, F.L.S.
G. S. Boulger, F.L.S.
George Brebner.
T. R. Archer Briggs, F.L.S.
James Britten, F.L.S.
Elizabeth G. Britton.
N. L. Britton.
W. Carruthers, F.R.S., Pres. L.S.
C. B. Clarke, M.A., F.L.S.
W. A. Clarke.
Alphonse DeCandolle.
G. C. Druce, F.L.S.
T. B. Flower, F.L.S.
H. O. Forbes, A.L.S.
David Fry.
Alfred Fryer.
H. D. Geldart.
Edward Lee Greene.
W. B. Grove, B.A.
Henry Groves.
James Groves, F.L.S.
F. J. Hanbury, F.L.S.
W. B. Hemsley, A.L.S.
Tokitaro Ito, F.L.S.
B. Daydon Jackson, Sec. L.S.

Bolton King, M.A.
E. F. Linton, M.A.
W. R. Linton, M.A.
E. S. Marshall, M.A., F.L.S.
J. J. Marshall.
George Massee, F.R.M.S.
Maxwell T. Masters, M.D., F.R.S.
J. Cosmo Melvill, M.A., F.L.S.
H. W. Monington.
Spencer Le M. Moore, F.L.S.
Baron F. von Mueller, K.C.M.G., &c.
G. R. M. Murray, F.L.S.
R. P. Murray, M.A., F.L.S.
Percy W. Myles, B.A., F.L.S.
T. A. Preston, M.A., F.L.S.
John Rattray, M.A., B.Sc.
W. Moyle Rogers, F.L.S.
R. A. Rolfe, A.L.S.
F. C. S. Roper, F.L.S.
N. J. Schultz.
Reginald Scully.
A. Sharland.
Worthington G. Smith, F.L.S.
S. A. Stewart.
R. F. Towndrow.
William West, F.L.S.
F. Buchanan White, M.D., F.L.S.
William Whitwell.
C. H. Wright.

## Directions to Binder.

| | | | | |
|---|---|---|---|---|
| TAB. 279 | . . . . . | to face page | **1** |
| ,, 280 | . . . . . . | .. | 33 |
| ,, 281 | , . . . . . | | 97 |
| ,, 282 | . . . . . . | .. | 129 |
| Portrait of ASA GRAY . | . . . | ,, | 161 |
| TAB. 284 | . . . . . . | ,, | 333 |
| ,, 285 | . . . . . . | ,, | 353 |

Or all may be placed together at the end of the volume.

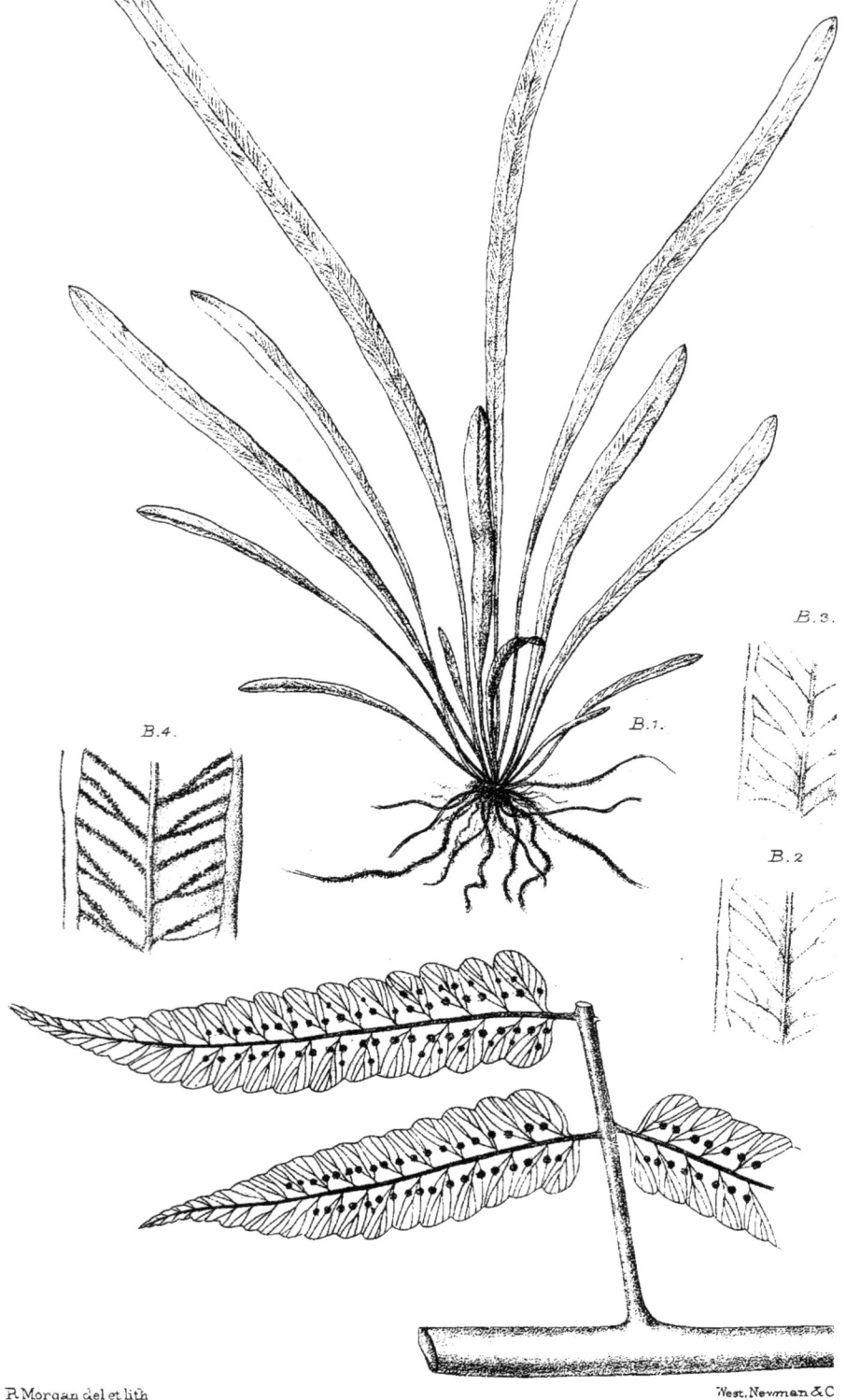

A. Alsophila dubia, *Bedd* B Gymnogramme Dayi, *Bedd*

THE

# JOURNAL OF BOTANY

## BRITISH AND FOREIGN.

## FERNS COLLECTED IN PERAK AND PENANG BY MR. J. DAY.

By Col. R. H. Beddome, F.L.S.

(Plate 279).

The species marked * are new to the Malay Peninsula.

*Gleichenia glauca* Hook.   Penang Hill, 1500 ft. elevation.
*G. flagellaris* Spr.   Penang and Perak, sea-level.
*G. dichotoma* Willd.   Perak, sea-level.
*Cyathea Brunonis* Wall.   Penang and Perak, up to 500 ft.
*Alsophila glauca* Bl.   Perak, 3000 ft.
*A. commutata* Mett.   Perak, 5000 ft.

*Alsophila dubia Bedd.   Stipes ? ; main rachis and rachis of pinnæ purple-brown, slightly furfuraceous above, glabrous below ; fronds subcoriaceo-membranaceous ; primary pinnæ 16–20 in. long, pinnate, with the apex only pinnatifid ; pinnules about 4 in. long by ⅝ in. broad on petioles 1–1½ lines long, more or less truncate at the base, much acuminate at the apex, pinnatifid only about one-sixth of the way to the costule, the very shallow lobes rather truncate ; costules scaly below or at length glabrous, furfuraceous above ; veins pinnate ; veinlets simple ; sori large, generally only in 1–2 rows, *i. e.*, on the 1–2 lower veinlets only, but sometimes in 3–4 rows, *i. e.*, on 3–4 veinlets and then near the base of the veinlets, and consequently parallel with the primary vein, and not shaped like an inverted V, as in *glabra*. Perak, 4000 ft. elevation. It is more allied to *podophylla* of Hooker than to *glabra*, and there is a specimen of it in the Kew packet of *podophylla* gathered by Curtis in Java ; which Mr. Baker now thinks distinct from the Chinese plant. Pl. 279 A.

*A. latebrosa* Hk.   Perak, 4000–5000 ft.
*A. Kingi* Clarke.   Perak, 1500 ft.
*A. obscura* Scortechini.   Perak, on the hill called Idjo, behind Taepeng, 4000–5000 ft.
*Dicksonia (Cybotium) Barometz* Link.   Perak, 1500 ft.

*Lecanopteris carnosa* Bl.  Perak, open places, Mt. Idjo, rare, 5600 ft.  Mr. Day states that ants live in the rhizome.

\*Hymenophyllum rarum R. Br.  Perak, Maxwell's Hill, 4000 ft.

*H. polyanthos* Sw.  Perak, 4000 ft.— Var. *Blumeanum.*  Perak, 4000 ft.

*H. javanicum* Spr.  Perak, 5000 ft.

*H. javanicum* Spr.. var *badium.*  Perak, 3000–4000 ft.

*H. Smithii* Hook.  Perak, 4000–5000 ft.

*H. Neesii* Hook.  Perak, 5000 ft.

*H. aculeatum* Van den Bosch.  Penang, 3000 ft.—(*sabinæfolium* Baker).  Perak, 4000 ft.

*Trichomanes parvulum* Poir.  Perak, 4000 ft.

*T. digitatum* Sw.  Perak.

*T. pallidum* Bl.  Perak, 5000 ft.

*T. bipunctatum* Poir.  Perak, 3000 ft.

*T. pyxidiferum* L.  Perak, 4000 ft.

*T. auriculatum* Bl.  Perak, 3000 ft.

*T. javanicum* Bl.  Perak, 500–1500 ft.

*T. rigidum* Sw.  Perak, 5000 ft.

*T. maximum* Bl.  Perak, 4000 ft.

*Davallia (Humata) angustata* Wall.  Perak and Penang, 3000 ft.

*D.*          ,,          *pedata* Sm.  Perak, 5000 ft.

*D. (Leucostegia) affinis* Hk.  Perak, 3000 ft.

*D. (Prosaptia) Emersoni* Hk. & Grev.  Perak & Penang, 3000 ft.

*D.*          ,,          *contigua* Sw.  Perak, 3000 ft.

*D. (Eudavallia) solida* Sw.  Perak, sea-level.

*D.*          ,,          *elegans* Sw.  Perak, sea-level.

*D.*          ,,          *epiphylla* Bl.  Perak, 3000 ft.

*D.*          ,,          *divaricata* Bl.  Perak, 2500 ft.

*D.*          ,,          *bullata* Wall.  Perak, 4000–5000 ft.

*D. (Microlepia) pinnata* Cav.  Perak and Penang, 3000 ft.

*D.*          ,,          *moluccana* Bl.  Perak, 4000 ft.

*D.*          ,,          *Speluncæ L.,* var.  Perak, 1500–3000 ft.

*D. (Stenoloma) tenuifolia* Sw.  Perak, sea-level.

*Lindsaya scandens* Hk.  Perak, 3000 ft.

*L. repens* Thw.  Perak, 3000–5000 ft.

*L. orbiculata* Lam.  Perak, 3000 -4000 ft.

*L. Lancea* L.  Perak, 1000–3000 ft.

*L. divergens* Wall.  Penang and Perak, 2000–3000 ft.

*Adiantum lunulatum* Burm.  Perak, sea-level.

*A. caudatum* L.  Perak, sea-level.

*Pteris longifolia* L.  Perak and Penang, sea-level.

*P. pellucida* Presl.  Perak, 3000 ft. (rare).

*P. patens* Hk.  Perak, 2500 ft.

*P. aquilina* L.  Perak, 5000 ft.

*P. (Campteria) biaurita* L.  Perak, 3000 ft.

*P. (Litobrochia) incisa* Thunb.  Perak, 3000 ft.

*P.*          ,,          ,, var. with entire pinnules.  Perak, 3000 ft.

*P.*          ,.          *marginata* Bory.  Perak, up to 3000 ft.

*Blechnum orientale* L.  Perak, sea-level.

*B. Finlaysonianum* Wall.  Perak, up to 1000 ft.

*Asplenium* (*Thamnopteris*) *Nidus* L.  Perak, up to 3000 ft., common.
*A. squamulatum* Bl.  Perak, near Taepeng, at no elevation.
*\*A. Scortechinii* Bedd.†  Perak, Caulfield's Hill, 3000–4000 ft.
*A. amboinense* Willd.  (*Fejeense* Syn. Fil.).  Perak, 3000–4000 ft.
*A. normale* Don.  Perak, 1500 ft.
*A. longissimum* Bl.  Perak, sea-level.
*A. tenerum* Först.  Perak, sea-level.
*A. borneense* Hk.  Perak, 3000 ft. (very rare).
*A. hirtum* Kaulf.  Perak, 1000–3000 ft.
*A. macrophyllum* Sw.  Perak, sea-level (rare).
*A. resectum* Smith.  Perak, 4000 ft.
*A. affine* Sw.  Perak, 3000 ft.
*A. nitidum* Sw.  Perak, 4000–5000 ft.
*A. Belangeri* Kze.  Perak, 3000–5000 ft., common.
*A.* (*Diplazium*) *subserratum* Bl.  Perak, 3000 ft.
*A.*       „      *pallidum* Bl.  Perak, 2000 ft.
*A.*       „      *porrectum* Wall.  Perak, 2000 ft.
*A.*       „      *bantamense* Baker.  Perak and Penang, 2000 ft.
*A.*       „      *sylvaticum* Presl.  Perak, 3000 ft.
*A.*       „      *tomentosum* Hk.  Perak, 3000–4000 ft.
*A.*       „      *speciosum* Mett.  Perak, 3000–4000 ft.
*A.*       „      *sorzogonense* Presl.  Perak, up to 2000 ft.
*A.* (*Anisogonium*) *cordifolium* Mett.  Perak, 5000 ft.
*A.*       „      *lineolatum* Mett.  Perak, 3000 ft.
*\*A.*       „      *decussatum* Sw.  Perak, in dense forests on Birch's Hill, 4000 ft.
*A.* (*Anisogonium*) *esculentum* Presl.  Perak, 1000–1500 ft.
*Didymochlœna lunulata* Desv.  Perak, Maxwell's Hill, 3000 ft.
*D. polycarpa* Bl.  Perak, Maxwell's Hill, 2000–3500 ft.
*Aspidium* (*Polystichum*) *aculeatum* Sw., var. *biaristatum*.  Perak, 3000 ft.
*A.* (*Euaspidium*) *singaporianum* Wall.  Perak, sea-level.
The rhizome is erect, not creeping, as described in Syn. Fil.
*A.*       „      ,      *polymorphum* Wall.  Perak, 2000 ft.
*A.*              *decurrens* Presl.  Perak, sea-level.
*A.*       „      ,:     *pachyphyllum* Kze.  Perak, sea-level.
*\*A.*       „      „      *multicaudatum* Wall.  Perak, at the Taepeng Waterfall, at no elevation.

---

† *Asplenium Griffithianum* does not appear to occur in Perak or Penang; the plant referred to at page 193, Syn. Fil., as having been collected by Mactier in Penang is an undescribed species, and as Mr. Baker has allowed me to describe it, I annex a description :—

Asplenium Mactieri, n. sp. — Caudex small, erect, scaly; scales dark brown, with a paler margin, lanceolate from a broad base, finely acuminated; stipes 6—9 in. long, pale yellow, whitish at the base; fronds 6—9 in. long, 1—1¼ in. broad, gradually narrowed below, gradually and finely acuminated at the apex, the margin crenate or serrate; texture subcoriaceous; veins distinct, usually once forked from near the base, occasionally again forked towards the apex, not quite reaching the margin; sori reaching from the midrib two-thirds towards the margin; indusium broad and very prominent.—Penang, gathered by Mactier. Allied to *Griffithianum*, but with a long slender stipe, rather more coriaceous in texture, and of a paler colour.

*A.* (*Lastræa*) *gracilescens* Hk.   Perak, 1500 ft.
*A.*     „     *immersum* Hk.   Perak, 3000 ft.
*A.*     „     *calcaratum* Hk.   Perak, near Taepeng, 2000 ft.
*A.*     „     *viscosum* Baker.   Perak, 5000 ft.
*A.*     „     *crassifolium* Hk.   Perak, 2000 ft.   The rhizome is wide-creeping.
*A.*     „     *Dayi* Bedd. in Scortechini's List.   Perak, Maxwell's Hill, 3000 ft.
*A.*     „     *fuscipes* Wall.   Perak.
*A.*     „     *syrmaticum* Willd.   Perak, 1500 ft.
*A.*     „     *Filix-mas* Rich., var. *elongata*.   Perak, Maxwell's Hill, 3000 ft.
*A.*     „     *flaccidum* Hk.   Perak, Maxwell's Hill, 3000 ft.
*A.* (*Nephrodium*) *unitum* R. Br.   Perak, sea-level.
*A.*     „        *aridum* Don.   Perak, sea-level.

\***Aspidium** (Nephrodium) **perakense** Bedd.   Caudex small, erect.   Stipes slender, villous; fronds pinnate, 12–14 in. long, 2–5 in. broad, oblong-lanceolate; central pinnæ the largest, lower ones gradually reduced to auricles, pinnæ $1\frac{1}{2}$–$2\frac{1}{2}$ in. long by $\frac{3}{8}$–$\frac{1}{2}$ in. broad, cut down about halfway to the rachis with close rather pointed lobes, texture softly herbaceous; veins 4–5 on each side, simple, the lower pair anastomosing with a long excurrent veinlet; stipes, rachis, and both surfaces of the frond copiously furnished with long whitish soft hairs; sori near the apex of the veins. Perak, Birch's Hill, 4000 ft., on exposed rocks.   In outline similar to *Lastrea Beddomei*, but of a very soft texture and densely hairy.

\**A.* (*Nephrodium*) *glandulosum* J. Sm.   Perak, sea-level.
*A.*     „       *eminens* Baker, Journ. Bot. 1880.   Perak and Penang Hill, 3000 ft.   Wall. cat. 353 in Linn. Herb.   3rd sheet from Penang is this plant, upper part of frond only.
*A.*     „       *pennigerum* Bl., var.   Perak, 3000 ft.
*A.*     „       *molle* Desv.   Perak, everywhere.
\**A.*     „       *ferox* Moore.   Perak, 3000–4000 ft.
\**A.*     „       *sagittæfolium* Moore.   Perak, 2000 ft., indusium often quite didymochlænoid.
*A.*     „       *truncatum* Presl.   Perak, 3000 ft.
*Nephrolepis exaltata* Schott.   Perak, sea-level.
*N. biserrata* Schott.   Perak, sea-level.
*N. acuminata* Hout.   Perak, 3000–5000 ft.
*Oleandra neriiformis* Cav.   Perak, 5000 ft.
*Polypodium* (*Phegopteris*) *punctatum* Thunb.   Perak, 4000 ft.
*P.* (*Dictyopteris*) *difforme* Bl.   Perak, sea-level.
*P.* (*Eupolypodium*) *subevenosum* Baker.   Penang Hill, 3000 ft.
*P.*     „       *sessilifolium* Hk.   Penang Hill, 3000 ft.; this and the last run one into the other, and are, I believe, only one species.
*P.*     „       *triangulare* (Scortechini).   Perak, Mt. Idjo, 5000 ft.
*P.*     „       *cornigerum* Baker.   Perak, Mt. Idjo, 5000 ft.
*P.*     *Khasyanum* Hk.   Perak, 4000 ft.
*P.*     „       *obliquatum* Bl.   Perak, 3000–4000 ft.
*P.*     „       *subfalcatum* Bl.   Perak, 3000–4000 ft.

*P.* (*Eupolypodium*) *fuscatum* Bl.   Perak, 3000–4000 ft.
*P.*   „   *decorum* Brack.   Penang Hill, 3000 ft.
*P.*   „   *papillosum* Bl.   Perak, 3000 ft.
*P.*   „   *tenuisectum* Bl.   Perak, 4000 ft.
*P.* (*Goniophlebium*) *verrucosum* Wall.   Perak, 4000 ft.
*P.* (*Niphobolus*) *adnascens* Sw.   Perak, sea-level.
*P.*   „   *acrostichoides* Forst.   Perak, sea-level.
*P.* (*Pleopeltis*) *accedens* Bl.   Perak. 3000 ft.
*P.*   „   *stenophyllum* Bl.   Perak, 5000 ft.   Very variable in breadth, and *P. Morgani* Zeiller (Bulletin de la Soc. Bot. de France) is only a synonym.
*P.*   „   *sinuosum* Wall.   Perak, sea-level.   Ants live in the rhizome.
*P.* (*Pleopeltis*) *longifolium* Mett.   Perak, 3000 ft.
*P.*   „   *superficiale* Bl.   Perak, 3000 ft.
*P.*   „   *angustatum* Sw.   Perak, 500 ft. ; Penang, 3000 ft.
*P.*   „   *rupestre* Bl.   Perak, 4000 ft.
*P.*   „   *platyphyllum* Sw.   Perak, 3000 ft.
*P.*   „   *Wrayi* Baker.   Perak, Mt. Idjo, on trees, 5000 ft.
*P.*   „   *irioides* Lam.   Perak, 3000 ft.
*P.*   „   *hemionitideum* Wall.   Perak, sea-level.
*P.*   „   *musæfolium* Bl.   Perak, 3000 ft.
*P.* (*Dipteris*) *Dipteris* Bl.   Perak, 5500 ft. ; Penang, 3000 ft.
*P.* (*Pleopeltis*) *laciniatum* Bl.   Perak, 5500 ft.   (= *P. macrochasma* Baker in Hook. Icones, tab. 1675).   Sori generally much sunk and forming pustules on the upper side of the frond, but in some fronds it is not so evident.
*P.*   „   *incurvatum* Bl.   Perak, 5000 ft.
*P.*   „   *Phymatodes* L.   Perak, sea-level.
*P.*   „   *nigrescens* Bl.   Perak, sea-level.
*P.*   „   *palmatum* Bl.   Perak, 3000 ft., and Penang Hill.
*P.* (*Drynaria*) *Heracleum* Kze.   Perak, 3000–4000 ft.
*P.*   „   *quercifolium* L.   Perak and Penang, low levels.
*P.*   „   *rigidulum* Sw.   Perak, 3000 ft.

*Gymnogramme** (Syngramme) **Dayi** Bedd.   Rhizome creeping, somewhat fibrillose; fronds narrow-linear, subentire or very obscurely crenated, 4–7 in. long by 1½ line broad, gradually attenuated at the base, but on a distinct stipe 1–1½ in. long; texture in age coriaceous; veins obscure in the old fronds, evident in the young, simple or once forked, the apices running into a submarginal transverse continuous or interrupted vein ; sori narrow-linear, thread-like, the length of the veins.   Perak, on quartz rocks, the pass between Kinala-Kansa and Kinta, about 2000 ft.   The affinity of this interesting little fern is with *G.* (*Syngramme*) *borneensis* Hook., though apart from its sori it has quite the aspect of a *Eupolypodium*.   Pl. 279 B.

*G.* (*Stegnogramme*) *aspidioides* Kze.   Perak, sea-level.   Var. with fertile fronds contracted; perhaps a new species.
*G.* (*Selliguea*) *Wallichii* Hk.   Penang, 3000 ft. ; Perak, 1500 ft.
*G.*   „   *involuta* Hk.   Perak, up to 2500 ft.
*G.*   „   *Feei* Hk.   Perak, sea-level.
*G.*   „   *Hamiltoniana* Hk.   Perak, 4000 ft.

*Meniscium salicifolium* Wall.  Perak rivers, up to 1000 ft.
*M. cuspidatum* Bl.  Perak, sea-level.
*Antrophyum plantagineum* Kaulf.  Perak, 3000 ft.
*A. reticulatum* Kaulf.  Perak, 2000 ft.
*Vittaria elongata* Sw.  Perak, 3000 ft.
*V. lineata* Sw.  Perak, 3000 ft.
*Vittaria scolopendrina* Thw.  Perak, 2000–3000 ft.
*Tænites blechnoides* Sw.  Perak and Penang, 3000 ft.
*Drymoglossum piloselloides* Presl.  Perak, sea-level.
*Acrostichum (Stenochlæna) sorbifolium* L.  Perak, 3000 ft.
*A.*          „          *palustre* L.  Perak, sea-level.  (The abnormal state, *Davallia achilleifolia* Wall. is also common).
*A. (Polybotrya) appendiculatum* Willd.  Perak, 4000 ft.
*A. (Gymnopteris) variabile* Hk.  Perak, 4000 ft.
*A.*          „          *subrepandum* Hk.  Perak, 4000 ft.
*A.*          „          *spicatum* L.  Perak, 3000–4000 ft.—Var. *latifrons.*  Perak, 3000–4000 ft.  Fronds 2 ft. 6 in. long, and $2\frac{1}{2}$ in. broad; the fertile apex 7–8 in.
\*A. *(Chrysodium) Blumeanum* Hk.  Perak, 4000 ft.
*A. aureum* L.  Perak, sea-level.
*A. (Photinopteris) rigidum* Wall.  Perak, 3000 ft.
*A.*          „          *drynarioides* Hk.  Perak, 2000 ft. (on tops of highest trees).
*Platycerium biforme* Bl.  Perak, sea-level.
*Lygodium scandens* Sw.  Perak, sea-level.
*L. flexuosum* Sw.  Perak, 500 ft.
*Angiopteris evecta* Hoffm.  Perak, sea-level.

EXPLANATION OF PLATE 279.—A. *Alsophila dubia,* a small portion of a pinna, life-size.  B 1. *Gymnogramme Dayi,* life-size.  B 2. venation showing a continuous intramarginal transverse vein.  B 3. The intramarginal vein interrupted. B 4. The lineal sori.

# THE NOMENCLATURE OF *NYMPHÆA*, &c.

## By James Britten, F.L.S.

Those who are working at synonymy on the only sound principle—that of priority—and who are anxious that the necessary changes should be made as promptly and as thoroughly as possible, will be interested in the latest discovery made and published by Mr. E. L. Greene in the 'Bulletin of the Torrey Botanical Club' for September last.  It is to be feared, however, that every one will not thank him for having demonstrated, as he has done, that the names of our two best known genera of *Nymphæaceæ* must be readjusted, and that in a manner which will cause some temporary inconvenience.

The discovery, as I have said, is due to Mr. Greene, but as in his paper he states that he has not seen all the books which establish it, and as the evidence which they afford is even stronger than he was aware of, I think it may be well to deal with it more fully in these pages.

Salisbury published in the 'Annals of Botany' (ii. 69—76) a "Description of the Natural Order of Nymphæēæ," in which he divided the Linnean genus *Nymphæa* into two. For one of these he retained the Linnean name; the other he styled *Castalia*. The volume is dated 1806; but internal evidence shows that this first part was issued in 1805. In the same year William Hooker published in the 'Paradisus' the plate lettered *Castalia magnifica* and dated Oct. 1, to which Salisbury supplied the letterpress.

In 1808 (or 1809), Smith (Fl. Græe. Prodr. i. 361) adopted Salisbury's division of the Linnean genus *Nymphæa*, but did not follow Salisbury's nomenclature. He restricts the name *Nymphæa* to Salisbury's *Castalia;* while he bestows upon the yellow-flowered species, for which Salisbury retained the name *Nymphæa*, a new name, *Nuphar*. He cites *Castalia* as a synonym of *Nymphæa* and says (under *Nuphar*), "Has, Nymphæam albam et luteam Linnæi, characterum ope in *Eng. Bot.* et *Fl. Brit.* evulgatorum ascitis insuper nectariis, in duo genera felicitèr disposuit D. Salisbury; at minus benè Nymphæam antiquorum veram, nomine, Castalia, ad novem et planè abnormem etymologiam formato, distinxit."*

The title-page of the 'Prodromus' is dated 1806; but, as Mr. Greene points out (Bull. Torr. Bot. Club, Dec., 1887, p. 257), the last part, containing *Nuphar*, did not appear until the end of 1808, or, more likely, the beginning of 1809. This is clear from the correspondence which took place between Goodenough and Smith regarding the name, extending from Nov. 17th to Dec. 14th of the former year.† Smith had already recognized the correctness of Salisbury's division of *Nymphæa*: "I believe," he says, "Mr. Salisbury's *Castalia* is well separated from *Nymphæa*"‡; and he writes to Goodenough (without, it must be admitted, displaying any animus against Salisbury personally), stating his wish to retain *Nymphæa* for the showy-flowered species, and to adopt *Blephara* for the yellow-flowered ones: he gives classical reasons for this course, which need not be referred to here. Goodenough, who certainly came as near hating Salisbury as a bishop could well do,§ promptly settles Smith's scruples. "You *must* and you do reject Salisbury's *Castalia* upon irrefragable [*i. e.*, on classical] ground‖": and he adds, by way of quieting any qualms of conscience which Smith may have had, "In your Introduction, you have pledged yourself, not to the name *Castalia*, but merely to the separation from *Nymphæa*."

Planchon (Ann. Sc. Nat. 3rd ser. xix. 59) demolishes the position of Smith and Goodenough in adopting the name *Nuphar*

---

* "Quasi ob pudicitiam, uterum totum petalis occultaut species hujus-generis, itaque *Castalias* dixi." Salisb in Ann. Bot. ii. 72.

† Mem. and Correspondence of Sir J. E. Smith, i. 576—582.

‡ Introd. to Botany (1807), 385.

§ See Smith's 'Correspondence,' i. 557, 575, 578, 587.

‖ In this latter case, L. C. Richard's name *Nymphosanthos*, proposed by him (Anal. du fruit, p. 68 (May, 1808) for the yellow-flowered species, in ignorance that Salisbury had already separated them, would take precedence of *Nuphar*.

on classical grounds; and shows a just appreciation of the claims
of Salisbury in connection with the establishment of the genus.
" Smith n'a en qu' à lui donner le nom aujourd'hui généralement
adopté. . . . On doit blâmer Smith d'avoir, probablement par
esprit d'antagonisme contre l'ingénieux Salisbury, bouleversé à
plaisir la nomenclature proposée par ce dernier botaniste. Il est
trop tard sans doute," he continues, though here I cannot follow
him, " pour revenir sur cette injustice qui fut en même temps une
maladresse: les termes resteront comme ils sont, à cause que
l'usage les a consacrés, mais on saura du moins de quel côté se
trouvaient le droit et la raison."

In Engl. Bot. t. 2292, published June 1, 1811, Smith established
*Nuphar minima* as a species, and says :—

" We take advantage of it to establish in our work the genus
*Nuphar*, first adopted from Dioscorides, in Prodr. Fl. Græc. v. i.
361, which embraces the yellow kinds of water-lily, and is clearly
distinguished by the above characters from the true *Nymphæa*
of that ancient author, to which the white and rose-coloured
kinds belong, as will appear in the new edition of the valuable
' Hortus Kewensis.' Mr. Salisbury determined that the nectary of
these last is a globe in the centre of the stigmas, while that of
*Nuphar* is at the back of the petals."

The arbitrary action of Smith cannot, of course, be defended,
although, as has been shown, Goodenough must take a large share
of the responsibility; and another tardy act of reparation will result
in the following restitution of Salisbury's names. It is certainly
inconvenient, for the time being, that we should have to style
*Nymphæa* what we have been accustomed to call *Nuphar*, while the
plants we know as *Nymphæa* will become *Castalia*; but this must
take place sooner or later, and it may as well be sooner.

The two genera will stand thus :—

CASTALIA Salisb. Ann. Bot. ii. = *Nymphæa*, Linn., in part; Smith,
    71. (1805); ' Paradisus,' t. 14      Prodr. Fl. Græc. i. 360
    (1805).                              (1808-9 ?).

NYMPHÆA Linn., in part; Salisb. = *Nuphar* Sm. Prodr. Fl. Græc. i.
    Ann. Bot. ii. 71 (1805).            361 (1808-9 ?).

Salisbury's two contributions above quoted seem to have
appeared almost simultaneously; but that in the ' Annals' takes
precedence, as is shown by his reference to the ' Paradisus.'

The following is the correct nomenclature so far as the species
enumerated by Salisbury are concerned. The transference of the
remaining species of *Nuphar* to *Nymphæa* (Linn., Salisb.) and of
those of *Nymphæa* (Linn., Smith) to *Castalia* is easy; but I do not
think it desirable that it should be made by one who has not worked
at the genus.

NYMPHÆA Linn., Salisb. = NUPHAR Smith (1808-9 ?).
NYMPHÆA LUTEA Linn. (*N. umbi-* = *Nuphar lutea* Sm.
    *licalis* Salisb.).

NYMPHÆA ADVENA [Soland. in] = *Nuphar advena* [Br. in] Ait.
Ait. Hort. Kew, ed. 1, ii. 226    Herb. Kew, ed. 2, iii. 295
(1789). *N. arifolia* Salisb.    (1811).

NYMPHÆA SAGITTIFOLIA Walt. Fl. = *Nuphar sagittæfolia* Pursh. Fl.
Car. 155 (1788).    Bor. Amer. ii. 370 (1814). *N. longifolia* Sm.

NYMPHÆA PÙMILA Hoffm.Deutschs.= *Nuphar minima* Sm. Engl. Bot. t.
Flora (1800), 241.    2292 (1811).

NYMPHÆA MICROPHYLLA Pers. En- = *Nuphar Kalmiana* [Br. in] Ait.
chirid. ii. 63 (1807). *N. Kal-*    Hort. Kew, ed. 2. iii. 295 (1811).
*miana* Sims, Bot. Mag. t.
1243 (1809).

      CASTALIA Salisb. = NYMPHÆA Linn. (in part), Smith.

CASTALIA PUDICA Salisb.    = *Nymphæa odorata* [Dryand. in]
    Ait. Hort. Kew, ed. 1, ii. 227
    (1803). [Kennedy in] Andr.
    Bot. Rep. t. 297 (same year.)

CASTALIA SPECIOSA Salisb.    = *Nymphæa alba* Linn.

CASTALIA SCUTIFOLIA Salisb.    = *Nymphæa cærulæa* [Kennedy in]
    Andr. Bot. Rep. t. 197 (Dec.
    1801), Dryand. in Bot. Mag. t.
    552 (Feb. 1802).

CASTALIA STELLARIS Salisb.    = *Nymphæa stellata* Willd. Sp. Pl.
(excl. plant. Austral.)    ii. 1153.

CASTALIA AMPLA Salisb.    = *Nymphæa ampla* DC. Syst. i. 54.

CASTALIA EDULIS Salisb.    = *Nymphæa edulis* DC. Syst. i. 52.

CASTALIA MAGNIFICA Salisb. Parad. = *Nymphæa rubra* " Roxb. MSS."
t. 14.    Andr. Bot. Rep. t. 503.

CASTALIA PYGMÆA Salisb. Parad. = *Nymphæa tetragona* Georgi, Reise
t. 68.    im Russ. Reich. i. 220 (1775);
    *N. pygmæa* Ait. Hort. Kew,
    ed. 2, iii. 293 (1811) et auct.

CASTALIA GIGANTEA (*C. stellaris* = *Nymphæa gigantea* Hook. Bot.
Salisb. Parad. Lond. quoad pl.    Mag. t. 4647.
Austral.)

There is one case in which some little difficulty as to synonymy arises. Salisbury in Ann. Bot. gives as the synonymy of his *Castalia mystica*—" *N. Lotus* Sims in Bot. Mag. n. 797, cum ic. *N. Lotus* Blandf. in Bot. Rep. n. 391, cum ic. *N. Lotus* Pl. Rar. Hung. v. 1, p. 13, t. 15. *Ambel* Rheed. Hort. Mal. v. ii. p. 51, t. 26. *Lotus* Alp. Exot. p. 214, cum figuris." In ' Paradisus,' how-ever, these citations, with others, are divided between two species, thus :—

CASTALIA MYSTICA Salisb. = "*Nymphæa Lotus* Sims in Bot. Mag.
p. 797, cum ic. *Nymphæa Lotus* Pl. Rar. Hung. v. i. p. 13.
f. 15. *Nymphæa Lotus* Willd. Sp. Pl. v. 2, p. 1153. *Nymphæa Lotus* Savign. in Ann. du Mus. v. 1, p. 366. *Nymphæa Lotus* Hasselq. Rcs. p. 471. *Lotus*, &c., Alp. Exot. p. 214, cum figuris."

He adds: " Though I have followed others in quoting the above synonyms, I am not absolutely certain that the Hungarian plant here taken up is the same with the Egyptian Lotus"; and De Candolle (Syst. i. 54), when establishing this as a distinct species, refers to this note of Salisbury's.

Salisbury's second species in the ' Paradisus ' is :—

CASTALIA SACRA Salisb. = " *Nymphæa Lotus* Marquis of Blandf. in Bot. Rep. n. 391, cum ic.   *Nymphæa pubescens* Willd. Sp. Pl. v. 2, p. 1154.   *Nymphæa Lotus* Roxb. MSS.   *Ambel* Rheed. Ital. Mal. v. 11, p. 51, f. 26."

It would seem that three plants were included in *C. mystica* of Ann. Bot., and that the names of these should stand as follows :—*

C. MYSTICA Salisb. = *C. mystica* Salisb. Ann. Bot. and ' Paradisus ' (in part) ; *Nymphæa Lotus* L., et auct. plur. (the Egyptian and African plant).

C. SACRA Salisb. = *C. mystica* Salisb. Ann. Bot. (in part); *Nymphæa pubescens* Willd. ; *N. Lotus* Roxb. Rep. 391 (the Indian plant).

C. THERMALIS = *C. mystica* Salisb. Ann. Bot. and 'Paradisus' (in part); *Nymphæa thermalis* DC.; *N. Lotus* Waldst. & Kit. (the Hungarian plant).

It may be remarked that, had the many writers who have recognised the priority of Salisbury's names, or have at least quoted them as synonyms, boldly faced the necessities of the case and restored them, our present inconvenience would have been largely obviated.   S. F. Gray's attempt to settle the matter is worth a note : he seems to have been anxious to recognise Salisbury's work, and at the same time to avoid changing Smith's name; so he boldly writes (Nat. Arr. Brit. Pl. ii. 706), " *Nuphar*, R. A. Salisbury," with the species, " *Nuphar luteum* Salisbury, Ann. Bot. 2, 69," and " *Nuphar minima*, Salisbury, Ann. Bot. 2, 69."   The latter species was not even *known* to the author to whom Gray attributes it!

If the date of the publication of *Anneslea* (Andr. Roxb., not of Wallich) given by Pfeiffer (Nomencl. p. 198) were correct, Salisbury's genus *Euryale* would have to give place to it.   But Pfeiffer's date, 1804, is certainly wrong; Andrews's plate (t. 618) is not dated, but t. 627 bears date Dec. 1, 1810 : *Euryale* dates from 1805. Roxburgh's name existed in MS. before this—at least so we may assume from Salisbury's remark (Parad. t. 64), "In the Annals of Botany I have called a plant, not before described by any botanist, *Euryale*, being ignorant that the friends of Lord Viscount Valentia in Hindostan had selected it to perpetuate his memory."   He goes on to say—"I am happy therefore now to offer him one from the same country," and establishes his genus *Anneslia*, the restitution of which will lead to further revolutions ; for not only must this supersede Bentham's *Calliandra* established for the same plant

---

* Caspary, however (Ann. Mus. Lugd. Bat. ii. 248), unites the three under *Nymphæa Lotus* L. (*C. mystica* Salisb.).

thirty-three years later (in 1840), but the Wallichian genus *Anneslea* (1824) must take another name, unless it is urged that the difference in termination is enough to justify the retention of the two names.

The authors of the ' Genera' were no doubt in this, as in so many other cases, actuated by a fear of the inconvenience which would arise from the restitution of the correct name. They were perfectly conscious that the name existed, for *Calliandra* stands thus in Gen. Pl. i. 596:—"CALLIANDRA Benth. in Hook. Journ. Bot. ii. 138 (*Anneslea* Salisb. Parad. Lond. t. 64, non Wall.)" And if we turn to Bentham's foundation of his genus, we read— " I propose the name of *Calliandra* for the genus indicated by DeCandolle under *Inga anomala* as the *Anneslea* of Salisbury, a name applied by Dr. Wallich to a very different East Indian genus." DeCandolle (Prodr. ii. 442), speaking of *Inga anomala?* Kunth. = *Anneslia grandiflora* Salisb., says, —"An genus cum 2 sequentibus et forsan præcedente proprium (*Anneslea* [*sic*] Salisb.) admittandum ? "

The case, then, is quite clear, and stands thus :—

ANNESLIA Salisb. Parad. t. 64 = CALLIANDRA Benth., Hook. (1807).                          Journ. Bot. ii. 138 (1840).
A. GRANDIFLORA Salisb. *l. c.*        = *C. grandiflora* Benth. *l. c.*

The other species of *Calliandra* (79 according to Gen. Pl.) must follow suit; but this must be the work of a future monographer.

It follows from what has been said that Wallich's genus *Anneslea* must be reduced to the rank of a synonym. Its author distinctly states (Plant. Asiat. Rar. i. 65) that he created it on the understanding that the plant so named by Roxburgh and Salisbury had been relegated to *Euryale* and *Acacia* respectively ; now that Salisbury's genus is allowed to stand, Wallich's *Anneslea* must disappear. The two species which it comprises are without a generic name; for *Anneslea* Wall. is in the rare position of never having received another generic appellation.

An innominate genus is so manifestly inconvenient that I venture to propose for it the name of DAYDONIA, in compliment to Mr. B. Daydon Jackson, whose services to Botany in his position as Secretary to the Linnean Society deserve recognition. His work upon the great 'Index of Plant-names,' which will necessitate many such changes of name, as well as the zeal he has shown in promoting the only sound principle which can govern botanical nomenclature, seem to render this commemoration specially appropriate. The specific names need not be changed.

DAYDONIA = ANNESLEA Wall.
D. FRAGRANS = *A. fragrans* Wall.
D. CRASSIPES = *A. crassipes* Hook. ex Chois.

# A SYNOPSIS OF *TILLANDSIEÆ*.

## By J. G. BAKER, F.R.S., F.L.S.

(Continued from Journ. Bot. 1887, p. 348).

**113. Tillandsia Sintenisii,** n. sp.—Leaves about a dozen in a rosette; dilated base ovate, 6 in. long, 3 in. broad; blade lanceolate, rigidly coriaceous, persistently thinly adpresso-lepidote, 1½ ft. long, an inch broad low down, tapering gradually into a convolute setaceous point. Inflorescence a lax very ample panicle; main axis ½ in. diam.; lower branches compound, a foot long; branch-bracts linear from an ovate base, 4–6 in. long; spikes 4–8 in. long, with lax erect flowers adpressed to the flexuose rachis; flower-bracts ovate, acute, ½ in. long. Calyx ¼ in. longer than the bract; sepals oblong, obtuse. Corolla not seen. Capsule-valves lanceolate, 1½–1¼ in. long, ⅙ in. broad.

Hab. Porto Rico; primæval woods of Mount Torito, alt. 3000 ft., *Sintenis* 2134!

**114. T. Swartzii,** n. sp.   *T. paniculata* Swartz in Herb. Mus. Brit., non Linn. — Basal leaves not seen.   Peduncle a foot long; lower bract-leaves with lanceolate free points 3–4 in. long.   Inflorescence a very lax panicle 2 ft. long, with ascending slender laxly-flowered flexuose branches; flower-bracts ovate, acute, ¼–⅓ in. long.   Calyx ⅜ in. long.   Capsule at least twice as long as the calyx.

Hab. Jamaica, *Swartz!*   Allied to *T. flexuosa* and *T. Sintenisii.*

**115. T. brassicoides,** n. sp. — Leaves 20 or more in a dense rosette; dilated base ovate, 3–4 in. long, 2 in. broad; blade lanceolate-acuminate, ½ ft. long, an inch broad at the base, tapering gradually to a short convolute point, rigidly coriaceous, densely persistently finely lepidote on both surfaces.   Peduncle rather longer than the leaves; lower bract-leaves with short free points; upper small, ovate, acute, not imbricated.   Inflorescence a lax subsecund spike 4–5 in. long, with a sulcate glabrous slightly flexuose rachis; flowers 5–6, lower spreading; bract-leaves coriaceous, broad-ovate, ascending, ¾ in. long.   Calyx 1–1¼ in. long; sepals obtuse.   Corolla not seen.   Capsule not longer than the calyx.

Hab. Rio Janeiro, *Burchell* 1393!   A remarkable plant, with bracts and calyx like those of *T. regina.*

Subgenus V. ANOPLOPHYTUM (Beer). — Leaves narrow, moderately firm in texture, acuminate or subulate, thinly or densely lepidote. Inflorescence more or less decidedly multifarious. Petals white, or red, or violet; blade lingulate, spreading at the tip only; claw not scaled. Stamens and style shorter than the petals.

## KEY.

Leaves linear-subulate.
    Spikes simple  .  .  Sp. 116–118.
    Spikes panicled  .  .  Sp. 119.
Leaves lanceolate-acuminate.
    Spikes simple  .  .  Sp. 120–123.
    Spikes panicled  .  .  Sp. 124–126.
Leaves lanceolate-acute  .  Sp. 127.

116. **T. plumosa,** n. sp. — Leaves densely rosulate, subulate from a clasping ovate scariose base, 3–4 in. long, 1-12th in. broad low down, not rigid, densely coated with large spreading white lanceolate scales. Peduncle shorter than the leaves ; bract-leaves crowded, with long densely lepidote erect free points. Flowers 12–20, aggregated in a dense head about ½ in. and broad ; flower-bracts ovate-cuspidate, strongly ribbed, ¼–⅓ in. long. Calyx reaching to the tip of the bract, laterally compressed, nearly flat on the inner face ; sepals deeply navicular, acute. Petals not seen.

Hab. Mexico, in the Province of Puebla, on trees, *Andrieux* 57 l

117. **T. rupicola,** n. sp. — Leaves 30–40 crowded on a stem an inch long, linear-subulate, reflexed, about two inches long, the channelled lanceolate base ⅛ in. broad, not rigid in texture, densely clothed all over with large persistent glittering erecto-patent white lanceolate scales. Peduncle about as long as the leaves ; bract-leaves crowded, all with long erect subulate points from a clasping ovate scariose base.. Flowers 8–12, in a very dense simple sub-globose very multifarious spike ½–¾ long ; flower-bracts oblong-navicular, acute or cuspidate, stramineous, thinly lepidote, ¼–⅓ in. long. Calyx ¼ in. long, glabrous ; sepals acute. Petal-blade deep violet, oblong, ⅛ in. long.

Hab. Ecuador ; forming a dense tuft on rocks in front of the village of Ona, alt. 8000 ft., *Col. Hall* ! Very distinct. I have not seen *Anoplophytum calothyrsus* Beer, Brom. 263 (name only), founded on Poppig 1224, from Peru ; nor *A. longebracteatum* Beer, gathered by Meyen ; *A. setaceum* Beer (Cuba, *Otto*) ; nor *A. Sprengelianum* Beer, all mentioned by name only in his ' Repertorium.'

118. **T. PULCHRA** Hook. Exot. Flora, 154 ; Roem. et Schultes, Syst. Veg. vii. 1208 ; Wawra in Oester. Bot. Zeit. 1880, 224 ; Itin. Prin. Sax. Cob. 173, t. 34. *T. pulchella* Hook. in Bot. Mag. t. 5229 ; Griseb. Brit. West Ind. 598 ; Belg. Hort. ix. 322, with figure. *T. pityphylla* Roem. et Schultes, Syst. Veg. *l. c. Anoplophytum pulchellum* Beer, Brom. 41. *Pourretia surinamensis* Hort. — Stem sometimes short, sometimes produced to a length of half a foot. Leaves densely crowded, ascending, linear, narrowed gradually into a long subulate point, 4–5 in. long, ¼ in. broad low down, firm in texture, thinly lepidote. Peduncle slender, erect, 3–4 in. long ; bracts lanceolate, with long subulate points. Flowers 6–12 in a dense spike ; flower-bracts ovate, acute, rose-tinted, ½–¾ in. long. Calyx ½ in. long ; sepals oblong, acute. Petals white, half as long again as the calyx. Stamens and style shorter than the petals.

Var. AMŒNUM Baker. *Anoplophytum amœnum* E. Morren in Belg. Hort. 1883, 265, t. 17. — Corolla violet. Leaves still more slender than in the type.

Hab. South Brazil, *Bowie & Cunningham* (year 1817) ! *Burchell* 2033 ! *Glaziou* 2730 ! 8025 ! 13257 ! 16456 ! *Gardner* 695 ! Dutch Guiana, *Kegel* 801. Mountains of Venezuela, alt. 4500 ft., *Fendler* 2576. Trinidad, *Baron Schacht* ! (Cultivated by Mr. Shepherd at Liverpool in 1824). Cuba, *Wright* 685 ! Var. *vaginata* Wawra is a form with a produced leafy stem.

119. **T. GLOBOSA** Wawra in Oester. Bot. Zeitsch. xxx. 222 ;

French transl. 72; Itin. Prin. Sax. Cob. 170, t. 32, fig. A. — Tufts nearly a foot in diameter. Leaves 30–40 in a dense rosette, linear-subulate, recurved, 6–9 in. long, ⅛ in. broad above the dilated base, moderately firm in texture, thinly grey-lepidote. Peduncle 3–4 in. long; bract-leaves many, erect, imbricated, subulate from a lanceolate base. Panicle dense, ovoid, about 2 in. long; lower branch-bracts lanceolate-subulate, 1½ in. long, tinged with red; flower-bracts lanceolate, red, as long as the calyx. Calyx above ½ in. long; sepals oblong-lanceolate. Petals violet, half as long again as the sepals, spreading at the tip only. Stamens and style shorter than the petals. Capsule-valves lanceolate, nearly an inch long.

Hab. South Brazil; Province of Entre Rios, &c., *Burchell* 1893! 3493! *Boog! Wawra & Maly*. Var. *crinifolia* Wawra differs from the type by its more slender leaves.

120. T. DIANTHOIDEA Rossi, Cat. Modoet. t. 1; Roem. et Schultes, Syst. Veg. vii. 1205; Regel, Gartenfl. t. 85; Ill. Hort. n. s. t. 322; Griseb. Symb. Fl. Argent. 1878, 333. *T. stricta* Lindl. in Bot. Reg. t. 1338. *Anoplophytum æranthos* and *dianthoideum* Beer, Brom. 40–41. *Pourrettia æranthos* Rossi, Herb. Gen. Amat. t. 307. *Amalia ærsincola* Bahi inedit.—Stems short or produced to a length of a few inches. Leaves crowded, falcate, lanceolate-acuminate, 3–4 in. long, ¼ in. broad low down, glaucous green, firm in texture, thinly lepidote. Peduncle 3–4 in. long; bract-leaves crowded, rigid, lanceolate-subulate. Flowers 5–10 in a moderately dense simple spike; bracts ovate, acute, reddish, ½–1 in. long. Calyx ½ in. long; sepals oblong. Petals bright lilac, half as long again as the calyx. Stamens a little longer than the calyx.

Var. *T. rosea* Lindl. in Bot. Reg. t. 1357. *T. recurvifolia* Hook. in Bot. Mag. t. 5246. *Anoplophytum roseum* Beer, Brom. 40.—Corolla smaller, white.

Hab. Uraguay, *Lorentz* 493! *King!* *Gibert* 1091! Buenos Ayres, *Tweedie!* Parana, *Christie!* Scarcely more than a variety of *T. stricta*. Mr. Im Thurn gathered lately on the summit of Mount Roraima (No. 315) a closely allied form with short rigid almost pungent leaves, large loose glossy brown bracts, and a large bright-coloured corolla.

121. T. STRICTA Soland MSS.; Sims in Bot. Mag. t. 1529; Rossi, Cat. Hort. Moedet. 1824, 82, t. 3; Roem. et Schultes, Syst. Veg. vii. 1206; Wawra, Bot. Reise. Maxim. 163; Oesterr. Bot. Zeit, 1880, 222; Itin. Prin. Sax. Cob. 173. *Anoplophytum strictum* Beer, Brom. 39; E. Morren in Belg. Hort. 1878, 188, t. 13. *T. bicolor* A. Brong. Voy. Coquille, 185, t. 86; Griseb. Pl. Lorentz. 224; Symb. Fl. Argent. 1878, 333. *A. bicolor* Beer, Brom. 41.—Tufts 6–9 in. diam. Leaves 30–50 in a dense rosette, lanceolate-acuminate, arcuate, 4–6 in. long, ⅙–¼ in. broad above the base, rigid in texture, glaucous green, thinly lepidote. Peduncle 3–4 in. long; bract-leaves many, lanceolate-subulate, erect. Spike dense, simple, erect, 10–20-flowered, 1–2 in. long; bracts reddish, ovate, the upper acute, the lower cuspidate; blade ½–¾ in. long. Calyx ½ in. long; sepals oblong, acute. Petals bright lilac, half as long

again as the calyx; blade lingulate, spreading only at the tip. Stamens and style a little longer than the calyx. Capsule-valves lanceolate, an inch long.

Var. CAULESCENS Baker. *T. subulata* Vell. Fl. Flum. iii. t. 197. *Diaphoranthema subulata* Beer, Brom. 155. — Leaves crowded on a produced stem, which is sometimes half a foot long, shorter, thicker, and less acuminate than in the type. Spike less dense; flowers fewer.

Hab. Southern and Central Brazil, *Lesson, Gardner* 696! *Miers* 3493! *Mosen* 3258! *Blanchet*! *Glaziou* 11689! Paraguay, *Balansa* 616! Tucuman, Catamarca and Oran, *Lorentz*! Andes of Bolivia, alt. 8000 ft., *Mandon* 1184! North Patagonia, *Capt. Middleton*! British Guiana, *Im Thurn*! *Jenman* 1845! There is a specimen without name in the herbarium of Linnæus. *β. caulescens;* Mountains of South Brazil, *Miers* 4610! *Glaziou* 8019! 15463! 15464! 16457! 16458!

122. **T. meridionalis,** n. sp.—Acaulescent. Leaves like those of *T. Duratii*, comparatively few in a rosette, lanceolate-acuminate, half a foot long, narrowed gradually from the base to the apex, $\frac{1}{2}$ in. broad low down, channelled deeply all down the face, thick and firm in texture, thinly persistently grey-lepidote. Peduncle rather shorter than the leaves, stiffly erect; bract-leaves many, imbricated, erect, with a long point from a broad clasping base. Spike dense, simple, oblong, 2 in. long; flower-bracts broad oblong-navicular, pointed; lower $\frac{3}{4}$ in. long; pedicel short, stout. Calyx above $\frac{1}{2}$ in. long; sepals oblong, obtuse or subacute. Petal-blade oblanceolate, $\frac{1}{8}$ in. long. Stamens and style not protruded beyond the calyx.

Hab. Uraguay, *Tweedie*!

123. **T. Benthamiana** (Klotzsch inedit?), n. sp. *T. vestita* Benth. in Pl. Hartweg. 25, non Cham. et Schlecht. *Anoplophytum Benthamianum?* et *vestitum* Beer, Brom. 263, 266. — Acaulescent. Leaves 30-40 in a dense rosette, lanceolate-acuminate, falcate, 6-8 in. long, $\frac{3}{4}$ in. broad at the dilated base, $\frac{1}{2}$ in. above it, deeply channelled down the face, thick and rigid in texture, persistently clothed with adpressed lepidote scales on both surfaces. Peduncle stout, erect, 3-6 in. long; bract-leaves crowded, similar to those of the rosette. Flowers in a dense oblong spike about 3 in. long, $1\frac{1}{2}$ in. diam.; flower-bracts very large, with a thin oblong boat-shaped blade about 2 in. long, 1 in. broad, acute, and in the lower with a long rigid cusp. Calyx much shorter than the bract. Corolla violet, much longer than the calyx, its narrow petals convolute in a tube, spreading at the very tip only. Stamens and style conspicuously exserted beyond the petals. Capsule-valves lanceolate, 1-1½ in. long, $\frac{1}{4}$ in. broad.

Hab. Central Mexico, on rocks and trees (Xalapa, Guadeloupe, and Balanos), *Hartweg* 223! *Coulter* 1579! *Bourgeau* 894! *Andrieux* 58! (a form with smaller bracts and narrower leaves). A very distinct species, connecting the sections *Anoplophytum* and *Pityro-phyllum*.

124. **T.** GEMINIFLORA A. Brong. in Voy. Coquille, 186. *T. rubida*

Lindl. in Bot. Reg. xxviii. t. 63. *Anoplophytum rubidum* Beer, Brom. 40. *A. geminiflorum* E. Morren in Belg. Hort. 1880, 191, t. 11. — Tufts 6-9 in. diam. Leaves 30-40 in a dense rosette, 4-5 in. long, ½ in. broad low down, narrowed gradually to a long point, moderately firm in texture, persistently grey-lepidote on both surfaces. Peduncle 2-3 in. long; bract-leaves large, lanceolate. Flowers 20-30 in a dense ovoid panicle 2-3 in. long; lower branch-bracts lanceolate-subulate, 1½-2 in. long, rose-red; flower-bracts lanceolate, about as long as the calyx. Calyx ½ in. long; sepals oblong-lanceolate. Petals bright red-purple, half as long again as the calyx, spreading only at the tip. Stamens shorter than the petals. Capsule-valves lanceolate, an inch long.

Hab. South Brazil; Rio Janeiro, *Lhotsky, Sello* 86, *Burchell* 2353! 3146! *Glaziou* 4263! Paranagua, *Platzmann*! Island of St.-Catherine, *Lesson.* Introduced into cultivation by Loddiges in 1842.

125. T. GARDNERI Lindl. in Bot. Reg. 1842, sub t. 63; Baker in Gard. Chron. 1878, ii. 461. *T. incana* Wawra in Oester. Bot. Zeitsch. xxx. 223; French trans. 72; Itin. Prin. Sax. Cob. 172. *Anoplophytum incanum* E. Morren in Belg. Hort. 1881, 209, t. 11. *A. Rollissoni* E. Morren inedit. *Tillandsia argentea* Hort. Angl. non Griseb.—Tufts a foot or more in diameter. Leaves 30-40 in a dense rosette, lanceolate-acuminate, 6-8 in. long, ½ in. broad low down, moderately firm in texture, densely clothed on both surfaces with glittering whitish lepidote scales. Peduncle 2-3 in. long; bract-leaves many, densely imbricated, lanceolate-subulate, lepidote. Inflorescence a dense panicle about 2 in. long and broad; lower branch-bracts lanceolate-subulate, 1½-2 in. long, densely lepidote on the back; flower bracts oblong-lanceolate, ½-¾ in. long. Calyx ⅓-½ in. long; sepals oblong-lanceolate. Corolla half as long again as the calyx; petal-blade lingulate, erect, bright red. Stamens and style not exserted beyond the tip of the petals. Capsule-valves rigid, 1½ in. long.

Hab. South Brazil, on the Serra Itatiaia, &c., *Burchell* 994! *Gardner* 134! *Dr. Cunningham*! *Boog*! *Glaziou* 16453! *Wawra & Maly* ii. 508. Trinidad, *Prestoe*! *Fendler* 822!

126. **T. didisticha,** n. sp. *Anoplophytum didistichum* E. Morren in Belg. Hort. 1881, 164; 1882, 335. — Leaves densely rosulate, lanceolate-acuminate, grey-lepidote, 6-8 in. long, channelled down the face. Peduncle longer than the leaves, arcuate; bract-leaves numerous, imbricated. Panicle congested; spikes dense, elliptical, distichous, 1½-2 in. long, 10-12-flowered; flower-bracts greenish brown, lepidote. Corolla white.

Hab. South Brazil. Introduced by M. Jacob-Makoy, of Liege, and flowered in 1880.

127. **T. brachyphylla,** n. sp. — Acaulescent. Leaves 30-40 in a dense rosette, arcuate inwardly, lanceolate, acute, not acuminate, narrowed gradually from the base to the apex, 3 in. long, ½ in. broad low down, thin in texture, flat, not at all rigid, densely persistently lepidote on both surfaces. Peduncle shorter than the leaves; bract-leaves crowded, linear from a broad base, erect, densely lepidote. Flowers in a dense simple spike about an inch

long; bracts ovate, acute, bright red, the lower ½–¾ in. long. Calyx glabrous, ½ in. long; sepals lanceolate, acute. Petals not seen. Capsule subcylindrical, twice as long as the calyx.

Hab. South Brazil, *Glaziou* 8018! Very distinct.

(To be continued.)

---

# NOTES ON THE FLORA OF EASTERNESS, ELGIN, BANFF, AND WEST ROSS.

## By G. Claridge Druce.

· For some years past the cliffs encircling Loch Ennich, in Easterness, which are well detailed in Ordnance Sheet No. 64, have appeared to me a desirable hunting-ground, as from their proximity to the Mac Dhu range, their great height, and precipitous character, in addition to their extent of six or seven miles, promised to repay one for a botanical raid. And then the great corries of Cairngorm and Braeriarch, which are such a prominent and beautiful object as one goes north by the Highland Railway, are within measurable distance of Aviemore, and had this attraction for me, that it was on their cliffs when a boy of fifteen that I first realised the beauty of the Scotch Flora. Aviemore unfortunately posseses no inn, so I made my head-quarters at the Boat of Garten, which is within a pleasant drive of ten miles of Glen More. I am greatly indebted to the owner and lessee of Glenmore and Rothiemurcus, for kind permission to visit the forests so late in the season. My first day was spent about Kingussie and yielded a large variety of roses with *Hieracium crocatum* and *corymbosum*, and *Equisetum sylvaticum* var. *capillare*, a variety which I believe I have also seen in Fife and East Perth. The next day I walked up to Glen More Lodge to make arrangement with the keepers, and found a great change had taken place in the condition of the country since my last visit. The following day I drove over and worked Corrie Sneachda, *Carex approximata*, *Saxifraga rivularis*, *Cerastium refractum* Allione, being the special prizes of the day. The Lochs Avinloch, Vaa, &c., were worked on the following day, *Nuphar minimum*, *Drosera obovata*. &c., being gathered. Lochs Garten and Mallachie gave *Carex limosa*, *Utricularia intermedia*, &c., and the Spey side was very gay with an abundant growth of *Hieracium corymbosum* in beautiful condition. *H. crocatum* was more scarce, but *Campanula rotundifolia* and *Galium verum* were abundant, while *Ribes petræum* was welcomed for other than botanical reasons, the fruit being very delicious. Near Kinchurdy an addition to the Scotch flora was made by discovering the variety *filifolium* of *Hieracium umbellatum*.

On my first visit to Glen Ennich I shortened the distance five miles by hiring a machine and driving through Rothiemurcus Forest, when we dismounted and walked six miles up the rather broken road, the last mile being by heaps of moraine of unusual

size. The loch is a fine piece of water more than a mile long, and surrounded on three sides by magnificent cliffs rivalling Canlochen, and surpassing Callater and the Dole. Overhanging the western side is the precipitous Sgoran Dubh, the corries of Braeriach being on the east; Cairntoul is on the south-east, but its summit is not visible from the glen. A lower range of cliffs, which are also less steep, are on the south side, and down them pours a pretty waterfall from the little Lochan nan Cnapan, situate on the table-land above, at an elevation of 2850 feet. Loch Ennich is about 1700 feet above the sea level, the summit of Sgoran Dubh being 3658 feet. So that in half a mile of actual distance a descent of 2000 feet is made, a degree of steepness difficult to match in Scotland. As may be imagined, the scenery is of a very wild and magnificent character. The golden eagle's nest was in the glen, and I saw some snow buntings on the cliffs. The southern side yielded *Cerastium alpinum, C. refractum, Carex pulla, C. atrata, C. vaginata, Saussurea, Saxifraga sponhemica, Deschampsia alpina, Poa alpina, Salix Lapponum, S. Myrsinites, Luzula arcuata, Lycopodium annotinum,* and a large variety of *Hieracia,* including *nigrescens, cæsium, senescens, pallidum, aggregatum, anglicum, lingulatum, holosericeum, eximium,* and *globosum.* I worked round the glen, and then scarcely relished the eleven miles walk back to Aviemore. I paid it another visit later on, but drove all the way to the loch, and was thus enabled to climb the cliffs of the Sgoran Dubh ; these, however, yielded little besides *Hieracia,* the rocks being too steep and dry. *Isoetes* was in the loch.

Another day was spent in the Corrie Leacainn of Cairngorm, which like Sneachda had plenty of *Saxifraga rivularis,* also *Alopecurus alpinus* and *Phleum alpinum.* I then climbed Ben McDhu and visited the upper end of Loch A'an. Glen Avon, and the mountains bordering the loch were quite sufficient for a whole day, which was enlivened by fine mountain grasses and many Hieracia, including *lingulatum,* as well as by a feast of ripe " evrons." The scenery by the Shelter Stone is very imposing and the place secluded enough for the eagles to nest. The snow was much less in quantity than for many years past.

A delightful day was spent in walking along the Findhorn from Dunphail to Forres. This yielded the same variety of *Melampyrum pratense,* which I first found in Wigtonshire, and which Father Reader has since gathered in Northumberland and Westmoreland.

After a few days spent among the lochs, I went north to Loch Maree, but here, as usual, the weather was wet and detestable. I climbed Ben Eay, *Arabis petræa* and *Arbutus alpina* rewarding the labour ; but the weather, the intolerable midges, and the asperities of a Scotch Sabbath, were enough to drive one eastward to Elgin.

I have to thank Mr. Arthur Bennett, Dr. Buchanan White, and Messrs. Bennett, Groves, Baker, Beeby, and Backhouse, for kind assistance.

The sign * means a new county record.    † means that personal authority is lacking in Top. Bot.

*Ranunculus Flammula* L. On the margins of Loch Morlich, Loch Garten, &c., the variety or form *pseudo-reptans* Syme, occurred.

*Nuphar minimum* Sm., E. Bot. 1811.—*N. pumilum* Sm., Eng. Fl. 1824. A plant which I doubtfully refer to this I gathered in Loch Avinlochy. The stigma was elliptic and the leaves hairy underneath. The Rev. McDougall informed me he had gathered it in a loch under Craig Ellachie. In Hooker's ' British Flora ' it is recorded for Loch Baladren, near Aviemore Inn, which is probably the same locality.

*Caltha palustris* L., var. *minor* DC. Tableland of Ben McDhu, Banff. Aberdeen S. and Easterness.

*Papaver dubium* L. (Lamottei), Forres, 95, and in West Ross, 105.

*Lepidium heterophyllum* Benth., var. *canescens* Gr. et Godr. (*Smithii* Hook). Loch Maree and West Ross, *105. In Journ. Bot. June, 1887, this is given (by a slip) as *heterostylum*.

*Subularia aquatica* L. Stony margin of Loch Morlich, † 96.

*Erophila verna* DC. Loch Maree, * 105.

*Cardamine flexuosa* With. Corrie Sneachda, * 96. Side of Loch A'an, * 94.

*Arabis petræa* Lamk. In 96 very rare ; a few small plants in Corrie Sneachda. * In 105, on the white quartzite screes of Ben Eay (Beinn Eigh), at an elevation of 2300 to 2800 feet, fairly plentiful. Some plants nine inches. In both counties the glabrous form alone noted.

*Erysimum cheiranthoides* L. A few small plants in cornfield near the Boat of Garten, 96.

*Viola tricolor* L. Margin of Loch Torridon, * 105.—*V. arvensis* Murray. Loch Torridon, * 105. — *V. lutea* Hudson, var. *amœna* (Symons) Hensl. The only form noticed about Kingussie, 96.

*Drosera anglica* Huds., var. *obovata* Mert. et Koch. This plant, which I found in West Ross in 1879, I again met with abundantly between Kenlochewe and Loch Torridon, and also by Loch Maree. There was a good deal of *anglica* near it, but not much *rotundifolia*. The large amount of viscid matter given off by it was very remarkable. There seems to be very little variation in its leaf-outline. In a small marsh near the Boat of Garten, * 96, I met with a good quantity of the same plant. Both the assumed parents occurred with it.

*Polygala serpyllacea* Weihe. The common plant of * Banff, * Easterness and * Elgin. I saw no plants which I could refer to *vulgaris*.

*Silene acaulis* L. On the Cairngorm, both on the † Banff and † Easterness side, in good flower near the snow patches.

*Lychnis alba* Mill. (*L. vespertina* Sibth.). * Waste ground near Forres, * 95. Aviemore and Boat of Garten, * 96. Little more than a colonist in these counties.—*L. Githago* Scop. Kenlochewe, * 105.

*Arenaria peploides* L. Loch Torridon, * 105.

*Spergula arvensis* L., var. *sativa* (Bœnng.). Frequent about Kingussie, Aviemore, Boat of Garten, * 96. Forres, Elgin, * 95, and Kenlochewe, * 105. I am induced to consider this a distinct species. Its grayish green foliage renders it at once noticeable, so different from the shining dark green foliage of *vulgaris*. The fruit characters appear to be constant.

*Stellaria media* Cyr., var. *neglecta* (Weihe). By the Findhorn, near Logie, 95 *; and Spey, 96 *.

*Cerastium refractum* Allione 1785, *C. trigynum* Vill. 1789. In Corrie Sneachda and Leacainn, Easterness, 96, glabrous form alone noted.—*C. alpinum* L. The var. *piloso-pubescens* Benth. (*pubescens* Syme) was the more frequent form. The aggregate plant was common in Corrie Sneachda, Leacainn and Glen Ennich.

*Trifolium agrarium* L. Cultivated fields near * 96, Boat of Garten, and near Logie, 95 *. Not native in either county. *T. minus* L., *T. dubium* Sibth. Kenlochewe, * 105.

*Vicia sativa* L. * 105, Kenlochewe.

*Prunus Padus* L. * 105, Kenlochewe.

*Rubus affinis* W. et N. * 95, near Forres.—*R. nessensis* Hall in Edin. Phil. Tr. (*R. suberectus* And.). Near Kinchurdy, * 96.—*R. macrophyllus*. Near Forres, * 95. — *R. rhamnifolius* Weihe et Nees. Near Forres, * 95.—*R. echinatus* Lindl. Near Altyre, * 95.

*Rosa mollis* Sm. Common about Kingussie, Aviemore and Boat of Garten, * 96, and Forres, * 95 ; with its var. *cærulea* (Woods). The latter variety also at Kenlochewe, * 105. — *R. tomentosa* Sm. Forres. Dunphail. Findhorn side, * 95. Spey side at Aviemore and Kinchurdy. Loch an Eilan and Kingussie, * 96. Dingwall, * 106. — Var. *scabriuscula* (Sm.) Hemsl. 1829.* Kinchurdy, 96.— Var. *subglobosa* (Sm.) Baker.* Boat of Garten, 96.—Var. *sylvestris* (Lindl.) Woods.* Kenlochewe, 105. * Forres, 95.—*R. Eglanteria* L. ed. i, and Huds. ed. i. (*R. rubiginosa* L. mant.). Near Forres, * 95. Spey side, * 96. This appears to me to have as good claim to being native in Scotland as it does in our Midland Counties.— *R. micrantha* Sm. Near Forres, * 95.—*R. involuta* Sm. Avinlochy. Auchgourish, * 96.—Var. *Sabini* (Woods) Baker. Loch Phitiulais, * 96.—*R. canina* L.—Var. *tomentella* Leman (Baker). Kingussie, * 96. — Var. *dumalis* (Bechst.) Dumort. Aviemore, with subglobose fruit, * 96. — Var. *dumetorum* (Thuill.) Hook. Boat of Garten, * 96.—Var. *sphærica* (Gren.) Dumort. Boat of Garten, * 96. — Var. *coriifolia* (Fries) Baker. Kingussie. Kinchurdy. With glandular sepals at Kingussie, and with hispid peduncle at Loch Phitiulais, * 96.—Var. *urbica* (Leman) Baker. Aviemore, * 96.—Var. *Watsoni* Baker. Kingussie, * 96.—Var. *subcristata* Baker. Forres, * 95.

*Ribes rubrum* L., var *petræum* (Sm.). Abundant and native by the Spey banks from Doune to Kinchurdy.

*Spiræa salicifolia* L. Near Kingussie, 96.

*Alchemilla alpina* L. Cairngorm, † 94.—*A. vulgaris* L., var. *hybrida* With. ed. 1 (*minor* Huds.) Glen More, 96.

*Pyrus Malus* L. Aviemore, * 96.

*Epilobium alpinum* L. 105 Lieuthgoch. A very pale-flowered form occurred in Corrie Leacainn, 96.—*E. obscurum* Schreb. 105, Kenlochewe.

*Circæa alpina* L., in a very small state, occurred by Loch Maree side, West Ross, and typical plants by Loch Carron in the same county. The var. *major* Hook. 1831, = *intermedia* (Ehrh.) Hensl. 1829. Findhorn, 95.

*Myriophyllum alterniflorum* DC. Findhorn, * 95. An abundant plant in the Spey and in many of the lochs in Easterness, * 96.

*Callitriche autumnalis* L. Loch Guinach, near Kingussie, * 96.— *C. hamulata* Kuetz. Near Logie, * 95.

*Scleranthus annuus* L. Kenlochewe, * 105.

*Sedum acre* L. Loch Maree, * 105.

*Saxifraga rivularis* L. Abundant in the corries of Sneachda, and Leachainn, Cairngorm, 96. — *S. sponhemica* Gmel. Glen Ennich, * 96.

*Galium boreale* L. A large branching form with acuminate leaves in Glen More, 96.

*Sherardia arvensis* L. Loch Maree, * 105.

*Valeriana officinalis* L., var. *montana* Milne and Gordon, 1793 (*Mikanii* Syme). Spey side, near Boat of Garten, * 96. There is some doubt as to the correct identity, as the plant was very dwarfed from the drought and grew on stony *débris*.

· *Scabiosa arvensis* L. Loch Maree. Probably introduced, * 105. —*S. Succisa* L. (*Succisa pratensis* Moench). I saw this at 2800 feet on the Cairngorms.

*Arctium minus* Schk. Forres, * 95. Kingussie. Boat of Garten * 96.—*A. intermedium* Lange. Kenlochewe, * 105.

*Saussurea alpina* DC. Corrie Sneachda, and fine specimen in Glen Ennich, * 96. The plant gives out a powerful odour of heliotrope. Also in Western Ross, † 105.

*Centaurea Cyanus* L. Kenlochewe, * 105.

*Gnaphalium dioicum* Gaertn. (Gaertner appears to be an earlier authority than R. Br.). Plentiful on moorland in Banff, † 94.

*Aster Tripolium* L. Loch Torridon, * 105.

*Hieracium lingulatum* Backh. * Glen Avon, 94. Glen Ennich and Cairngorm Corrie, * 96; and on the slopes of Ben Eay very rare, * 105.—*H. anglicum* Fr. Cliffs of Sgoran Dubh, * 96. Glen Avon, * 94.—*H. aggregatum* Backh. Rare. Cliffs of Sgoran Dubh, * 96. — *H. vulgatum* Fries. Kingussie. Rothiemurcus, &c., * 96. Logie, * 95.—*H. Eupatorium* Griseb. Plentiful by Spey side and at Kingusssie, 96. Near Grantown, * 95. — *H. crocatum* Fr. Spey side, near Kinchurdy, * 96. — *H. melanocephalum* Tausch. Glen Avon, * 94. Glen Ennich and Corrie Sneachda, * 96. — *H. eximium* Backh. Glen Avon, * 94. Glen Ennich, * 96. — Var. *tenellum* Backh. Glen Avon, * 94. Corrie Leacainn and Glen Ennich, * 96. Ben Eay, * 105. — *H. holosericeum* Backh. Glen Avon, * 94. Corrie Sneachda and Glen Ennich, * 96.—*H. globosum* Backh. South side, Glen Ennich, * 96. — *H. nigrescens* Willd. Glen Avon, * 94. Not rare on Corries of the Cairngorm and Glen Ennich, 96. Ben Eay, rare, * 105. — *H. senescens*, Backh. Cliffs of Sgoran Dubh, * 96.—*H. murorum* L. p. p. Glen Ennich, * 96. —*H. cæsium* Fries. Glen Ennich, * 96.—*H. pallidum* Biv. Glen Avon, * 94. A frequent and variable form in Glen Ennich, 96. —*H. umbellatum* L. Glen Ennich, * 96. Findhorn side, near Logie, * 95.—Var. *filifolium* Backh. Kinchurdy, * 96.

*Solidago Virgaurea* L., var. *cambrica* Sm. (Huds.). Plentiful in the corries of Cairngorm and in Glen Ennich, 96.

*Lobelia Dortmanna* L.   Loch Clare, &c.   West Ross, † 105. Abundant in Loch Garten, Loch an Eilan, Lochs Phitiulais, Mallachie, and Loch Guinach near Kingussie.   The plant fades much less quickly when removed from the water than many of our aquatics.

*Arctostaphylos alpina* Spreng.   I was extremely pleased to meet with this plant in good condition on the north slopes of Beinn Eigh, at about 4200 feet.   West Ross was placed as a doubtful record in Top. Bot.   The authority for its occurrence was Dr. Lightfoot, who says " he gathered it to the south of Little Loch Broom and between that lake and Loch Mari."   Mr. Watson thought there might have been some confusion of this species with *Uva Ursi.* — *A. Uva Ursi* Spreng.   Lieuthgoch, † 105.

*Vaccinium uliginosum* L.   Glen Shiel, † 105.   Loch Aan, * 94.

*Pyrola secunda* L.   A few plants occurred in a ravine near Glen More Lodge, 96.—*P. rotundifolia* L.   Glen More, 96.

*Gentiana campestris* L.   Loch Torridon, * 105.

*Veronica alpina* L.   A variety of this occurred on the Cairngorms with a suppressed spike almost hidden in the leaves.—*V. scutellata* L.   This occurred with pale blue flowers on the margin of Loch Morlich, 96.   I saw nothing of the hairy form var. *pubescens* Gray, 1821, *parmularia* (Turp. et Poir.). — *V. agrestis* L.   Kenlochewe, * 105.—*V. serpyllifolia* L., var. *humifusa* (Dicks.).   Cairngorm. Ben McDhu, and Loch Aan, 92, * 94 and 96.

*Melampyrum pratense* L.   Loch Maree, * 105.—*M. pratense* var. *hians* Druce.   This occurs in plenty by the Findhorn near Logie, * 95.   The corolla lips were generally open, the colour deep golden yellow, even to the tube ; the capsules were as frequently erect as deflexed, and the position of the flowers often suberect.   The size of the flower was the same as those of var. *montanum* (Johnst.), which was a common moorland plant in 95 and 96.   The flowers darkened in spirit, and very much in drying.   In the ' Norge Flora ' Dr. Blytt includes a var. *luteum* which I thought might be the same plant, but the Rev. F. Woods informs me that *luteum* is a large plant with deeply-toothed bracts and very numerous flowers.

*Mentha rotundifolia* Huds.   Boat of Garten, on waste ground,* 96.

*Calamintha Clinopodium* Benth.   Spey side, near Kinchurdy, * 96.

*Lamium intermedium* Fries.   Boat of Garten, * 96.   Forres, † 96, and Kenlochewe, * 105.

*Pinguicala lusitanica* L.   Loch Maree, 105.

*Utricularia intermedia* Hayne.   Near Boat of Garten.   Loch Mallachie, * 96.   Loch near Forres, * 95.

*Littorella juncea* Bergh.   Loch Mallachie, Morlich, &c., Easterness.

*Chenopodium album* L. var. *incanum* Moq. Tand.   Kingussie, Aviemore, * 96.   Forres, * 95.

*Atriplex angustifolia* Sm.   Forres * 95.   Boat of Garten, * 96.

*Oxyria digyna* Hill.   Loch A'an, † 94.

*Polygonum Convolvulus* L.   Loch Torridon, * 105.

*Euphorbia helioscopia* L. Kenlochewe, * 105. — *E. Peplus* L. Kenlochewe, * 105.

*Quercus sessiliflora* Salisb. Logie. Findhorn, * 95.—*Q. Robur* L. *Q. pedunculata* Ehrh.). Rothiemurchus * 96. Findhorn, Logie and Altyre, * 95.

*Betula alba* L. (*B. verrucosa* Ehrh.). Kingussie. Rothiemurchus, * 96. Dunphail, 95.

*B. glutinosa* Fries. Kingussie, * 96. Logie. Findhorn side, * 95.

*Pinus sylvestris* L. Native in Glen Avon, 94 *.

*Salix rubra* L. Near Forres, * 95. Very difficult to speak with certainty if the osier willows are indigenous in the north.—*S. rugosa* Leefe. In a hedge near Kenlochewe, * 105. — *S. caprea* L. So far I have been unable to meet with *caprea* as a generic name in pre-Linnean botany. It is written with a capital in L. C. This is native in Glen Avon, * 94.—*S. phylicifolia* L. By the Spey side, between Aviemoor and Rothiemurchus, † 96.—*S. repens* L. On Ben Eay, * 105. Apparently rare.—*S. Myrsinites* L. On the south side of Glen Ennich, not uncommon but very much disfigured with galls, 96 — *S. herbacea* L. Ben Eay, † 105. — *S. nigricans* L. By the Spey side near Kinchurdy, 96, apparently rare. Two forms were noticed : one near *cotinifolia*, named, I presume, after *Rhus Cotinus: cotonifolia* in L. C. appears to be a misprint.

*Myrica Gale* L. Glen Avon, † 94.

*Juniperus communis* L. Glen Avon, † 94. — *J. nana* Willd. Cairngorm, * 96. Loch A'an, * 94.

*Listera cordata* Br. Loch Maree side, † 105.

*Orchis incarnata* L. Near Boat of Garten, * 96.

*Habenaria bifolia* Br. Balblair, * 96. Dunphail, * 95. — *H. viridis* Br. Glen Ennich, 96.

*Malaxis paludosa* Sw. Near Boat of Garten, growing with *Hypnum lycopodioides*, † 96. Plentiful by Loch Maree side, † 105.

*Potamogeton natans* L. Loch near Forres, * 95.— *P. polygonifolius* Pourr. Both floating and heath state, near Dunphail, * 95 ; and in Glen Avon, † 94.

**Sparganium minimum* Fries. Loch Mallachie, * 96.—*S. natans* L. (*S. affine* Schn.). Avinloch, 96. Loch on Ben Eay, * 105. — *S. ramosum* Huds. By the Spey, Inverdruie, * 96. Loch near Forres, * 95.—*S. neglectum* Beeby. Not seen.

*Typha latifolia* L. Near Dingwall, * 106. Splendid specimens.

*Juncus supinus* Moench., var. *fluitans* Fries. In the Spey near Boat of Garten, * 96.—Var. *uliginosus* Fries. Loch Morlich, * 96. —*J. trifidus* L. Abundant on the quartzite of Ben Eay, † 105.

*Luzula erecta* Desf.—Var. *congesta* (Lej.). Common on the moorland, * 94, * 95.—*L. arcuata* Wahl. Cliffs about Loch A'an, † 94. Splendid specimen, seven or eight inches, in Corries Sneachda, Leacainn and Glen Ennich.

*Carex lagopina* Wahl, 1803, *C. approximata* Hoppe, 1800. Abundant over a limited area in one of the Cairngorm corries. It has a very restricted perpendicular range. The specimens were in good condition and in fine fruit, * 96. — *C. dioica* L. Glen Avon, † 94. A form with fruit rather deflexed, near Boat of Garten, 96. —*C. aquatilis* Wahl., var. *elatior* Bab., 1845 (*Watsoni* Syme). I do not know the reason for retaining Syme's name when that of

Babington is not only earlier but more appropriate. If Mr. Watson's name (it is worthy of any honour) is to be connected with *C. aquatilis* L., it should be with the small alpine variety, which is now assumed to be the type: for if I am not mistaken, it was this which Mr. Watson found on the Clova tableland. The lowland plant which must bear Babington's name, I found in three or four stations by Spey side, sparingly at Kinchurdy, but abundantly near Downe, * 96. In Top. Bot. Moray 96 is given instead of Moray 95.—*C. atrata* L. Very fine specimens occurred on the south side of Glen Ennich, 96.—*C. Goodenowii* J. Gay, var. *juncella* (Fries). By the Spey near Auchgourish, * 96, and near Altyre, * 95. — *C. curta* Good. Near Aultnancaber, Glen More, * 96. —Var. *alpicola* (Wahl). Glen Ennich, * 96.— *C. limosa* L. Marshy ground near Loch Mallachie, * 96.—*C. saxatilis* L. (*C. pulla* Good.). South end of Glen Ennich in fair quantity, 96.—*C. flava* Loch A'an side, * 94. — Var. *lepidocarpa*, Top. Bot., which is, I suppose, equal to *minor* Towns. Common by Loch Morlich, Loch Vaa, Loch Phitiulais, &c., * 96. Glen Avon, * 94, and near Dunphail, * 95. Also in Deeside, 92 †.— *C. vaginata* Tausch. Plentiful and variable in Glen Ennich and the corries of Cairngorm, † 96. Glen Avon, 94. — *C. echinata* Murray. — A small darker-glumed form occurred in Glen Ennich, and on the Cairngorms, 94, 96, but Mr. Arthur Bennett would scarcely refer it to var. *grypus* (Schk.). — *C. pilulifera*. Typical; Glen Avon, * 94.— *C. filiformis* L. Rare. Loch Mallachie. Loch Gahmna, 96.— *C. vesicaria* L. Near Downe, 96.—*C. binervis* Sm. —Abundant in 94, 95, 96 and 105, and very variable. On the higher cliffs of the Cairngorms and in Glen Ennich the dark-glumed form was prevalent. In W. Ross a tall plant, three to four feet high, with long thin spikes and pale glumes, occurred. — *C. xanthocarpa* Degl. Near Loch Clare, * 105.

*Scirpus setaceus* L. Kenlochewe, * 105. —*S. pauciflorus* Lightf. Marsh near Dunphail * 95.

*Phleum alpinum* L. A form with more cylindric panicle than the Clova plant secured in Corrie Sneachda, Glen Ennich, 96, and round Loch A'an, * 94. These were all *commutatum* Gaud.

*Alopecurus alpinus* Sm. This occurred also with a more lax and cylindric panicle in Corrie Sneachda, Leacainn and Glen Ennich; it is probable the var. *Watsoni* Syme.

*Phragmites communis* Trin. The form or variety *nigricans* Gren. et Godr., var. *uniflorus* Boreau, near Loch Garten, Loch Phitiulais, * 96, and near Dunphail, * 95.

*Deschampsia alpina* Roem. et Schult. Rock above west side of Loch A'an, † 94. Typical plants occurred also in Corrie Sneachda, 3600 feet; Corrie Leacainn, 3900 feet; and in Glen Ennich, 96.— *D. cæspitosa* Beauv., var. *altissima* (Lamk.) = *pallida* (Koch). Cairngorms, 96.—Var. *brevifolia* (Parn.). A common form on the corries, above 2000 feet. It occurred with whitish as well as with dark purple florets. In Corrie Leachainn it occurred with *D. alpina*, and its range in altitude exceeded that plant, * 92, 94, 96. —Var. *pseudo-alpina* (Syme). Corrie Sneachda. Glen Ennich,

96. Loch A'an, 94. Viviparous and dark coloured. Dr. Buchanan White has, I believe, found it recur to type in cultivation. The var. *brevifolia* appears to be joined to the type by a series of forms.—*D. flexuosa* Trin., var. *montana* (Huds.), non Linn. ed. 1. In beautiful condition on the Cairngorm ridge, and by the waterfall in Glen Ennich 96; also on rocks at head of Loch A'an 94, and on Ben McDhu, 92.—*D. discolor* Röem. et Schultz., *D. Thuilleri* Gr. et Godr., *Aira setacea* Huds., *A. uliginosa* Weihe. I was unaware this interesting grass had been found in Easterness by Mr. Groves at Loch an Eilan, until Mr. Arthur Bennett reminded me of it upon my return. I found it on the borders of Loch Phitiulais, by Loch an Eilan, Loch Gamhna, and very abundantly by Loch Dallas in Easterness. Also by a loch between Forres and the Findhorn, in the Altyre policies, * 95. In all cases the plant was in beautiful condition, the lovely green tufts of capillary leaves being noticeable for a considerable distance. This year was a very favourable one for it, the dry weather having reduced the level of the lochs considerably, so that the plant, which grows upon their margin, was more easily reached, was not so much crowded by aquatic vegetation, and doubtless flowered more frequently. At Loch Dallas *D. flexuosa* grew with it, and at Loch an Eilan it was close by. *D. discolor* appears to me a distinct species from *flexuosa*. The different habit, the paler, more nearly equal glumes, and the broader florets, well distinguish it. Moreover, the panicle branches of *flexuosa* are glossy, while those of *discolor* are dull, from the numerous scarious scales with which they are thickly covered. Under a 2-inch glass the branches of *flexuosa* are seen to be also clothed with scales, but they are much smaller, and more thinly scattered than in *discolor*.

*Molinia varia* Schrank. Principally as a small plant with short leaves and almost simple panicle, probably the var. *breviramosa* Parnell (of *cærulea*).

*Poa alpina* L. Rocks at head of Loch A'an, * 94. Abundant in Glen Ennich and Corrie Sneachda, † 96. Both viviparous and normal plants occurred.—*P. nemoralis* L. Spey side. Kinchurdy, * 96. Glen Tilt, * 89. In both cases this appears to be the var. *Parnellii* Bab.

*Agrostis palustris* Huds. (*A. alba* L.). A large form of this occurred in a marsh by the Spey, near Aviemore and Inverdruie. It is, I suppose, the variety (of *alba*) *palustris* Parnell, principally distinguished by its larger spikelets. This at first suggested a starved form of *Digraphis*.

*Festuca sciuroides* Roth. Loch Torridon, * 105.—*F. rubra* L. This occurred as the form *pruinosa* Hack., by Loch Torridon, * 105.

*Agropyron repens* Beauv., var. *barbatum* (Duval Jouve). Boat of Garten, &c., * 96. Near Forres, * 95. A purple-spiked form of *barbatum* occurred at Kingussie, 96, and a very glaucous form of *repens*. Near Aviemore, 96. The name *barbatum* is, I think, clearly preceded by that of *Leersianum* Gray, 1821.

*Lastrea Filix-Mas* Presl, var. *paleacea* Moore. Lieuthgoch, * 105.

*Polystichum Lonchitis* Roth. Glen Ennich. Cairngorm, 96.

Also a new record to * 105.—*P. lobatum* Presl. The small form (*genuinum* Syme) by the Findhorn, † 95.

*Athyrium Filix-fœmina* Roth., var. *convexum* (Newm.). Loch Maree, † 105. — *A. alpestre* Milde, var. *obtusatum* Syme. Corrie Leacainn, * 96.

*Osmunda regalis* L.  * 105.

*Botrychium Lunaria* L.  † 105.

*Lycopodium clavatum* L.  † 94.—*L. complanatum* L., "Hook. fil." Cairngorms, * 94, * 96, * 105.—*L. alpinum* L.  † 94.--*L. Selago* L. Loch A'an, † 94.—*L. Selago* L., var *recurvum* Syme. Ben Slioch, * 105. Loch Brandy, 90. — *L. selaginoides* L. Glen Avon, 94.

*Isoetes lacustris* L. Loch Ennich, large form. Loch Garten. And a small form in Loch Morlich, 96 †.

*Equisetum sylvaticum* L., var. *capillare* Wahl. Kingussie. Ardvroilach, * 96. I also saw this in * 93, North Aberdeen.—*E. hyemale* L. Near Boat of Garten, 96.

*Nitella flexilis* Agardh. Spey at Aviemore, and Loch Mallachie, 96 *.

*Chara fragilis* Desv. Loch Phitiulais, 96 *.

Several critical Graminæ, &c., are in Mr. Arthur Bennett's hands.

---

# ON A NEW *SELAGINELLA* FROM NEW GUINEA.

## By Baron von Mueller and J. G. Baker.

The following very distinct new species of *Selaginella* has lately been collected by the botanist of the Cuthbertson Expedition (Mr. W. Sayer) on the mountains of New Guinea. It belongs to the subgenus *Stachygynandrum*, to the series *Caulescentes*, and to the group *Flabellatæ* near *S. usta* and *caulescens*.

*229. **Selaginella angustiramea**, F. M. et Baker, n.sp. — Stem slender, continuous, stiffly erect, under a foot long, simple in the lower quarter, tripinnate in the upper three-quarters; pinnæ erecto-patent, oblong-deltoid, 2–3 in. long; ultimate branches sometimes 1–1½ in. long, including the leaves not more than 1-16th in. diam. Leaves of the lower plane crowded, firm in texture, erecto-patent, ovate, subobtuse, 1-24th in. long; of the upper plane rather shorter, more ascending, oblique, ovate, acute. Spikes short, terminal on the branchlets, 1-24th in. diam.; bracts ovate, acute, strongly keeled.

New Guinea; Mount Obree, alt. 7000 ft., *W. Sayer.*

---

# SHORT NOTES.

Botany of the Steep Holmes. — Staying at Weston-super-Mare a few days during the past summer, I took the opportunity of visiting the Steep Holmes, an island in the Severn. It is a rock of about one mile and a half in circumference, and in many places overhangs the water, being inaccessible except by two narrow passages very difficult of access, rising from the small pebbly beach on the north-east and south-western sides. The island is an outlying mass of the mountain limestone of the Mendip range, on the axis of the chain prolonged under the sea, the one being connected with Crook's Peak by the links of Brean Down, Uphill, and Bleadon Hill; the other with Banwell Hill by those of Bearn Back and Worle Hill. The summit is a sandy unfruitful soil, bearing little grass or any vegetable except those that seem peculiar to such situations. The Steep Holmes, whose summit rises about 400 feet above the level of the Channel, is known to botanists as the habitat of the single peony (*Pæonia corallina* Retz.), first added to the British Flora by the late Mr. F. Boucher Wright, of Hinton-Blewett, Somerset, in 1803, then growing in great profusion in the rocky clefts of the island, where it is conjectured to have grown for ages, but of late years, owing to the rapacity of collectors, it has become very scarce; but was glad to find on this visit, as well as others made in recent years to the island, the peony was gradually increasing. In 1848 and 1853 it was scarcely obtainable. The following plants were observed on my visit:—*Fumaria officinalis* L.; *Brassica oleracea* L.; *Silene maritima* With.; *Hypericum montanum* L.; *Lavatera arborea* L., on the north side of the island; *Erodium maritimum* L'Her.; *Smyrnium Olusatrum* L.; *Coriandrum sativum* L. (naturalised); *Crithmum maritimum* L.; *Hedera Helix* L.; *Sambucus nigra* L.; *Inula crithmoides* L.; *Statice occidentalis* Lloyd; *Ligustrum vulgare* L.; *Euphorbia Lathyris* L. (naturalized on the declivities of the island); *Allium Ampeloprasum* L. From its great abundance in the island Ray gave this the specific name of " *Allium Holmense spherico capite* "—the great round-headed garlick of the Holm Islands; of late years it has become less plentiful. *Suæda fruticosa* Forsk. has not been observed on the island for many years. It rests on the authority of Lobel. I would add, the Flat Holm is about three miles to the northward of the Steep Holm, and about one mile and a half in circumference. Being under cultivation it affords little or no interest whatever to the botanist.—T. BRUGES FLOWRR.

HIERACIUM GIBSONI Backh. AND CAREX IRRIGUA Hoppe IN WESTMORELAND.—In July, 1883, I gathered, in the company of the Rev. R. P. Murray, a specimen of the former plant on mountain limestone, near Kirkby Stephen. When remounting it, this year, I was struck by its appearance; and, on careful comparison with specimens from Settle, and with Mr. Backhouse's 'Monograph,' feel confident about the name. A few days previously we found a sedge in a wet sphagnous bog above the Mazebeck, between Caldron Snout and Highcup Scar, which we thought to be *C. limosa*, but when I was looking through my sedges lately it seemed more properly referable to *C. irrigua*. Mr. Arthur Bennett has kindly examined a plant, and confirms the name. These appear to be additions to the county flora.—EDWARD S. MARSHALL.

EQUISETUM SYLVATICUM L., *var*. CAPILLARE Hoffm., IN W. SUSSEX. —In August last I found this beautiful form in clayey copses near

Fernhurst. Mr. W. H. Beeby agrees in the naming, and tells me that it grows in Surrey, near Haslemere, with other forms. The species is not recorded for Vice-county 13 in 'Topographical Botany,' ed. 2.—EDWARD S. MARSHALL.

CRACKLING SOUND OF UTRICULARIA.—I write to ask an explanation of the distinct crackling sound produced by *Utricularia vulgaris* when it is disturbed. I had for some time supposed it was to be heard only from fronds removed from the water and beginning to dry, but I find the same phenomenon when the plants,—old ones filled with sacs, still in the water,—are disturbed. On shaking such a stem the rattle is distinct, the separate clicks being as loud as those made by slowly winding a watch. After the first series of clicks, the plant must rest some time before a second disturbance will produce a second fusillade.—D. S. KELLICOTT (in Bot. Gazette, Nov. 1887, p. 276), Buffalo, N. Y.

---

## NOTICES OF BOOKS.

*A Manual of Orchidaceous Plants cultivated under Glass in Great Britain.* James Veitch & Sons, Royal Exotic Nursery, Chelsea. Part II., *Cattleya, Lælia,* &c.

THE second part of this work deals with *Cattleya, Lælia,* and some small closely-allied genera. The two former have been excessively subdivided in horticultural works, and those botanists who have tried to keep pace with the horticulturists in this matter, and have almost given it up in disgust, will surely be surprised to find in a horticultural work, that *Cattleya* has but 18 species (with 9 supposed natural hybrids in addition); for something like 50 or 60 have been fully described, and the garden names are yet more numerous. *Cattleya labiata* absorbs something like 20 described and so-called " species," the numerous forms being simply treated as varieties and subvarieties, a method of treatment with which few botanists will be disposed to quarrel. Besides a full statement under each species of the habitat, five elaborate maps are devoted to the geographical distribution of the different species of *Cattleya, Lælia,* and *Odontoglossum.* The key to the situation is admirably supplied in the following extract:—" The position of some of the names on the maps illustrating the geographical distribution of *Cattleya* and *Lælia* must be accepted as approximately correct only. In such cases, the true habitat of the species has either been too vaguely recorded, or it has been purposely withheld for trade objects, to which the interests of science are, unfortunately, often regarded as altogether subordinate."

A word as to hybrids. In *Cattleya* 13 artificial hybrids, and 9 supposed natural ones are given, while in *Lælia* the numbers are 15 and 7 respectively; and it is curious to note that with all these artificial hybrids they only appear to have solved the parentage of one of the supposed natural ones. Of *L. lilacina* it is remarked:— " A supposed hybrid between *Lælia crispa* and *L. Perrinii,* the same two species from which Dominy raised *L. Pilcheri.* The natural hybrid differs from the artificially raised one chiefly in the form of the labellum and in the season of flowering." These artificial hybrids requires from eight to nineteen years to arrive at the flowering stage, so that we can scarcely hope to have the parentage of all the supposed natural hybrids solved at present. Singularly enough, 2 of the artificially raised *Cattleyas*, and no less than 11 of the *Lælias*, were raised by crosses between the two genera ; a matter which furnishes food for reflection as to why these bigeneric hybrids should pertain to the former genus in the one case and to the latter one in the other. They adopt *Sophro-cattleya* for a hybrid between *Cattleya intermedia*. Why was not this hybrid either a *Cattleya* or a *Sophronitis?* *Sophronitis* and *Lælia* are two exactly parallel genera ; both have eight pollen-masses ; and *Cattleya* differs from both in having but four. The fact is we have either a true bigeneric hybrid in both cases or in neither. What is sauce for the goose will have to be sauce for the gander, for *Sophronitis* and *Lælia* are both more distinct from *Cattleya* than they are from each other.

It seems a pity that the Messrs. Veitch should have commenced a fresh pagination for the second part, as a consecutive pagination of the volume would have been more handy for citation. A considerable number of excellent woodcuts greatly enhance the value of the work. R. A. ROLFE.

THE last part of the 'Icones Plantarum,' dated November, 1887, contains the following new genera :—*Polydragma* Hook. f. (Euphorbiaceæ, Crotoneæ) ; *Sphyranthera* Hook. f, (Euphorbiacæ) ; *Scortechinia* Hook. f. (Euphorbiacæ, Phyllantheæ ?) ; *Megaphyllæa* Hemsl. (Meliaceæ, Trichilieæ) ; *Lophopyxis* Hook. f. (Euphorbiaceæ?) : *Petrocosmea* Oliv. (Gesneraceæ, Cyrtandreæ).

THE BARON F. VON MUELLER has made some progress with the third of the series of illustrated monographs with which, through the liberality of the Victorian Government, he has been enabled to enrich botanical literature. The iconography of *Eucalyptus* and *Myoporineæ* is now succeeded by an 'Iconography of Australian species of *Acacia*,' of which four quarto parts, each containing ten species, have been issued at the moderate cost of three shillings each. The text is confined to an explanation of the plates, which have been prepared by Mr. Robert Graff. The term indefatigable has often been applied to Baron von Mueller, and we are glad to see that it still remains one of his most striking attributes.

NEW BOOKS.—H. E. F. GARNSEY & I. BAYLEY BALFOUR, 'Lectures on Bacteria' (Oxford, Clarendon Press : 8vo, pp. xii. 193 : 20 cuts : price 6*s.*). — L. DOSCH, 'Excursions-Flora . . . des Grossherzogtums Hessen' (Giessen, Roth, "1888" : ed. 3, 8vo, pp. cviii. 616).—E. G. CAMUS, 'Catalogue des Plantes de France, de Suisse, et de Belgique' (Paris, Dupont, "1888" : 8vo, pp. vii. 325).— A. KERNER, 'Pflanzenleben . . . 1 Band, Gestalt und Leben der Pflanze' (Leipzig : large 8vo, pp. x. 734 : 20 coloured plates, 553 cuts).—E. COSSON, 'Compendium Floræ Atlanticæ,' vol. (Ranunculaceæ—Cruciferæ) (Paris : 8vo, pp. cviii. 367).

ARTICLES IN JOURNALS.

*American Naturalist* (Nov.). — E. L. Sturtevant, 'History of Garden Vegetables.' — A. P. Morgan, 'The Genus *Geaster*' (*G. campestris, G. delicatus*, spp. nn.).

*Annals of Botany*, No. 2 (Nov.).—J. D. Hooker, '*Hydrothrix*, a new genus of *Pontederiaceæ* (*H. Gardneri*, sp. unica: 1 plate). — F. W. Oliver, 'Obliteration of sieve-tubes in *Laminarieæ*' (2 plates). M. Treub, 'On the life-history of Lycopods.'—F. O. Bower, 'Modes of climbing in *Calamus*.' — Id., 'On the terms "Phyllome" and "Caulome."' — J. R. Vaizey, 'Absorption of Water in Mosses.'— D. Morris, 'On certain plants as Alexipharmics or snake-bite antidotes.'—B. L. Robinson, 'On the Genus *Taphrina*.'

*Bot. Centralblatt.* (No. 48).—R. Keller, 'Bildungsabweichungen der Blüten angiospermer Pflanzen' (1 plate).—(No. 49). J. Wollheim, 'Untersuchungen über den Chlorophyllfarbstoff.'

*Bot. Gazette* (Nov.). — J. M. Coulter & J. N. Rose, 'Notes on *Umbelliferæ* of E. United States' (1 plate). — A. J. Stace, 'Plant Odours.'—W. J. Beal & C. E. St. John, '*Silphium perfoliatum* and *Dipsacus laciniatus* in regard to insects.'

*Bot. Zeitung* (Nov. 25). — H. Hoffmann, 'Culturversuche über Variation.' — (Dec. 2, 9, 16). J. Wostmann, 'Zur Kenntniss der Reizbewegungen.' — (Dec. 9). O. Loew, 'Ueber die Formose in pflanzenchemischer Hinsicht.'

*Bull. Soc. Bot. France* (xxxiv.: Comptes rendus 6: Dec. 1).— D. Clos, '*Stachys germanica, intermedia*, and *biennis*.' — P. A. Dangeard & —. Barbé, 'La polystélie dans le genre *Pinguicula*.'— M. Gandoger, 'Plantes de Gibraltar.' — —. Granel, 'Sur l'origine des suçoirs de quelques phanérogames parasites' (tt. 2).—P. Sagot, 'Sur le genre Bananier.' — A. Chatin, 'Flore Montagnarde.'— E. Mer, 'Sur la formation du bois parfait dans les essences feuillues.'

*Bull. Torrey Bot. Club* (Dec.). — G. N. Best, 'On the Group *Carolinæ* of the genus *Rosa*.' — E. L. Greene, 'Bibliographical Notes' (*Nymphæa* and *Nuphar*).—K. B. Claypole, 'Colour of *Caulophyllum thalictroides*.'

*Flora* (34–36: Dec.). — R. Diez, 'Ueber die Knospenlage der Laubblätter.'

*Gardeners' Chronicle* (Dec. 3).—*Oxalis imbricata*, fl. pl. (fig. 129). *Nepenthes Curtisii* Mast. (fig. 133). *Dendrobium trigonopus* Rchb. f., *Mormodes vernixium* Rchb. f., *Cryptophoranthus maculatus* Rolfe, spp. nn. — (Dec. 10). *Kniphofia Kirkii* Baker, *Alocasia marginata* N. E. Br., *Masdevallia sororcula* Rchb. f., *Octomeria supraglauca* Rolfe, spp. nn.—Germination of Carica (figs. 138, 139).—*Athrotaxis* (figs. 140–145). — (Dec. 17). *Dendrobium rutriferum* Rchb. f., *Phalænopsis Regnieriana* Rchb. f., spp .nn.—H. N. Ridley, *Solanum cornigerum* (fig. 148).

*Journal de Botanique* (Dec. 11). — P. van Tieghem, 'Structure de la racine des les Centrolépidées, Joncées, etc.' — P. Vuillemain, 'Sur une maladie des Cerisiers et des Pruniers.' — (Dec. 15). E.

Wasserzug, 'Principes procedés de coloration des Bactéries.' —
D. Bois, 'Herborisations dans la département de la Manche.'—
E. Bondier, ' *Tremella fimetaria* Schum.'

*Journ. Linn. Soc.* (Bot.: xxiv., No. 160: Nov. 28).—H. Trimen,
' Hermann's Ceylon Herbarium and Linnæus' 'Flora Zeylanica.' '
—R. A. Rolfe, ' Bigeneric Orchid Hybrids ' (1 plate). — H. Bolus,
' Contributions to S. African Botany' (*Melhania griquensis, Celastrus
maritimus, Lotononis foliosa, Crotalaria griquensis, Argyrolobium
megarhizum, Senecio sociorum, S. namaquanus, S. albopunctatus, S.
Rehmanni, Erica tetrastigmata, E. Baurii, E. Cooperi, F. Missionis,
F. urna-viridis, F. adenophylla, E. hæmantha, E. Tysoni, E.
Lerouxiæ, E. aspalathifolia, E. trichadenia, E. trachysantha, E.
caffrorum, E. Brownleeæ, E. eriocodon, E. inops, E. natalitia,
Philippia tristis*, spp. nn.). — D. H. Scott, ' On Nuclei in *Oscillaria*
and *Tolypothrix* ' (1 plate).—T. Ito, ' A *Balanophora* new to Japan '
(*B. dioica* Wall. (1 plate). — (No. 161: Nov. 30). H. N. Ridley,
' A New Genus of *Orchideæ* from the Island of St. Thomas ' (*Orestia:
O. unica*, sp. unica : 1 plate). — S. le M. Moore, ' The influence of
light upon protoplasmic movement.' — M. C. Potter, ' An Alga
(*Dermatophyta radicans*) growing on the European Tortoise ' (1 plate).
—C. Spegazzini & T. Ito, ' Fungi Japonici nonnulli ' (*Fusarium
oidioide* Speg., *Phyllosticta Tokutaroi* Speg., *Tuberculina japonica*
Speg., spp. nn.). — J. G. Baker, ' Ferns from West Borneo '
(*Matonia sarmentosa, Davallia pinnatifida, D. nephrodioides, Asplenium
crinitum, Nephrodium subdigitatum, Polypodium subasboreum, P. quin-
quefurcatum,, Gymnogramme chrysosora, G. campyloneuroides, Acrostichum oligodictyon*, spp. nn.). — F. B. Forbes & W. B. Hemsley,
Index Floræ Sinensis, part iv. (Dec.) (*Astilbe polyandra, Saxifraga
tabularis, Hydrangea longipes, Deutzia discolor, Sedum filipes, D. polytrichoides* (t. 7), *Eugenia fluviatilis, Thladiantha? Henryi, T. nudiflora* (t. 8), *Begonia Henryi, Acanthopanax diversifolium, Cornus
hongkongensis*, spp. nn. : all of Hemsley).

*Oesterr. Bot. Zeitschrift* (Dec.).—R. v. Wettstein, ' Ueber einen
abnormen Fruchtkörper von *Agaricus procerus*.'—O. Stapf, ' Ueber
einige Iris-Arten des bot. Gartens in Wien.' — B. Blocki, ' *Rosa
Herbichiana*, sp. n.'

---

## LINNEAN SOCIETY OF LONDON.

*Dec.* 1st.—William Carruthers, F.R.S., President, in the chair.
The following gentlemen were elected Fellows of the Society :—
Mr. K. H. Bennett (Sydney, N.S.W.), Lord Egerton of Tatton,
Mr. W. Francis, the Rev. F. W. Galpin, Mr. W. S. McMillan, Mr.
A. J. North (Melbourne, Victoria), Mr. J. Ogilby (Sydney, N.S.W.),
Mr. A. S. C. Stuart (Madras), Mr. G. Swainson, Mr. I. C. Thompson, and Mr. C. Topp (Melbourne, Victoria). There was exhibited
for Mr. O. Fraser, F.R.S., of the India Museum, Calcutta, a
specimen of what was supposed to be a weather-worn seed of a
palm. In a letter read at the meeting from Mr. Fraser, the object
in question was stated to have been picked up on the Madras

coast, but could not be satisfactorily identified by the Indian authorities. The specimen having been examined by Dr. J. Anderson (late of Calcutta), and Mr. Denny of the British Museum, they inclined to regard it as possibly the consolidated roe of a fish, while Prof. C. Stewart surmised the substance as vegetable in structure; decision was left *sub judice*, pending further microscopic and chemical investigation. A paper was read by Mr. C. B. Clarke, " On a new Species of *Panicum*, with remarks on the Terminology of the *Gramineae*."

---

OBITUARY.

On Friday, the 14th October, 1887, Mr. JOHN PRICE, M.A., died at his residence in Chester. Mr. Price was eighty-four years of age, but long years ago he was familiarly known by a wide circle of friends as " Old Price "—a title self-bestowed, but which seemed quite naturally to incorporate the many aspects of his character as he was known, not only to those who came into more immediate contact with him as teacher, friend, or philanthropist, but to those fortunate enough to be familiar with some prose essays and odd papers and poems bound together in a now scarce volume, entitled, with characteristic fancy, "Old Price's Remains." Mr. Price was born at Pwllycrochon, North Wales. He received his earlier education at Chester, and afterwards went to Shrewsbury School, where, under Dr. Butler, he had Charles Darwin as a school-fellow. He went afterwards to St. John's College, Cambridge, where he became a tutor. After a residence extending over many years at Birkenhead, he eventually made Chester his home. Mr. Price was an honorary member and chairman of the botanical section of the Chester Natural Science Society, and was the second recipient, Mrs. Kingsley being the first, of the Kingsley memorial medal, given by the Society to those who have contributed materially to the advancement of Science in the District. Mr. Price took special delight in biological study, and more particularly in botanical research. As a teacher and lecturer his methods were unique, and full of an originality and refined humour peculiarly his own. His eagerness and devotion to any subject which commended itself to him ended often in the announcement of unexpected discoveries and a propounding of various theories, which set numbers of his hearers thinking and working in new fields. In his garden he had his beds filled with large plants of the giant *Heracleum*, which always yielded abundant matter for speculative reasoning and close examination; he carried with him frequently for weeks in a time, leaves of *Cardamine* in a small phial of water to watch the process of development in leaf-propagation. His lectures and letters were always characteristic and delightful, marked invariably with that strong individuality which he never lost.—EDWARD J. BAILLIE.

coast, but could not be satisfactorily identified by the Indian authorities. The specimen having been examined by Dr. J. Anderson (late of Calcutta), and Mr. Dendy of the British Museum, they inclined to regard it as possibly the consolidated roe of a fish, while Prof. C. Stewart surmised the substance as vegetable in structure; decision was left *sub judice*, pending further microscopic and chemical investigation. A paper was read by Mr. C. B. Clarke, "On a new Species of *Panicum*, with remarks on the Terminology of the *Gvamineæ*."

## OBITUARY.

On Friday, the 14th October, 1887, Mr. JOHN PRICE, M.A., died at his residence in Chester. Mr. Price was eighty-four years of age, but long years ago he was familiarly known by a wide circle of friends as "Old Price"—a title self-bestowed, but which seemed quite naturally to incorporate the many aspects of his character as he was known, not only to those who came into more immediate contact with him as teacher, friend, or philanthropist, but to those fortunate enough to be familiar with some quaint essays and odd papers and poems bound together in a now scarce volume, entitled, with characteristic fancy, 'Old Price's Remains.' Mr. Price was born at Pwllycrochon, North Wales. He received his earlier education at Chester, and afterwards went to Shrewsbury School, where, under Dr. Butler, he had Charles Darwin as a school-fellow. He went afterwards to St. John's College, Cambridge, where he became a tutor. After a residence extending over many years at Birkenhead, he eventually made Chester his home. Mr. Price was an honorary member and chairman of the botanical section of the Chester Natural Science Society, and was the second recipient (Mrs. Kingsley being the first) of the Kingsley memorial medal given by the Society to those who have contributed materially to the advancement of Science in the district. Mr. Price took special delight in biological study, and more particularly in botanical research. As a teacher and lecturer his methods were unique, and full of an originality and refined humour peculiarly his own. His attention and devotion to any subject which commended itself to him ended often in the announcement of unexpected discoveries and a propounding of curious theories, which set numbers of his hearers thinking and working in new fields. In his garden he had his beds filled with huge plants of the giant *Heracleum*, which always yielded abundant matter for speculative reasoning and close examination; he carried with him, frequently for weeks at a time, leaves of *Cardamine* in a small phial of water to watch the process of development of leaf-propagation. His lectures and letters were always characteristic and delightful, marked invariably with that strong individuality which he never lost.—EDMUND J. BAILLIE.

Tab. 280.

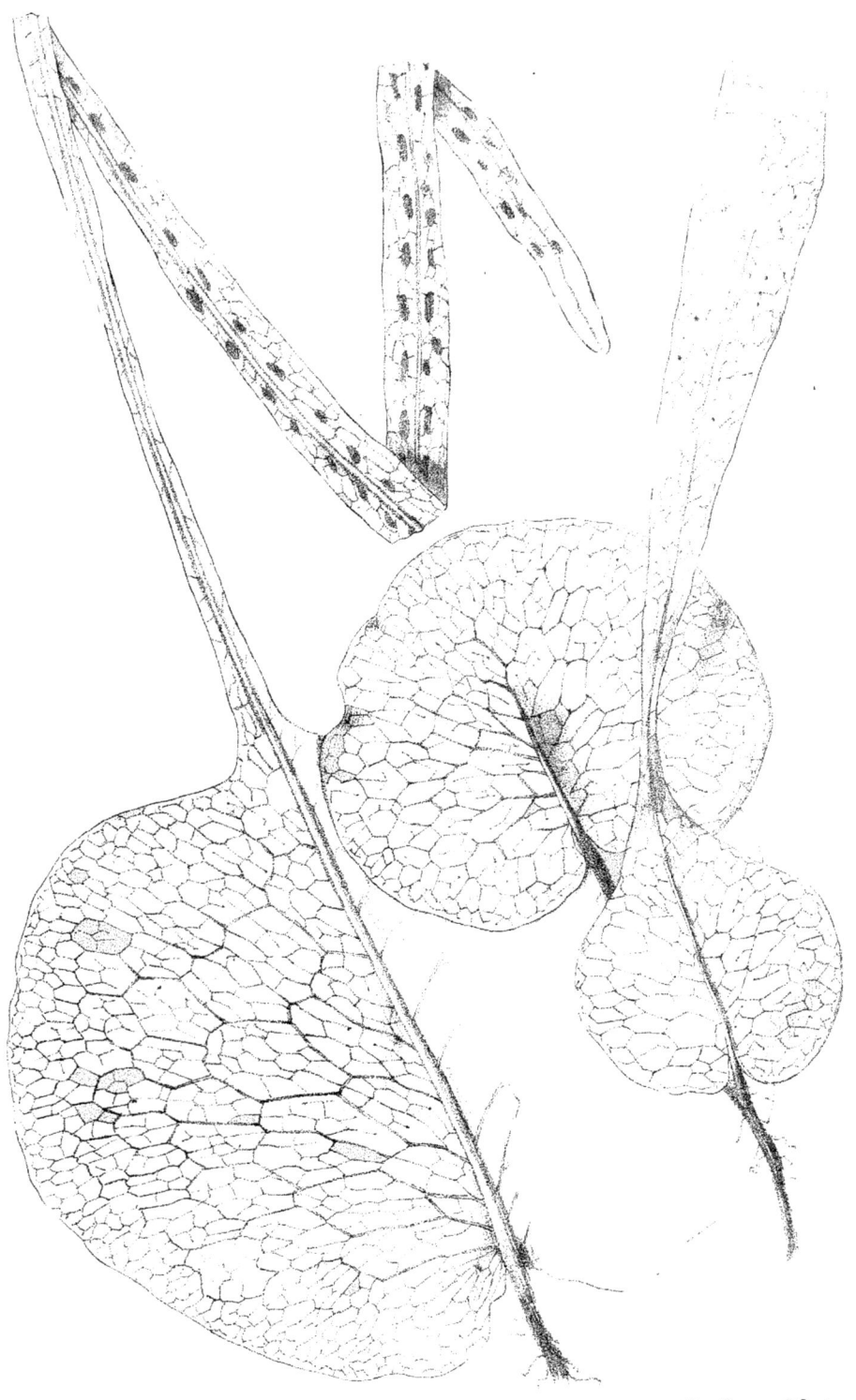

R. Morgan del et lith.

West, Newman & Co. imp.

Polypodium annabellæ, H.O Forbes.

# A NEW FERN FROM NEW GUINEA.

## By H. O. Forbes, F.R.G.S., A.L.S.

### (Plate 280).

**Polypodium Annabellæ**, sp. n. — Rhizome slender, clothed with scattered, brown, membranaceous, lanceolate-acuminate scales, creeping on tree-stems, and giving off branching rootlets densely covered with rich brown hairs. Fronds sessile on the creeping rhizome, at intervals so that the base of one is overlapped by the apex of the one below, simple, membranaceous, the barren fronds and the sterile portion of the fertile fronds orbicular, with entire margins; veins elevated, repeatedly branching and enclosing large oblique irregular areolæ, with small inclosed areolæ, and a few free veins; the veins terminate a little within the margin of the frond, and, being looped, form an irregular intramarginal line, with an occasional free veinlet extending outwards. In the barren frond the midrib disappears in the venation about the middle. The fertile portion is produced from the apex of the barren portion of frond, the midrib being continued beyond this portion with a narrow wing, for two or three inches, and then expanding into a linear-lanceolate lamina, with similar venation to the rest of the frond. The sori are somewhat oblong, and are borne in a single or double series at the union of the veinlets within the meshes.

Found at an elevation of 2000 ft., above the Murray River, on the stems of trees, on a coast-trending outlier of (perhaps) the Owen Stanley Range. The upper surface of the frond is close adpressed to the tree.

[Note.—This interesting fern, discovered by Mr. Forbes, and named by him in honour of his brave and devoted wife, belongs to the section *Drynaria* of *Polypodium*, and to the small group of that section in which the sorus-bearing portion of the frond is only an extension of the lower barren part: it is a much simpler form than any yet known.—W. Carruthers.]

---

# ON A COLLECTION OF FERNS MADE BY BARON EGGERS IN ST. DOMINGO.

## By J. G. Baker, F.R.S.

The following is a list of the Ferns and other Vascular Cryptogamia collected by Baron Eggers during his recent exploration of St. Domingo, with descriptions of the novelties:—

2035, 2173. *Gleichenia pubescens* H.B.K., two varieties.
2735, 1854? *Cyathea arborea* Sm.
2060. *C. Schanschin* Mart., or a near ally.

2738. *Hemitelia horrida* R. Br.
2180. *Alsophila pruinata* Kaulf.
2750. *Hymenophyllum hirsutum* Sw.
2753. *Trichomanes muscoides* Sw.

1590. *T. pusillum* Sw.
2684. *T. Krausii* H. & G.
2746. *T. scandens* L.
2749. *T. tenerum* Spreng.
2735 b. *Dicksonia cicutaria* Sw.
2653. *D. rubiginosa* Kaulf.
2653 b. *Davallia inæqualis* Kunze.
2027. *D. aculeata* Sw.
2507. *D. clavata* Sw.
2588. *Adiantum deltoideum* Sw.
2502. *A. cristatum* L.
2243. *Pellæa ternifolia* Fée.
2822. *Cheilanthes microphylla* Sw.
2526. *Pteris longifolia* L.
2818. *P. mutilata* L.
2172. *P. aquilina* L.

2805. *P. grandifolia* L.
2782. *P. aculeata* Sw.
2041. *Lomaria procera* Spreng.
2268. *Asplenium monanthemum* L.
2537. *A. dentatum* L.
2665. *A. auriculatum* Sw.
2092. *A. præmorsum* Sw.
2774. *A. rhizophorum* L.
2639. *A. cicutarium* Sw.
2788. *A. arboreum* Willd.
1866 a, b. *Nephrodium patens* Desv.
1698. *N. Sloanei* Baker.
2780. *N. sanctum* Baker.
2775. *N. conterminum* Desv.
2306. *N. Filix-mas* Rich.
1585. *N. hirtum* Hook.

1575 (142*). **Nephrodium myriolepis**, n. sp.—Caudex erect. Stipes 8–9 in. long, tufted, densely crinite up to the top with spreading hair-pointed brown paleæ, as is the main rachis. Lamina ovate, tripinnate, 1½ ft. long, moderately firm in texture, green and glabrous on both surfaces. Many lower pinnæ ovate- or oblong-lanceolate, 2 in. broad, the lowest pair rather shortened, produced on the lower side. Pinnules many, close, sessile, lanceolate, ⅓ in. broad, cut down to the rachis into entire linear-oblong parallel obtuse tertiary segments ½ line broad. Veins very few and distant. Sori one at the base of each tertiary segment. Indusium small, glabrous, persistent.—Allied to *N. amplum* and *Grisebachii*.

2733. *Nephrodium scolopendrioides* Hook.
1866. *N. molle* Desv.
2513. *N. cicutarium* Baker, var. Identical with a small little-cut form frequent in Jamaica.
2529. *Aspidium semicordatum* Sw.
2821. *A. plantagineum* Griseb.
2560. *A. trifoliatum* Sw.
2779, 1829, 2801. *Nephrolepis exaltata* Schott., forms.

1577. *Polypodium reptans* Sw.
2806.      var. *asplenioides*.
2595. *P. pectinatum* L.
2736. *P. sororium* H.B.K.
2789, 2307. *P. angustifolium* Sw.
2546. *P. repens* L.
2300. *P. lycopodioides* L.
2766. *Gymnogramme sulphurea* Desv.
1594. *Antrophyum lineatum* Kaulf.
2203. *Acrostichum viscosum* Sw.

2201 (52*). **Acrostichum** (ELAPHOGLOSSUM) **Eggersii**, n. sp.— Rhizome short-creeping, densely clothed with small spreading lanceolate dark brown paleæ of firm texture. Stipes of the sterile frond slender, stramineous, 4–5 in. long, clothed throughout with ascending whitish brown ovate membranous paleæ. Sterile frond lanceolate, 5–6 in. long, ¼–⅓ in. broad at the middle, narrowed gradually to the base and apex, subrigid in texture, green above, with a few bleached paleæ, densely persistently coated all over beneath with ovate membranous fimbriated pale brown paleæ. Fertile frond narrower, with a much longer stipe.—Lamina like that of *A. simplex*; paleæ of the under surface most like those of *A. squamosum*.

2604. *Acrostichum sorbifolium* L.  1655. *A. adiantifolia* Sw.
2755. *A. nicotianæfolium* Sw.  2820 b. *Lygodium venustum* Sw.
2046. *Anemia Phyllitidis* Sw.

2536 (7*). **Lygodium gracile,** n. sp. — Stem very slender, sarmentose, glabrous. Sterile pinnæ long-petioled, deltoid, tripinnate, glabrous, moderately firm .in texture; lower pinnules deltoid, with compound lower tertiary segments; ultimate segments lanceolate, ½–1 in. long, ⅛ in. broad. Veins lax, distinct, forked, erecto-patent. Ultimate fertile segments deeply pinnatifid; spikes of fruit placed at end of the lobes, not above 1-12th in. long.—Most resembles *L. microphyllum* R. Br.

2737. *Danæa nodosa* Smith.  2158. *L. clavatum* L.
2055. *Lycopodium reflexum* Lam.  2270. *L. complanatum* L.
2161. *L. dichotomum* Jacq.  2611. *Selaginella albo-nitens*
2174. *L. subulatum* Desv.  Spring.

So far, therefore, it would seem that the ferns of St. Domingo correspond closely with those of Jamaica. Here, as in Jamaica, we obtain characteristically temperate types (e. g. *Nephrodium Filix-mas* and *Lycopodium clavatum*) on the high mountains. Besides the three novelties, this collection adds to the West Indian flora *Pellæa ternifolia** and *Lycopodium subulatum,* two widely spread Mexicano-Andine species.

---

# ON NOMENCLATURE.

## By W. H. Beeby, A.L.S.

In Mr. B. D. Jackson's concluding paper on the ' Nomenclature of the London Catalogue' (Journ. Bot. 1887, p. 334), he proposes an alteration in the statement of the name " *Sparganium ramosum* Curtis"; and in the same number of this Journal (p. 349), Mr. Druce asks, —" Why write *Sparganium ramosum* Curtis, why *S. affine* Schniz.? " Neither of these gentlemen has brought forward any evidence to show that *S. ramosum* Hudson can be regarded sa other than an aggregate name (*Cf.* Journ. Bot. 1885, p. 193, and 1886, p. 142); and if Mr. Jackson's proposition (which consists of the citation of Hudson's name with the supplementary quotation of a plate) be accepted, we must also concede to South European botanists the right to continue to apply Hudson's name to their *S. neglectum;* for there is not one word in Hudson's description which is not equally applicable to both plants, nor is there anything in their distribution to give a clue to either plant as being the one intended by him. As the above proposals have a somewhat wide bearing on the subject of nomenclature, I may perhaps be permitted to make a few observations. Although heartily in accord with the movement, of late so active in this country, to ascertain what are really the names of our plants, I think that something more is required than the hunting-up of the oldest name ever applied, but sometimes applicable only in the most

---

* Since this was written *Pellæa ternifolia* has also been received from Jamaica.

1655. *A. adiantifolia* Sw.
2820 b. *Lygodium venustum* Sw.

grace, n. sp. — Stem very slender,
nnæ long-petioled, deltoid, tri-
m in texture; lower pinnules
rary segments; ultimate segments
bad. Veins lax, distinct, forked,
ments deeply pinnatifid; spikes
ot above 1-12th in. long.—Most

2158. *L. clavatum* L.
2270. *L. complanatum* L.
2611. *Selaginella albo-nitens*
          Spring.

that the ferns of St. Domingo
Jamaica. Here, as in Jamaica,
te types (e. g. *Nephrodium Filix-*
high mountains. Besides the
to the West Indian flora *Pellæa*
, two widely spread Mexicano-

# NOMENCLATURE.

### W. H. BY, A.L.S.

n's concluding paper on the ' Nomenclature
e' (Journ. Bot. 1887, p. 334), he proposes
tement of he name " *Sparganium ramosum*
same number of this Journal (p. 349),
Why write *parganium ramosum* Curtis, why
either of the gentlemen has brought forward
that *S. ramum* Hudson can be regarded sa
egate name (. Journ. Bot. 1885, p. 193, and
if Mr. Jackso's proposition (which consists of
son's name with the supplementary quotation of a
d, we must ao concede to South European
t to continue apply Hudson's name to their
there is not on word in Hudson's description
ally applicable both plants, nor is there any-
stribution to give clue to either plant as being
by him posals have a some-
g on I may perhaps
heartily in
ntry, to
ink that
oldest

general way ; the far more difficult task remains of finding out the oldest name which is sufficiently exact in meaning to be applicable in a strict sense to the plant it is intended to represent.    One example of an aggregate which has found its way into the Catalogue is seen in the name " *Viola canina* L.," a name which has no meaning in an exact sense.    In the previous edition of the Catalogue this name stood " *Viola canina* auct."—a statement perhaps preferable to that of the present edition, but wanting in the preciseness which is found in " *Viola canina* Reich."    In the beautiful fasciculus of Swedish Violæ recently issued by Drs. Neuman, Wahlstedt, and Murbeck there occurs the following :— " *Viola canina* Reich. Plant. Crit. i. p. 59.—Linné Spec. plant. &c. p. min. p. ! "—and I think that the testimony of three Swedish botanists on such a point will be generally accepted.    Mr. Druce has called attention (Journ. Bot. 1887, p. 312) to another aggregate in *Rhinanthus Crista-galli* L. ; but I am at a loss to harmonise his remarks on that name with the apparent intention of his note on the *Spargania*.    For these latter Mr. Druce does not indicate his alternative names ; and with regard to *S. affine* Schniz. I am unable to imagine what alternative he would suggest.

I think that most botanists will agree with Mr. Druce in the protests he has from time to time made against the practice of making a man " say what he has not said " in the matter of varieties.    It is not apparent why a rule should be enforced for the transfer of species, yet be ignored in the transfer of varieties.    It is easy to show (as I have shown elsewhere) that the injustice inflicted on authors is quite as great in the latter as in the former case.    It is of course only too certain that a long time must elapse before all our varieties are properly adjusted in this respect; still it does seem desirable that the principle should be acknowledged, and carried out as far as possible, in order to prevent the making of more work which must eventually be undone.    That some botanists are alive to the necessity of applying the ordinary rule to varieties is shown by the Messrs. Groves' statement of the var. *e.* of *Chara vulgaris* in the last edition of the Catalogue.

In citing the name for restricted *Sparganium ramosum* as " *S. ramosum* Curtis " I did not make any new departure as regards the form of name, but merely followed what appeared to me to be the best course,—often, but not uniformly adopted.    Among other similarly restricted names I may mention that of *Viola canina* Reich., quoted above ; *V. silvestris* Reich. (*Reichenbachiana*) ; *Polygonum nodosum* Reich. (*maculatum*) ; *Sparganium natans* Fries (*Friesii* Beurling) ; to which I would propose to add *Juncus conglomeratus* Smith, or earlier authority.    The last-named plant has been called *J. Leersii* by Marsson (Fl. v. Neu-Vorpommern, 1869), on account of the uncertainty attaching to the old Linnean name, and the objection to the latter is sustained by a no less authority than Dr. F. Buchenau, who now (Krit. Zusammenst. der europ. Juncaceen, 1885, &c.) adopts *J. Leersii* Marss.    All the new names proposed for the above plants are open to the, as it appears to me, fatal objection that they do not refer the student to the earliest adequate description of the plant intended; it is, moreover, a manifest advantage to be able to retain such widely-known names as *Juncus conglomeratus*, when this can legitimately be done by merely amending the authority for such name. Had Marsson renamed the plant before any adequate description had been published under the old name, his name would then undoubtedly have claimed adoption. I might have accepted the name for restricted *Sparganium ramosum*, kindly proposed by Dr. Boswell for the 'London Catalogue'; but in the face of Curtis's sufficient description and plate, I did not feel able to do so.

*Note.*—Since this paper was communicated in November last, several articles on Nomenclature have appeared; as well as one by Mr. Druce, on Scotch plants, in the January number of this Journal. In the last-named paper the mystery of Mr. Druce's earlier name for *Sparganium affine* is explained. Apparently Mr. Druce does not know that *S. natans* is a distinct species,—far more distinct, indeed, from *S. affine* than is *S. simplex* in its floating forms. Mr. Druce's proposed adjustment of these names affords the best possible confirmation of the justness of the opinion expressed above, that "something more is required than the hunting-up of the oldest name ever applied." I have not gone to this length merely for the sake of pointing out that a mistake has been made about the name of a *Sparganium;* far wider issues are involved, affecting largely the nomenclature of our plants, through the, as I think, unsound principles on which that nomenclature has been, and is being, manipulated.

---

## " ENDOSPERM."

### By G. S. Boulger, F.L.S.

No one can deny the value, both from the point of view of the teacher and from that of the student wishing for clear general principles to guide him in original work, of a uniform system of descriptive terminology for the whole Vegetable Kingdom based upon ascertained homologies. This alone is one great reason for botanists to welcome the appearance of Prof. Goebel's 'Outlines of Classification and Special Morphology' in an English dress. As hinted in the author's preface to the work, there are two classes of sins against such a desirable uniformity: first, the use of several terms for structures, in various groups, now known to be homologons, as in the cited example of "placenta," "receptacle," and "columella"; and secondly, the use of one term for structures, in different groups, now known not to be homologous. This second class of misleading terms, of which the cases cited are "frons" and "pro-embryo," seems to be by far the more dangerous; and it is to be regretted that Prof. Goebel and his translators seem to have perpetuated one particularly striking case of it in what appears to the present writer to be a wholly unnecessary manner. This case is that of the term "endosperm" in Gymnosperms and Angiosperms. In the "Explanation of Terms," to the maintenance of which the translators, it must be stated, do not pledge themselves, this word appears with three, if not four, somewhat disparate significations, thus: "Endosperm. (*a*) In *Selaginella:* tissue formed in the cavity of the macrospore below the prothallium. (*b*) In Gymosperms: prothallium within the embryo-sac (macrospore); secondary endosperm may be formed as a nutritive tissue after the prothallium is absorbed. (*c*) In Angiosperms: tissue formed within the embryo-sac (macrospore) after fertilisation (commencing by division of the secondary nucleus), and serving for the nutrition of the embryo."

The confusion, however, appears in the body of the work; and, as it is at first apparently recognised and avoided, must be deliberate, Speaking of the *Selaginelleæ*, especially *Selaginella*, and basing its description on Pfeffer's in Hanstein's Bot. Abhand. iv. (1871), it is stated (pp. 285–6) that, "While the macrospores . . . . . are still lying in the sporangium, their apical region is occupied by a small-celled meniscus-shaped tissue, formed probably during the maturing of the spores by the breaking up of a quantity of protoplasm collected there. It is this tissue which subsequently produces the archegonium, and is therefore the true prothallium; but some weeks after the dispersion of the spores free cell-formation begins beneath this earlier tissue in the cavity of the spore, which results in the filling up of the entire cavity and the production of a large-celled tissue, a secondary prothallium as it may be termed."

This description, without the concluding words, is virtually given in the first English edition of Sachs' 'Text-book' (1875), where it is illustrated, as in Goebel, with Pfeffer's now familiar figure showing the well-marked "diaphragm" between the two tissues. There, however, Sachs adds that the large-celled tissue, "Pfeffer, supported by considerations with which I also agree, compares to the endosperm of Angiosperms, and, following this analogy, calls by the same name" (*op cit.*, p. 404). Goebel, however, merely adds (*loc. cit.*) the following note:—

"Pfeffer compared this tissue with the endosperm of the Angiosperms, and gave it that name; but since the homology of the two formations must be doubtful as long as the processes in the macrospore of the *Selaginelleæ* are not better known than they now are, a more definite term is preferable. It is probable that the contents of the macrospore divide into two primordial cells, one of which moves to the apex of the macrospore, and there produces the primary prothallium; while the other remains at first at the base of the macrospore, and subsequently produces the secondary prothallium."

This distinction in terms is not clearly maintained in the later part of the work. It appears, in fact, to be more clearly expressed in Sachs (p. 422), where, after acknowledging that the analogy of the endosperm of Gymnosperms with the prothallium of the higher Cryptogamia was first shown by Hofmeister, the author continues:—

"The processes which take place in the embryo-sac of Mono-cotyledons and Dicotyledons appear somewhat different, and bear a greater resemblance to what takes place in the macrospore of *Selaginella*. In this genus, besides the prothallium which produces the archegonia, there arises subsequently by free cell-formation another tissue, which fills up the rest of the space of the macro-spore; to this tissue the endosperm of Monocotyledons and Dicotyledons, which is formed by free cell-formation only after fertilisation, appears to correspond; the prothallium of *Selaginella* does not appear to have anything to correspond to it in Angiosperms." In a note, however, it is added that "the 'Antipodal cells' . . . may probably be considered as the last occasional occurrence of the rudiment of the true prothallium."

This latter identification is fully adopted by Goebel (p. 300). "The antipodal cells," he says, "are to be considered as a rudimentary prothallium. Here, too [among Angiosperms], a tissue, the endosperm, is formed in and fills the embryo-sac after fertilisation, but the cells of the rudimentary prothallium do not take part in its formation; this commences with the division of the nucleus of the embryo-sac, which is still present along with the six cells. We must not therefore consider the endosperm of the Angiosperms as equivalent to the endosperm of the Gymnosperms, which, as has been said, is simply the tissue of the prothallium in the macrospore; whereas the endosperm of the Angiosperms, as compared with the Vascular Cryptogams, is probably to be regarded as a new formation."

There could hardly be a more complete abnegation of the principle of terminology based upon homology than this. Of course, from a merely physiological standpoint we might term any storehouse tissue within the embryo-sac or macrospore "endosperm"; but this is not in accordance with modern custom; nor does it seem desirable to extend the already loosely applied term "prothallium" to structures which bear neither archegonia nor antheridia. Might it not then be well to apply the term "archisperm" to those structures formed before fertilisation, or at an early stage, in the macrospore, viz., the meniscus-shaped "primary" (female) prothallium above the diaphragm in *Selaginella*, the so-called "endosperm" in Gymnosperms, and the antipodal cells of Angiosperms, and either to reserve the term "endosperm," or to use "metasperm," for those formed at a later stage, viz., the large-celled "secondary prothallium," below the "diaphragm" in *Selaginella*, the "secondary endosperm" formed as a nutritive tissue after the prothallium is absorbed in Gymnosperms, and the endosperm originally so called, formed after fertilisation by the division of the secondary nucleus of the embryo-sac, in Angiosperms?

---

## A SYNOPSIS OF *TILLANDSIEÆ*.

By J. G. BAKER, F.R.S., F.L.S.

(Continued from p. 17.)

Subgenus VI. PITYROPHYLLUM (Beer). — Leaves narrow, rigidly coriaceous, densely rosulate, densely lepidote. Flowers arranged in a nearly sessile capitulum in the centre of the rosette of leaves. Petal-blade long, erect, lingulate; claw not scaled. Stamens and style longer than the petals. Sp. 128, 129.

128. TILLANDSIA IONANTHA Planch. in Flore des Serres, t. 1006 ; Hook. fil. in Bot. Mag. t. 5892. *T. erubescens* Hort. *Pityrophyllum erubescens* Beer, Brom. 79. *T. Scopus* Hook fil., *loc. cit.*—Tufts crowded, 2–3 in. long and broad. Leaves 30–40 in a rosette, linear-acuminate, recurved, 2–3 in. long, $\frac{1}{6}$–$\frac{1}{4}$ in. broad at the base, thick in texture, densely lepidote all over, channelled all down the face. Flowers few, arranged in a nearly sessile capitulum in the centre of the rosette of leaves ; bracts lanceolate, reaching to the top of the calyx. Calyx green, $\frac{1}{2}$ in. long ; sepals oblong. Petals bright violet, twice as long as the calyx ; blade erect, lingulate. Stamens and style longer than the petals.

Hab. Mexico. Introduced into cultivation at the Herrenhausen Garden before 1857. I cannot from the brief description separate *Pityrophyllum gracile* Beer (= *Tillandsia Quesneliana* and *Pourrettia stricta* Hort.).

129. T. BRACHYCAULOS Schlecht. in Linnæa, xviii. 422 ; E. Morren in Belg. Hort. 1878, 185, t. 11 ; Hemsl. in Bot. Cent. Amer. iii. 319.—Tuft. 8–9 in. broad, 5–6 in. high. Leaves densely rosulate, lanceolate-acuminate, 6–9 in. long, $\frac{1}{3}$ in. broad low down, recurved, thick and rigid in texture, lepidote all over, tinged more or less with reddish brown, channelled all down the face. Flowers 10–12 in a central capitulum overtopped by the recurved similar inner leaves. Calyx $\frac{1}{3}$ as long as the corolla ; sepals subacute. Petals bright violet, above an inch long ; blade erect, lingulate. Stamens and style exserted beyond the tip of the petals.

Hab. Central Mexico ; gathered by Schiede and Karwinsky. Introduced into cultivation by Roezl in 1876, and flowered in the collection of Prince Furstenberg at Donauschingen.

Subgenus VII. ALLARDTIA (Dietrich = *Platystachys* K. Koch non Beer = *Vriesea* Beer et Griseb. ex parte).—Differs from *Platystachys* by its thin flat flexuose subglabrous lorate or lanceolate leaves, and from *Vriesea* by its smaller flowers, without any scale on the claw of the petal. Acaulescent, with leaves in a dense utricular rosette. Spikes distichous, simple, or forming a distichous panicle.

KEY.

Inflorescence a simple spike  .    .    . Sp. 130–135.
Inflorescence panicled.
    Whole plant not more than 2–3 feet long   Sp. 136–149.
    Whole plant much longer  .    .    . Sp. 150–154.

130. **T. brachycephala,** n. sp. — Leaves few in a rosette, linear, acute, glabrous, thin in texture, closely ribbed, above a foot long, $\frac{1}{3}$ in. broad at the middle, $\frac{3}{4}$–1 in. at the dilated base. Peduncle slender, as long as the leaves ; lower bract-leaves with long erect points. Flowers in a dense globose capitulum ; flower-bracts oblong, obtuse, $\frac{1}{2}$ in. long. Calyx reaching to the tip of the bract. Petals not seen. Capsule-valves linear, $1\frac{1}{4}$ in. long.

Hab. Peru ; St. Gavan, *Lechler* 2409 !

131. **T. gymnophylla,** n. sp. *T. heliconioides* Griseb. in Gotting. Nachtrag. 1864, 18, non H. B. K.—Leaves few in a rosette, lanceolate from a slightly dilated base, a foot long, ½–¾ in. broad at the middle, thin, flat, naked, acute, not acuminate. Peduncle arcuate, shorter than the leaves. Inflorescence a moderately dense spike 3–4 in. long; flowers ascending, adpressed to the axis; flower-bracts oblong-lanceolate, glabrous, ½ in. long. Calyx reaching nearly to the tip of the bract. Petals not seen. Capsule-valves lanceolate, an inch long.

Hab. Mountains of Venezuela, *Fendler* 2615!

132. **T. drepanocarpa,** n. sp.—Leaves about 20 in a rosette; dilated base 1½ in. broad; blade lanceolate, acute, thin, bright green, glabrous, 6–8 in. long, ⅓–½ in. broad at the middle. Peduncle as long as the leaves; lower bract-leaves with lanceolate free points. Inflorescence a simple erect lax spike 5–6 in. long; flowers ascending; flower-bracts lanceolate, scariose; lower 2 in. long. Calyx glabrous, ½–⅝ in. long. Corolla not seen. Capsule-valves lanceolate, 1–1¼ in. long, spreading like a sickle after they dehisce.

Hab. South Brazil; province of St. Paulo, *Burchell* 3596!

133. **T. complanata** Benth. Bot. Sulphur, 173; Walp. Ann. i. 839, non E. Morren in Belg. Hort. 1872, 23.—Leaves about 20 in a dense rosette, lanceolate from an ovate dilated base above an inch broad, 9–10 in. long, an inch broad at the middle, thin, flexible, subglabrous, acute. Peduncles many to a rosette, slender, shorter than the leaves; bract-leaves small, lanceolate, entirely adpressed. Inflorescence a dense simple distichous spike 1½ in. long, ½ in. broad; flower-bracts oblong-lanceolate, acuminate, much compressed, ¾ in. long. Calyx ½ in. long, falling short of the bract. Petal-blade narrow, reddish, ½ in. long.

Hab. Columbia, on trees, near the River Machalay, October, 1836, *Barclay* 525! Also *Sinclair*! *Edmonstone*! *Cuming* 1190!

134. **T. axillaris** Griseb. in Gotting. Nachtrage, 1864, 17; Flora Brit. West. Ind. 597.—Leaves lorate from a large oblong base above 2 in. broad, above a foot long, 1¼–1½ in. broad at the middle, thin, flexible, subglabrous, deltoid at the tip. Peduncles many to a spike, much shorter than the leaves; bract-leaves small, scariose, entirely adpressed. Inflorescence a simple distichous spike 2–3 in. long, ¼ in. diam.; flower-bracts oblong-lanceolate, cuspidate, much compressed, ¾ in. long. Calyx reaching nearly to the tip of the bract; sepals acute. Petal-blade narrow, reddish, ⅓ in. long. Capsule-valves lanceolate, above an inch long.

Hab. Jamaica; St. Andrews Mountains, *Purdie*! Venezuela; mountains of Tovar, *Fendler* 1512! 1513! Ecuador; Pasto, *Lehmann*! Very near *T. complanata;* probably a mere variety.

135. **T. virginalis** E. Morren in Belg. Hort. 1880, 238. *T. heterophylla* E. Morren in Belg. Hort. 1873, 138.—Leaves about 20 in a rosette, lorate from a dilated base, 1½ ft. long, 1½–2 in. broad, flexible, pale green, glaucous beneath, especially towards the base. Peduncle 2½–3 ft. long, including the spike; spike simple, large, distichous; bracts large, conduplicate, glossy, green,

farinaceous.  Calyx not protruded beyond the tip of the bract.
Corolla white, above 3 in. long.  Stamens not exserted beyond the
tip of the petals; anthers very large.

Hab.  Mexico; province of Cordova.  Introduced into cultivation about 1870 by M. Omer de Malzine.  Flowered in 1880 at the
Botanic Garden at Liege, and by Mons. F. Massange at Louvrex.

136.  **T. triticea** Burchell MSS.—Leaves lorate from a very
large dilated ovate base 4 in. long, 3 in. broad, a foot long, 1¼ in.
broad at the middle, copiously branched and spotted, like *T. guttata*
and *splendens*, with claret-purple, flexible, subglabrous, deltoid-
cuspidate at the apex.  Peduncle slender, as long as the leaves;
bract-leaves small, lanceolate, adpressed.  Panicle a foot long;
spikes 10–12, arcuate-ascending, 2–3 in. long, all simple; lower
branch-bracts lanceolate, 1–1½ in. long; flowers numerous,
ascending, not dense; flower-bracts oblong, obtuse, ¼ in. long.
Calyx ¼ in. long; sepals obtuse.  Capsule-valves lanceolate, nearly
three times as long as the calyx.

Hab.  South Brazil; swampy woods near Santos, in province
of St. Paulo, October, 1826, *Burchell* 3217 !

137.  **T. Parkeri**, n. sp.—Leaves lorate from an ovate dilated
base 1½ in. diam., a foot long, an inch broad at the middle, flexible,
subglabrous, deltoid-cuspidate at the apex.  Peduncle slender;
upper bract-leaves small, lanceolate, adpressed, scariose.  Panicle
6–8 in. long; branches about 6, shortly peduncled, erecto-patent;
spikes 1–2 in. long; flowers many, erecto-patent, not very close;
flower-bracts ovate-oblong, subobtuse, ¼ in. long.  Calyx reaching
to the tip of the bract; sepals obtuse.  Capsule-valves lanceolate,
three times as long as the calyx.

Hab.  British Guiana, *Parker* !  Nearly allied to *T. triticea*
Burchell.

138.  **T. spiculosa** Griseb. Gott. Nachtrag. 1864, 17.—Leaves
lanceolate from a dilated ovate base 1¼–1½ in. broad, ½ ft. long,
thin, subglabrous, narrowed gradually to an acute point.  Peduncle
as long as the leaves; bract-leaves small, lanceolate,· entirely
adpressed, not imbricated.  Panicle composed of two or more
erecto-patent distichous spikes; flower-bracts ovate, ¼ in. long.
Calyx as long as the bract; sepals obtuse.  Petal-blade small,
violet.  Capsule-valves lanceolate, under an inch long.

Hab.  Mountains of Venezuela; Tovar and Maya, alt. 4000–
7000 ft., *Fendler* 1511 ! 1518, 2446.

139.  **T. compacta** Griseb. in Gott. Nachtrag. 1864, 18.—Leaves
above a dozen in a rosette, lorate from a large dilated oblong base,
3 in. long, 1½ in. broad, a foot long, an inch broad at the middle,
deltoid-cuspidate at the apex, thin, flexible, subglabrous.  Peduncle
as long as the leaves; bract-leaves many, small, imbricated,
entirely scariose and adpressed.  Panicle short, dense; spikes
oblong, erecto-patent, 1–1½ in. long, 1 in. diam.; branch-bracts
ovate, acute, nearly as long as the spikes; flower-bracts oblong-
navicular, glossy, cuspidate, ¾ in. long.  Calyx ¾ in. long; sepals
acute.  Petals and capsule not seen.

Hab.  Mountains of Venezuela; Tovar, *Fendler* 1508 !

140. T. cyanea E. Morren in Belg. Hort. 1879, 297. *Allardtia cyanea* Dietrich in Berl. Gartenzeit. 1852, xx. 241. *Platystachys cyanea* K. Koch in Ind. Sem. Hort. Berol. 1854, App. 2; Walp. Ann. vi. 68.—Leaves lanceolate from a dilated ovate base, a foot or more long, 1–1½ in. broad above the base, green, flexible, glabrous. Peduncle including the panicle 2–3 ft. long; lower bract-leaves with free points. Inflorescence a dense panicle; lower branches compound; spikes distichous, 1–1½ in. long, under ½ in. diam. ; flower-bracts oblong-navicular, glabrous, ½–⅝ in. long. Calyx ½ in. long; sepals pointed. ʻPetal-blade oblong, bright violet, ¼ in. long. Stamens and style not protruded beyond the tip of the petals. Capsule-valves lanceolate, ¾ in. long, ⅙ in. broad.

Hab. Guatemala, *Warcewicz*! Described from a specimen in the herbarium of Dr. Karl Koch. As Professor Morren has already pointed out, it is quite different from *T. Lindeni*.

141. T. tetrantha Ruiz & Pavon, Fl. Peruv. iii. 39, t. 265 ; Roem. et Schultes, Syst. Veg. vii. 1228. *Billbergia tetrantha* Beer, Brom. 127.—Leaves a dozen or more to a rosette, lanceolate from a large dilated oblong base 2 in. broad, a foot or more long, an inch broad at the middle, narrowed gradually to an acute point, thin, flexible, subglabrous, maculate with purple on both sides. Peduncle rather longer than the leaves. Inflorescence a lax panicle ; branch-bracts ovate-cuspidate, very convex, 1–1½ in. long : spikes few-flowered, secund. Calyx under ½ in. long, yellow ; sepals oblong. Petal-blade small, oblong, violet. Stamens not protruded.

Hab. Peru ; Andes of Muna, on rocks and trees. I have not seen authentic specimens of any of these three species of Ruiz and Pavon's, and have not been able to match them with plants of recent collectors.

142. T. maculata Ruiz et Pavon, Fl. Peruv. iii. 40, t. 267 ; Roem. et Schultes, Syst. Veg. vii. 1223. *Vriesea maculata* Beer, Brom. 98.—Leaves a dozen or more to a rosette, lorate from an ovate base, thin, flexible, glabrous, deltoid-cuspidate at the apex, copiously spotted with red-brown. Peduncle above a foot long ; bract-leaves small and distant. Inflorescence a panicle a foot or more long, with a red rachis ; lower branches compound ; spikes distichous, 1–2 in. long ; flower-bracts oblong-lanceolate, ¾ in. long. Calyx reaching to the tip of the bract. Petal-blade small, oblong, violet.

Hab. Andes of Peru at Muna, &c., *Ruiz & Pavon*.

143. T. rubra Ruiz et Pavon, Fl. Peruv. iii. 40, t. 266 ; Roem. et Schultes, Syst. Veg. vii. 1222 ; Griseb. Symb. Fl. Argent. 1878, 232. *Vriesea rubra* Beer, Brom. 98. *Phytarhiza rubra* E. Morren in Belg. Hort. 1879, 370.—Leaves 10–12 in a rosette, lorate from an ovate base, nearly 2 ft. long, an inch broad at the middle, flexible, subglabrous, deltoid-cuspidate at the apex. Peduncle as long as the leaves ; bract-leaves entirely adpressed. Inflorescence a panicle above a foot long, with erecto-patent distichous dense spikes, the lower 3 in. long ; rachises red ; branch-bracts small, ovate ; flower-bracts oblong-lanceolate, ¾ in. long. Calyx as long as the bract ; sepals acute. Petal-blade small, oblong, violet.

Hab.   Andes of Peru at Tarma, *Ruiz & Pavon*.   Oran, on trees and rocks on the banks of the Rio Blanco, *Lorentz*, teste *Grisebach* ("taller and leaves larger than in Peruvian type").

144. **T. caracasana,** n. sp. — Leaves lorate from an oblong dilated base 3 in. long, 2 in. broad, thin, flexible, subglabrous, an inch broad at the middle, deltoid-cuspidate at the apex.   Peduncle shorter than the leaves; bract-leaves all adpressed.   Inflorescence a panicle with 8–9 erecto-patent distichous spikes 1–1½ in. long, ⅜–¾ in. diam.; lower branch-bracts as long as the spikes; flower-bracts oblong-cuspidate, glossy, ¾ in. long.   Calyx ½ in. long, reaching nearly to the tip of the bract; sepals acute.   Petal-blade oblong, ¼ in. long.

Hab.   Caracas, *Moritz* 448!   Described from a specimen in the British Museum.   Very near *T. rubra* R. & P.

145. **T. rubella,** n. sp. — Root-leaves lorate from an ovate dilated base, acute, a foot long, above an inch broad at the middle, thin, flexible, subglabrous.   Peduncle much longer than the leaves; bract-leaves ovate-lanceolate, imbricated, almost entirely adpressed and scariose, the upper bright red.   Inflorescence a panicle 6–8 in. long; spikes numerous, simply distichous, erecto-patent, 1–1½ in. long, ⅝ in. diam.; branch-bracts ovate-cuspidate, shorter than the spikes, bright red; flower-bracts ovate-oblong, obtuse or subacute, bright red, ½ in. long.   Calyx ½ in. long, reaching to the tip of the bract; sepals glossy, much imbricated.   Petal-blade narrow, acute, ¼ in. long.   Stamens not protruded beyond the petals.

Hab.   Andes of Bolivia, near Sorata, alt. 8000 ft., *Mandon* 1187!   Nearly allied to *T. rubra* R. & P.

146. T. FENDLERI, Griseb. Gott. Nacht. 1864, 17.—Root-leaves lorate, acute, with minute cross-ridges.   Lower bract-leaves with leafy-points.   Inflorescence a panicle with a few long distichous branches; flower-bracts ovate-oblong, acute.   Calyx reaching to the tip of the bract.   Petal-blade small.

Hab.   Venezuela; between Caracas and Tovar, alt. 6000 ft., *Fendler* 1515.   This I have not seen.   It is said to be nearly allied to *T. rubra* R. & P.

147. T. ROEZLII E. Morren in Belg. Hort. 1877, 272, t. 15. *Allardtia Roezlii* E. Morren, Cat. 1873, 3.—Leaves 15–20 in a rosette, lorate from a dilated oblong base, 1–1½ ft. long, an inch broad at the middle, deltoid at the apex, thin, flexible, subglabrous, pale green blotched with darker green.   Peduncle as long as the leaves; bract-leaves small, adpressed.   Inflorescence a lax panicle ½ ft. long, with about 4 distichous spikes 2–4 in. long, an inch broad, the lower spreading; branch-bracts lanceolate, 1½–2 in. long; flower-bracts oblong-lanceolate, green, 1–1¼ in. long.   Calyx 1 in. long, reaching to the tip of the bract; sepals acute.   Petal-blade reddish, oblong-lanceolate, ½ in. long.   Stamens shorter than the petals.

Hab.   Andes of Northern Peru; discovered by Roezl in 1871 and brought into cultivation.

148. **T. rigidula,** n. sp.—Basal leaves lorate from an ovate dilated base 1½ in. diam., a foot long, an inch broad at the middle,

subrigid, green, naked, deltoid at the apex. Peduncle with inflorescence 3 ft. long ; bract-leaves small, lanceolate, entirely scariose and adpressed. Spikes 2, distichous, 1½ in. diam., the end one 6-8 in. long, the side one distant, peduncled, ascending ; lower flowers subpatent, spaced out ; flower-bracts ovate, obtuse, ½-⅝ in. long. Calyx glabrous, protruded ⅛ in. beyond the tip of the bract ; sepals obtuse. Petal-blade small.

Hab. British Guiana, *Appun* 840 !

149. **T. Kalbreyeri,** n. sp.—Leaves lanceolate, rigid, glabrous, pale green, 1½-2 ft. long, 3-4 in. broad low down, 2 in. broad at the middle, narrowed gradually to an acute point. Peduncle with panicle 2-2½ ft. long ; branches many, not distinctly peduncled, the side ones about half a foot long, above an inch broad, the end one longer ; flowers crowded, erecto-patent ; flower-bracts oblong, 1¼-1½ in. long, 1 in. broad. Calyx just protruded beyond the tip of the bract. Petal-blade small, violet.

Hab. New Granada ; between Ocana and Pamplona, alt. 3500 ft., *Kalbreyer* 1013 !

150. **T. martinicensis,** n. sp.—Basal leaves not seen. Lowest stem-leaf lorate, flexible, glabrous, a foot long, above an inch broad. Peduncle stout (½ in. diam.), above 2 ft. long ; bract-leaves with large lanceolate free points, growing gradually shorter in the upper ones. Panicle 1½ ft. long ; branches very numerous, spreading or ascending, the central ones sometimes slightly compound ; spikes dense, the longest 2-3 in. long ; branch-bracts coriaceous, ovate, acute, shorter than the spikes ; flower-bracts coriaceous, suborbicular, ⅓ in. long. Calyx ½ in. long ; sepals obtuse. Petals and capsule not seen.

Hab. Martinique, on the Montagne Pelie, *Hahn* 521 ! 523 !

151. **T. penduliflora** Griseb. Fl. Brit. West. Ind. 597.—Leaf lanceolate, rigidly coriaceous, subglabrous, 2-3 ft. long, 2 in. broad at the middle. Inflorescence a huge panicle, the lower branches of which are a foot long, with small ovate bracts at the base, a flexuose rachis, and 7-8 nearly sessile orbicular deflexed distichous congested many-flowered spikes 1½-2 in. long, an inch broad, with a rigid ovate basal bract ; flower-bracts lanceolate-navicular, 1-1¼ in. long. Calyx an inch long, reaching to the tip of the bract ; sepals acute. Petals not seen.

Hab. Dominica, *Imray* 107 ! Martinique, *Hahn* 676 ! A very distinct and fine species, apparently 4-5 ft. high.

152. **T. excelsa** Griseb. Fl. Brit. West. Ind. 597 ; Sauvalle, Cat. Cub. 169.—Leaf lorate from a dilated oblong base 2-3 in. broad, 1½ ft. long, 2 in. broad at the middle, thin, flexible, subglabrous. Peduncle stout, as long as the leaves ; lower bract-leaves with large lanceolate free points. Inflorescence a lax panicle a foot long ; many lower branches spreading and compound ; lower branch-bracts lanceolate, 3 in. long ; ultimate spikes 1-2 in. long, ½-⅔ in. broad ; flower-bracts oblong-lanceolate, ¾ in. long. Calyx ½ in. long ; sepals acute. Petal-blade oblong-lanceolate, ¼ in. long. Capsule-valves lanceolate, an inch long.

Hab. Jamaica ; mountains of Manchester and Westmoreland, *Purdie* ! Eastern Cuba (near Monte Verde), *Wright* 1517 !

153. **T. elata,** n. sp.—Leaves 2½–3 ft. long, 2 in. broad at the middle, thin, flexible, subglabrous, deltoid-cuspidate at the apex. Peduncle stout.   Panicle 2½–3 ft. long; central branches copiously compound; branch-bracts oblong-lanceolate, 2–3 in. long; ultimate spikes erecto-patent, 1½–2 in. long, ¾ in. broad; flower-bracts oblong-lanceolate, acute, ¾–1 in. long.   Calyx ½ in. long; sepals acute.   Petal-blade narrow, acute, ¼ in. long.

Hab.   Santa Moarta; Sierra Nigra, *Purdie*! Closely allied to the West Indian *T. excelsa.*

154. **T. megastachya,** n. sp.   *T. foliosa* Griseb. Fl. Brit. West. Ind. 597, non Mart. et Galeotti. — Basal leaves lorate, 2–2½ ft. long, 1½ in. broad at the middle, flexible, glabrous. Peduncle stout (½ in. diam.); bract-leaves with lanceolate free points ½–1 ft. long.   Panicle above a foot long; branches very numerous, subsessile; branch-bracts coriaceous, ovate, acute, or cuspidate, the upper shorter than the spikes, the lower much longer; spikes about 3 in. long and nearly as broad; flowers dense, erecto-patent; flower-bracts narrow, acute, an inch long. Calyx reaching nearly to the tip of the bract; sepals acute. Capsule-valves lanceolate, ¼ in. longer than the calyx.

Hab.   St. Vincent, *Rev. L. Guilding*!

Subgenus VIII. WALLISIA (Regel).—Leaves thin, acuminate, nearly naked.   Spikes distichous, simple or panicled.   Blade of petals large, spreading, orbicular; claw not scaled.   Stamens and pistil not exserted from the calyx.   .   Sp. 155–158.

155. T. LINDENI E. Morren in Belg. Hort. xix. 321, t. 18; Gard. Chron. 1879, ii. 460; Ill. Hort. 1869, t. 610; Floral Mag. 1872, t. 44; Rev. Hort. 1878, 300, with figure; Garden, 1876, 466, with figure.   *Phytarhiza Lindeni* E. Morren in Belg. Hort. 1879, 297.   *T. Morreniana* Regel in Gartenfl. 1870, 40. — Acaulescent. Leaves 40–60 in a dense rosette, lanceolate from a dilated base, 1½–2 in. broad, 1–1½ ft. long, an inch broad at the middle, tapering gradually to a long point, thin, hardly at all lepidote, dull green, with vertical stripes of red-brown.   Peduncle a foot long; bract-leaves lanceolate, erect, imbricated.   Inflorescence a simple dense distichous spike 4–6 in. long; flowers about 10 on a side; flower-bracts oblong-navicular, 1½–2 in. long, bright red in the type. Calyx 1½ in. long; sepals oblong, acute.   Petal-blade orbicular, spreading, bright violet, an inch broad.   Stamens and pistil not longer than the calyx.

Hab.   Andes of Peru, discovered by Wallis.   First shown by Linden at the Paris Exhibition of 1867.   Var. *intermedia* Morren (Floral Mag. 1871, t. 529) is a form with a longer peduncle and greenish red bracts.   Var. *Regeliana* Morren (*T. Lindeniana* Regel, Ind. Sem. Hort. Petrop. 1868, 92; Gard. Chron. 1879, ii. 461, fig. 72) is a form with a long peduncle, green bracts, and a white eye to the petal; and vars. *luxurians* and *major* are luxuriant garden forms.

156. T. HAMALEANA E. Morren in Gard. Chron. 1869, ii. 460. *Wallisia Hamaleana* E. Morren in Belg. Hort. 1870, 97, t. 5.

*T. Lindeni* var. *Hamaleana* André in Ill. Hort. 1877, 188.   *T. Com-melyna* E. Morren.   *Phytarhiza Hamaleana* E. Morren in Belg. Hort. 1879, 297, 370. — Acaulescent.   Leaves 15–20, densely rosulate, lanceolate, acuminate, thin, subglabrous, pale green, 6–9 in. long, an inch broad at the middle.   Peduncle shorter than the leaves; bract-leaves small, scariose, adpressed, not imbricated. Spikes 3–4 in a small deltoid panicle, distichous; branches erecto-patent; branch-bracts small, ovate; flower-bracts ovate, purplish green, shorter than the calyx.   Calyx 1½ in. long; sepals oblong-lanceolate, green, glabrous.   Petal-blade orbicular, bright violet with a white eye, ¾ in. diam.   Stamens and style not exserted from the calyx.

Hab.   Andes of Peru, discovered by Wallis.   First exhibited by Linden at Paris in 1867.   Probably this is a mere variety of *T. Lindeni*, but I have not seen it.

157.  **T. platypetala**, n. sp. — Leaves few in a rosette, erect, lorate from a large dilated ovate base 2½ in. diam., above a foot long, 1½–1¾ in. broad, obtuse, thin, flexible, subglabrous.   Peduncle as long as the leaves; bract-leaves small, scariose, entirely adpressed. Inflorescence a dense panicle 3–4 in. long and broad; branches 4–5, dense, erecto-patent; branch-bracts small, scariose, ovate; flower-bracts ovate-oblong, ¾ in. long.   Calyx reaching to the tip of the bract; sepals oblong, very convolute and imbricated.   Petal-blade obovate-cuneate, ½ in. long and broad.   Stamens and style not exserted from the calyx.

Hab.   Ecuador; mountains of El Cisne, *Hartweg*!

158.  T. UMBELLATA André in Rev. Hort. 1886, 60, with coloured figure. — Leaves densely rosulate, lanceolate-acuminate, a foot or more long, ⅓–½ in. broad low down, bright green, glabrous, faintly streaked vertically with claret-brown on the back.   Peduncle much shorter than the leaves.   Spikes dense, simple, 5–6-flowered; flower-bracts oblong-cuspidate, bright green, ¾ in. long.   Calyx as long as the bract.   Petal-limb obovate-cuneate, an inch broad, bright blue, with a white eye.   Stamens and style not longer than the calyx.

Hab.   Ecuador; Cordillera of Cisné, temperate region, gathered by M. Poortman in May, 1882.

Subgenus IX. VRIESEA (Lindl.).—Acaulescent, with leaves in a dense utricular rosette, thin, flexible, nearly or quite without lepidote scales.   Spikes distichous, simple or panicled.   Flowers large.   Petals white or yellow or greenish, never violet, with a broad claw, appendiculate with a pair of scales at the base.

KEY.

| | |
|---|---|
| Spikes simple, dense .   .   .   . | Sp. 159–173. |
| Spikes simple, lax. | |
|    Spikes erect; flowers erecto-patent . | Sp. 174–188. |
|    Spikes erect; flowers spreading   . | Sp. 189–193. |
|    Spikes drooping   .   .   .   . | Sp. 194–195. |
|    Garden hybrids   .   .   .   . | Sp. 196–197. |

Spikes panicled.
Flower-bracts nearly or quite as long
　　as the calyx　　.　　.　　.　　.　　Sp. 198-206.
Flower-bracts shorter than the calyx　Sp. 207-217.
Imperfectly-known species　　.　　.　　.　　Sp. 218-220.

159. T. BILLBERGIÆ Baker. *Vriesea Billbergiæ* Lemaire in Ill. Hort. xvi. Misc. 91.—Habit of a *Billbergia*. Leaf-bases utriculate in a globe 4-5 in. high ; blade lorate, 6-8 in. long, 1½ in. broad, deltoid-cuspidate at the apex. Peduncle much longer than the leaves. Spike dense, distichous. Petals green, red and white. Stamens exserted.

Hab. Mexico, in oak forests. Sent by Ghiesbreght to Verschaffelt, about 1865.

160. T. DUVALIANA Baker. *Vriesea Duvaliana* E. Morren in Belg. Hort. 1884, 105, t. 7-8.—Leaves about twenty in a rosette, lorate from a dilated base 1½ in. diam., under a foot long, an inch broad below the middle, plain green on the face, tinged with purple on the back, flexible, subglabrous, deltoid-cuspidate at the tip. Peduncle rather exceeding the leaves ; bract-leaves small, broad, scariose, adpressed. Spike dense, distichous, about 20-flowered, 5-6 in. long ; flowers all erecto-patent ; flower-bracts oblong-navicular, acute, 1¼-1½ in. long, bright red and yellow at the base, green at the tip. Calyx reaching to the tip of the bract. Petals greenish yellow, half an inch longer than the calyx. Stamens a little longer than the petals.

Hab. South Brazil, in woods near Rio Janeiro, *Glaziou* 14344 ! Introduced into cultivation about 1875, probably by Binot. Named after M. Duval, of Versailles, who exhibited it in flower at Paris in 1883.

161. T. HELICONIOIDES H.B.K., Nov. Gen. i. 234 ; Roem. et Schultes Syst. Veg. vii. 1226. *Vriesea heliconioides* Lindl. in Bot. Reg. 1843, sub. t. 10; E. Morren in Ill. Hort. n. s. t. 490; Antoine Brom. ii. t. 8 ; Gard. Chron. 1884, ii. 140, fig. 26. *V. bellula* and *Falkenbergii* Hort. *T. disticha* Willd. herb. No. 6327. *Platystachys disticha* Beer, Brom. 264.—Leaves 15-20 in a rosette, lorate from an ovate base 1½ in. diam., thin, flexible, plain from above, tinged with purple beneath, an inch broad at the middle, narrowed gradually to the point. Peduncle much shorter than the leaves ; bract-leaves small, adpressed. Flowers 6-10 in a simple erect spike 4-6 in. long, 2 in. diam. ; flower-bracts ovate-navicular, acute, recurved, 1½ in. long, bright red at the base, green at the tip. Calyx a little over an inch long; sepals obtuse. Petals white, twice as long as the calyx ; blade ¼ in. broad. Stamens as long as the petals.

Hab. Valley of Rio Magdalena; first gathered there by Humboldt, Introduced into cultivation by the Continental Company of Horticulture, and exhibited at Ghent in 1883.

162. T. PACHYCHLAMYS Baker. —Leaves lorate from an ovate dilated base 2 in. broad, 3 ft. long, 2 in. broad at the middle, flexible, subglabrous, deltoid-cuspidate at the apex. Peduncle

stout, stiffly erect, above a foot long; bract-leaves scariose, imbricated, entirely adpressed. Spike dense, erect, $\frac{1}{2}$–1 ft. long, $1\frac{1}{4}$ in. diam.; flower-bracts ovate, coriaceous, very ascending, 2 in. long, $1\frac{1}{4}$ in. round low down. Calyx $\frac{1}{2}$ in. shorter than the bract. Petals not seen.

Hab. British Guiana, *Parker*! *Jenman* 2044!

163. T. SCHLECTENDAHLII Baker. *T. cæspitosa* Cham. et Schlecht. in Linnæa, vi. 54; Hemsl. in Biol. Cent. Amer. Bot. iii. 319, non Leconte. *Vriesea cæspitosa* E. Morren in Bourg. Pl. Mex. Exsic. No. 2960.—Leaves lanceolate from a large ovate dilated base $1\frac{1}{2}$–2 in. broad, flexible, subglabrous, a foot long. 1–$1\frac{1}{2}$ in. broad at the middle, narrowed gradually to the point. Peduncles 4–5 in. long, sometimes 3 from one rosette; bract-leaves small, ovate, scariose. Inflorescence a dense simple distichous spike 4–6 in. long, $1\frac{1}{2}$–2 in. broad; flower-bracts ovate, acute or cuspidate, $1\frac{1}{2}$–2 in. long. $1\frac{1}{4}$ in. round at the base, upper red. Calyx $1\frac{1}{4}$ in. long, glabrous. Petals twice as long as the sepals; blade oblanceolate. Stamens slightly longer than the petals.

Hab. Mexico, in the provinces of Xalapa and Orizaba, *Pavon*! *Schiede & Deppe*! *Bourgeau* 2960! *Hahn*! Very near *T. incurvata*.

164. T. INCURVATA Baker. *Vriesea incurvata* Gaudich, Atlas Bonité, t. 68; Beer, Brom. 92; E. Morren in Belg. Hort. 1882, 52, t. 2. *T. inflata* Baker in Bot. Mag. t. 6882. *Vriesea inflata*, Wawra Itin. Prin. Cob. i. 162; Antoine Brom. 28, t. 18. *V. carinata* var. *inflata* Wawra in Œster. Bot. Zeitsch. xxx. 183; French transl. 64.—Leaves about 15 in a rosette, lorate from an ovate dilated base 2 in. broad, thin, flexible, plain green, subglabrous, 8–12 in. long, 1–$1\frac{1}{2}$ in. broad at the middle, deltoid-cuspidate at the tip. Peduncle much shorter than the leaves; bract-leaves ovate, scariose, imbricated. Inflorescence a simple dense distichous spike $\frac{1}{2}$–1 ft. long, 2 in. broad; flower-bracts ovate, acute, $1\frac{1}{2}$–2 in. long, $\frac{1}{2}$ in. round at the base, the upper bright red. Calyx $1\frac{1}{4}$ in. long, yellowish white. Petals bright yellow, half as long again as the calyx; blade lingulate. Stamens half an inch longer than the petals. Capsule as long as the calyx.

Hab. South Brazil, on the Corcovado and other mountains of Rio Janeiro, and St. Paulo, *Burchell* 2282! 3488! 3864/2! *Gaudichaud, Longman*! *Raddi* 105, *Wawra* 86, 95 Introduced into cultivation by Binot, in 1880. I cannot follow Wawra in separating *inflata* as a species from *incurvata*.

165. T. CARINATA Baker. *Vriesea carinata* Wawra in Œster. Bot. Zeit. 1862, 349, 1880, 183; Bot. Ergeb. 154, t. 26; Itin. Prin. Cob. 157; Antoine Brom. 9, t. 7. *V. brachystachys* Regel. Gartenfl. 1866, 258, t. 518; Bot. Mag. t. 6014. *V. psittacina* var. *brachystachys* E. Morren in Belg. Hort. 1870, 161, t. 8. *V. psittacina* var. *carinata* E. Morren in Belg. Hort. (1882, 28), 1882, 287, t. 10–12, fig. 1.—Leaves 15–20 in a rosette, lorate from an ovate dilated base $1\frac{1}{2}$ in. broad, thin, flexible, plain green, under a foot long, under an inch broad at the middle, deltoid-cuspidate at the apex. Peduncle rather exceeding the leaves; bract-leaves all small and adpressed. Flowers 10–12, in a dense distichous spike 2–3 in.

long, all erecto-patent; flower-bracts oblong-navicular, keeled, about an inch long, bright red at the base, green at the tip, with a small incurved cusp.  Calyx rather shorter than the bract.  Petals bright yellow, $\frac{1}{2}$ in. longer than the calyx.  Stamens a little longer than the petals.  Capsule-valves above an inch long.

Hab.  South Brazil about Rio Janeiro and in the province of St. Paulo, *Burchell* 2308! 3326! *Wawra, Longman*! *Glaziou* 8026! St. Catherina, *Tweedie* 529! Introduced into cultivation about 1865 by Messrs. Booth of Hamburg.

(To be continued.)

---

# BIOGRAPHICAL INDEX OF BRITISH AND IRISH BOTANISTS.

By JAMES BRITTEN, F.L.S., AND G. S. BOULGER, F.L.S.

HAVING often in the course of our own work felt the want of some reference-list of byegone workers in Botany, the compilers have thought that the present Index may be of use to others.  Their plan has been to be liberal in including all who have in any way contributed to the literature of the science, who have made scientific collections of plants, or who are known to have otherwise assisted in the progress of Botany, exclusive of pure Horticulture.  Where known, the name is followed by the years of birth and death, and in other cases an approximate date is given.  Then follows the place and day of birth and death, chief titles, dates of election to the Linnean and Royal Societies, or chief University degrees.  In conclusion, reference is made to the chief sources of further information in which Pulteney, Rees, Pritzel, Jackson, and the Royal Society Catalogue are first quoted, and then the fullest known record, with a note of any portrait and of genera dedicated to the various persons catalogued.  The following abbreviations indicate the most frequently quoted of the many works we have had to consult :—

*Ann. & Mag.*  'Annals and Magazine of Natural History,' 1841–86.
*Bot. Misc.*  'Botanical Miscellany,' London, 1830–33.
*Cash.*  'Where there's a Will there's a Way,' by Jas. Cash, 1873.
*Cott. Gard.*  'The Cottage Gardener,' 1849–1860.
*Dict. Nat. Biog.*  'Dictionary of National Biography,' edited by Leslie Stephen.  London, 1885, and in progress.
*Encycl. Gard.*  'An Encyclopædia of Gardening,' by J. C. Loudon. London, 1850.
*Felton.*  'Portraits of English Authors on Gardening,' 1830.
*Fl. Midd.*  'Flora of Middlesex,' by H. Trimen & W. T. T. Dyer. London, 1869, 8vo.
*Friends' Books.*  'Catalogue of Friends' Books,' by Joseph Smith, 1867.
*Gard. Chron.*  'The Gardeners' Chronicle.'  London, 1841, and in progress, fol.
*Gorham.*  'Memoirs of John and Thomas Martyn,' 1830.

*Jacks.* ' Guide to the Literature of Botany,' by B. D. Jackson. London, 1881.

*Johnson.* ' History of English Gardening,' 1829.

*Journ. Bot.* ' The Journal of Botany,' 1834–1886, including ' Hooker's Journal of Botany,' 1834–1842 ; ' The London Journal of Botany,' 1842–48; ' Hooker's Journal of Botany and Kew Gardens Miscellany,' 1849–1857 ; and ' The Journal of Botany,' 1863, and in progress.

*Journ. Hort.* ' The Journal of Horticulture, 1860–1886.

*Linn. Letters.* ' Correspondence of Linnæus and other Naturalists.' Edited by Sir J. E. Smith. 2 vols. London, 1821.

*Loud. Gard. Mag.* ' Loudon's Gardeners' Magazine,' 1826–43.

*Mag. Nat. Hist.* ' The Magazine of Natural History.' Conducted by J. C. Loudon. 9 vols. 1829–1836. New Series, conducted by E. Charlesworth, 4 vols., 1837–4 .

*Mag. Zool. Bot.* ' Magazine of Zoology and Botany,' conducted by Sir W. Jardine, P. J. Selby, and Dr. Johnson, 1837–1838. Continued as ' Annals of Natural, History ; or Magazine, &c.,' 1838–1840, vols. i.–v.

*Martyn.* ' Plantæ Cantabrigienses,' by Thomas Martyn, M.A. London, 1763.

*Munk.* ' The Roll of the Royal College of Physicians,' 2nd ed., 1878.

*Nich. Anec.* ' Nichols' Literary Anecdotes of the 18th Century,' 1812–15.

*Nich. Illust.* ' Nichols' Literary Illustrations of the 18th Century,' 1817–58.

*Phyt.* ' The Phytologist,' 1842–1868.

*Pritz.* ' Thesaurus Literaturæ Botanicæ,' by A. Pritzel, 1872.

*Proc. Geol. Soc.* ' Proceedings of the Geological Society.'

*Proc. Linn. Soc.* ' Proceedings of the Linnean Society,' 1838, and in progress.

*Pult.* ' Historical and Biographical Sketches of the Progress of Botany in England,' by Richard Pulteney, 1790.

*Ray Lett.* ' Correspondence of John Ray.' Ray Society, 1848.

*Ray Mem.* ' Memorials of John Ray.' Ray Society, 1846.

*Rees.* ' Rees' Cyclopædia,' 1819–1820. (The biographies by Sir J. E. Smith).

*Rich. Corr.* ' The Correspondence of Richard Richardson, M.D.' Yarmouth, 1835.

*R. S. C.* ' The Royal Society's Catalogue of Scientific Papers,' 1867–1879.

*Semple.* ' Memoirs of the Botanic Garden at Chelsea,' 1878.

*Smith Lett.* ' Memoir and Correspondence of Sir J. E. Smith,' by Lady Smith. 2 vols., 1832.

*Trans. Bot. Soc. Edin.* ' Transactions of the Botanical Society, Edinburgh,' 1844, and in progress.

*Weston.* ' Catalogue of English Authors,' &c., 1773.

As the list of botanists is necessarily incomplete, and many dates and other facts are yet unknown to us, we shall be much obliged for any corrections or additions that may be sent us. Living workers are omitted.

**Abbot, Charles** (1761?–1817): b. Winchester? 1761?; d. Bedford, October, 1817. Clerk. D.D., Oxon, 1802. Vicar of Oakley Raynes and Goldington, Beds. F.L.S., 1793. 'Flora Bedfordiensis,' 1798. Herbarium at Turvey Abbey, Beds. Gent. Mag. 1817, ii. 378; Journ. Bot. 1881, 40; Dict. Nat. Biog. i. 3.

**Abbot, Robert** (fl. 1630), of Hatfield, Clk. "Excellent and diligent herbalist," Pult. i. 137; Johnson's Gerard, 166, 175, 216.

**Abel, Clarke** (1780–1826): d. at Cawnpore, 14th November, 1826. M.D., F.L.S., 1818. In China with Lord Macartney, 1816. Chinese plants in Brit. Mus. Pritz. 1; Dict. Nat. Biog. i. 32; Gent. Mag. xcvii. pt. II. (1827), 644. Portr. at Kew. *Abelia* R. Br.

**Abercrombie, John** (1726–1806): b. Edinburgh, 1726; d. London, 30th April or 1st May, 1806. Pritz. (ed. 1), 1; Dict. Nat. Biog. i. 36; Johnson, 219; Cott. Gard. iv. 65; Journ. Hort. lv. (1876), 469, portr.; Felton, 153; Biog. in his 'Gardener's Pocket Journal,' ed. 35, 1857. Best portr. in his 'Universal Gardener,' ed. 1783; full-length, æt. 72, in ed. 16, 1800.

**Acton, Frances Stackhouse,** *née* KNIGHT (1793?–1881): d. Acton Scott, Salop, 24th January, 1881. Eldest d. of T. A. Knight. m. Thomas Pendarves Stackhouse Acton, 1812. Shared in her father's experiments. Artist. Gard. Chron. 1881, i. 182.

**Adams, Francis** (1796–1861): b. Lumphanan, Aberdeen, 13th March, 1796; d. Banchory Ternan, 26th February, 1861. M.A., Aberdeen; Hon. M.D., 1856; LL.D., Glasgow, 1846. Greek scholar. Translated Hippocrates, 1849. Dict. Nat. Biog. i. 95; Murray, 'Northern Flora,' xvii. and Append. I. Bust at Univ., Aberdeen.

**Adams, George** (1720–1786? or 1773): b. London, 1720; d. London, 5th March, 1786? 'Micrographia Illustrata'; Pritz. 1; Jacks. 219; Dict. Nat. Biog. i. 97.

**Adams, George,** *jun.* (1750–95): b. London, 1750; d. London, 14th August, 1795. F.L.S., 1788. Son of above. Pritz. 1; Jacks. 219; Dict. Nat. Biog. i. 97; Gent. Mag. lxv. 708.

**Adams, John** (fl. 1793). F.L.S., 1795. Of Pembroke. Correspondent of Sowerby. Drowned off Pembrokeshire, 183–. E. B. 111, 248, 462; Lees' 'Botanical Looker-out.'

**Adams, John** (fl. 1785). Of Edmonton. 'Account of a Variegated American Aloe,' 1785; Pritz. (ed. 1), 2.

**Aikin, Arthur** 1773–1854): b. Warrington, 19th May, 1773: d. London, 15th April, 1854. F.L.S., 1818. Original Member Geological Soc., 1807; Sec., 1811; Sec. to Society of Arts, 1817–1840. Dict. Nat. Biog. i. 184; Proc. Geol. Soc. xi. (1855), p. xli.; Wall. Pl. Asiat. iii. 65. *Aikinia* Br. = *Epithema* Bl.

**Aikin, John** (1747–1822): b. Kibworth Harcourt, Leicester, 15th January, 1747; d. Stoke Newington, London, 7th December, 1822. M.D., Leyden, 1784; F.L.S., 1795. Memoir by Lucy Aikin, 1823, portr., engr. by Englehart; Jacks. 213, 244; Dict. Nat. Biog. i. 185; Munk, ii. 421; Gent. Mag. 1823, i. 85; Portr. by J. Donaldson, engr. by Knight. *Aikinia* Salisb.

**Ainslie, Whitelaw** (c. 1788–c. 1835). M.D. 'Materia Indica,'

1826; Pritz. 3; R. S. C. i. 30; Dict. Nat. Biog, i. 190.
*Ainsliæa* DC.

**Aiton, William** (1731-1793): b. near Hamilton, N.B, 1731;
d. Kew, 1st February, 1793. Director of Kew, 1759-1793.
'Hortus Kewensis,' 1789; Rees; Pritz. 3; Jacks. 412; Dict.
Nat. Biog. i. 207; Cott. Gard. v. 263; Oil portr. and engr.,
Kew; Johnson, 298; Proc. Linn. Soc. ii. 82. *Aitonia* Thunb.

**Aiton, William Townsend** (1766-1849): b. Kew, 2nd February,
1766; d. Kensington, 9th October, 1849. F.L.S., 1797.
Director of Kew, 1793-1841, Pritz. 3; Jacks. 412; Dict. Nat.
Biog. i. 208; Proc. Linn. Soc. ii. 82; 'Hortus Kewensis,' ed. 2,
v. 531. Portr. by L. Poyet, Kew.

**Alchorne, Stanesby** (d. 1799 or 1800). Apothecary. Assay-
master in the Mint. Hon. Demonstrator at Chelsea, 1771-1773.
Semple; Linn. Letters, ii. 4-7. *Alchornea* Martyn.

**Alcock, Randal Hibbert** (1833-1885): b. Gatley, Cheshire, 21st
July, 1833; d. Didsbury, Lanc., 9th November, 1885. F.L.S.,
1876. 'Botanical Names for English Readers,' 1876; Jacks.
9, 499; Journ. Bot. 1886, 160; Proc. Linn. Soc. 1885-6, 137·

**Alderson, John** (1757 ?-1829): b. Lowestoft, 1757?; d. Hull,
16th September, 1829. M.D. One of founders of Hull Garden.
'Essay on *Rhus Toxicodendron*,' 1793; Dict. Nat. Biog. i. 243;
Corlass & Andrews, ' Sketches of Hull Authors,' 6; Gent. Mag.
Nov. 1830, 451. Two statues at Hull.

**Alexander, William** (fl. 1820), M.D., of Halifax. Trans. Hort.
Soc. i. 328.

**Alfred the Philosopher** (? *de Sarchel*, or *Sereshel*) : d. 1270. "An
Englishman, much respected at home." ' On Vegetables.' Felton, 4; Dict. Nat. Biog. i. 285.

**Allan, James** (fl. 1853). Ph.D.; A. M.: Prof. Chemistry, Manchester. 'Botanist's Word-book,' 1853, with G. Macdonald;
Jacks. 9; Gard. Chron. 1853, 791.

**Allcard, J.** F.L.S. (d. 1844), of Stratford Green, Essex. His
garden described, Gard. Chron. 1841, 119, 599; 1842, 271, 855;
1843, 559.

**Allman, William** (1776-1846): b. Kingston, Jamaica, 7th Feb.,
1776; d. Dublin, 8th December, 1846. M.D., Dublin, 1804;
Prof. Botany, Dublin, 1809-1844. MS. on 'Mathematical
Connection between Parts of Vegetables,' 1811, in Herb. Mus.
Brit. Pritz. 4; Jacks. 17, 40, 65; Dict. Nat. Biog. i. 335; R. S. C.

**Alston, Charles** (1683-1760): b. Eddlewood, W. Scotland, 1683;
d. Edinburgh, 22nd November, 1760. M.D., Leyden ? 1716;
Prof. of Botany, Edinburgh, 1719-1760. 'Tirocinium Botanicum
Edinburgense,' 1753. Pult. ii. 9-17; Pritz. 5; Gent. Mag.
1760, 544; Jacks. 16, 32, 411; Rees; Dict. Nat. Biog. i. 346;
Rich. Corr. 275. *Alstonia* Linn. fil.

**Anderson, Alexander** (d. 1811): d. St. Vincent, W. Indies, 8th
September, 1811. M.D.; F.L.S., 1808. In America from 1775.
Curator, St. Vincent Garden, from 1785. Demerara pl. in Mus.
Brit. Pritz. 6; Dict. Nat. Biog. i. 372; Cott. Gard. viii.;
Lambert, ' Pinus,' ii. 14; Gard. Mag. i. (1826), 194; Banksian
Corresp. 3rd May, 1789, and 30th March, 1796. *Andersonia* R. Br.

**Anderson, George** (d. 10th January, 1817). F.L.S., 1800. Of West Ham and Leadenhall Street. 'Pæonia,' Linn. Trans. v. 12, 283. Had a salicetum.

**Anderson, J.** Collector on H.M.S. 'Adventurer,' 1825-1830. Pl. in Herb. Mus. Brit.

**Anderson, James** (d. 1838). Director, Sydney Garden, N.S.W. Friend of the Cunninghams. Woolls, 'Lectures on Veg. Kingdom,' 1879, p. 58.

**Anderson, James** (1739-1808) : b. Hermiston, Edinburgh, 1739; d. Isleworth, 15th October, 1808. LL.D., Aberdeen, 1780. Pritz. 6; Jacks. 34; Dict. Nat. Biog. i. 381; Cott. Gard. v. 1.

**Anderson, James** (d. 1809): d. Madras, 5th August, 1809. Physician to East India Company. Dict. Nat. Biog. i. 382.

**Anderson, Robert.** Brother of Thomas and John. 'Catalogue of Calcutta Plants,' 1862.

**Anderson, Samuel** (d. 1878). F.L.S., 1854. Of Whitby. Contributed to 'Sphagnaceæ Britannicæ Exsiccatæ.' Journ. Bot. 1878, 64; Gard. Chron. 1878, i. 178.

**Anderson, Thomas** (1832-1870): d. Edinburgh, 26th October, 1870. M.D. Edin., 1853; F.L.S., 1859. Superintendent, Calcutta Garden. Singapore pl. in Mus. Brit. Pritz. 6; Jacks. 384, 388, 451; Journ. Bot. 1870, 368; Gard. Chron. 1870, 1478; Dict. Nat. Biog. i. 392; R. S. C. i. 65; vii. 33; Trans. Bot. Soc. Ed. xi. 1875, 41.

**Anderson, William** (1766-1846): b. Easter Warriston, Edinboro', 1766; d. Chelsea, 6th October, 1846. A.L.S., 1798; F.L.S.. 1815. Gardener and Curator at Chelsea from 1814. Pritz. 6; Semple, 119, 203; Dict. Nat. Biog. i. 393; Proc. Linn. Soc. i. 331.

**Anderson, William** (d. 1778): d. off Anderson's Island, 3rd Aug. 1778. Surgeon R.N. on Cook's voyages. MSS. in Banksian Library. New Caledonia pl. in Herb. Mus. Brit. Pritz. 6; Dict. Nat. Biog. i. 393; 'Cook's Voyages,' ii. 440; Brown, Prodr. Fl. Nov. Holland. 553; Comp. Bot. Mag. ii. (1836), 227. *Andersonia* R. Br.

**Anderson-Henry, Isaac** (d. 1884): b. Caputh, Perthshire; d. 21st September, 1884. F.L.S., 1865. President Bot. Soc. Edin., 1867-8. Gard. Chron. 1873, 399, portr.; 1884, ii. 400; Trans. Bot. Soc. Ed. ix.

**Andrewes.** Apothecary. Of Sudbury. Friend of Ray, Dale, &c. Rich. Corr. 184; Ray. Syn. 114.

**Andrews, Henry C.** (fl. 1796-1828). Knightsbridge, London, 'Botanists' Repository.' 'Heaths.' Pritz. 6; Jacks. 131, 132, 142, 471; Dict. Nat. Biog. i. 406. *Andreusia* Vent. = *Myoporum*.

**Andrews, James** (1801?-1876): d. Penrose Street, Walworth, 17th December, 1876. Botanical artist. 'The Parterre,' 1841. 'Floral Tableaux,' 1846. Jacks. 39, 41; Gard. Chron. 1877, i. 24.

**Andrews, William** (1802-1880): b. Chichester, 1802; d. Dublin, 11th March, 1880. President, Dublin N. H. Soc. Discovered *Trichomanes radicans Andrewsii.* Journ. Bot. 1880, 256; 1883, 181; Proc. Roy. Irish Acad. iii. (1880); Dict. Nat. Biog. i. 409; Ann. N. H. vi. (1841), 382.

**Anglicus, Gilbertus** [*see* LEGLE].

**Ansell, John** (b. Hertford). Afterwards of Chislehurst, Kent. On Niger Expedition with Vogel in 1841. *Ansellia* Lindl.

**Aram, William** (fl. 1770). Of Norwich (?). Wrote list of Norfolk pl. in 'Description of England and Wales,' 1769–70, vol. vi. Jacks. 503.

**Archer, John** (fl. 1660–1684). Physician to Charles II. Of Knightsbridge. 'Complete Herbal,' 1673. Jacks. 199; Granger, iv. 5; Dict. Nat. Biog. ii. 71. Engr. portr. in his 'Secrets Disclosed,' 1684.

**Archer, Thomas Croxen** (1817–1885): b. Northamptonshire, 1817; d. Edinburgh, 19th February, 1885. Surgeon. In Liverpool Custom House, 1841–1860. Director, Edinburgh Museum of Science and Art, 1860–1885. Pres. Bot. Soc. Edinb. 1862. 'Popular Economic Botany,' 1853. Pritz. 8; Jacks. 66, 192; R. S. C. i. 85; vi. 567; vii. 42; Trans. Bot. Soc. Ed. xvi. (1886), 272.

**Arden, Lady** (fl. 1796). Correspondent of J. E. Smith. Fungologist. E. B. 461.

**Ardern, John** (fl. 1349–1370). Surgeon. Of Newark. 'De re herbaria . . . . .,' Sloane MS. 56, 335, 341, 2002, 3844, 1991. Pult. i. 23; Dict. Nat. Biog. ii. 76; Friend, Hist. Physic; Tanner, Bibl. Brit.-Hib. 48; Haller, i. 229. *Ardernia* Salisb.

**Argyle, Archibald, Duke of** [*see* CAMPBELL, ARCHIBALD].

**Armistead, Wilson** (fl. 1838–1865). Of Leeds. Meteorologist and entomologist. 'To my Botanical . . . Friends,' 1865, 8vo. Herbarium. 'Friends' Books,' i. 131; Hall, 'Flora of Liverpool,' vii.

**Arnold, Joseph** (1782–1818): b. Beccles, 28th December, 1782; d. Padang, Sumatra, July or August, 1818. M.D., Edin., 1807; R.N.; F.L.S., 1815. Memoir by Dawson Turner, Ipswich, 1849; Dict. Nat. Biog. ii. 110; Linn. Trans. xiii. 201. Col. medallion portr. Kew. *Rafflesia Arnoldi* R. Br.

**Arnott, George Arnott-Walker** [*see* WALKER-ARNOTT].

**Artis, Edmund Tyrrell** (fl. 1825). House-steward to Earl Fitzwilliam. 'Antediluvian Phytology,' 1825. Pritz. 9; Jacks. 176, 182.

**Arviel, Henry** (c. 1280). Resided at Bologna. 'De Botanica, sive Stirpium Varia Historia.' Pult. i. 22; Tanner, Bibl. Brit.-Hib. ; "Varia itinera susceperat," Haller, i. 219.

**Ascham, Anthony** (fl. 1550). Clerk. Physician. Vicar of Burniston, near Bedale, Yorkshire. 'A Little Herbal,' 1550. Pult. i. 50–1; Pritz. 9; Jacks. 25; Dict. Nat. Biog. ii. 149; Baker, 'Fathers of Yorkshire Botany.' *Aschamia* Salisb. = *Hippeastrum* (in part).

**Ashby, John** (1754 ?–1828): d. Bungay? 24th November, 1828. Grocer and draper. Of Bungay. Contributor to Smith's 'Flora Britannica.' Mag. Nat. Hist. ii. (1829), 120.

**Ashfield, Charles James** (fl. 1860). R. S. C. i. 107; vii. 53.

**Ashmole, Elias** (1617–1692): b. Lichfield, 23rd May, 1617; d. 18th May, 1692. Hon. M.D., Oxon, 1690. Dict. Nat. Biog. ii. 172; Mag. Nat. Hist. n. s. i. 272; Cott. Gard. iv. 269; Athen. Oxon. iii. 354.

**Atherstone, W. Guyton.** R. S. C. i. 109.

**Atkins, A.,** *née* CHILDREN (d. 1871): d. Halstead, Kent, June, 1871. d. of J. G. Children, of Brit. Mus. Collection of Brit. pl. and 3 vols. of impressions of Algæ in Herb. Mus. Brit. Jacks. 242.

**Atkins, Sarah or Lucy** [*see* WILSON].

**Atkinson, William** (1821–1875). F.L.S., 1860. Of Calcutta. Photo. portr., Kew.

**Atthey, Thomas** (d. 1880): d. at Gosforth, 1880. A.L.S., 1875. Contributor, on *Diatomaceæ*, to Ann. & Mag. N. H. Journ. Bot. xviii. 224.

**Attwood, E. Marcus.** R. S. C. i. 111.

**Aubrey, John** (1626–1697); b. Easton Perry, Kingston, Wilts., 12th March, 1626; d. Oxford, June, 1697. Nephew to Henry Lyte. Dict. Nat. Biog. ii. 244; Biogr. by J. Britton, London, 1845.

**Ayres, Philip Burnard** (d. 1863): d. St. Louis, Mauritius, 1863. M.D. Cryptogamist. Contributor to 'Phytol.' Pritz. 10; R. S. C. i. 129; Fl. Maurit. 10*. MSS. at Kew.

**Ayton, John** (fl. 1776). Of Kew. *Aytonia* Forst.

(To be continued.)

---

# SHORT NOTES.

WEST CORNISH PLANTS.—In May, 1886, Rev. R. P. Murray and myself gathered under trees at Antron House, near Helston, some specimens of a *Poa*, which Mr. Beeby inclined to name *sudetica* Haenke. Prof. Häckel now reports on it:—"*Poa Chaixi* Vill. (1785), var. *remota*. *P. remota* Fr. *P. sudetica* Haenke (1791), var. *remota*, auct. pl." I brought away from the Lizard coast a turf containing *Scilla autumnalis*, &c. Last summer a *Festuca* flowered in it, which differed from the forms I had previously seen. Through Mr. Beeby's kindness I have the following definition from Prof. Häckel:—"*F. ovina* var. *vulgaris* Koch, subvar. *hispidula*." It is remarkably glaucous and broad in the flower.—*Polygala vulgaris* L. (*fide* Ar. Bennett), from the south shore of the Looe Pool, and *Carex vesicaria* L., from Gunwalloe, are additions to the records for v.-c. 1 in Top. Bot., ed. 2.—*Scirpus pauciflorus* Lightf., queried there, occurs above Kynance Cove and in Kynance Vale. — EDWARD S. MARSHALL.

CAREX TRINERVIS Degl. IN IRELAND. — In the summer of 1885 (August 12th) I gathered a *Carex* in the neighbourhood of Round-stone, Co. Galway, which I took for a form of *C. Goodenowii*, but thought at the time a peculiar form. It grew in a damp part of an extensive seaside field, which was in grass, and for the most part was of a sandy character. There was not much of the *Carex* to be seen, and it was only noticed by me in one spot. I had not at that time seen the English form of *C. trinervis*, and the specimen in question was laid aside among other remnants of my Irish gatherings till the autumn of 1887, when, looking out Irish plants for a friend,

I came across it. I was struck with its general aspect, and, being by then familiar with the appearance of our Norfolk *trinervis*, I quickly examined the nervation of the fruit. I was soon satisfied that I had a form of *trinervis* Degl. before me, and on comparison with other specimens I found it agreed best with one which Mr. Arthur Bennett had given me of the variety *laxa* of Lange. Mr. Bennett on seeing the specimen did not feel justified in pronouncing the Irish plant to be Lange's variety. This question is, however, of minor importance : the interesting features of the discovery are, the fact that Ireland has another *Carex* to add to its list, and that one only lately known for Britain ; the westward extension of the geographical area of this species, Oporto being (according to Nyman) the point farthest south and farthest west for which it has hitherto been reported ; and an additional link, of no little interest, between the flora of S. and W. of Ireland and the flora of the Spanish Peninsula.—EDWARD F. LINTON.

GLAMORGANSHIRE PLANTS.—The following species, none of which are recorded for Glamorganshire in the last edition of Top. Bot., were observed in that county during last August by Mr. D. Morris, of Kew, Mr. R. V. Sherring, and myself. Some of them were found by Mr. Morris and Mr. Sherring when together ; others by Mr. Sherring and myself :—*Raphanus maritimus* Sm. Very sparingly on a shingly beach at Penclawdd. — *Viola Curtisii* Forst. Burrows near Kenfig ; only in small quantity.—*Rubus plicatus* W. & N. (name endorsed by Mr. T. R. Archer Briggs). Clive Common, near the Mumbles. — *R. affinis* var. c. *cordifolius* W. & N. (so named by Mr. J. G. Baker). At and near the Mumbles ; apparently rather frequent.—*Apium nodiflorum* Reichb., var. c. *ochreatum* DC. (considered such by Mr. Arthur Bennett). Clive Common, near the Mumbles. —*Juncus obtusiflorus* Ehrh. Oxwich Marsh. — *J. acutiflorus* Ehrh. Oxwich Marsh.—*Scirpus Tabernæmontani* Gmel. Oxwich Marsh.— *Aira caryophyllea* L. Near the Mumbles. — *Sieglingia decumbens* Bernh. Near the Mumbles. — The *Viola Curtisii* from Kenfig appears precisely similar to that which grows so abundantly on some parts of Braunton Burrows, N. Devon. Dr. Boswell, in Eng. Bot., says that *V. Curtisii* probably occurs on Crumlin Burrows, Glamorganshire, but that he had not seen any specimens from that locality. When I visited the Crumlin Burrows in August last I could not find it there ; but I was able to examine a portion only of that extensive tract, in some part of which the plant may very likely still exist. Glamorganshire is noted in the last edition of Top. Bot. as a comital exception for *Juncus acutiflorus*, *Aira caryophyllea*, and *Sieglingia decumbens ;* hence the reason for here recording those not uncommon and widely distributed species.— DAVID FRY.

ON LEAF-BEARING STIPULES IN POTAMOGETON. — Last summer my friend Mr. Arthur Bennett sent me a root of his *Potamogeton Griffithii*, which I planted in a large tub, so that its growth might be more carefully watched than it could have been in a pond. It soon produced two or three small shoots, which made but little progress ; but it was evidently pushing out strong stolons in all

directions in the mud in which it grew.  In September these stolons
sent up several vigorous stems, on each of which the lowest stipule
was closely clasping, and furnished on its back with a narrow
*linear-spathulate coriaceous leaf*.  This year, when putting a fine
series of *Potamogeton plantagineus* in the press, I was much struck
with the resemblance its early state bore to that of *P. Griffithii*,
and was therefore induced to examine my specimens more closely
to see if similar adnate stipules were present; I soon found some
few examples, but they occur more sparingly than in *P. Griffithii*.
In the latter species they seem to be always present at the base of
each shoot.  This hurried notice is written to call the attention of
observers to this new form of stipule, which is not the exact
analogue of that of the *pectinatus* group.  They should be sought for
at the base of the stem, as they seem only to be produced immediately above the surface of the mud.  Probably they may occasionally
occur in other species of *Potamogeton*, and should be looked for as
early in the year as possible, as they soon decay.  The collection of
early states of all our Pondweeds is well worth the attention of
botanists, and they make very beautiful specimens for the herbarium.
—ALFRED FRYER.

## NOTICES OF BOOKS.

*A Flora of Hertfordshire.*  By the late ALFRED REGINALD PRYOR,
   B.A., F.L.S.  Edited for the Hertfordshire Natural History
   Society, by Benjamin Daydon Jackson, Sec. L.S.: with an
   introduction on the Geology, Climate, Botanical History, &c.
   of the County, by John Hopkinson, F.L.S., F.G.S., and the
   Editor.  London : Gurney and Jackson.  1887.  8vo, pp.
   viii. 588.

AFTER a series of delays, unavoidable although regrettable, the
posthumous work of the botanist known to the readers of this
Journal as "R. A. Pryor," sees the light.  Its author died early in
1881 ; and the task of completing his work, or preparing it for the
press, has passed through various hands before it came into those
of Mr. Daydon Jackson. who has brought it to a successful issue.
It was originally intended that I should edit the work, as Mr.
Jackson states in his Preface ; but he does not state that this undertaking of mine was contingent on a promise of the late Mr.
Newbould to transcribe Mr. Pryor's MS. for the press.  It will
surprise no one who knew Mr. Newbould's anxious conscientious-
ness, that this part of the work progressed but slowly ; nor was it
to be wondered at that the Herts Natural History Society became
impatient, and requested, at the end of two years, that the
MS. should be returned to them.  I was quite prepared to do all
that I had promised ; and I may further say that, although the
public are doubtless the gainers, it was a regret to me to be
deprived of the opportunity of carrying out what I know would
have been the wish of my late friend.  Mr. John Hopkinson then
undertook the task, but his progress was hardly more rapid ; and

this and other reasons render it a matter for satisfaction that the work was taken up by Mr. Jackson in the autumn of 1885.

There are many reasons which render it difficult to criticise the volume now before us. Mr. Jackson has conscientiously followed Mr. Pryor's MS.; but this was in many cases incomplete, while in others localities were given with a fullness and detail which, although representing the intentions of the author at the time they were committed to paper, would, there is reason to think, have been considerably modified before printing. Considering the small extent of the county, it may fairly be questioned whether the localities for the commoner species need have been given at such length. *Papaver Argemone*, for example, occupies a whole page; so does *Arenaria leptoclados*; *Helianthemum Chamæcistus* has nearly two pages devoted to it; the Resedas have nearly three pages between them. Still more open to criticism is the prominence given to such annual casuals as *Papaver somniferum* and *Camelina sativa*; and it is to be regretted that such plants as *Silene nutans* (one or two plants on a garden wall), and *S. conica* ("three plants in the middle of a fifty-acre field"), should be admitted to the honours of thick type, numbered as part of the native flora of the county, and included in the "tabular statement" of distribution. Other species, as it seems to me, should have been referred to incidentally rather than as actually forming part of the Flora, on the ground of needing verification. Such are *Pyrola rotundifolia*,—surely a mistake for *P. minor*,—*Malaxis paludosa* (on Parkinson's authority only), and *Cephalanthera ensifolia* (only given in Gibson's Camden).

As Mr. Jackson remarks in his Preface, it is matter for sincere regret that Mr. Pryor did not live to complete his work. It would, under his hands, have taken rank with Mr. Briggs's 'Flora of Plymouth,' in the critical notes with which the author would have enriched it. Unfortunately, they were hardly ever committed to paper, although carefully stored up in Mr. Pryor's accurate mind. In some cases his views were published, as, for example, on *Epipactis latifolia* (Journ. Bot. 1881, 71), Bobart's green Scrophularia (Id. 1877, p. 238), and the Hertfordshire Carices (Id. 1876, 365); and an editorial note calling attention to these would have been desirable. He had made a special study of some plants, such as the Poppies and Water Buttercups, and had hoped to publish notes on these in his book. A careful collation of Mr. Pryor's published papers in this Journal with the Flora would probably result in many corrections and additions: thus, *Typha angustifolia* is recorded for Hatfield (Journ. Bot. 1874, 22), which seems to answer the question raised at p. 509 of the Flora, and removes it from the list of plants (p. xxxiii.) peculiar to the Thame district.

In the main, however, the work has been carried out as the author would have wished. Among its distinctive features are the arrangements and naming according to Nyman, with such exceptions as Mr. Jackson believes Mr. Pryor would have made; the reference following the name of each genus and species to the place (I think Mr. Pryor meant also to have added the date) of publication; and an index of species as well as of genera. For the last

we have to thank Mr. Jackson, who is also responsible for the full
and interesting list of contributors and the "botanical history of
the county." Mr. Hopkinson contributes the geographical and
geological notes. At the end are lists of Mosses, Algæ, Hepatics,
Lichens, and Fungi, which "must be regarded only as an attempt
to gather together the Hertfordshire records of cellular cryptogams."

Mr. Pryor was a strong advocate of the division of districts in
accordance with the river-basins, and this plan, originally proposed
by Coleman in the previous Flora, is here more fully adopted. As
a natural result, the extent of the various districts is very different:
Thame and Brent being very small (though the latter is very
interesting), while Colne and Lea between them include nearly the
whole of the county. The six divisions are grouped under two
main heads, Ouse and Thames, the former including the small
divisions of Cam and Ivel, the latter the four named above.

It is much to be regretted that the Hertfordshire Society, which
was materially benefitted at Mr. Pryor's decease, has not published
a fuller biography than the sketch given at pp. xliv—xlvi. A copy
of an excellent portrait, taken not long before his death, might well
have faced the title page; and an example of his characteristic
letters might have been added. No reference is made to the careful
manner in which Mr. Pryor personally examined the various
districts, nor is even a list of his publications appended. These seem
to me serious omissions.

It was my privilege to accompany Mr. Pryor on three of the
"jaunts," as he used to call them, undertaken for the purpose of
examining the plants of the country. His custom was to make some
place a centre for two or three days, taking walks in different
directions, and carefully noting what was seen. In the evening his
memoranda were transcribed into one or more of his numerous
note-books, and doubtful plants examined. Recalling some of
these rambles, and the plants met with, I am inclined to think that
some of those books must have been lost, or that their contents were
not again entered by him in the quarto MS. books which contained
the Flora. Be this as it may, there is no record in the volume of
localities of certain plants which we noted together—of *Cerastium
arvense*, for instance, which we tracked at Aldbury Owers from
Buckinghamshire just into Herts, on an occasion when we vainly
endeavoured to find *Pulsatilla* on the Buckingham side of the
boundary; of *Myosurus minimus*, which we picked up in a cornfield
near Tring station, during an after-dinner stroll on a bleak bright
evening of May, 1876; and many more. Mr. Pryor's herbarium
was very small, so that specimens of his gatherings are nowhere
very largely represented; a considerable number, however, are in
the British Museum Herbarium.

The appendix of "additional published localities with a few
hitherto unpublished" (for which, of course, Mr. Pryor is in no
way responsible) contains some matter for remark. The locality
given for *Pulsatilla vulgaris* is Herts in New Bot. Guide, as
"Ashley" is annotated "probably Ashley Green in Bucks." If so,
it is new to the county. An old Rayan plant—"Alsine montana

minima "—which " has remained doubtful for nearly two hundred years " is cleared up by Mr. Jackson's consultation of the Sloane Herbarium, and shown to be a "compact" (? young) form of *Moehringia trinervia*. *Pyrola* "*media*" of the New Bot. Guide figures here, although the authors of the 'Flora Hertfordiensis' say that the station was in Bucks, and there can be no doubt but that *P. minor* was meant; so that the query may be removed from the word "error?," which Mr. Jackson has appended to the record.* *Euphorbia stricta* is surely an error. *Blysmus compressus* is here added to the Flora on the authority of Dr. De Crespigny, but without locality; a reference to the British Museum Herbarium shows specimens collected by that botanist on Rickmansworth Common Moor in 1877; Middlesex is queried for this species, but the same herbarium contains it from Harefield, collected by Dr. Forbes Young. *Cystopteris fragilis* surely has no claim to thick type and a number.

The list of "additions and corrections," independently of the above, extends to five pages, but is certainly by no means complete; *e. g.*, on p. 151, "Journ. Bot. 1874, 272," should be 1875, 212; " *Couringia* " (p. 31) should be *Conringia*, and " *Vaccinum* " (p. 269), *Vaccinium ;* the last error might well be taken for a " recurrence to primitive type," in these days of restoration of old names and spellings.                    JAMES BRITTEN.

---

*Lectures on Bacteria.* By A. DE BARY. Translated by Henry E. F. Garnsey, Revised by Isaac Bayley Balfour. (Oxford : Clarendon Press, 1887). Pp. xii. 193 ; 20 cuts. Price 6s.

ABOUT a year ago it was said in these pages that the study of Bacteriology was rapidly becoming an affair of pots and pans,—apparatus, staining media and the like,—that the Bacteria themselves were being lost sight of. No naturalist could survey the literature of the subject in our language without a misgiving that true words were then spoken in jest. Since they were printed, however, two remarkable additions have been made to our literature, *viz.*, the section on Bacteria in De Bary's 'Comparative Morphology and Biology of Fungi, Mycetozoa and Bacteria,' which has been published in English form, and the same author's 'Lectures on Bacteria,' now under notice. These books have been made part of our literature, accessible to all; and they exhibit to us exactly the state of our knowledge of the natural history of Bacteria. The 'Lectures on Bacteria' are remarkable, not only on account of the survey of the subject and admirable arrangement of the matter, but in an equal degree for the style of exposition, which the translators have rendered very happily. The book not only

---

* In Fl. Hertfordiensis we read, " *P. media* is certainly a native of Bucks." This is an error ; and we learn from Mr. Watson's MS8. (now in the British Museum), that " the certainty rests on Mr. Pamplin's authority, who says that he knows both species (*P. media* and *minor*), and that he found the former two or three miles west of Tring, which would certainly be in Bucks.—Rev. R. H. Webb, in letter of April 10, 1849."

contains an account of Bacteria, which may be read with the greatest profit by those already familiar with the whole subject, but it is written in such a manner that no cultivated reader who desires information about Bacteria will turn to it in vain.

It would be needless to attempt a calculation of the numbers of botanical and other scientific books which have been designed for the general educated public. The almost invariable result is, that when an enlightened member of the public sits down to peruse such a book with thankfulness in his heart that now at last he will learn something fundamental about so-and-so, he begins exceedingly comfortably and all goes well for a page or two. Then a sentence is reached which has to be read again; if he be persevering, index or glossary are consulted and—well, the result is that he feels a few minutes later he "would like his money back." The present book not only contains, as has been said, the whole matter from the author's point of view; it is not only a severely and profoundly accurate book, but it is intelligible to the merest chemist who needs biological information on the subject. With this book, and with Dr. Klein's 'Micro-organism and Disease,' especially that portion of it which deals with the methods of research, the student may equip himself, for a few shillings, with the necessary fundamental literature of Bacteriology.

Prof. de Bary may be heartily congratulated on the successful form of the English edition of his lectures, which conveys them to all here "who are not strangers to the elements of a scientific training."

The book contains, besides a good index, an excellent conspectus of the literature, with notes on it.

G. Murray.

### ARTICLES IN JOURNALS.

*Bot. Centralblatt.* (No. 1).—J. Jankó, *Equisetum albo-marginatum.* — (No. 3). A. Hansgirg, 'Einige Bemerkungen zum Aufsatze A. Tomaschek's 'Ueber *Bacillus muralis.*'' — (No. 4). J. Murr, 'Ueber die Einschleppung und Verwilderung von Pflanzenarten im mittleren Nord-Tirol.'

*Bot. Gazette* (Dec.). — B. D. Halsted, 'Three nuclei in pollengrains' (1 plate).— C. Robertson, 'Fertilisation of *Calopogon parviflorus.*' — J. M. Coulter & J. N. Rose, ' *Umbelliferæ* of E. United States' (1 plate).—A. Gray, *Coptis laciniata*, sp. n.

*Bot. Zeitung* (Dec. 23). — J. Wortmann, ' Zur Kenntniss der Reizbewegungun.' — (Dec. 30). O. Loew & Th. Bokorny, 'Ueber des Vorkommen von activen Albumin im Zellsaft und dessen Ausschiedung in Körnchen durch Basen.'—(Jan. 6, 13). M. W. Beyerinck, 'Ueber das *Cecidium* von *Nematus Capreæ* auf *Salix amygdalina.*' — (Jan. 20, 27). E. Zacharias, 'Ueber Kern- und Zelltheilung.' — W. Detmer, ' Ueber physiologische Oxydation im Protoplasma der Pflanzenzellen.'

*Bull. Torrey Bot. Club* (Jan.).— T. Morong, ' Studies in *Typhaceæ.*'—F. L. Scribner, ' New or little-known Grasses' (*Muhlenbergia arizonica* Scrib., *Sporobolus interruptus* Vasey, *Deyeuxia Suksdorfii* Scrib., *Bromus Pampellianus* Scrib., spp. nn. : 1 plate). — T. F. Allen, ' *Nitella* (not *Tolypella*) *Macounii.*'—G. Vasey, ' New Western Grasses' *Poa macrantha* Vasey, *P. argentea* Howell, *Alopecurus Howellii* Vasey, *A. Macounii* Vasey, *A. californicus* Vasey, spp. nn.).

*Gardeners' Chronicle* (Dec. 24). — *Anthurium acutum* N. E. Br., sp. n. — (Jan. 7). *Ficus Canoni* N. E. Br., *Albuca Allenæ* Baker, *Catasetum pulchrum* N. E. Br., spp. nn. — *Stachys tuberifera* (fig. 1). N. E. Brown, ' *Veronica cupressoides* ' (figs. 3–7).—(Jan. 14). *Laelia Gouldiana* Rchb. f. ("n. sp. or n. hyb." ).— (Jan. 28). *Oncidium chrysops* Rchb. f., sp. n.—Fasciated Petunia (fig. 21).

*Journal de Botanique* (Jan. 1).— L. Mangin, ' Sur de développement des fleurs dans les bourgeons.'—E. Rose, ' La Flore Parisienne aux commencement du xvii. siècle.' — N. Patouillard, ' La classification des Champignons.' — (Jan. 15). E. Bornet, ' Algues du voyage au golfe de Tadjoura.'—L. Morot, ' Sur l'identité spécifique du *Polyporus abietinus* Fr. et de l'*Irpex fusco-violaceus* Fr.'

*Notarisia* (Jan.). — ' Algæ novæ quæ ad litora scandinaviæ indagavit H. F. G. Stroemfelt' (*Microcoryne, Phycocelis,* genn. nov. : 1 plate).—F. Castracane, ' Saggio sulla flora diatomacea delle cosi dette muffe delle terme di Valdieri.' — E. Bornet & C. Flahault, Concordance des ' Algen Sachsens et Europás.' — A. Hansgirg, ' Algæ Novæ aquæ dulcis.'

*Oesterr. Bot. Zeitschrift* (Jan.).—F. Krašan (memoir and portrait). — L. Celakovsky, *Lathyrus spathulatus*, sp. n. — J. Bornmüller, *Ptilotrichum Uechtritzianum* sp. n. — O. Stapf, ' Ueber einige Iris-Arten des botanischen Gartens in Wien.' — B. Blocki, *Viola roxalanica*, sp. n. — P. Conrath, ' Zur Flora von Bosnien.'—J. Ullepitsch, ' Neue Pflanzenarten.' — E. Formánek, ' Flora von Nord. Mahren.'—P. G. Strobl, ' Flora des Etna.'

*Pharmaceutical Journal* (Jan. 14).—D. Hooper, ' Bark of *Michelia nilagirica.*'

---

## OBITUARY.

ALEXANDER DICKSON, M.D. (Edin. & Dubl.), LL.D. (Glasgow), Professor of Botany in the University of Edinburgh, died suddenly, on 30th Dec. last, from heart-disease, when he was entering in his note-book the state of a game of curling in which he was one of the leaders. Prof. Dickson was born on the 21st February, 1836, in Edinburgh. He was the second son of David Dickson, of Hartree, and by the early death of his elder brother he became the heir and then the owner of the estates of Hartree and Kilbucho. His father also died suddenly from heart-disease. He studied at the University of Edinburgh, and graduated in Medicine in 1860. A year of his medical curriculum was spent at the University of Wurzburg. He early manifested a great love for Botany, was one of the most distinguished of the late Dr. Balfour's students, and gained a gold medal at graduation for his thesis on " The Development of the

Seed-vessel in *Caryophyllaceæ*," an abstract of which was published in the Transactions of the Edinburgh Bot. Soc., to which Society he had already contributed a paper "On a Monstrosity in *Silene inflata*," which suggested some generalisations on placentation, and a dissertation "On the Compound Nature of the Cormophyte." He had qualified himself to practise medicine, but his heart was in botanical studies, and in the hope that some suitable opening might present itself to enable him to devote himself to his favourite science he delayed for some time taking any steps to begin the practice of medicine. At length in the beginning of 1862 he was, as he himself wrote, " at last reduced to the dire necessity of announcing himself as a servant of the public, by way of a door-plate." Before, however, any practical issue came of this he was called to act as deputy for Prof. Dickie, then in bad health, in the University of Aberdeen. His investigations into the morphology and development of the flower occupied all his spare time. He made endless preparations, and a great series of the most careful drawings; and his greatest pleasure was to get some appreciative listener to hear his demonstration of his specimens. His little black wicker-work basket containing his precious slides was his constant companion in all his travels. At the close of 1866 he was appointed to succeed Prof. Harvey in the chair of Botany at Trinity College, Dublin, and shortly after he became also Professor in the Royal College of Science, Dublin. His stay in Ireland was short, for in 1868 he was appointed Professor of Botany in the University of Glasgow, in succession to Prof. Walker-Arnott. He found fresh opportunities for prosecuting his investigations on the morphology of plants, and his published memoirs are all characterised by the singularly careful statement of the facts, and by the cautious, judicious, and philosophic generalisations based on the facts. He has at different times communicated to the world some of these memoirs through the pages of this Journal. In 1879 the chair of Botany became vacant through the resignation of Prof. Balfour, and Dr. Dickson was appointed his successor both as Professor of Botany and as Regius Keeper of the Royal Botanic Gardens, Edinburgh. The duties of the immense class of students in Edinburgh are very heavy, though happily limited to the short summer session. Dr. Dickson went heartily into his work. His masterly knowledge of his subject, his happy illustrations, his facile use of the black-board, and his genuine sympathy with his students, who, even the laziest of them, fully appreciated his warm heart and conscientious work, made him universally beloved. No grief at his sudden death surpassed that of the students of Edinburgh University. In Edinburgh, at Hartree, and wherever he was known, Prof. Dickson will be sorrowed for : he was a true friend, and a good man.

WILLIAM CARRUTHERS.

WE regret to record the death of Prof. DE BARY, which took place on Jan. 19th. We hope to give a portrait and memoir of this distinguished botanist at an early date, as also of Prof. ASA GRAY, whose death, like that of Dr. BOSWELL, occurred on Jan. 31st.

## HEINRICH ANTON DE BARY.

ANTON DE BARY was born in Frankfurt on the 26th January, 1831, and died at Strassburg on the 19th January, 1888. In a 'Journal of Botany' it seems hardly necessary to say more. There are no readers of these pages to whom his name is not familiar—to most his works are familiar. The weekly issue of the 'Botanische Zeitung' brought it freshly before our eyes, the almost daily use of one or other of his books will keep it there for many of us. To a smaller circle in this country no such summons to memory is necessary—I mean those who have known him. One is accustomed to the platitude that a man's works are his monument. In de Bary's case they are more—they are the story of his life in one sense. When one surveys the work and measures the short span -of the years of its accomplishment, the reflection is obvious that there was little room for more than a man's domestic life. De Bary found room, however, for mental culture of a very wide kind. His delight in art was at all times fresh and strong, and the last day which the present writer passed in his company was largely spent in our National Gallery and among the classical antiquities of the British Museum. He seemed to forget there the great bodily pain I know he was suffering and the mental anxiety I believe he then felt.

The reproach of ignorance of systematic Botany is often deservedly laid on morphologists and physiologists. It would be the greatest mistake to lay it on de Bary. He had a remarkable knowledge of the plants of the regions in which he lived and of the Alps,—not only the flowering plants but the cryptogams as well—and he had a humorous contempt for the botanist who cared for none of these things. In connection with this I may be allowed to recal the only rebuke I ever received from him. I had declared, in the course of a Sunday walk,—with the rashness of youth,—my opinion of the monotonous character of the grasses. The only reply was the suggestion that it would be a good thing to bring him one or two named every morning, adding that the season of the year was very favourable. Soon afterwards, in the course of some experiments, I had infected a number of crucifers in the Botanic Garden with *Cystopus candidus*. The disease spread among the few allied plants in the old garden, and there was discovered one morning a dearth of material for illustrating a lecture including that group. With evident amusement de Bary asked me if there were any other natural orders to which I owed a grudge.

One recollection of his own boyhood will have interest for readers in this country. His curiosity was greatly excited by hearing his elders talk of the great disaster of the potato disease at the first serious outbreak.

The following extract from an obituary notice of him written by me for the 'Academy' relates shortly the story of his life:—

"Having entered the University of Berlin he came under the influence of the celebrated Alexander Braun. He began at once the work of original research; and, in his 'Untersuchungen über die Brandpilze,' published when twenty-two years of age, there is no trace of a prentice hand. The next memoir of note was his 'Untersuchungen über die Familie der Conjugaten'—an investigation full of interest to the student of the development of sexuality in lower organisms. These researches established his reputation for brilliant work; but when, in the year following (1859), the publication of the last memoir, there appeared 'Die Mycetozoen' (second edition 1864), de Bary came at once into the front rank of biologists. In this remarkable paper there was told the life-history of these organisms, which have continued to fascinate every one since. There is hardly a biologist of note of the present generation who has not at some time or other "taken up" the Mycetozoa. Are they animals or plants? Or is it profitable to put the question in that form at all? They had been considered fungi of high organisation, until at one stroke they sank so low in the scale of classification that the botanist likes to think of them as beyond the frontier line altogether. Next followed the 'Recherches sur les développement de quelques champignons parasites,' in which our knowledge of Peronosporeae especially was much extended. Next 'Die Fruchtentwickelung der Ascomyceten' gave rise to much discussion—limited, however, to botanists. In the meantime de Bary and Woronin had established the 'Beiträge zur Morphologie und Physiologie der Pilze,' consisting of a series of memoirs coming out at uncertain times and continued down to a few years ago. In 1866 his handbook, 'Die Morphologie und Physiologie der Pilze, Flechten und Myxomyceten,' represented the first serious attempt to establish order in the vast literature of mycology. It was a splendid performance; and the impetus it gave to research, and, better still, the direction cannot be overvalued. Numerous memoirs followed. De Bary became editor of the 'Botanische Zeitung,' a weekly journal, in addition to his other labours, and enriched it with much of his own work. Among the papers published during this time was the account of his own investigation of the potato-disease, which attracted much notice in this country. A great labour was carried on during these years and finally saw the light in 1877—his 'Vergleichende Anatomie der Vegetations organe der Phanerogamen und Farne'—a book representing enormous labour as well as insight of the highest order. In 1878 he published his charming primer of Botany; and another period followed in which papers now and then appeared—for example, that on apogamy—and during which he was perfecting what was nominally a second edition of the great book on fungi, but turned out to be in point of fact a new work. In 1884 appeared the 'Vergleichende Morphologie und Biologie der Pilze, Mycetozoen und Bacterien,' which, in many respects, stands not only above his own previous work but well in advance of anything in the contemporary literature of botany. In 1886 his 'Vorlesungen über Bacterien' came as an especial pleasure to those who wished to see this group dealt with by an accomplished naturalist.

It would be interesting to point out in greater detail than these columns permit the direct influence of de Bary's work on agriculture and on medicine, as well as on the progress of botany. His method of *cultivation* of disease organisms has been the one by which all true progress has been made in the study of them.

"During these years of productive labour de Bary held the post of Professor of Botany, first at Freiburg, then at Halle, and, since the war, at the new German University of Strassburg. Both in Germany and in this country numerous pupils are striving to carry on his work in the spirit of their master. His remarkable personal kindliness and delightful humour inspired those who have had the privilege of working under his direction with feelings of devotion not only to botany but also to Anton de Bary."

GEORGE MURRAY.

---

# DE DUABUS ROSIS BRITANNICIS

SCRIPSIT N. J. SCHEUTZ.

REVERENDUS E. F. LINTON illas Rosas, novas floræ Britannicæ cives, mihi examinandas nuper misit, de quibus pauca disserere in animo habeo.

R. MOLLIS, Sm., var. *glabrata* Fries Novitiæ fl. Suecicæ, ed. 2 p. 151. Foliolis utrinque glabris vel glabrescentibus, subtus glanduloso-punctatis vel rarius ferè eglandulosis.

Habitat in Scotia, Strome Ferry, Ross. E. F. Linton legit.

Specimina scotica cum suecicis congruunt. Primo quidem adspectu crederes, formam in Scotia lectam pertinere ad *R. tomentosam*, Sm.; sed rami breves, erecto-patentes, forma et serraturæ foliolorum, fructus præcoces hispidi, sepala adscendentia et persistentia sunt omnino ut in *R. molli*, cujus forma *glabrata* certissime est. Formam analogam in Anglia septentrionali lectam et sub nomine *R. tomentosæ* missam Crèpin commemorat in 'Primitiis Monographiæ Rosarum,' fasc. vi. p. 108.

Sunt qui saltem in Suecia habuerint *R. mollem* var. *glabratam* pro *R. marginata* Wallr., quia descriptio a Wallrothio data convenire videbatur. Sed falso; nullo enim modo *R. marginata* Wallr., pertinet ad *R. mollem*. Déséglise, qui in herbario De Candollei examinavit typum *R. marginatæ* Wallr., in 'Catalogue raisonné des espèces du genre Rosier,' p. 251, docuit hunc typum pertinere ad Caninas hispidas. Exstat tamen alia *R. marginata* Auctt. non Wallr., quam Déséglise l. c. dixit esse lectam in Britannia et Gallia. Diagnosis, quam Wallroth dedit in 'Rosarum pl. gen. historia,' tam ambigua et vaga est, ut ex ea nihil certi concludi possit, quare rhodologi alii aliter de *R. marginata* Wallr. cogitant. Sic Baker, in 'Monogr. of British Roses,' *R. marginatam* collocavit inter Caninas subrubiginosas. Christ autem in 'Die Rosen der Schweiz' existimavit *R. marginatam* esse formam *R. trachyphyllæ* Rau. In Primitiis monagr. Rosarum, fasc. ii. p. 33, Crèpin dixit *R. marginatam* Wallr. videri esse formam hybridam ex una Caninarum atque una alterave forma R. Gallicæ L. ortam.

F 2

Koch in Synopsi fl. Germ. et Helvet., atque Grenier in " Flore de
la chaine Jurassique' alias sententias amplexi sunt. Utcumque res
se habet, certum est *R. mollem* var. *glabratam* non esse *R. marginatam* Wallr.

R. CORIIFOLIA Fries var. *Lintoni*, nov. var., foliolis pubescentibus,
subtus plus minusve glandulosis, duplicato-serratis dentibus cum
1–3 denticulis glandulosis ; pedunculis brevibus nudis; receptaculis
fructiferis subglobosis eglandulosis ; sepalis post anthesin erecto-
patentibus persistentibus, dorso eglandulosis.

Habitat in Scotia, ad flumen prope Braemar, Aberdeen.  E. F.
Linton legit.

Forma notabilis, foliis subtus glandulosis vergens ad Tomen-
tellas, ad quas retulissem, si sepala essent reflexa et decidua.
Si quis sequatur dispositionem, quam Cel. J. G. Baker dedit in
opere jure laudato ' Monograph of British Roses,' hanc varietatem
referat ad seriem, quæ] appellatur Subrubiginosæ, et inter *R. caninam* var. *Borreri* (Woods) et var. *Bakeri* (Déséglise) collocet.
Videtur proxime accedere ad var. *Bakeri*, quam Crépin in ' Primitiis monogr. Rosarum,' fasc. VI. p. 58, considerat ut varietatem
*R. coriifoliæ*. Inter Rosas Scandinavicas occurrit forma, quæ,
varietati *Lintoni* affinis sed habitu et nonnullis characteribus
diversa, appellata est *R. gothica* Winslow (in Botaniska Notiser
1879, Herb. Rosar. Scandin. No. 29).

Hoc loco haud alienum esse arbitror legentium plerosque admo-
nere, nomen *Rosæ coriifoliæ* Fr. apud complures hujus ætatis rhodo-
logos latins patere atque plures formas comprehendere quam apud
Fries in Novitiis fl. Suecicæ, ubi Fries expressis verbis dixit
*R. coriifoliam* præditam esse *foliolis subtus eglandulosis* subæqualiter
serratis, *serraturis simplicibus eglandulosis*. Recentiore tempore haud
paucæ formæ et varietates *Rosæ coriifoliæ* in Scandinavia sunt
lectæ quæ a descriptione Friesii in Novitiis fl. Suec. et in Summa
Veget. Scandin. data plus minusve discrepant. Hæc enim Rosa in
Suecia multum variat atque aliis locis alias formas sibi induit, quæ
una cum forma typica *R. coriifoliæ* ad *R. dumetorum* Thuill. eodem
fere modo se habent, quo *R. Reuteri* (God. = *R. glauca* Vill.)
ejusque formæ ad *R. caninam* L. sensu strictiore.

Examinanti mihi var. *Lintoni* in mentem venit, formas *R. coriifoliæ* foliolis subtus *glandulosis*, ut var. *gothicam* (Winslow), var.
*Bakeri* (Déségl.), var. *Lintoni* aliasque, fortasse ita se habere ad
cæteras formas *R. coriifoliæ* foliolis subtus *eglandulosis*, ut *R. tomentellam* Lém. ejusque formas ad *R. dumetorum* Thuill. ejusque
formas.  *R. tomentellam* credo propius accedere ad *R. dumetorum*
quam ad sectionem Subrubiginosas.  Hæ formæ *R. coriifoliæ* foliis
subtus glandulosis recedentes, quæ possunt in unam seriem conferri
atque appellari *Subtomentellæ*, præcipue sepalis erectis persistentibus
a Tomentellis veris differunt.

# THE MOSS FLORA OF SUFFOLK.

## By the Rev. E. N. Bloomfield, M.A.

Since writing the paper under the above title in this Journal for 1885, pp. 233-238, I have obtained much additional information which it may be well to put on record.

On sending a copy to Sir Charles Bunbury, of Barton Hall, near Bury St. Edmunds, he wrote, " Mr. Eagle's collection of which you speak is now in my possession, having been bought by my father after Mr. Eagle's death, and I have incorporated it with my own."*

My friend, the Rev. W. M. Hind, has carefully examined Sir Charles Moss's Herbarium, and has sent me a list of all the Suffolk species contained in it, with their localities. Several of Mr. Eagle's Suffolk specimens have notes and drawings appended by Mr. W. Wilson (the author of the 'Bryologia Britannica'), while the names of others have been corrected by him. Besides this, Mrs. Skepper, of Bury St. Edmunds, has very kindly forwarded me over twenty letters from Mr. Wilson to Mr. F. K. Eagle, ranging from 1842 to 1855, which throw light on many of the specimens in Mr. Eagle's collection. I shall therefore quote freely from these letters.

Sir C. Bunbury died at Barton in 1886. He has left his Herbarium to the University of Cambridge, but it will, I believe, remain for the present in Lady Bunbury's possession at Mildenhall.

I propose to incorporate with this supplementary list the additions made to the Moss Flora of the county by Mr. H. N. Dixon, in the Journal for 1885, p. 311, and for 1886, p. 283. He very kindly sent me specimens of most of these additions which have been examined and confirmed by Mr. Boswell, who kindly looked them over, as he had done my former Suffolk specimens.

Mr. Eagle's species are given on his own authority, except so far as they are confirmed by Mr. W. Wilson; while Sir C. Bunbury's specimens were doubtless named by himself after comparison with specimens from Messrs. Dickson, Turner, &c.

I have employed as far as possible the nomenclature of the London Catalogue of 'Mosses and Hepatics.'

In order to indicate the distribution, I have added E. East Suffolk, W. West Suffolk, wherever these are not already given in my former paper.

*Sphagnum cymbifolium* Ehrh. var. *squarrulosum* Nees. Lound, *H. M. Dixon*. E.

*Systegium crispum* Hedw. Barton, *Sir C. B.* Newmarket Heath, *Eagle*. W.

---

* " I have long had great pleasure in the study of Mosses, and have collected them in various countries. Mr. Eagle's collection, which I have incorporated with my own is, I believe, rich as a British collection, for Mr. Eagle had received a great number of species from Dickson, Dawson, Turner, and other early authorities, as well as from Sir William Hooker, and at a later time from Mr. Wilson, of Warrington; moreover Mr. Eagle continued till late in his life to collect from time to time, whenever he had an opportunity, and especially in Suffolk."—Extract from Letter of Sir C. Bunbury, October 1st, 1885.

*Gymnostomum microstomum* Hedw.  Bury ; Newmarket Heath, *Eagle*.

*Dicranella cerviculata* Hedw.  Wangford, *Eagle*, W.—*D. rufesens* Turn.  Suffolk, *Eagle*.

*Campylopus flexuosus* Brid.  Walberswick, *H. N. Dixon*.  E.— *C. paradoxus* Wils.  Walberwick, *H. N. Dixon*.  E.—*C. fragilis* B. & S.  Lound, *H. N. Dixon*.  E.

*Pleuridium nitidum* Hedw.  Brandon, *Eagle*.  W.

*Sphærangium muticum* Schreb.  Rougham ; Bury, *Eagle*.  Barton, *Sir C. B.*

*Phascum curvicollum* Hedw.  Barton, *Sir C. B.*—*P. rectum* Sm., Newmarket, *Eagle*.  W.

*Pottia crinita* Wils.  Southwold, *H. N. Dixon*.  E.

*Didymodon rubellus* B. & S.  Rushbrooke ; Wangford ; Mildenhall *Eagle*.—*D. luridus* Hornsch.  Dunwich, *H. N. Dixon*.  E.

*Trichostomum tophaceum* Brid.  Wangford ; Eriswell, *Eagle*.  W.

*Barbula rigida* Schultz. (enervis).  Thetford, *Eagle*.  Barton, *Sir C. B.*  W.  Of specimens sent by Mr. Eagle, Mr. Wilson writes, " The leaves are particularly broad and obtuse." — *B. marginata* B. & S.  Blythburgh Church, *H. N. Dixon*.  E.— *B. muralis* L. var. *æstiva*.  Woodbridge, *Eagle*.  E. — *B. unguiculata* Dill. var. *apiculata*.  Suffolk, *Eagle*.—*B. rigidula* Dicks.  Wangford, *Eagle*.  W.—*B. vinealis* Brid.  Road to Whepstead, *Eagle*.  W.  Mr. Eagle speaks of this as common, and Mr. Wilson rejoins, " As common with us in a barren state as with you."—*B. latifolia* B. & S.  Halesworth ; Mendham ; Groton, *Eagle*.  W.E.  Mr. Eagle sent many fruiting and other specimens to Mr. Wilson, who answers, " This is a good species, very rare in fruit, but found sparingly in that state in Sussex."  Mr. W. and Mr. E. both failed to detect any male inflorescence.—*B. convoluta* Hedw.  Wangford ; Rougham, *Eagle*.—*B. intermedia* Brid.  Blythburgh, *H. N. Dixon*.  E.—*B. papillosa* Wils.  Bury ; Halesworth, &c., *Eagle*.  Wrentham, *H. N. Dixon*.  W.E.

*Ulota phyllantha* Brid.  A specimen was sent from Suffolk to Mr. Wilson.

*Orthotrichum saxatile* Brid.  Burgh Castle, *H. N. Dixon*.  E.  This is probably the *O. cupulatum* of the Hist. Yar.—*O. tenellum* Bruch.  Bury, *Eagle*.  W.  On receipt of specimens Mr. Wilson wrote, " The yellow calyptra seem to be peculiar to this species. I believe it to be *O. tenellum*."

*Ephemerum serratum* Schreb.  Barton, *Sir C. B.*

*Physcomitrella patens* Hedw.  Wangford, *Eagle*.  Mr. Wilson acknowledges the receipt of examples of this species and of *Systegium crispum*.

*Amblyodon dealbatus* Dicks.  Tuddenham, *Eagle*.  W.  Mr. Wilson writes : " This is a moss which one would not expect to find in Suffolk ; but your specimens are genuine."

*Leptobryum pyriforme* L.  Barton Mills, *Eagle*.

*Bryum lacustre* Brid.  Wangford, *Eagle*.  Mr. Wilson writes :— " From a memorandum given to Mr. Spruce by Mr. Borrer, I infer that *Bryum lacustre* was gathered by you at Wangford in 1804, in the same year that Mr. Turner published his work on Irish Mosses, and Smith his Fl. Brit. Amongst the specimens sent, I was particularly interested with an example of a double capsule on the same seta, of which I enclose a sketch. I never before met with a similar instance, except that I have seen two capsules on the same receptacle in *Sphagnum contortum*, and a forked seta of a species of *Bryum*." The specimen and sketch referred to are in Sir C. Bunbury's Herbarium. — *B. intermedium* W. & M. Mildenhall, *Eagle*. This was determined by Mr. Wilson, and there is in the Herbarium an enlarged drawing of the capsule by him.

*Mnium cuspidatum* Hedw. Lakenheath; Lackford, *Eagle*. Barton, *Sir C. B.* Fritton, *H. N. Dixon*.

*Aulacomnium androgynum* L. Ipswich, *Eagle*. E.—*A. palustre* L. var. *ramosum*. Suffolk *Eagle*.

*Pogonatum urnigerum* L. Lackford, *Sir C. B.* W.

*Antitrichia curtipendula* L. Lowestoft; Framlingham, near the Countess Well, *Eagle*. The latter specimen was growing luxuriantly on a tree.

*Camptothecium nitens* Schreb. Suffolk, *Eagle*. Lackford bog, *Sir C. B.* W.

*Scleropodium ? illecebrum* Schwg. (*blandum*). Thetford, *Eagle*. Between Rougham and Bury, *Sir C. B.* W. This may be *S. cæspitosum*.

*Eurhynchium pumilum* Wils. Lound, *H. N. Dixon*. E.

*Rhynchostegium tenellum* Dicks. Eye churchyard, *Eagle*. — *R. Murale* Hedw. Bury ; Risby church ; Harleston, *Eagle*. W.

*Plagiothecium sylvaticum* L. Barton, *Sir C. B.* W.—*P. undulatum* L. Fritton, *H. N. Dixon*. E.

*Amblystegium radicale* P. Beauv. Icklingham Sluice, *Eagle*. W. Mr. Wilson writes, " The barren specimen sufficiently agrees with the previous fertile one of *Hypnum radicale* from Icklingham Sluice."

*Hypnum Cossoni* Schpr. Redgrave Fen, *E. M. Holmes*.—*H. uncinatum* Hedw. Wangford, *Eagle*.—*H. palustre* L. Barton Mills ; The Priory, Bury, *Eagle*.—*H. elodes* Spruce. Redgrave Fen, *E. M. Holmes*. W.—*H. polygamum* B. & S. Lakenheath Fen, *Eagle*.— var. *stagnatum* Wils. Lakenheath Fen, *Eagle*. W. These were determined for Mr. Eagle by Mr. Wilson.—*H. stramineum* Dicks. Suffolk, *Turner*. This specimen was probably from Belton, near Yarmouth, whence it is recorded in the History of Yarmouth.

---

# NOTES ON SOME KERRY PLANTS.

## By Reginald Scully.

Last summer I spent the greater part of July and a week or so of August botanising in Kerry. Most of the time was devoted to the Lakes of Killarney and their surroundings, the rest to Tralee and the coast-line thence to Ballyheigne. Several interesting 'finds' rewarded my endeavours, the most important being *Utricularia neglecta*, an addition to the flora of Ireland.

I made my head-quarters at the Muckross Hotel in Cloghereen, a small village about three miles from Killarney on the Glengariffe road, and from this centre explored the three Killarney lakes with their connecting Long Range, and Loch Guitane, a large lake some three miles to the east of the hotel. On the suggestion of my friend Mr. A. G. More, I provided myself with a long-handled garden rake, and by making extensive use of boats I dragged with this instrument nearly every day and grubbing spot in these lakes. In addition to these boating trips, I made my way round the entire Lower Lake, with a shore line of over 20 miles, most of the Middle and part of the Upper Lake, with numerous excursions into the surrounding bogs and woods.

Instead of taking my work day by day, it may be better perhaps to take each lake, &c., in turn, mentioning the more interesting plants as they occur, before combining in a tabular form those which are new to, or rare in, the County Kerry.

On the 8th July I started for my first trip from the hotel, and arrived in the bay at the Muckross boat-house in the Lower Lake. *Eriocaulon* ... *Littorella* ... and *Lobelia* ... in several lakes, while in the surrounding ... in the ... *Allium* *Scorodoprasum* occurred in one spot. There seems some doubt as to the origin of this plant here, and the presence of *Iris* *foetidissima*, a plant certainly introduced in the west of Ireland, if not in the east, ... a few miles of the *Allium* lends strength to this doubt. Around the boggy margin of the boat-harbour *Eleocharis acicularis* grows abundantly; it occurs again in several spots as far round as the mouth of the Lower Flesk, being especially abundant near the ... boat-house; this plant is new to Dist. 1 of the "Cybele Hibernica." In a ... near the boat-harbour *Limosella* ... occurs sparingly with *Lobelia* ... After several unsuccessful hauls in the bay just outside our starting-point, the rake came up once loaded with *Naias flexilis*, and looking over the side of the boat I could see this plant growing in great abundance in about six feet of water. I found two other stations, some three or four miles apart, for this rare plant. So far, *Naias* is known from two localities in Kerry: Killarney ... first found by Rev. E. F. Linton ... 1886, and Caragh Lake A. G. More ... Muckross shore and Castlelough Bay grow *Sedum* ... on both sides of the Killarney ... with *Sedum* ... and many ... forms of *Ranunculus* *Flammula*. Mr. Charles Bailey ... the most extreme of my Killarney forms near the ... boat-house ... from which it differs in having almost straight peduncles, and in giving ... very robust and numerous roots from the nodes; ... of the Killarney forms he places midway between ... and ... Around this bay I noted *Samolus Valerandi*, *Teucrium Scordium*, and *Smyrnium Olusatrum*, the remains ... cultivation. Near ... boat-house *Scutellaria minor* occurs in great plenty, and near by I gathered *Sedum* ... *Ranunculus* ... seems the commonest ... of the Killarney ... Off the shore here I found *Potamogeton* ... previously only known in

Ireland from another Kerry station, Castlegregory. Mr. Arthur Bennett, who has kindly looked over some of my plants, considers the Killarney *nitens* to be midway between the type and var. *aucuphilus*. At the mouth of the Flesk, *Sanicaria* again occurs with *Wahlenbergia hederacea* scarred but plentiful, and *Carex verticillata* sparingly. Growing here in the river were some curious forms of *Hisine Pantava*: a small tuft of similar leaves with several long filiform stems springing directly from this tuft and ending in a small root-bearing leaf. The Mine holes in Ross Island gave me a most interesting series of these forms, which probably have been often mistaken in Ireland for *Hisine nitans*. *Rarviia lingalis* occurs in sandy hollows at Tan-hole Bay, and *Lathraea Squamaria* in the woods on Ross Island, which here forms one side of this bay. Neither *Wahlenbergia* nor *Lathraea* have been previously recorded from Kerry. The Mine holes on Ross Island yielded *Sparganium minimum*, more *Potamogeton nitens*, and a fine selection of the various forms of *Hisine repens*. Within a few feet of each other were growing the submerged form with grass-like leaves and creeping suckers, the form in a few inches of water where these grass-like leaves are supplemented by a few others which reach the surface and give out narrow floating leaves, passing into typical *Hisine repens*, which also grows here and elsewhere in great profusion on the muddy slopes of the pools. Around grew *Silene maritima* and *Armeria maritima*, a well-known instance of maritime plants surviving far inland not many feet above sea-level. Round the shores of Ross Island, *Elatine hexandra* occurs in nearly every suitable bay; in fact this is a very common plant in all three lakes and in the Long Range. *Hieracium pallidum* and *Galium boreale*, the latter very luxuriant, occur here on the shores with *Rhamnus catharticus, Tolmenta Salvia* and *Galium palustre* var. *Graenmani*. In Ross Bay the rake brought to the surface *Naias flexilis* in several spots along the south side, with plenty of *Callitriche autumnalis*; it was here that the Rev. E. F. Linton, as recorded in the "Journal of Botany," March, 1886, found *Naias*, then new to Killarney. Among my *Characeae* kindly looked over by the Messrs. Groves, I find *Chara fragilis* var. *capillacea* and *Nitella translucens* were gathered here. Around the bay *Equisetum Wilsoni, Polygonum minus* and *Nasturtium palustre*, the tall upright form, were mixed with *Plantago major* var. *intermedia*; a small island in the bay here is full of *Lysimachia vulgaris*, a very rare Kerry plant.

Leaving Ross Bay on our northward course, *Hieracium umbellatum* takes the place of *H. pallidum* along the broken limestone shore with *Thalictrum flexuosum* and *Rubus saxatilis*. On entering Wicaria Bay I came across a small pool, shut off from the lake, owing to the excessive drought, by a ridge of rocks; growing this time within reach of my arm were several fine tufts of *Naias* with plenty of *Chara aspera* and *C. fragilis*. Along the wooded shore here *Carex pallescens* was mixed, while about Whitarr's Point *Carex verticillata* again occurs rather plentifully, with *Rarviia nitens* sparingly. Two curious forms of *Mentha sativa* I gathered near

I made my head-quarters at the Muckross Hotel in Cloghereen, a small village about three miles from Killarney on the Glengariffe road, and from this centre explored the three Killarney lakes with their connecting Long Range, and Loch Guitane, a large lake some three miles to the east of the hotel. On the suggestion of my friend Mr. A. G. More, I provided myself with a long-handled garden rake, and by making extensive use of boats I dragged with this instrument nearly every bay and promising spot in these lakes. In addition to these boating trips, I made my way round the entire Lower Lake, with a shore line of over 20 miles, most of the Middle and part of the Upper Lakes, with numerous excursions into the surrounding bogs and woods.

Instead of taking my work day by day, it may be better perhaps to take each lake, &c., in turn, indicating the more interesting plants as they occur, before combining in a tabular form those which are new to, or rare in, the County Kerry.

On the 6th July I started for my first trip from the hotel, and noticed on the way to the Muckross boat-house on the Lower Lake, *Pimpinella major*, *Calamintha officinalis* and *Carex divulsa* in several places, while in the shrubberies on the right of the roadside *Allium Scorodoprasum* occurred in one spot. There seems some doubt as to the origin of this plant here, and the presence of *Iris fœtidissima* (a plant certainly introduced in the west of Ireland, if not in the east) within a few yards of the *Allium* lends strength to this doubt. Around the boggy margin of the boat-harbour *Eleocharis acicularis* grows abundantly ; it occurs again in several spots as far round as the mouth of the River Flesk, being especially abundant near the Cahernane boat-house ; this plant is new to Dist. 1 of the ‘ Cybele Hibernica.’ In a damp copse near the boat-harbour *Lastrea Thelypteris* occurs sparingly with *Carex vesicaria*. After several unsuccessful hauls in the bay just outside our starting-point, the rake came up quite loaded with *Naias flexilis*, and looking over the side of the boat I could see this plant growing in great luxuriance in about six feet of water. I found two other stations, some three or four miles apart, for this rare plant. So far, *Naias* is known from two localities in Kerry; Killarney (first found by Rev. E. F. Linton, Journ. Bot., p. 83, 1886), and Caragh Lake (A. G. More, Journ. Bot., p. 350, 1877). Muckross shore and Castlelough Bay gave *Galium boreale* on both sides of the Bilrook stream, with *Stachys Betonica* and many curious creeping forms of *Ranunculus Flammula*. Mr. Charles Bailey places the most extreme of my Killarney forms near the Ullswater *pseudo-reptans*, from which it differs in having almost straight internodes, and in giving out very robust and numerous roots from the nodes ; another of the Killarney forms he places midway between *pseudo-reptans* and *suberectus*. Around this bay I noted *Saponaria officinalis*, *Fœniculum officinale*, and *Smyrnium Olusatrum*, the remains of former cultivation. Near Cahernane boat-house *Subularia aquatica* occurs in great plenty, and near by I gathered *Stellaria media* var. *neglecta*. *Ranunculus peltatus* seems the commonest of the Killarney Batrachian Ranunculi. Off the shore here I found *Potamogeton nitens*, previously only known in

Ireland from another Kerry station, Castlegregory. Mr. Arthur Bennett, who has kindly looked over some of my plants, considers the Killarney *nitens* to be midway between the type and var. *curvifolius*. At the mouth of the Flesk, *Subularia* again occurs with *Wahlenbergia hederacea*, stunted but plentiful, and *Carum verticillatum* sparingly. Growing here in the river were some curious forms of *Alisma Plantago*; a small tuft of subulate leaves with several long filiform stems springing directly from this tuft and ending in a small ovate floating leaf. The Mine holes on Ross Island gave me a most interesting series of these forms, which probably have been often mistaken in Ireland for *Alisma natans*. *Radiola linoides* occurs in sandy hollows at Tan-hole Bay, and *Lathræa Squamaria* in the woods on Ross Island, which here forms one side of this bay. Neither *Wahlenbergia* nor *Lathræa* have been previously recorded from Kerry. The Mine holes on Ross Island yielded *Sparganium minimum*, more *Potamogetens nitens*, and a fine selection of the various forms of *Alisma repens*. Within a few feet of each other were growing the submerged form with grass-like leaves and creeping suckers, the form in a few inches of water where these grass-like leaves are supplemented by a few others which reach the surface and give out narrow floating leaves, passing into typical *Alisma repens*, which also grows here and elsewhere in great profusion on the muddy slopes of the pools. Around grow *Silene maritima* and *Armeria maritima*, a well-known instance of maritime plants surviving far inland not many feet above sea-level. Round the shores of Ross Island, *Elatine hexandra* occurs in nearly every suitable bay; in fact this is a very common plant in all three lakes and in the Long Range. *Hieracium pallidum* and *Galium boreale*, the latter very luxuriant, occur here on the shores with *Rhamnus catharticus Orobanche Hederæ* and *Caltha palustris* var. *Guerangerii*. In Ross Bay the rake brought to the surface *Naias flexilis* in several spots along the south side, with plenty of *Callitriche autumnalis*; it was here that the Rev. E. F. Linton, as recorded in the 'Journal of Botany,' March, 1886, found *Naias*, then new to Killarney. Among my *Characeæ* (kindly looked over by the Messrs. Grove) I find *Chara fragilis* var. *capillacea* and *Nitella translucens* were gathered here. Around the bay *Equisetum Wilsoni*, *Polygonum minus* and *Nasturtium palustre*, the tall upright form, were noted, with *Plantago major* var. *intermedia*; a small island in the bay here is full of *Lysimachia vulgaris*, a very rare Kerry plant.

Leaving Ross Bay on our northward course, *Hieracium umbellatum* takes the place of *H. pallidum* along the broken limestone shore with *Thalictrum flexuosum* and *Rubus saxatilis*. On entering Victoria Bay I came across a small pool, shut off from the lake, owing to the excessive drought, by a ridge of rocks; growing this time within reach of my arm were several fine tufts of *Naias* with plenty of *Chara aspera* and *C. hispida*. Along the wooded shore here *Carex pallescens* was noted, while about Mahony's Point *Carum verticillatum* again occurs rather plentifully, with *Bartsia viscosa* sparingly. Two curious forms of *Mentha sativa* I gathered near here are named vars. *paludosa* and *subglabra* by Mr. Bennett. On the shore below Lakeview House I was surprised to meet *Mimulus luteus*, well established and very abundant for about a mile. Mr. Ross O'Connell, of Lakeview, has kindly given me some interesting particulars of the introduction of this plant into Kerry, from which I gather that the *Mimulus* was brought from Virginia to the O'Connells of Derrynane Abbey by a French officer about 1757, and was planted by them in a small lake near the Abbey, some forty miles from here. An ancestor of the present owner of Lakeview brought a few plants to Killarney in 1820 and planted them on the lake shore below his house, where they have flourished and spread ever since. I understand the *Mimulus* has died out at its original Kerry locality, though there seems little likelihood of its doing so at Killarney for years to come.

Crossing the River Laune, which here leaves the lake, *Carex pallescens* again occurs plentifully with *C. filiformis* close by. I gathered here a curious form of *C. remota* with several of the spikelets branched; Mr. A. G. More, to whom I showed the plant, had not seen this state of *C. remota* before. From the Laune for more than a mile to Benson's Point *Carum verticillatum* occurs in great plenty; this is the headquarters of this plant on the lake. *Milium effusum* and *Carex paniculata* were next noted in several spots from O'Sullivan's Cascade to Glena Bay; at this latter place *Carex pendula* grows above the salmon haul with *Pinguicula lusitanica* and *Hymenophyllum Wilsoni*, much rarer about these lakes than *H. tunbridgense*. Along the rocky shore which extends from Brickeen Bridge to our starting-point at the Muckross Boat-house, the only interesting plants noted were *Hieracium pallidum* and various forms of the *Saxifraga umbrosa* and *Geum* group.

The Middle Lake afforded *Rubus saxatilis*, *Cladium germanicum*, *Ranunculus peltatus*, *Elatine hexandra* in several places, while along the south side, under Torc Mountain, *Carex pallescens* again occurs with luxuriant specimens of *Saxifraga hirsuta* and *S. Geum*. Growing along the roadside just above the lake, I found *Neottia Nidus-avis*; it occurs again a mile or so nearer Cloghereen.

The Black Channel gave me, near the Meeting of the Waters, fine specimens of *Isoetes echinospora*, the var. *linearis* of *Potamogeton polygonifolius*, discovered here by Mr. A. G. More, *Carex acuta* and *Utricularia intermedia*. Above the Old Weir Bridge *P. linearis* quite fills the channel, and is the most abundant of the Long Range pond-weeds. I gathered here also a *Potamogeton* which Mr. Bennett considers to be Syme's var. *pseudo-fluitans*; unlike the 'linearis' this occurred very sparingly. Beyond the Eagle's Nest, I found *Subularia* again, *Callitriche autumnalis*, *Isoetes echinospora*, *Elatine hexandra*, and *Carex vesicaria* in several places. In some of the boggy ditches hereabouts more *Utricularia intermedia* occurs; I could find no flowers. Near the Upper Lake, *Sparganium affine* was flowering plentifully in the channel. My rakings in the Upper Lake were poorly rewarded; besides that, the only two days of broken weather which I had during my stay in Killarney were those which I spent on this lake, and in few places in the British

Isles do wind and rain do their work better. *Isoetes echinospora, Subularia, Elatine, Callitriche hamulata,* and *Nitella translucens* were the most interesting plants noticed. Near the Hunting Tower *Lastrea Thelypteris* was gathered growing with *L. Oreopteris.* About a mile south of the Upper Lake lies Loch Beg, *Utricularia intermedia* and *Cladium germanicum* occur in this pond, while in the bogs near I found *Rynchospora fusca,* the station probably referred to in the ' Cybele Hibernica,' " Bogs near the Upper Lake of Killarney, Flor. Hib."

I spent a day examining Loch Guitane and its shores. This lake lies some three miles to the east of Muckross and is about the size of the Middle Lake. No subaqueous rarities were found, but its shores gave me *Microcala filiformis* in several spots with *Lycopodium inundatum,* two of our rarest Irish plants ; both have been recorded from here by Mr. A. G. More (Journ. Bot., p. 373, 1876). *Ranunculus peltatus, Cladium,* the three *Droseræ,* and on the lower slopes of Mangerton *Lycopodium clavatum,* a rare Kerry plant, were also noted. A little north of Loch Guitane lie the Doo loughs ; around these *Rynchospora fusca* occurs in great plenty ; I found it again in several of the bogs between these lakes and the village of Cloghereen. It is abundant in the Sherehee bog about a mile N.E. of the village, where it grows with *Carex limosa,* a rare Kerry sedge, and *Vaccinium Oxycoccos,* another Kerry rarity. In the bog holes here and elsewhere, *Utricularia minor* was flowering plentifully, while *Mentha Pulegium* forms a handsome fringe round the narrow stretch of water which almost encircles this bog. Near here I gathered the rayed form of *Centaurea nigra.* *Subularia* occurs with *Elatine hexandra* in Ardagh Pond. Two days spent in the Woodlawn and Glenflesk demesne woods gave *Sisymbrium Thalianum, Wahlenbergia,* very luxuriant, *Stachys Betonica, Luzula pilosa, Carex lævigata, Lemna trisulca, Equisetum hyemale, Bromus giganteus,* and on a wall several plants of *Poterium muricatum,* evidently introduced, with *Chelidonium majus* and *Linaria viscida.*

Leaving Killarney, a run of twenty miles by rail brings you to Tralee. Here I spent a week or more, examining the coast-line to Ballyheigue and a few spots inland. Following the Tralee Canal to where it enters the sea, I noted at Blennerville, *Nasturtium palustre,* the prostrate stunted form, *Lepigonum neglectum, Senebiera didyma* and *Coronopus* ; thence to the Spa, a small seaside resort, *Salsola Kali, Suæda maritima, Eryngium maritinum, Crithmum maritimum,* and *Glaucium flavum.* On the soft strand below Seafield House, I found *Zostera nana* in its second known Irish locality, Mr. More's station at Baldoyle, near Howth, being the other ; nearer the house, *Marrubium vulgare,* a solitary plant was noted. About Fenit, the maritime form of *Solanum Dulcamara* grows among the shingle ; it is noticed by Mr. H. C. Hart as frequent along the opposite shore of this bay. *Hyoscyamus niger* is plentiful beyond the new pier with *Silybum Marianum.* Near the northern head of the Fenit peninsula, *Senecio Jacobæa* var. *flosculosus* is very plentiful, while near it, in a boggy hollow, *Trifolium fragiferum* is abundant. This seems a common clover in Kerry wherever sand-

vars. illosa and subglabra by Mr. Bennett. On
Lan vi House I was surprised to meet *Mimulus*
... I ad very abundant for about a mile. Mr.
of Lake ow, has kindly given me some interesting
... lution of this plant into Kerry, from which
M was was brought from Virginia to the
... Abbey by a French officer about 1757,
i v them in a small lake near the Abbey, some
... A ancestor of the present owner of Lake.
w ant to Killarney in 1820 and planted them
... b house, where they have flourished and
I understand the *Mimulus* has died out at its
... though there seems little likelihood of its
... r oars to come.
... r Lane, which here leaves the lake, *Carex*
... plentifully with *C. filiformis* close by. I
... fm of *C. remota* with several of the spike-
Ir. A. GM re. to whom I showed the plant, had
... *mota* before. From the Laune for more
... hint *Carum verticillatum* occurs in great
... quarters of this plant on the lake. *Milium*
... ata were next noted in several spots from
... to Glena Bay; at this latter place *Carex*
... dmon haul with *Pinguicula lusitanica* and
... much rarer about these lakes than *H.*
... rocky shore which extends from Brickeen
... it at the Muckross Boat-house, the only
... 1 wre *Hieracium pallidum* and various forms
... ad *Geum* group.
... noed *Rubus saxatilis*, *Cladium germanicum*,
... *ita hexandra* in several places, while along
il r TorMountain. *Carex pallescens* again occurs
... us f *Saxifraga hirsuta* and *S. Geum*. Grow-
... jus above the lake. I found *Neottia Nidus-*
... in a m or so nearer Cloghereen.
annel gave me. near the Meeting of the Waters,
*Isoetes echinospora*, the var. *linearis* of *Potamogeton*
overed lre by Mr. A. G. More, *Carex acuta* and
... Above the Old Weir Bridge *P. linearis* quite
an is le most abundant of the Long Range
athered here also a *Potamogeton* which Mr.
s to be lyme's var. *pseudo-fluitans*; unlike the
urred vav sparingly. Beyond the Eagle's Nest,
again, *callitriche autumnalis*, *Isoetes echinospora*,
and *Carex vesicaria* in several places. In some
es hereabouts more *Utricularia intermedia* occurs;
... bar the Upper Lake, *Sparganium affine*
plentifully in the channel. My rakings in ... r
rdl; besides that
which I ad during m
nt on th lake, ar

Isles do wind and rain do their won better. *Isoetes echin* ?  ? ?.
*Subularia, Elatine, Callitriche hamulata* and *Nitel'a tr ms'*  ? ?
the most interesting plants notic l. Near the Hunting I  ?
*Lastrea Thelypteris* was gathered gr wg with *L. Ore pteris*. Al ut
a mile south of the Upper Lake li s och Beg. *Urtica* ? ? ?
*media* and *Cladium germanicum* occur i this pond, while in the l  ?
near I found *Rynchospora fusca*, the ation probably referred to in
the ' Cybele Hibernica,' " Bogs near to Upper Lake of Killarney,
Flor. Hib."

I spent a day examining Loch Gitane and its sh res. This
lake lies some three miles to the east f Muckross and is ab it the
size of the Middle Lake. No subaquous rarities were f ui.l, but
its shores gave me *Microcala filiforn* is n several sp ts with *I
podium inundatum*, two of our rarest Ish plants ; both have b n
recorded from here by Mr. A. G. Mor (Journ. Bot., p. 873, 1876 .
*Ranunculus peltatus, Cladium*, the thre *Drosera*, and on the l  ?
slopes of Mangerton *Lycopodium clavatum*, a rare Kerry plant, were
also noted. A little north of Loch uitane lie the Doo l  l  ;
around these *Rynchospora fusca* occurin great plenty ; I f und it
again in several of the bogs between tese lakes and the villa  (
Cloghereen. It is abundant in the Sherehee bog about a  mile
N.E. of the village, where it grows wh *Carex limos* i, a rare Kerry
sedge, and *Vaccinium Oxycoccos*, anothr Kerry rarity. In th  bug
holes here and elsewhere, *Utricularia minor* was flowering pl ou
fully, while *Mentha Pulegium* forms a andsome fringe round the
narrow stretch of water which almt encircles this bog. Near
here I gathered the rayed <sub>fo</sub>rm of *Centaurea nigra*. *Subu aria*
occurs with *Elatine hexandra* in Ardaa Pond. Two days el nt in
the Woodlawn and Glenflesk demme woods gave *Sisymbrium
Thalianum, Wahlenbergia*, very luxurnt, *Stachys Betmica, Luula
pilosa, Carex lævigata, Lemna trisulca Equisetum hyemale, E  ?
giganteus*, and on a wall several plnts of *Poterium muricat m*,
evidently introduced, with *Chelidoniu majus* and *Linaria ruci li*.

Leaving Killarney, a run of twen miles by rail brings you to
Tralee. Here I spent a week or mor examining the coast-line to
Ballyheigue and a few spots inland. Following the Tralee Canal
to where it enters the sea, I note at Blennerville, *Nasturtium
palustre*, the prostrate stunted form, *epigonum neglectum, Senebiera
didyma* and *Coronopus* ; thence to the Spa, a small seaside resort,
*Salsola Kali, Suæda maritima, Eringium maritimum, Crithmum
maritimum*, and *Glaucium flavum*. On the soft strand below Seafield House, I found *Zostera nana* in a second known Irish locality,
Mr. More's station at Baldoyle, near Howth, being the other ;
nearer the house, *Marrubium vulgare* solitary plant was noted.
About Fenit, the maritime form of *Solanum Dulcamara* grows
among the shingle ; it is noticed b Mr. H. C. Hart as frequent
along the opposite shore of this bay. *Hyoscyamus niger* is plentiful
beyond the new  —ith *Silebum Marianum*. Near the northern
head of

hills meet a boggy flat.    On rocks at both sides of the entrance to Barrow Harbour, *Statice occidentalis* occurs, a plant probably new to district 1 of the Cyb. Hib.    Around the harbour I gathered *Statice rariflora* and *Œnanthe Lachenalii* in several places with *Silybum*, *Brassica nigra, Rubia peregrina, Lepigonum marginatum, L. rupestre, Juncus obtusiflorus, Carex extensa, Lepturus filiformis,* and *Asplenium marinum*, while among the crops I noticed *Papaver dubium, Caucalis nodosa,* and *Scandix Pecten-Veneris.    Asperula cynanchica* and *Senecio Jacobæa* var. *flosculosus* are common plants all along this coast. Resuming our walk at the east side of the entrance to Barrow Harbour, *Orchis pyramidalis* was noted in three spots between this and Rahaneen, where more *Statice rariflora* and *Silybum* occur near an old castle.    Rahaneen is a wide, but shallow, bay, nearly empty at low tide, but intersected by dangerous quicksands.    Rounding this, a nearly straight line of sandhills runs all the way to Bally-heigue, about six miles of heavy tramping.    Along the land side of these sandhills, *Trifolium fragiferum* occurs abundantly in many spots, *Viola Curtisii, Calystegia Soldanella, Chenopodium rubrum, Orchis pyramidalis,* and *Phleum arenarium* were also noted.    Four miles of this dreary sandhill waste brought me to a very curious and unexpected find, namely, *Cuscuta Trifolii* growing on stunted *Lotus corniculatus;* it occurred abundantly over the space of an acre or two, on a low-lying sandy flat, not a stone's throw from the sea.    The *Lotus* and sandy grass on this storm-swept spot rarely exceeded an inch or two in height.    Hitherto this plant has been known in Ireland only as a colonist near Bray ; here it is probably an escape from a shipwreck, or from some crop on the Ardfert farms about four miles distant.    Beyond this lies Akeragh, a semi-tidal lake about two miles round.    *Blysmus rufus*, a northern plant first shown to extend to Kerry by Mr. H. C. Hart, occurs here luxuriantly ; *Trifolium fragiferum* is again abundant, with *Œnanthe Lachenalii, Chenopodium rubrum,* and *Carex extensa*.    Nothing of interest was noted hence to Ballyheigue, except the profusion of *Senecio flosculosus.    Diotis* and *Peucedanum* are recorded from here in Smith's Kerry ; I could find neither ; they are probably errors for perhaps *Eryngium marinum* and *Œnanthe Lachenalii,* both of which are very abundant in this locality.    On my way back to Tralee, I noted near Ardfert *Inula Helenium* and abundance of a hybrid thistle, unfortunately too far past flowering to be satis-factorily identified.

A day spent in the Clogherbrian bogs, a mile north of the Spa, gave me *Pimpinella major,* abundant *Galium uliginosum,* a very rare Irish plant, new to this district, *Epipactis palustris, Juncus obtusi-florus, Carex dioica* and *C. teretiuscula* in the lower bog, while in the upper more *Galium uliginosum* occurs, with *Pinguicula grandi-flora, P. lusitanica, Bartsia viscosa, Carex dioica,* and *C. paniculata.* Between Clogherbrian and the Spa *Althæa officinalis* and *Inula Helenium* were noticed near some houses, with *Radiola, Verbascum Thapsus,* and *Carex riparia.*

Another day spent inland to the east of Tralee, gave *Carex strigosa,* by the roadside near Chute Hall, a plant new to this district.   In Ballycarty demesne *Pimpinella major* is very abundant, and in a small lake here I discovered *Utricularia neglecta* in its first Irish locality growing abundantly with *U. vulgaris*, from which it seems best distinguished by the smallness of the leaf bladders, which at first sight appear to be entirely wanting.   Mr. Arthur Bennett has confirmed the correctness of my Ballycarty plant. *Callitriche autumnalis*, *Ranunculus trichophyllus* and *Zannichellia palustris* were also gathered in this lake.   In a small bog close by, *Galium uliginosum* again occurs sparingly.   Near the roadside leading back to Tralee, *Carex pendula* and *Pimpinella major* are abundant.

In a trip south of Tralee to the lower slopes of the Slieve Mish Mountains, *Ægopodium Podagraria* in several spots near houses, *Bartsia viscosa*, *Habenaria viridis*, *Carum verticillatum* sparingly, and *Carex lævigata* were noted; near Blennerville, on the muddy slopes of a tidal ditch, a large form of *Cochlearia* with all its leaves stalked, was abundant; Mr. Bennett refers this plant doubtfully, in the absence of fruit, to *C. officinalis*.

A short excursion to Derrymore, a peninsula which juts from the south side of Tralee Bay, gave me *Chenopodium rubrum* var. *pseudo-botryodes*, a very rare Irish plant; it was pretty plentiful in stony hollows and was the only form which occurred.   The maritime variety of *Solanum Dulcamara* was also noticed here abundantly.

In the following list of the more interesting plants recorded above, the affix κ denotes that the plant is new to the County Kerry; ɪ, that it is an addition to District 1 of the ' Cybele Hibernica and its Supplement.

Ranunculus trichophyllus
   *Chaix.* κ.
R. peltatus *Schrank.* ɪ.
Caltha palustris, *var.* Guerangerii
   *Bor.* ɪ.
Nasturtium palustre *DC.*
Sisymbrium Thalianum *Hook.* κ.
Subularia aquatica *Linn.*
Brassica nigra *Koch.* κ.
Stellaria media, *v.* neglecta
   *Weihe.* ɪ.
Trifolium fragiferum *Linn.*
*Poterium muricatum *Spach.* ɪ.
Callitriche hamulata *Kuetz.*
C. autumnalis *Linn.*
Carum verticillatum *Koch.*
Pimpinella major *Huds.*
Galium boreale *Linn.*
G. uliginosum *Linn.* ɪ.
*Inula Helenium *Linn.*
*Silybum Marianum *Gaertn.*
Senecio Jacobæa, *v.* flosculosus
   *Jord.*

Hieracium pallidum *Biv.*
H. umbellatum *Linn.*        [κ.
Wahlenbergia hederacea *Reichb.*
Vaccinium Oxycoccos *Linn.*
Statice rariflora *Drej.*
S. occidentalis *Lloyd.*
Lysimachia vulgaris *Linn.*
Microcala filiformis *Link.*
*Cuscuta Trifolii *Bab.* ɪ.
Hyoscyamus niger *Linn.*
*Linaria viscida *Mœnch.* κ.
*Mimulus luteus *Linn.* κ.
Bartsia viscosa *Linn.*
Orobanche Hederæ *Duby.*
Lathræa Squamaria *Linn.* κ.
Utricularia neglecta *Lehm.*   See
   above.
U. intermedia *Hayne.*
Mentha Pulegium *Linn.*
*Marrubium vulgare *Linn.* κ.
Chenopodium rubrum *Linn.* κ.
   Var. pseudo-botryodes *H. C.*
   *Wats.* ɪ.

Polygonum minus *Huds.*  K.
Neottia Nidus-avis *Rich.*
Epipactis palustris *Crantz.*  K.
Orchis pyramidalis *Linn.*
Allium Scorodoprasum *Linn.*
Juncus obtusiflorus *Ehrh.*
Luzula pilosa *Willd.*  K.
Sparganium affine *Schnizl.*
Lemna trisulca *Linn.*  K.
Potamogeton polygonifolius,
    *var.* pseudo-fluitans *Syme.*
    ,, linearis *Syme.*
P. nitens *Web.*
Zostera nana *Roth.*  I.
Naias flexilis *Rostk.*
Eleocharis acicularis *Sm.*  I.
Scirpus rufus *Wahlb.*
Rynchospora fusca *R and S.*

Carex dioica *Linn.*
C. teretiuscula *Good.*
C. acuta *Linn.*
C. limosa *Linn.*
C. pallescens *Linn.*
C. pendula *Huds.*  K.
C. strigosa *Huds.*  I.
C. filiformis *Linn.*
C. riparia *Curtis.*  K.
Milium effusum *Linn.*
Phleum arenarium *Linn.*
Lastræa Thelypteris *Presl.*
Equisetum hyemale *Linn.*
Lycopodium inundatum *Linn.*
Isoetes echinospora *Dur.*
Chara fragilis *v.* capillacca *Coss*
    *and G.*
Nitella translucens *Agardh.*

---

# ON *POTENTILLA REPTANS* AND ITS ALLIES.

## By W. H. BEEBY, A.L.S.

LAST winter Herr Svanté Murbeck, of the Botanical Museum, Lund., who has studied the various forms of *Potentilla* which are intermediate to *P. reptans* L., *P. procumbens* Sibth., and *P. Tormentilla* Sibth., asked me to send him as large a series as possible of the British forms, in the expectation that he might find among them, besides *P. mixta* Nolte, possibly *P. suberecta* Zimm. and *P. Gremlii* Zimm. Accordingly I forwarded the whole of my collection, together with a number of examples from the herbaria of Messrs. Arthur Bennett, Alfred Fryer, W. F. Miller, and Rev. E. S. Marshall. The result was that various examples were identified by Herr Murbeck as *P. suberecta*, but *P. Gremlii* was not detected.

Among the plants lent by Mr. Bennett was one from a number of specimens sent out as " *Tormentilla reptans?* " by Watson through Bot. Soc. Lond. in 1849 ; these were cultivated specimens from a root originally collected at Ockshott, Surrey. In the 'Phytologist' (1849, p. 485) Mr. Watson expresses himself with considerable doubt as to the proper name for this plant, and it is therefore interesting to note that it is one of those determined by Herr Murbeck as *P. suberecta;* and I may say that of my own examples so identified several had been referred by me to *P. procumbens* with doubt ; it seems therefore that there is room for the name. The *P. suberecta* is generally accepted as a hybrid *P. Tormentilla* × *procumbens;* while *P. Gremlii* is considered a hybrid *P. reptans* × *Tormentilla.* If the latter be really a hybrid, it should be found in Britain, as its two supposed parents are both of them much more common than *P. procumbens,* one of the parents of *P. suberecta.*

I append the more interesting of Herr Murbeck's determinations. The separation of the different forms does not appear to me to be always easy, and I think that a good knowledge of the group is needed before they can be accurately named.

*P. reptans* L., var. *microphylla* Tratt. — Chalky hillocks in the Fens, Cambridge (*Bennett*).

*P. mixta* Nolte.—Surrey! (*Marshall*).   Maresfield, E. Sussex!

*P. suberecta* Zimm. — Surrey! (*Watson*, 1849; various other localities).  Cambridge (*Hb. Fryer*).  Lake Lancashire (*F. A. Lees, Hb. Bennett*).  Kirkcudbright (*J. McAndrew, Hb. Bennett*).

Short diagnoses of these and allied forms are given in 'Die Europ. Arten der Gattung *Potentilla*,' by A. Zimmeter, published at Steyr, 1884.

---

# A SYNOPSIS OF *TILLANDSIEÆ*.

## By J. G. BAKER, F.R.S., F.L.S.

(Continued from p. 50.)

166. T. CHRYSOSTACHYS Baker in Bot. Mag. t. 6906.  *Vriesea chrysostachys* E. Morren in Belg. Hort. 1881, 87, 1882, 335.—Leaves about 30 in a rosette, lorate from an ovate base, thin, flexible, glabrous, pale green on the face, tinged with brown on the back towards the base, 1–1½ ft. long, 2–2½ in. broad at the middle, deltoid-cupidate at the apex.   Peduncle, including infloresence, 2–2½ ft. long; bract-leaves small, greenish, ovate, adpressed. Spikes one or two, dense, distichous, 6–12 in. long, about an inch broad; flower-bracts ovate, acute, lemon-yellow, an inch long, Calyx ½ in. long; sepals oblong-lanceolate.   Petals bright yellow, as long as the bract; blade lingulate.   Stamens as long as the petals.

Hab. Andes of Peru, *Davis*!  Introduced by Messrs. Veitch about 1880.  Our plant at Kew bore a couple of spikes in 1884.

167. T. BARILLETI Baker.   *Vriesea Barilleti* E. Morren in Belg. Hort. 1883, 33, t. 3; Antoine Brom. 20, t. 13.—Leaves about 16 in a rosette, lanceolate from an ovate base, acute, flexible, bright green, 1½ ft. long, 2 in. broad at the middle.   Peduncle, including the spike, about as long as the leaves; bract-leaves all small, scariose, adpressed.   Flowers in a simple dense spike 8–12 in. long, 2½ in. broad; flower-bracts oblong, truncate at the end, with a small incurved point, 1½ in. long, yellowish, with dark red dots. Calyx as long as the bract.   Petal-blade pale yellow, lingulate, ½ in. long.   Stamens a little longer than the petals.

Hab. Andes of Ecuador; discovered by M. Barillet Deslongchamps.  Introduced about 1877.  Flowered by Professor Morren at Liège in 1883.

168. T. SPLENDENS A. Brong. in Flore des Serres, May 1846, t. 4.  *Vriesea splendens* Lemaire in Flore des Serres, vi. 162, with woodcut.   *V. speciosa* Hook. in Bot. Mag. t. 4382; Beer Brom. 91; Walp. Ann. iii. 622; Antoine Brom. 18, t. 12.   *T. picta* and *zebrina* Hort.—Leaves 12–20 in a rosette, lorate from a dilated ovate base,

thin, flexible, nearly naked, 1–1½ ft. long, 1–1½ in. broad at the middle, dull green with very distinct irregular cross-bands of purple, rounded at the apex to a cusp. Peduncle ½ ft. long; bract-leaves also fasciated with purple. Flowers 12–30, in a flat lanceolate spike ½–1 ft. long, 1½–2 in. broad at the middle; flower-bracts 1½–2 in. long, acute, all bright red or the lower green. Calyx under an inch long; sepals acute. Petals yellow, lanceolate, cernuous, three times as long as the calyx, ¼ in. broad. Stamens about as long as the petals. Capsule-valves lanceolate, an inch long.

Hab. French Guiana, *Poiteau*! (1824). Introduced into cultivation by Melinon and Leprieur about 1842. British Guiana, *Schomburgk*!

169. T. GLADIOLIFLORA Wend. in Hamb. Gartenzeit. 1863, 31. *Vriesea gladioliflora* E. Morren in Belg. Hort. 1880, 87, 216; Antoine in Wiener Gartenzeit. 1880, 97, with figure; Brom. 23, t. 15. — Leaves about 20 in a rosette, lorate from a dilated ovate base, thin, flexible, recurved, 1½–2 ft. long, 3 in. broad, plain green on the face, claret-purple on the back, deltoid-cuspidate at the apex. Peduncle as long as the leaves; bract-leaves many, small, adpressed. Inflorescence a simple dense lanceolate spike a foot long, 1½ in. broad at the middle; bracts green, ovate, acute, 1½ in. long, an inch broad at the base. Calyx reaching to the tip of the bract; sepals oblong, obtuse. Petal-blade white, orbicular, half as long as the calyx. Stamens reaching to the tip of the petals.

Hab. Costa Rica, received alive by Dr. Wendland at Herrenhausen in 1863. Sinteni's 2792, from Porto Rico, seems from an imperfect specimen to be either this or near it.

170. T. VIMINALIS Hemsl. Biol. Cent. Amer. iii. 323. *Vriesea viminalis* E. Morren in Belg. Hort. 1878, 257, t. 14–15; Antoine, Brom. 21, t. 14. *T. viridiflora* Hort.—Leaves about 15 in a rosette, lanceolate from a dilated base, 2 in. diam., flexible, plain green, subglabrous, acute, a foot or more long, 1½–2 in. broad at the middle. Peduncle 1¼ ft. long; bract-leaves many, small, imbricated, adpressed. Flowers in a simple dense lanceolate spike 5–6 in. long; flower-bracts oblong, acute, about an inch long, bright green. Calyx reaching to the tip of the bract; sepals oblong, obtuse. Petals white, half as long again as the calyx; blade broad and very obtuse. Stamens reaching to the tip of the petals. Capsule 1½ in. long.

Hab. Costa Rica, near Cartago, sent alive by Wendland to Liege in 1873.

171. **T. longicaulis**, n. sp.—Leaves lorate from an ovate dilated base 3–4 in. long, 2 in. broad, thin, flexible, subglabrous, 1¼ ft. long, an inch broad at the middle, deltoid-cuspidate at the tip. Peduncle stiffly erect, 2½–3 ft. long; bract-leaves small, scariose, imbricated, adpressed. Inflorescence a dense flat simple spike 6–8 in. long, an inch broad; flower-bracts very ascending, ovate, acute, 1½–2 in. long, an inch round at the base. Calyx ¼ in. shorter than the bract. Petals not seen.

Hab. South Brazil, *Glaziou* 8988! Nearly allied to *T. gladioliflora*.

172. T. VIRIDIFLORA Baker. *Platystachys viridiflora* Beer, Brom. 81. — Leaves a dozen or more in a dense utricular rosette, lorate from an ovate base 4 in. long, 2½–3 in. broad, thin, flexible, subglabrous, above a foot long, an inch broad at the middle, deltoid-cuspidate at the tip. Peduncle stiffly erect, 1½–2 ft. long; bract-leaves imbricated, scariose, the lower with small lanceolate coriaceous free points. Inflorescence a dense flat simple spike 8–12 in. long, 2 in. broad, with sometimes a small second one; flower-bracts oblong-lanceolate, acute, 2–2½ in. long, an inch broad low down, pale green, glabrous. Calyx half an inch shorter than the bract, 1¼ in. long. Petals green, fugitive. Stamens longer than the petals.

Hab. Mexico; Province of Cordova, *Bourgeau* 2274! Described by Beer in 1857 from plants sent alive by Carl Heller to Count Attems at Gratz, which flowered in November, 1854. Allied to *T. gladioliflora*.

173. **T. longibracteata**, n. sp. — Leaf thin, flexible, lorate, subglabrous, 1½–2 ft. long, 1½ in. broad, and narrowed to the point in the Venezuelan plant, 2 in. broad and rounded to a cup in the Trinidad plant. Peduncle stiffly erect; bract-leaves small, scariose, adpressed, imbricated. Inflorescence a broad flat dense simple spike a foot long, 3 in. broad; flower-bracts oblong-lanceolate, 2–2½ in. long, ¾–1 in. round low down, very ascending. Calyx ¾ in. long; sepals oblong, obtuse. Petals not seen. Capsule-valves lanceolate, 1¼ in. long.

Hab. Venezuela; Mountains of Tovar, *Fendler* 2449! Trinidad, *Fendler* 830! 831!

174. T. PSITTACINA Hook. in Bot. Mag. t. 2841; Roem. et Schultes, Syst. Veg. vii. 1225. *Vriesea psittacina* Lindl. in Bot. Reg. xxix. t. 10; Walp. Ann. iii. 622; Wawra in Oesterr. Bot. Zeit. xxx. 182; French trans. 62; Itin. Prin. Cob. 154; E. Morren in Belg. Hort. 1882, 287, t. 10–12; Antoine, Brom. 8, t. 6. *T. simplex* Vill. Fl. Flum. iii. t. 130 ? *V. simplex* Beer, Brom. 97.— Leaves 12–20 in a rosette, lanceolate from a dilated base, 2 in. broad, a foot or more long, 1–1½ in. broad at the middle, narrowed gradually to the point, plain green, flexible, subglabrous. Peduncle shorter than the leaves; bract-leaves small, scariose, adpressed. Inflorescence a simple lax erect spike 6–9 in. long; flowers 6–10, erecto-patent; rachis red, flexuose; flower-bracts oblong, ¾ in. round, 1½ in. long, bright red or red and yellow. Calyx rather longer than the bract, yellowish; sepals obtuse. Petal-limb oblanceolate, ½ in. long, yellow. Stamens a little longer than the petals.

Hab. Forest round Rio Janeiro, *Burchell* 2540! *Miers* 3766! 3874! *Wawra* 216. Figured by Hooker from a plant grown by Mr. Shepherd at the Liverpool Botanic Garden in 1827. Var. *decolor* Wawra, Itin. Prin. Cob. 156, t. 33A; Antoine, Brom. t. 6, has green flower-bracts and a green calyx. Morren, *l. c.*, unites with this *Morreniana*, *carinata*, and *brachystachys* as varieties.

175. T. LAXA Griseb. in Gott. Nachtrag. 1864, 18, non Fl. Brit. West. Ind. 596. — Leaves lorate, with a dilated base. Peduncle longer than the leaves; bract-leaves small, adpressed. Spike lax,

simple, 6–8-flowered; flower-bracts ovate-oblong, obtuse.  Calyx
slightly longer than the bract.  Corolla 2 in. long.  Capsule as long
as the calyx.

Hab.   Venezuela; Mountains of Tovar, alt. 3000 ft., *Fendler*
2166!  Allied to *T. psittacina*.

176. T. PARABAICA Baker.    *Vriesea parabaica* Wawra, Itin. Prin.
Sax. Cob. 160, t. 33, 36 B; Antoine, Brom. 4, t. 4.   *V. carinata* var.
*constricta* Wawra in Oester. Bot. Zeitsch. xxx. 183; French trans.
64.—Leaves about 20 in a dense rosette, lorate from an ovate base,
6–8 in. long, an inch broad at the middle, thin, flexible, subglabrous,
deltoid-cuspidate at the apex.   Peduncle much shorter than the
leaves; bract-leaves small, adpressed, imbricated.  Spike moderately
dense, simple, 4–5 in. long; flowers 12–15, all erecto-patent;
flower-bracts ovate-lanceolate, $1\frac{1}{4}$–$1\frac{1}{2}$ in. long, bright red, green
towards the tip.  Calyx yellowish green, reaching to the tip of the
bract; sepals oblong-lanceolate.  Petals yellow, twice as long as
the calyx; limb oblong.  Stamens longer than the petals.

Hab.   South Brazil; Woods of Juiz de Fora, *Wawra & Maly*,
ii. 184.  Very near *T. psittacina*.

(To be continued.)

---

## THE LATE DR. BOSWELL.

DR. JOHN THOMAS IRVINE BOSWELL was born in Queen Street,
Edinburgh, on Dec. 1st, 1822, in the house now occupied by the
Philosophical Institution.  His father was Patrick Syme, an artist
who paid much attention to Natural History, and who published an
illustrated work on the British Song-birds.  His mother was a
Miss Boswell, a daughter of Lord Balmuto, and she also was an
excellent artist and very fond of Botany.  Patrick Syme took an
appointment as teacher of drawing to the academy of Dollar, and
removed there when his son was very young, and it was at this
academy that the future Dr. Boswell was educated.

From his earliest years he showed a decided taste for collecting
not only plants, but also insects and shells.  After leaving school
he was apprenticed to a firm of engineers in Edinburgh, and after
serving his time with them was engaged as a land-surveyor for a
few years.  Whilst travelling about in the exercise of his profession
he took advantage of every opportunity of botanical exploration.
The Scotch counties for which he checked lists for Mr. Watson are
85 Fife, 87 West Perth, 91 Kincardine, and 111 Orkney.  I believe
it was in 1850 that he undertook the curatorship of the Edinburgh
Botanical Society.  In February of that year he read a paper before
a meeting of the Society on the plants which he collected during a
visit paid to his relatives in Orkney in the summer of 1849.  It was
printed in the 'Transactions' of the Society (vol. iv. p. 49), and it
led to a correspondence with Mr. Watson, the result of which was
that in 1851 he undertook the curatorship of the Botanical Society

of London and removed to town. He lived first at Provost Road, and afterwards in Adelaide Road, Haverstock Hill.

In 1852 and 1853 he explored carefully the neighbourhood of London, and saw growing a large number of south-country plants he had never met with before. Two papers on his London explorations will be found in the 4th volume of the 'Phytologist.' In 1854 Mr. Syme was elected a Fellow of the Linnean Society, and undertook the Botanical Lectureship at the Charing Cross School of Medicine, and afterwards that of Westminster, where he did duty for many years. I became a member of the London Botanical Society in 1852, and I remember that at this period the parcels we received consisted largely of London and Scotch plants gathered by Watson and Syme. The London Society was broken up in 1857, and from that year till 1866 the specimens were distributed from Thirsk. The earlier editions of the 'London Catalogue' were mainly or entirely the work of Mr. Watson. Mr. Syme shared with him the editing of the fifth edition, which came out in 1857, and also in the sixth and seventh. The third edition of 'English Botany' began in 1863, and it is upon this his reputation as a botanist mainly rests.

He had by this time accumulated an extensive herbarium, both of British and European plants, and had seen growing in Scotland and England a large proportion of the species he undertook to describe. The accuracy and carefulness of his descriptions are known far too well to most of the readers of this Journal, from daily practical experience of their usefulness, to need any commendation from me here now. I will only venture in this connection to extract a few words from a letter I have received since his death from Mrs. Boswell, who shared from his early years in all his botanical work and interests. " I who acted as his amanuensis, and to whom he dictated the whole of the text of the third edition of 'English Botany,' can testify to the pains he took to make it complete, never resting whilst anything remained possible to be done in the way of comparison and research." And it is not alone the fulness and accuracy of the descriptions that make the book so valuable, but the power he shows in grasping the relationship of the types and the acute sense of proportion shown in their arrangement. This was the first time that the British plants were classified on Darwinian lines, and I never cease, when I use the book, to admire the skill which is shown in dividing out the types into species, subspecies, and varieties, a task that was done so thoroughly well that when Sir J. D. Hooker, with all his wide experience, went over the same ground shortly after, in his 'Student's Flora,' he found extremely little to change.

The eleven volumes of 'English Botany' came out between 1863 and 1872. The first volume was just finished in time to be reviewed by Dr. Seemann in the first monthly issue of this Journal. In 1868 he left London for his ancestral home of Balmuto, and in 1875, on the death of his uncle, he became the head of that branch of the Boswell family, and took his mother's maiden name. In 1875 he received the honorary degree of LL.D. from the University of

St. Andrew's. In his later years it was always a source of regret that he was not able to carry out a plan which had been proposed of revising the text of 'English Botany' and printing it as a book without the plates. From 1870 to 1875 he managed the distributions of the Exchange Club and drew up the Annual Reports, all of which will be found in the pages of this Journal. About 1875 his health began to fail, and he gradually felt his botanical correspondence more and more of an effort. He had two slight attacks of paralysis, and for the last two years was a complete invalid.

Dr. Boswell died on the 29th of January, and his coffin was carried, covered with snowdrops and Christmas roses, the last his favourite flower, to the ancestral vault of the Boswell family at Kinghorn. A portrait of him appeared in the 'Illustrated London News' on February 11th.         J. G. BAKER.

[Mr. Charles Bailey kindly forwards us, with permission to print, the following copy of what must have been one of the last botanical letters ever written by Dr. Boswell. It is interesting as showing that the writer retained to the last his interest in botanical studies :—

"Balmuto, Kirkcaldy, Fife, N.B.,
"19th January, 1888.

"Dear Mr. Bailey,—Thanks for the 'Botanical Record Club Report'; I was much pleased to read Dr. Lees's note on the Cannock Chase *Vaccinium*. I wish some one would send me a specimen, but now I have no claim. I never ask a discoverer for a specimen, as I am not doing anything for Botany. . . . If among the duplicates of the 'Botanical Exchange Club' there should ever be *Carex trinervis* (much wanted), *C. salina* var. *Kattegattensis*, and *Calamagrostis strigosa*, I should much like to have them. *Cerastium arcticum* var., from Shetland, in fruit, and *C.* '*longirostre*' are, I suppose, hopeless. I got a number of garden plants brought in, and dried them, but have not been able to get them labelled. Who is distributor this year? Perhaps I could get *Senecio spathulifolius* and *Pyrus fennica* done by the end of the month. I have a number of others from garden, and escaped or even naturalised; also a lot of Sparganiums in fruit (the most troublesome plant I ever dried)— fruit more dense and with less shoulder than the South English plant; but Mr. Beeby says they are certainly *S. ramosum*, so useless I suppose. Prof. Archangeli, of Pisa, author of new 'Moss Flora of Italy,' writes to me wishing to exchange Italian Mosses for Scotch and English. Can you help him? I cannot. I am confined to my bedroom, mostly to bed, with ulcerated leg,—two years,— from valvular disease of the heart, producing complete bodily and mental failing.

"Yours very truly,
"(*Signed*) JOHN T. BOSWELL."]

# BIOGRAPHICAL INDEX OF BRITISH AND IRISH BOTANISTS.

## By James Britten, F.L.S., and G. S. Boulger, F.L.S.

(Continued from p. 56.)

**Babington, Joseph** (fl. 1798). M.D. Lichenologist. E. B. 450. Discovered *Lichen punctatus* = *Lecanora lacustris* With. f. *punctata*.

**Backhouse, James** (1794?–1869): b. 1794?; d. Holgate House, York, 20th January, 1869. Nurseryman. Missionary Friend in Norway and the Southern Hemisphere. Correspondent of J. E. Smith and W. J. Hooker. Found *Trichomanes radicans* and *Viola arenaria*. 'British Hieracia,' 1856. Pritz. 11; Journ. Bot. 1869, 51; Journ. Hort. xli. 1869, 32; Gard. Chron. 1869, 136; R. S. C. i. 147; vi. 573; vii. 65. *Backhousia* Hook.

**Badham, Charles David** (1806–1857): b. London, 1806; d. 14th July, 1857. M.D. F.R.C.P. Clerk. 'Esculent Funguses,' 1847. Pritz. 11; Jacks. 244; Dict. Nat. Biog. ii. 387.

**Baikie, William Balfour** (1820 or 1825–1864): b. Kirkwall, Orkney, 27th August, 1825; d. Sierra Leone, W. Africa, 30th November or 12th December? 1864. M.D., Edin. R.N. Surgeon to Niger Expedition, 1854. 'Historia Naturalis Orcadensis, 1848. Jacks. 217; Journ. Bot. 1865, 71; Trans. Bot. Soc. Edin. viii. 336; Gard. Chron. 1858, 622, 734; Dict. Nat. Biog. ii. 406; Gent. Mag. March, 1865. Monument in Kirkwall Cathedral.

**Baily, Miss** [*see* Kane, Lady].

**Baines, Henry** (c. 1800–c. 1880): b. Halifax, Yorkshire, c. 1800; d. York, c. 1880. Sub-curator, York Philosophical Society; Curator of the garden. 'Flora of Yorkshire,' 1840. Formerly in the employ of Mr. Backhouse. Pritz. 12; Jacks. 262.

**Baines, Thomas** (1822–1875): b. King's Lynn, Norfolk, 1822; d. Durban, Natal, 8th May, 1876. F.R.G.S. Cape Colony, 1842. Artist in Kafir War, 1848–1851; in N.W. Australia in 1855; to Zambesi Expedition, 1858, &c. Second edition of Lindley & Paxton's 'Flower Garden,' 1880. Jacks. 408; R. S. C. i. 161; Dict. Nat. Biog. ii. 441. Vignette photo. at Kew. Biog. sketch in his 'Explorations in S.W. Africa.' Journ. R. G. S. xli. 100; xlv., cxli.

**Baker, H. C.** (fl. 1836). Captain, Bengal Artillery. 'List of specimens of wood from India,' 1836. Pritz. 12.

**Baker, Henry** (1698–1774): b. Chancery Lane, London, 8th May, 1698; d. Strand, London, 25th November, 1774. F.R.S., 1740. Sloane, 152, 252; Nich. Illust. iv. 762; Biogr. Britann. i. 525; Nich. Anecd. v. 172; Sloane MSS., 4435, 4436; Dict. Nat. Biog. iii. 10. Introduced *Rheum palmatum*.

**Baker, William Lloyd.** Clerk. F.L.S., 1793. Of Stout's Hill, Gloucester. Discovered *Cephalanthera rubra*. Sowerby Letters in Bot. Dep. Brit. Mus. Oil-painting at Hardwicke, Gloucester.

**Bakewell, Robert** (1768–1843): b. 1768; d. Downshire Hill,

Hampstead, London, 15th August, 1843. Geologist. 'Pollen,' Mag. Nat. Hist. ii. 1. R. S. C. i. 167; Dict. Nat. Biog. iii. 23·

**Balfour, Sir Andrew** (1630-1694): b.˙ Denmiln, Fife, 18th January, 1630; d. London, 10th January, 1694. M.D., Caen, 1661. Founded Edinburgh Bot. Garden, c. 1680. Pult. ii. 3; 'Memoria Balfouriana,' by Sir Robert Sibbald, 1699; Dict. Nat. Biog. iii, 48. *Balfouria* R. Brown.

**Balfour, John Hutton** (1808-1884): b. Edinburgh, 15th September, 1808; d. Inverleith House, Edinburgh, 11th February, 1884. M.A. and M.D., Edin., 1832. F.L.S., 1844. F.R.S. Prof. Bot., Glasgow, 1845; Edinburgh, 1845-1879. 'Flora of Edinburgh,' 1863. 'Manual of Botany,' 1848. 'Class-book of Botany,' 1852. Pritz. 13; Jacks. 518; Proc. Linn. Soc. 1883-4, 30; Gard. Chron. 1884, i. 220; Dict. Nat. Biog. iii. 56. *Balfourodendron* Méllo.

**Ball, Anne E.** (fl. 1840). Of Youghal. Algologist. Journ. Bot. 1840, 191. *Ballia* Harv.

**Balls, Matthew** (fl. 1869). Of Hitchin and Hertford. Phyt. n. s. ii. (1869), 202; Pryor, 'Flora of Hertfordshire,' xlvii.

**Banckes, Richarde** (fl. 1526). "Here begynnyth a new Mater ye whiche sheweth . . . . Vertues and Properties of Herbes," London, 1526. Jacks. 23.

**Bancroft, Edward Nathaniel** (fl. 1829-1841). M.D. Of Jamaica. R. S. C. i. 75. *Bancroftia* Macfadyen.

**Banister, John** (d. 1692 or 1696 ?): d. Virginia, U.S.A., 1692. Clerk. Missionary in Virginia. Correspondent of Ray, Lister, and Compton. 'Cat. of Pl. observed in N. America,' in Ray, 'Hist. Pl.' ii. 'Herbarium Virginianum,' 1767. Previously at Oxford, Phil. Trans. xxviii. 1713, 188. Letters in Sloane MSS. Herb. in Sloane's. Pult. ii. 55-7; Pritz. 13; Phil. Trans. xvi. 667; Ray, 'Hist. Pl.' ii. 1928; Loudon, 'Arboretum,' 44; Dict. Nat. Biog. iii. 119. *Banisteria* Houston.

**Banks, George** (fl. 1823-1832). F.L.S., 1824. Lecturer on Botany at Devonport. 'Introduction to . . . . English Botany,' 1823. 'Indigenous Flora of London and Plymouth,' in Mag. Nat. Hist. 1829, 265. 'Plymouth and Devonport Flora,' 1830-2. Pritz. 13; Jacks. 234, 258.

**Banks, Sir Joseph** (1743-1820): b. Argyle Street, London, 13th February, 1743; d. Spring Grove, Isleworth, Middlesex, 19th June, 1820. M.A., Oxon, 1763. F.R.S., 1766. P.R.S., 1778. D.C.L., Oxon, 1771. Bart. 1781. K.C.B., 1795. Round the world with Cook, 1768-1771; to Iceland with Solander, 1772. Purchased Cliffort's Herbarium. Herbarium and library in Brit. Mus. Pritz. 13; Jacks. 518; Lives by Duncan, Brougham, &c.; Dict. Nat. Biog. iii. 129; Gent. Mag. xc. i. 574, 637; Nich. Anec. vii. 20, 509; Cott. Gard. iv. 169; Felton, 181; Linn. Letters, ii. 574-580. Portr. at Hortic. Soc.; engr. in 'European Magazine,' 1795. Oil-painting and busts, one by Chantrey at Linn. Soc. Statue by Chantrey at Brit. Mus., and portr. at Kew. *Banksia* L. fil. *Josephia* Brown.

**Barclay, George** (fl. 1835): b. Huntley, Aberdeenshire; d. Buenos

Ayres. Kew gardener and collector. H.M.S. 'Sulphur,' 1835, to Chili, Peru, Panama, Sandwich Isles, Nootka, &c. Plants in Herb. Mus. Brit. Gard. Chron. 1882. i. 305.

Barclay, Robert (1757-1830): b. 1757; d. 22nd October, 1830. F.L.S., 1788. Of Clapham and Buryhill. Pritz. 14; Bot. Misc. ii. 122. Portr. at Kew. *Barclaya* Wall.

Barker, John Theodore (fl. 1852). Of Bath. 'Beauty of Flowers,' 1852. Jacks. 43.

Barnard, Alicia M. (fl. 1845). Of Norwich. Grand-niece (?) of Sir J. E. Smith. Pryor, 'Flora of Hertfordshire,' xlvii.

Barnard, Edward (1786-1861). F.L.S., 1818. Proc. Linn. Soc. 1862, lxxxv.

Barrington, Hon. Daines (1727-1800): b. 1727; d. in the Temple, London, 11th March, 1800. F.R.S. 'Naturalists' Calendar.' Pritz. 14; Jacks. 219; Felton, 177; Baillon, i. 372; Diet. Nat. Biog. iii. 286; Nich. Illust. v. 582; portr. vii. 4; Nich. Anec. ii. 553; iii. 3; viii. 424; Gent. Mag. lxx. 291. *Barringtonia* Forst.

Barrow, Sir John (1764-1848): b. Dragley Beck, Ulverston, Lancashire; d. London, 23rd November, 1848. Secretary to the Admiralty, 1804. Baronet, 1835. Autobiography, 1847. Diet. Nat. Biog. iii. 305. *Barrowia* Dcne.

Barter, Charles (d. 1859): d. Rabba, W. Africa, 15th July, 1859. Niger Expedition, 1858. A.L.S. 1858. Pritz. 14; R. S. C. i. 196; Proc. Linn. Soc. v. xx. *Barteria* Hook. f.

Barton, Benjamin H. F.L.S., 1835. Of Great Bissenden. 'British Flora Medica,' 1837, with Thos. Castle. Pritz. 15; Jacks.

Barton, John (fl. 1812-1830). Of Chichester. Friend. Brother of Bernard Barton. 'Lecture on the Geography of Plants,' 1827. Pritz. 15; Jacks. 221; R.S.C. i. 200.

Bartram, John (1699 or 1701-1777): b. Marple, Co. Delaware, Penn., 1699 or 1701; d. 1777. "King's botanist in America." "The greatest natural botanist in the world," Linnæus. Correspondent of Sloane, Hill, Ellis, and Collinson, Phil. Trans. 1740-1763. Pritz. 15; Medical & Phys. Journ. i. (1804), 115; Memorials, by W. Darlington, 1849. Loudon, 'Arboretum,' 85. *Bartramia* Hedw.

Baskerville, Thomas (1812-1840): b. 26th April, 1812; d. London, 1840. M.R.C.S., 1835. Practised at Canterbury. 'Affinities of Plants,' 1839. Pritz. 16; Jacks. 17; Diet. Nat. Biog. iii. 369. *Baskervilla* Lindl.

Bateman, John (?) (fl. cire. 1700). Clerk. List of Faversham plants in Blackstone's 'Specimen.' Pult. ii. 272; Jacob, 'Plantæ Favershamienses.'

Bauer, Ferdinand (1760-1826): b. Feldsberg, Austria, 20th January, 1760; d. Hietzing, Vienna, 17th March, 1826. Accompanied Sibthorp to Greece, 1784. Flinders' Expedition, 1801-5. 'Illustrationes Floræ Novæ Hollandiæ,' 1813. Lambert's 'Pinus,' 1810-1837. Lindley's 'Digitalis.' Pritz. 17; Jacks. 398; Nich. Illust. vi. 838; Ann. & Mag. iv. (1840), 67; Proc.

Linn. Soc. i. 1839, 39; Journ. Bot. 1843, 106.  *Bauera* Banks ex And. Bot. Rep.

**Bauer, Francis** (1758-1840): b. Feldsberg, Austria, 4th October, 1758; d. Kew, 11th December, 1840; bur. at Kew. F.L.S., 1804. F.R.S., 1820. Came to England, 1788. Employed as artist by Banks. 'Delineations of Exotic Plants' . . . , 1796. 'Strelitzia depicta,' 1818. 'Ergot,' Linn. Trans. 1840. Ann. & Mag. v. 1840, 47; Pritz. 17; Jacks. 519; Gard. Chron. 1841, 22; Proc. Linn. Soc. i. 1841, 101; Proc. Roy. Soc. iv. (1843), 342; Ann. & Mag. vii. 1841, 77, 439. Portr. at Kew. *Bauera* Banks ex Andr. Bot. Rep.

**Baxter, William** (1787-1871): b. Rugby, Warwick, 15th January, 1787; d. Oxford, 1st November, 1871. A.L.S., 1817. Curator, Oxford Bot. Garden, 1813-1851. 'British Phænogamous Botany,' 1834-1843. Pritz. 18; Jacks. 235; Journ. Bot. 1871, 380; Gard. Chron. 1871, 1426; Loud. Gard. Mag. x. (1834), 110; Dict. Nat. Biog. iii. 438; Druce, 'Flora of Oxfordshire,' 392. Engr. at Kew, by J. Whessell, from drawing by A. Burt.

**Baxter, William** (fl. 1823-1830). Collector in South Australia. Plants in Herb. Mus. Brit. R. Brown, Prodr. Fl. Nov. Holl. Suppl. 1830. *Baxteria* R. Brown.

**Beaton, Donald** (1802-1863): b. Urray, Ross-shire, 8th March, 1802; d. Surbiton, Surrey, October, 1863. Cott. Gard. xiii. 153, portr.; Journ. Hort. v. (1863), 349 and 415, portr. *Beatonia* Herbert.

**Beattie, James** (1735-1803): b. Laurencekirk, Kincardine, 25th October, 1735; d. Aberdeen, 18th August, 1803. M.A., Aberdeen, 1753. A.L.S., 1807. Prof., Aberdeen. Added *Linnæa* to British Flora, E. B., 433. Life by Sir W. Forbes, 1806; Smith, Lett. i. 441-3; Dict. Nat. Biog. iv. 23.

**Beaufort, Mary, Duchess of** [*see* SOMERSET].

**Beaumont, John** (d. 1731): d. Stone-Easton, Somerset, March, 1731; bur. Stone-Easton. Surgeon. F.R.S., 1685. 'Rock-plants in the Lead Mines of Mendip Hills.' Sloane MS. 4037, 128-32; Dict. Nat. Biog. iv. 60; Ray Lett.

**Beaumont, Lady Diana** (fl. 1826). Contributed European plants to the Calcutta Bot. Garden. *Beaumontia* Wall. Tent. Fl. Nepal., not as in Pritz. 18.

**Becker, John Thomas** (1770-1848): b. 1770; d. Hill House, Southwell, Notts., 3rd January, 1848. Clerk. M.A., Oxon, 1795. Prebendary of Southwell, 1818. Determined *Crocus nudiflorus*. E. B. 491; Gent. Mag. April, 1848; Dict. Nat. Biog. iv. 75.

**Bedford, John, Duke of** [*see* RUSSELL, JOHN].

**Beeke, Henry** (1751-1837): b. Kingsteignton, Devon, 6th January, 1751; d. Torquay, Devon, 9th March, 1837. Clerk. B.A., Oxon, 1773. D.D., 1800. F.L.S., 1800. Dean of Bristol, 1814. Gent. Mag. n. s. vol. vii.; Smith MSS.; Turner & Dillwyn, 'Bot. Guide,' p. 527, 8; Dict. Nat. Biog. iv. 124; Mag. Nat. Hist. (1837), 61, 392. Distinguished *Lotus pilosus*.

**Beeston, William** (1672?-1732): d. Bentley, Suffolk, 4th

December, 1732. M.B., Camb., 1692. M.D., 1702. Of Ipswich. "Very curious and knowing in plants," Sherard, Rich. Corr. 184; Nich. Illustr. vi. 879; Loudon 'Arboretum,' 62.

Bell, G. (d. 1784): b. Manchester; d. Manchester, 1784. Doctor. ';De physiologiâ plantarum,' 1777. Mem. Manchester Soc. ii. 394 ; Baillon, i. 395.

Belt, Thomas (1832–1878): b. Newcastle-on-Tyne, 1832; d. Kansas City, Colorado, 21st September, 1878. Traveller. 'Naturalist in Nicaragua,' 1874. Gard. Chron. 1878, ii. 478 ; Proc. Geol. Soc. 1878–9, 48; Dict. Nat. Biog. iv. 204 ; Jacks. 368.

Bennett, Frederick Debell (fl. 1840). F.R.C.S. F.R.G.S. ' Catalogue of Plants collected during the Tuscan's Voyage ' in ' Narrative of a Whaling Voyage, 1833–1836,' 1840. Pritz. 21.

Bennett, John Joseph (1801–1876): b. Tottenham, Middlesex, 8th January, 1801 ; d. Maresfield, Sussex, 29th February, 1876, bur. M.R.C.S., 1825. F.L.S., 1828. Sec. L.S., 1840–1860. F.R.S., 1841. Assistant-Keeper, Bot. Dep. Brit. Mus., 1827. Keeper, 1857–1870. ' Plantæ Javanicæ,' 1838–1852. Pritz. 21 ; Jacks. 521 ; R. S. C. i. 275 ; Journ. Bot. 1876, 97, with portr. ; Dict. Nat. Biog. iv. 246. Oil-portr. by Eddis and bust by Weekes at Linn. Soc., Kew, and Bot. Dep. Brit. Mus. *Bennettia* R. Br. = *Cremostachys* Tulasne. *Bennettia* Gray = *Saussurea* DC. *Bennettia* Miquel. *Bennettites* (fossil) Carruthers.

Bennett, John Whitchurch (fl. 1830). F.L.S., 1828. 'Coconut Palm,' 1831. Jacks. 208.

Benson, Thomas (1802–1887): b. Cockermouth, Cumberland, October, 1802; d. Great Fambridge, Essex, 9th June, 1887. B.A., Cantab., 1824. Clerk. Vicar of Great Fambridge, 1832. Contributed to Gibson's ' Flora of Essex.' Herbarium bequeathed to Essex Field Club.

Bentall, Thomas (fl. 1847–1862). Of Halstead, Essex. Invented a drying paper. Contributed to Gibson's ' Flora of Essex.' R. S. C. i. 280.

Bentham, George (1800–1884): b. Stoke, near Plymouth, 22nd September, 1800 ; d. Wilton Place, London, 10th September, 1884. LL.D., Camb. C.M.G., 1878. F.L.S., 1826. Pres., 1861–1874. Sec. R. Hort. Soc. 1829–1840. F.R.S., 1862. 'Plantæ Hartwegianæ,' 1857. 'Flora of Hongkong,' 1861. 'Flora Australiensis,' 1863–1878. ' Genera Plantarum,' with J. D. Hooker, 1862–1883. Herb. at Kew. Pritz. 21 ; Jacks. 21 ; R. S. C. i. 280 ; vii. 240 ; Proc. Linn. Soc. 1884, 90, with bibliography; Journ. Bot. 1884, 353, portr.; Gard. Chron. 1884, ii. 336, 368 ; Dict. Nat. Biog. iv. 263. Oil-portr. by Dickinson at Linn. Soc. Photo. and engr. at Kew. *Benthamia* Lindl.

(To be continued.)

---

## SHORT NOTES.

NOTE ON MENTHA PRATENSIS Sole. — In a letter of Aylmer Bourke Lambert to the late Sir James Smith, which is contained in his collections at the Linnean Society, is the following note :—

" Mr. Lambert informs Sir James Smith that he ascertained that *Mentha pratensis*' (Sole) was thrown out of the Roebuck Inn garden on Alderbury Common, and was merely a single plant ; this Mr. Sole dug up, and the original specimen is at the Linnean Society."—Thos. Bruges Flower.

The Summit Flora of the Grand Tournalin. — In an ascent of the Grand Tournalin of Val Tournanche made on the 3rd July last, I gathered the following species in full flower on the summit-ridge of the mountain, at a height of 3400 mètres (say 11,150 feet), as determined by the Survey of the Italian État-Major :—*Ranunculus glacialis* L., *Thlaspi rotundifolium* DC., *Draba Wahlenbergii* Hartm., *Saxifraga oppositifolia* L., *S. planifolia* Lap., *Artemisia spicata* DC., *Linaria alpina* DC., *Androsace glacialis* Hoppe. To the best of my knowledge no details of the summit flora of the Tournalin have hitherto been published, so that the foregoing list may not be altogether without interest, if only as affording one more example of the striking uniformity in the groups of species occurring at heights of 10,000 feet and upwards in the European Alps. Three days earlier (30th June), in an ascent of the Rympfischhorn (Zermatt), I found *Sempervivum arachnoideum* L. growing on the highest point of Rympfischgrat, marked 3314 mètres (10,850 feet) in General Dufour's Swiss Survey map, an elevation to which, I believe, this species very seldom attains in the Swiss Alps ; and the day after discovered *Woodsia hyperborea* (R. Br,) on the rocks above the Boden Glacier, near the base of the Riffelhorn, perhaps a new Swiss station for the species. Mr. Nicholson, of Kew, has kindly compared two of my specimens with the materials in the Herbarium, and confirms me in the opinion that they are *Draba Wahlenbergii* Hartm. and *Saxifraga planifolia* Lap.—N. Colgan.

The Name Conringia.—In the January number of the ' Journal of Botany,' p. 61, attention is called to the fact of *Conringia* being mis-spelled in Pryor's ' Flora of Hertfordshire.' This was due to extreme regard to the author's views, rather than editorial oversight, for on turning to the place of publication I found it *Couringia*, and the index to the same volume of DeCandolle's ' Systema ' has it spelled the same way. At that time I had not access to the whole literature of the genus, and therefore had to let the matter rest, as I could not thoroughly investigate it. Having recently looked up the question, it may be summarised thus :—

Conringia Heist. Ind. Pl. Rar. n. 1. p. 34 (1730) ; Linn. Syst. ed. 1 (1735).

Couringia Adans. Fam. ii. 418 (1763) ; A. Juss. Dict. xi. 244 (1818) ; DC. Mém. Mus. vii. 239 (1821) ; Syst. ii. 507, 508 (1821).

Hermann Conring was a noted jurisconsult of the previous century, and a collected folio edition of his works was in course of issue at the time that Heister was publishing his little tracts on the Helmstadt garden plants.—B. Daydon Jackson.

NOTICES OF BOOKS.

*The Botanical Record Club: Phanerogamic and Cryptogamic.* Report for the years 1884, 1885, 1886, by the Editor, F. ARNOLD LEES, M.R.C.S. Manchester: Printed by James Collins & Co., King Street, 1887. 8vo, pp. 81–156, 77, 78.

THIS latest addition to the Reports of the Record Club in no way differs in general plan from its predecessors, which have been duly noticed in these pages; and it is open to the same objections which, as it seems to us, have characterised former issues. It must always be borne in mind that these Reports are not in the market, and their circulation is confined to the members of the Club, and to the privileged " public Institutions and Journals to which the ' Reports' are sent—21 in number: so that the new names published therein are inaccessible to the botanical world at large.

The plan of including as " new county records" all such as are " additional to ' Topographical Botany,' ed. 2, and to previous Reports of the Botanical Record Club," involves the repetition of much that has already appeared in our own columns and elsewhere. This would be the case even if this non-natural use of the phrase " new county record" were adhered to by those who employ it: but this is not so. *Ononis antiquorum*, for example, is duly entered for Beds in ' Top. Bot.'—on old authority, it is true, but Abbot's testimony has hitherto been considered worthy of credence. Now it is entered as a " new county record," with the remark, " No record since that in Abbot's *Flora*." This may be a desirable confirmation, but a " new county record" it certainly is not. *Vicia lutea*, from the railway near Bedford, although marked as a colonist, has certainly no business here, nor would Mr. Watson have sanctioned its insertion: the same objection might be raised to the inclusion of *Anthoxanthum Puelii*.

Another point which strikes us as of doubtful benefit is the portentous list of Roses and Rubi which swells the Report—more than three pages of " new county records " alone being allowed to each, plus three of Roses and four of Rubi in the "new locality list." Is it disrespectful to express a doubt as to whether the collectors of these prickly subjects mean the same thing by the names attached to them? Is it not rather certain that they do not by any means recognise the same limit of species in these genera? It is true that Mr. Baker is given as the authority for the Roses, but we find no indication that the specimens of the " new county records " have passed through his hands; while the note on one specimen that the name was " certified by Mr. J. G. Baker" seems rather to imply that for the others he is not to be held responsible. He has, however, furnished many notes in the " new locality list." As to the Rubi—well, on most of them Prof. Babington has given an opinion, and no better can be obtained. For the novelties under these genera, however, Mr. Lees himself is responsible. These are *Rubus Bakeri* and *Rosa tomentosa* var. *uncinata*—the former " = the *R. nitidus* var. *hamulosus* P. J. Mull. of L. C. 8th"; the latter is a new form found near Llanfairfechan by Mr. Charles Bailey.

The "County Catalogues" begin with a list of Breconshire "new records" observed by Mr. W. Bowles Barrett in 1884, which for the most part correspond with the more complete enumeration of the plants of that county printed in this Journal for 1885. Mr. Lees, on the principles of the Record Club, is fully justified in including them in this Report, but it would, we think, have been convenient to refer his readers to the more detailed paper published in these pages. Is Mr. Barrett responsible for the note on p. 143, which runs: "The following, also, are additional Breconshire Records, not heretofore published, communicated to me by the botanists named in each case. They were included in my paper in *Journ. Bot.*, with names of finders attached.—W. B. B." If "not heretofore published," how could they have been "included in [his] paper in Journ. Bot."? It is certain, moreover, that "Lady Wilson" and "Sowerby," whose names are attached to two of the plants in question, did not themselves communicate with Mr. Barrett. Mr. Ley's Radnorshire list has much in common with Mr. Ridley's, published in this Journal in 1881, although it includes two noteworthy additions, *Potentilla rupestris* and *Allium Schœnoprasum.*

It is matter for satisfaction that the voucher-specimens on which these records of the Club are based are now accessible to the public in the Natural History Museum. No doubt anyone engaged on a local flora would consult these specimens, and this would be desirable, as, in spite of the great care exercised by Mr. Lees, errors will creep in. Thus the specimen on faith of which *Scirpus acicularis* is recorded in the present Report for Cardigan is not that species, but *S. setaceus;* and other examples of misnaming might be adduced. On the other hand, Mr. Druce's specimen of *Schoenus nigricans*, as to the occurrence of which in Glen Shee Dr. White expresses some doubt, is certainly that species.

May we utter a word of protest against the peculiar phraseology in which these Reports are couched? "Gracile" (pp. 134, 135, &c.) seems needless when we have so good a word as slender ready to hand.

_____

*The Elements of Botany for beginners and for schools.* By ASA GRAY. Revised edition. New York: Ivison and Co. London. Trübner. 1887. 8vo, pp. 226. 589 cuts.

THIS new edition of an old favourite should have been noticed before ; but the delay has given an additional interest to the work, for we now know it as the last which will come from the hand of the venerable and kindly writer. It takes the place, as the author tells us in his preface, of his 'Lessons in Botany,' published over a quarter of a century ago, and " is a kind of new and revised edition of that successful work." It is meant to occupy in the higher schools the position which " How Plants Grow " fills in the common schools, and " is intended to ground beginners in Structural Botany and the principles of vegetable life, mainly as concerns Flowering or Phanerogamous plants, with which botanical instruction should always begin."

There are some of us old-fashioned enough to be thankful for a book which begins somewhat in the good old way to which we in England were accustomed in Oliver's ' Lessons,' and tells us about things which we can see and handle and examine without the aid of a microscope. " No mention has been made," says Dr. Gray, " of certain terms and names which recent cryptogamically-minded botanists, with lack of proportion and just perspective, are endeavouring to introduce into phanerogamic botany, and which are not needed nor appropriate, even in more advanced works, for the adequate recognition of the ascertained analogies and homologies." It is certain that the fondness for new terms has been indulged to an inconvenient extent, and it may be doubted whether the actual knowledge of the plants themselves is always promoted by the system of instruction now in vogue. The story of a student who had mastered the structure of *Peziza*, but who did not at all recognise our common scarlet species as the plant she had been studying with creditable results is, we believe, perfectly true ; and it is certain that the present mode of teaching does not encourage what used to be known as " field botany."

There is no need to describe Dr. Gray's little book in detail. Beginning with a description of flax as a type plant, he then takes the different organs and their modifications in detail, with a chapter on fertilization and an interesting section on " Vegetable Life and Work."—the volume ending with an excellent Glossary and Index of Terms. Dr. Gray's work, throughout his long and active botanical career, has been marked by clearness and consistent usefulness ; and these characteristics are as noticeable in a work like the present as in the greater undertakings by which his position among botanists has been secured.

In an interesting little illustrated volume entitled ' The Vegetable Lamb of Tartary ; a Curious Fable of the Cotton Plant ' (Sampson Low & Co. : 8vo, pp. xi. 112), Mr. Henry Lee has established to his own satisfaction the identity of the " Scythian Lamb," which was first brought into prominence by Sir John Mandeville, with the cotton plant ; and he certainly makes out a good case in support of his view. In spite of the general belief which identified the " Scythian Lamb " with the rhizome of a fern, which, in allusion to this belief, was styled *Dicksonia Barometz*, Mr. Lee considers that these rhizomes and the " Lambs " made from them " had no more to do with the origin of the fable of the ' Barometz ' than the artificial mermaids so cleverly made by the Japanese have had to do with the origin of the belief in fish-tailed human beings and divinities." For the evidence which the author brings in support of his theory, and which has been carefully and exhaustively compiled from a large number of writers, we must refer our readers to the volume itself, which is further enriched by a readable sketch of the history of cotton and the cotton trade.

THE first part of the ' Transactions and Proceedings of the Perthshire Society of Natural Science' contains an interesting paper on the Flora of Woody Island, near Perth, by Mr. W. Barclay; and two papers by Mr. R. H. Meldrum, one on *Mnium riparium* in Scotland, the other on localities for Perthshire plants.

THE recent issue (vol. v. pt. ii.) of the 'Proceedings of the Bristol Naturalists' Society' contains "Notes supplemental to the Flora of the Bristol Coal-field," by Mr. J. W. White, which contains the gratifying intelligence that "the Cheddar Pink is likely to hold its ground as long as the world shall last"; and a continuation of Mr. Cedric Bucknall's "Fungi of the Bristol District," in which a new species, *Lasiosphæria fulcita*, is described and figured.

---

### ARTICLES IN JOURNALS.

*Bot. Centralblatt.* (Nos. 6, 7).—J. Murr, 'Ueber die Einschleppung und Verwilderung von Pflanzenarten im mittleren Nord-Tirol.'—(No. 6). A. N. Lundström, 'Ueber Mykodomatien in den Wurzeln der Papilionaceen' (1 plate).—A. Pater, 'Ueber die Pleomorphie einiger Süsswasseralgen aus der Umgebung Münchens.'— (No. 7). C. O. Harz, 'Ueber vergleichende Stickstoffdüngungsversuche.n'—Id., *Agaricus lecensis*, sp. n.—C. J. Johanson, 'Ueber die Pilzgattung *Taphrina*.' — (Nos. 8, 9). C. Dünnenberger, 'Bacteriologische-chemische Untersuchung über die beim Aufgeben des Brotteiges wirkenden Ursachen.'—G. Beck, 'Geschichte des Wiener Herbariums.'

*Bot. Gazette* (Jan.).—D. H. Campbell, 'The Botanical Institute at Tübingen.' — J. W. Moll, 'The application of the paraffin-imbedding method in Botany.'—*Erigeron Tweedyi* Canby, sp. n.

*Bot. Notiser* (häft. 1).—F. W. C. Areschoug, 'Om *Rubus affinis* och *R. relatus*.' — G. Anderson, 'Redogörelse för senare tiders undersökningar af torfmossor, kalktuffer och sötvattensleror, särdeles med hänsyn till den skandinaviska vegetationens invandringshistoria.' — E. Ljungström, 'En Primula-exkursion till Möen.'—S. Berggren, 'Om apogami hos prothalliet af *Notochlæna*.' F. Areschoug, 'Om *Trapa natans* var. *conocarpa*.'—A. N. Lundström, 'Om Jenissej-strändernas Salixflora.' — J. A. Leffler, 'Ofversigt af den skandinaviska halfons anmärkningsvärdare Rosaformer.'

*Bot. Zeitung* (Feb.). — A. F. W. Schimper, 'Ueber Kalkoxalatbildung in den Laubblättern.'

*Bull. Bot. Soc. France* (xxxiv. Comptes rendus, 7 : Feb. 1).— E. Mer, 'Récherches sur la formation du bois parfait.'—C. Degagny, 'L'hyaloplasma ou protoplasma fondamental, son origine nucléaire.' ——. Hue, 'Lichens du Cantal.'—A. Battandier, 'Plantes d'Algérie' (*Trifolium Juliani, Lathyrus numidicus, Typha Maresii, Romulea Rouyana,* spp. nn.). — L. Trabut, 'Additions à la Flore d'Algérie' (*Aristida sahelica,* sp. n.).

*Bull. Torrey Bot. Club* (Feb.).—J. Schrank, 'Histology of Vegetative Organs of *Brasenia peltata*' (2 plates). — G. Vasey, 'New or Rare Grasses.'

*Gardeners' Chronicle* (Feb. 4).—*Esmeralda bella* Rchb. f., *Maxillaria Hübschii* Rchb. f., *Catasetum tapiriceps* Rchb. f., spp. nn.—*Abies numidica* (fig. 23).—(Feb. 25). *Dendrobium chryseum* Rolfe, sp. n.

*Journal de Botanique* (Feb. 1). — W. Nylander, 'Note sur le *Farmelia perlata* et quelques espèces affines.' — C. Flahault, 'Les herborisations aux environs de Montpellier.' — M. Gomont, 'Sur les envelopes cellulaires dans les Nostocacées filamenteuses.'—(Feb. 16). N. Patouillard, 'Fragments mycologiques' (*Camillea*: 1 plate). — E. Roze, 'La Flore parisienne au commencement du xviii. siècle.'

*Magyar Növénytani Lapok.* (Feb.). — L. Simonkai, 'Fiume Florája.'

*Nuovo Giornale Bot. Italiano* (Jan. 31).—A. N. Berlese, 'Monografia dei generi *Pleospora, Clathospora, e Pyrenophora.*'—O. Beccari, 'Nuove specie di Palme recentemente scoperte alla Nuova Guinea' (*Ptychandra Muelleriana, P. Obriensis, Ptychosperma Sayeri, P. Ridleyi, Calamus Cuthbertsoni*, spp. nn.).

*Oesterr. Bot. Zeitschrift.* (Feb.).—A. Hausgirg, 'Zur Algenflora Böhmens.'—V. Borbás, *Cynoglossum paucisetum* sp. n.—L. Celakovsky, 'Orientalische Pflanzenarten' (*Lathyrus brachypterus*, sp. n.).—B. Blocki, *Hieracium pseudobifidum*, sp. n. — P. Conrath, 'Zur Flora von Bosnien' (*Corydalis leiosperma*, sp. n.).

---

## LINNEAN SOCIETY OF LONDON.

*Feb. 2nd*, 1888. — W. Carruthers, F.R.S., President, in the chair.—Dr. William Schlich, Mr. Isaac Thompson, and Mr. W. S. McMillan were formally admitted Fellows of the Society. The President called attention to the loss which the Society had sustained by the death of Professor Asa Gray, Professor Anton de Bary, and Dr. Irvine Boswell (formerly Syme), which had occurred since the date of the last meeting, and gave a brief review of the life and labours of each. Mr. C. T. Druery exhibited a collection of abnormal British Ferns, and made some remarks on the extraordinary number of named varieties which had been recognised, and which now required to be carefully examined and compared with a view to some systematic arrangement of them. A discussion followed, in which the President, Mr. J. G. Baker, Dr. Murie and others took part. Dr. Amadeo exhibited and made some observations on a new species of *Tabernæmontana*. A long and interesting paper was then read by Mr. Henry T. Blanford, F.R.S., on the Ferns of Simla, based upon a collection which he had himself made there, "not much below 4500 feet, nor above 10,500 feet." His remarks were illustrated by a map, and by the exhibition of a number of the more noticeable Ferns collected, many of which were extremely beautiful. Criticisms were offered by Mr. C. B. Clarke, F.R.S., Mr. J. S. Gamble (Conservator of Forests, Northern Circle, Madras), and Dr. William Schlich (Inspector-General of Forests to the Government of India.) A paper was then read by Mr. H. J. Veitch, F.L.S., on the fertilization of

*Cattleya labiata*, var. *Mossiæ*, in which the author detailed an elaborate series of observations, undertaken with the object of detecting, if possible, the act of fertilization of the ovules, to determine the time that elapses between pollination and that event, and to trace the development of the ovules into perfect seeds.  After explaining the structure of the sexual apparatus of *Cattleya labiata*, with the aid of drawings, showing the separate parts, the processes following pollination were then dealt with, first from the development of the rudiment into the perfect ovule, and then the ripening of the ovules into seeds, these processes being also illustrated by drawings made of particular stages.  A discussion followed, in which Mr. J. G. Baker, Mr. H. N. Ridley and others took part.

*February* 16*th.*—William Carruthers, F.R.S., President, in the chair.—Announcement was made of a donation of books to the Library by the widow of the late Dr. John Millar, Fellow of the Society, recently deceased. — Mr. Spencer Moore exhibited and made some remarks upon specimens illustrative of the *Palmella* state of *Draparnaldia glomerata.*—Mr. D. Morris exhibited a species of wood of *Hieronyma alchornioides* received from Trinidad, showing in its fissures mineral deposits which on chemical analysis proved to be calcic carbonate.  For comparison Mr. Morris also exhibited and made some observations upon some deposits of calcic phosphate in teak.  Some of these (described by Sir Fredk. Abel, Quart. Journ. Chem. Soc. xv. 91) are 6 ft. in length, 6 in. in breadth, and from $\frac{1}{8}$ in. to $\frac{3}{8}$ in. in thickness.  Deposits in bamboo known as *tábasheer* (silicate) were shown, as also pearls (carbonate of lime) from cocoa-nuts, received from Dr. Sydney J. Hickson (see 'Nature,' vol. xxxvi. p. 157). — Dr. Burn Murdock exhibited and offered remarks upon the intramarginal (so-called) veins in the section *Areolata* of the genus *Erythroxylon*, of which *E. Coca* is the most familiar species.  These lines are due to a thickening of the parenchymatous tissue which takes place in the bud-stage, and are in no way connected with the venation of the leaf.—Mr. G. F. Sherwood exhibited a collection of photographs taken in Samoa illustrating the scenery and people, together with a number of necklets formed with strings of various bright-coloured seeds. — The first paper of the evening was by Mr. H. N. Ridley, · On Self-Fertilisation and Cleistogamy in Orchids.'  Three common methods of self-fertilisation were explained : (1) By the breaking up of the pollen mass, and falling of the dust either directly upon the stigma or into the lips, whence it comes into contact with the stigma ; (2) by the falling of the pollen masses as a whole from the clinandrum into the stigma ; and (3) by the pulling forward of the pollinia from the clinandrum or the anther cap, the caudicle and gland remaining attached to the column.  An interesting discussion followed, in which Prof. Marshall Ward, the Rev. G. Henslow, and Mr. A. W. Bennett took part.

THE vacancy caused in the Botanical Chair at Edinburgh, caused by the death of the late Prof. Dickson, has been filled by the appointment of Prof. I. B. Balfour, of Oxford.

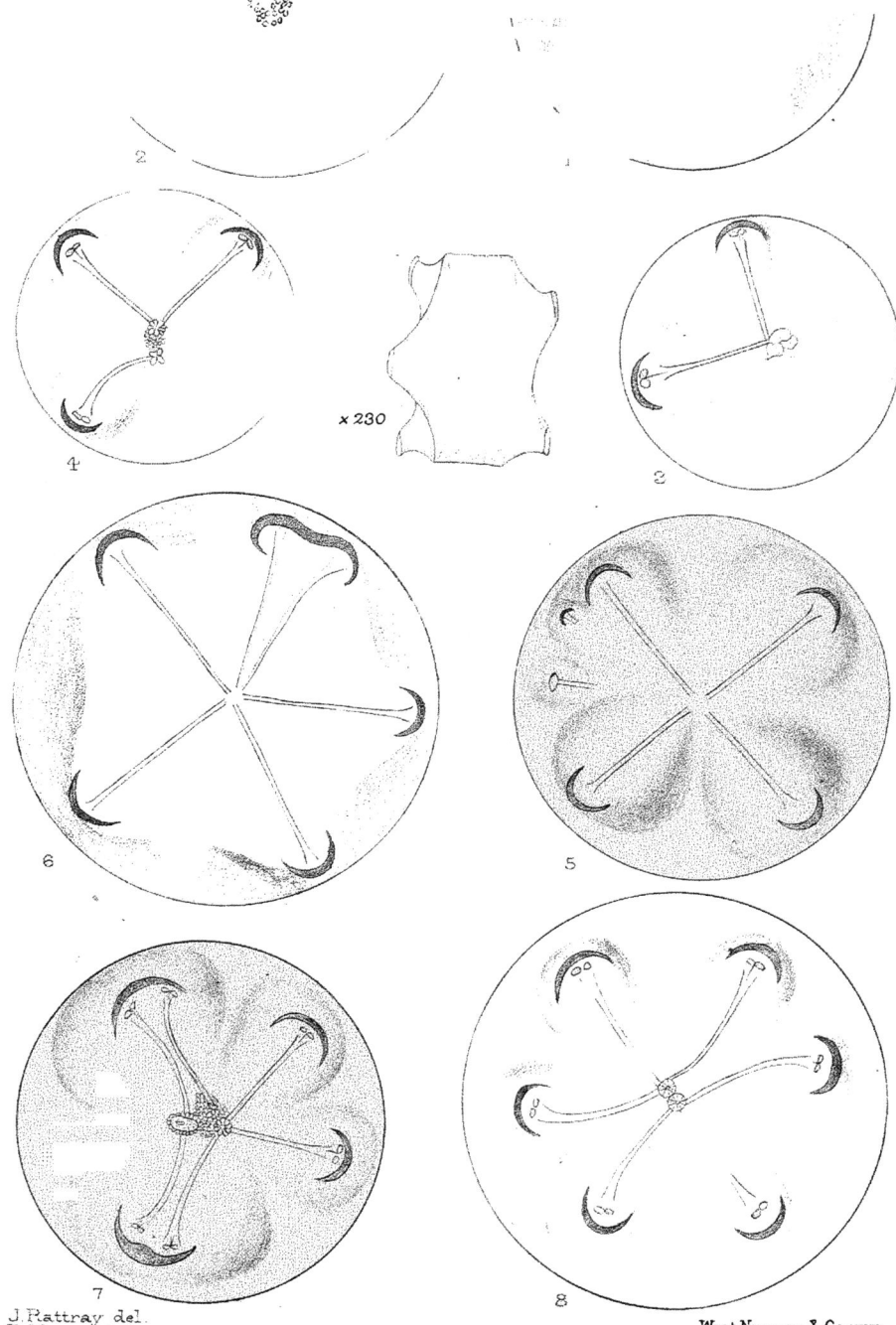

× 230

J. Rattray del.
R. Morgan lith.

West, Newman & Co. imp.

Aulacodiscus Kittoni, *Arnott*.
*Abnormal Valves.*

# NOTES ON SOME ABNORMAL FORMS OF *AULACODISCUS* Ehrb.

## By John Rattray, M.A., B.Sc., F.R.S.E.

### (Plate 281).

In the course of investigating this genus for the Monograph which I have just laid before the Royal Microscopical Society, I have met with a number of abnormal forms, which it seems to me desirable to record and describe. The genus is characterised by the circular outline of its valves, by the surface—almost flat or with a somewhat elevated band in the course of the radius; by the free circular or closely approximated angular markings, and by the presence of distinct processes at the ends of the primary rays, which proceed outwards from a hyaline or punctate central space or from a central rosette.

The abnormalities which I have observed affect the outline, the undulations of the surface, the colour, the character of the central space, the markings, the number and direction of the primary rays, and the processes.

1. Outline.—With the single exception of *A. polygonus* Grun., which, as its name implies, is many-sided, all the species of this genus have a circular outline. In valves of *A. Kittoni* Arnott, however, the margin sometimes becomes more or less straight, so as to produce an obtusely angular form, suggesting the normal more rectilinear-sided specimens of *A. polygonus;* and in a considerable number of species I have met with valves that are more or less regularly elliptical, e. g., *A. Barbadensis* Ralfs, *A. angulatus* Grev., *A. elegans* Grove & Sturt, *A. dispersus* mihi, *A. amœnus* Grev., *A. affinis* Grun., *A. formosus* Arnott, *A. aucklandicus* Grun., *A. Petersii* Ehrb., *A. Kittoni* Arnott, *A. Rattrayii* Grove & Sturt, *A. Sollittianus* Norman. A more local interference with the circular outline sometimes occurs in *A. Petersii* Ehrb. by the formation of one, two, or three lobes which proceed sometimes suddenly from the margin or by a more gradual extension.

2. Surface.—In only a few species is the surface almost flat from the centre to the border, e. g., *A. exiguus* Witt., *A. Barbadensis* Ralfs; in several it is flat to the zone of the processes, but beyond this slopes distinctly to the border, e. g., *A. Comberi* Arnott, *A. Beeveriæ* Johnson; in many there is an elevated band either at or within the zone of the processes, and this may be associated with the presence of inflated areas along the primary rays, e. g., *A. decorus* Grev., *A. Archangelskianus* Witt., *A. superbus* Kitton, &c., or it may not, as in *A. Sollittianus* Norman, *A. margaritaceus* Ralfs, *A. Lahuseni* Witt., &c.; distinct inflations without a well-defined narrow highest zone are normal in *A. formosus* Arnott, *A. inflatus* Grev., *A. mammosus* Grev., *A. Jamischii* Grove & Sturt, *A. Carruthersianus* Kitton & Grove, *A. quadrans* Sch., &c.

In some species the depression of the centre below the general surface becomes considerable, and may either be accompanied by slight inflations along the primary rays, e. g., *A. margaritaceus* Ralfs, *A. secedens* Sch., or without them, e. g., *A. Kilkellyanus* Grev. Sometimes several distinctly defined concentric zones occur in forms normally devoid of them, e. g., *A. Sollittianus*, and apiculi may also appear over the general surface, e. g., *A. Sollittianus*, or be absent from certain well-defined areas, e. g., *A. Barbadensis*, or only become visible when the lower stratum of the siliceous valve is exposed by removal of the upper, e. g., *A. Comberi*. Finally, in abnormal specimens of *A. Kittoni*, from Vera Cruz, the outer edge of the elevated central area may be concave, almost straight, or more irregular, and at the centre the elevation may be somewhat greater upon one side of the valve than on the other. By the approximation of primary rays the outer edge of the intervening elevated area is closer to the border (Pl. 281, fig. 8); and the same is observed when a single ray is replaced by two occurring on the same elevated area and surrounded at their outer ends by a single elongated medially concave curved band representing the outer edge of the siliceous process (Pl. 281, figs. 6, 7, & 8). When the primary rays and processes are confined to one half of the valve, as in specimens of *A. Kittoni* from Monterey, the outer edge of the elevated central area is concave between the primary rays, but convex outwards on that half of the valve from which the rays are absent (Pl. 281, fig. 2). Accessory inflations of small size in the interval between or coalescent with large normal inflations have been found in valves of *A. Kittoni* from Vera Cruz (Pl. 281, fig. 5).

Specimens of *A. formosus* from Iquique have been observed with four inflations, of which two, almost at right angles, were normal; the remaining two wider, with their adjacent sides more indistinct, and merging gradually into the intervening area. Other specimens of the same species found in Palillos guano, and provided with four inflations, have only one of these normal; a second subnormal, but uniting at the side distant from the preceding, with a similar imperfect inflation of equal breadth, but devoid of a primary ray. The surface concavity between the united edges of these inflations is shallow. The fourth inflation is unsymmetrical, having one side wider than the other. Valves of *A. formosus* from the same guano have sometimes the four inflations confined to one half of the valve, and at almost equal distances apart. Their edges are then less sharply defined than in the type, and there is a crescentic elevation at the centre between the inner ends of the most distant inflations. This elevation is concave towards the central space, more convex on its outer side, and highest at the middle, its ends sloping downwards towards the inflations.

In lobate valves of *A. Petersii* Ehrb., from Vera Cruz, the inflations opposite the lobes are either unaffected, as in the largest valves, or are, together with their primary ray, somewhat longer. In smaller valves the inflations opposite the lobes are longer than the others. In Colon specimens of this species the surface sometimes exhibits an irregularly convex sharp line between the adjacent angles of inflations or nearer their outer ends; this line is sometimes continued along the sides of the inflations and the outer edge of the elevated central area.

A valve of *A. excavatus* Sch., from Sysran deposit, has the sur-face sloping uniformly outwards from the process placed farthest from the margin; the compartments between this and the two remaining processes are normal, but the third is devoid of the shallow basin-like depression characteristic of the species.

3. COLOUR.—Differences of colour in mature valves depend on the thickness of the valve, those having the superficial layer absent from certain portions being lighter there than elsewhere, e. g., *A. rotulus* mihi and *A. Comberi*, on the size and proximity of the puncta, on the spaces between the markings and on the central space, e. g., *A. orientalis* Grev., and on the delicacy of the markings, e. g., *A. exiguus*, *A. crux* Ehrb. In an abnormal specimen of *A. margaritaceus* Ralfs, from Pisagua, the surface is much mottled in appearance, the intensity of the bluish to dark greyish hue varying in different places. In abnormal valves of *A. Kittoni*, as in normal specimens of *A. Rattrayii*, darker irregular radial areas are very evident. Hyaline valves of normally darker frustules are not uncommon in *A. Oregonus* Harv. & Bail. and *O. orientalis*.

4. CENTRAL SPACE.—In some species the central space is uniform in outline and dimensions, e. g., *A. Oregonus. A. orientalis*, but in others the form is variable, e. g., *A. margaritaceus*. In *A. exiguus* it normally has narrow extensions outwards between the primary rays, and wider lobate extensions have sometimes been observed in *A. formosus*. The limiting markings may be regular or irregular in the same species. In abnormal Vera Cruz specimens of *A. Kittoni* two minute irregular subcentral spaces separated by a single irregular row of markings sometimes occur, the one being at the inner ends of the widest adjacent primary rays, the other at those of the remaining rays. Another specimen of *A. Kittoni* from the same locality (Pl. 281, fig. 7) has two single rays, and two double ones united in the same crescent and occupying one elevated area. The single rays and one of the double rays meet in a subcentral rosette, another of the double rays stops short of the rosette, and the remaining rays originate in an elliptical excentric clear space.

A further modification occurs in the specimen represented in Pl. 281, fig. 8, also from Vera Cruz, where there are two equal sub-central rosettes in contact with each other and placed in line with the two opposite straight primary rays. On the other hand, the two rosettes may be inconspicuous, separated by a few irregular markings without interspaces (Pl. 281, fig. 4), or a distinct rosette and central space may be absent and replaced by an irregular excentric area crossed by irregular curved bands (Pl. 281, fig. 2, 3).

In Pisagua specimens of *A. margaritaceus*, and in others found by the 'Gazelle' expedition, two subcentral spaces have been observed. In the former these are markedly different from one another, the one elongate and placed at the inner end of six of the primary rays, the other smaller, quadrangular at the inner end of the seventh ray.

A specimen of *A. affinis* Grun., found by Macrae in Holothurians from China and now in the collection of Kitton, has the central space replaced by seven large and a few intermediate smaller hyaline or minutely punctate areas separated by the inner ends of irregularly bent flexuous strands of striæ, sometimes continuous with the space between the rows in the primary rays. An excentric space of normal form and size is sometimes found at the inner ends of the primary rays in *A. excavatus* from the Sysran deposit.

5. MARKINGS.—Variations are occasionally met with in the degree of distinctness of the individual markings arising from their greater or less elevation above the general surface. The examination of a series of individuals of *A. formosus* and *A. margaritaceus*, embracing mature and immature valves, well illustrate these variations. The arrangement of the markings sometimes shows striking irregularities. In some abnormal Vera Cruz specimens of *A. Kittoni* the rows between the approximated primary rays are almost parallel instead of radial, in others a greater irregularity in size and arrangement is found about the centre than towards the periphery. Occasionally, too, considerable areas are entirely devoid of them, e. g., *A. neglectus*, and the interspaces, which afford a character of some importance, in the discrimination of species varying in size and form, but, generally speaking, are most inconstant in the vicinity of the central space, e. g., *A. margaritaceus, A. mammosus, A. intumescens* mihi. Sometimes the markings are entirely absent from a narrow clear band, of which the inner margin is irregular, the outer regular ; and this is sometimes associated with changes in the position of the adjacent markings, as in *A. Comberi*, or is unconnected with any further modifications, as in *A. margaritaceus* and *A. affinis*.

In the abnormal Chinese valve of *A. affinis*, already referred to (p. 99), the markings become of normal form only on the outer portion of the disc. Here they form short straight or regularly curved rows, and are irregular towards the inner ends of the rows. The outer edge of the elevated area is formed by a sharp dark irregularly angular or lobate line, and the border is at some points separated from the outer ends of the rows of markings by unequal punctate areas, the largest bicrescentic at opposite ends of one diameter, the smallest lunate at opposite ends of the diameter at right angles to the former.

6. THE PRIMARY RAYS.—The primary rays are an important character in the present genus, yet their number is extremely variable, —from entire absence in abnormal forms of *A. Kittoni* to forty-five in *A. orientalis*,—and they are far from constant, even in the same species. Normally they proceed from the centre towards the circumference in straight lines. In *A. Kittoni* the rays are four or eight on the surface of each valve; they are formed of two contiguous and conspicuous rows of angular markings, and terminate in a considerable elevation provided with a siliceous cap that produces a well-defined dark crescent when the valve is next the observer. Among the Vera Cruz specimens of this common species are some with six rays, two being approximated and curved in opposite directions, the rows are irregular on the central third of their length, and only the more curved are traceable to the central space ; the markings in one row are sometimes larger than those in the other (Pl. 281, fig. 6). A second valve (Pl. 281, fig. 5) with

six rays had four of these normal; of the two remaining, that on the larger accessory inflation is distinct with the rows in contact, that on the smaller hardly differentiated. A third valve, also with six primary rays, had two of these normal, but the remaining in two unsymmetrical pairs (Pl. 281, fig. 7): a beautiful case of symmetrical arrangement was, however, observed on a fourth valve, shown on fig. 8 of Pl. 281, where two of the six rays standing opposite to each other are straight and end in contiguous rosettes, whilst the two others passing to each rosette are similarly curved in opposite directions, the crescents and inflations at the outer ends of all being of similar size and form, and quite normal.

A specimen with three primary rays (Pl. 281, fig. 4) showed two of these straight at right angles, and ending in an inconspicuous rosette; the third curved away from that adjacent to it, and ending in a second similar rosette. Valves with two rays were observed: the rays being straight and at right angles, but proceeding from an excentric point (Pl. 281, fig. 3). Only one valve has been seen with a single ray, which also started from an excentric point (Pl. 281, fig. 2). Of the two valves seen with no rays, the one showed the regularly quadrangular outer edge of the elevated central portion indicating the position of the absent rays (Pl. 281, fig. 1); on the border of the other was an elongated elevation with convex surface, and irregularly bent linear markings.

Among the specimens procured by Kitton from Monterey, the gaps in this curious series from Vera Cruz are filled up. Thus one valve with five primary rays had three on one half of the valve, meeting at a central rosette; two on the other half were unsymmetrical, and met at an obtuse angle in a second excentric rosette. Another valve with four rays confined to one of its halves showed three meeting in a common rosette, and the central portion of the fourth curving to an irregular excentric clear space; and finally, a valve with three rays confined to one half, but in other respects normal.

In *A. margaritaceus* somewhat similar abnormalities have been found. Thus a Pisagua valve with seven rays had four of these in pairs towards opposite sides of the valve, whilst a 'Gazelle' specimen showed two contiguous rays separated by but a single row of markings. It has also been found that a Yokohama valve with seven rays had two in one half of the valve contiguous, and formed by three almost parallel rows.

On displaced inflations of *A. formosus* from Iquique the primary ray is sometimes sigmoid, with the central flexure sharper than the peripheral; and in a specimen from Palillos guano one of the primary rays, though connected with a process, is without a subjacent inflation—one of the inflations adjacent being without its ray and process. Finally, in a valve of *A. Rattrayii* from the Oamaru deposit with four primary rays, two are almost at right angles, a third is at a greater, and the fourth at a less acute angle; a fifth fissure-like ray similar to the other four is present, but has no process or clear space at its outer end.

7. PROCESSES.—These are absent in *A. apedicellatus*, *A. suspectus*,

and abnormal forms of *A. Kittoni*. The chief abnormality arises from variations in their distance from the circumference. In an abnormal Chinese specimen of *A. affinis* with a diameter of ·14 mm., the unsymmetrical processes are inserted at distances varying from one-fourth to one-ninth of the radius from the circumference.

It may further be pointed out that in the flat forms of *A. sus-pectus*, *A. Beeveriæ*, *A. Comberi*, and other species, in the absence of primary rays in specimens of *A. Kittoni*, and of processes in this as well as in *A. suspectus* and *A. apedicellatus*, this genus approaches *Coscinodiscus*. The affinity of the two is still further shown by the occurrence of, *a*, a central rosette in *Coscinodiscus symmetricus* Grev., *C. asteromphalus* Ehrb., and *C. omphalanthus* Ehrb.; *b*, a central space in *C. gigas* Ehrb. and *C. diorama* Sch.; *c*, compartments on the valves resulting from fasciculate arrangement of the markings in *C. denarius* Sch. and *C. curvatulus* Grun.; *d*, distinct but single radial rows of markings representing primary rays in *C. denarius* Sch. and *C. extravagans* Sch.; *e*, marginal processes in *C. extrava-gans* Sch. and *C. tuberculatus* Grev.; *f*, clear spaces near the border, corresponding to those at the base of the processes in *Aulacodiscus*, in *C. armatus* Grev., var. Sch.; *g*, occasional interspaces between the outer ends of the radial rows of markings and the border in *C. biradiatus* Grev.; *h*, irregularities of surface in *C. excavatus* Grev. The markings are often similar in the two genera.

EXPLANATION OF PLATE 281.—Figs. 1—8, Abnormal valves of *A. Kittoni* Arnott, showing variations in the arrangement of the primary rays, and central spaces or central rosettes. Fig. 9, Normal valve viewed in its girdle aspect.

## THE LATE JOHN SMITH, A.L.S.

I SHOULD like to see a biography of John Smith, written in detail from the same point of view as that of his fellow Scotchmen, Robert Dick and Thomas Edward. He was born at Aberdour, in Fifeshire, where his father was a gentleman's gardener, on the 5th of October, 1798. The story of his early life and of his struggles to educate himself in the rudiments of Botany will be found, told by himself and accompanied by an excellent portrait, in Gard. Chron. n. s. vol. v. p. 363 (1876). His whole school education did not cost more than five pounds. When he was employed as a journeyman at the Edinburgh Botanic Garden, in 1818, four of them lived together in a one-roomed bothy, and his wages were nine shillings a week, out of which he saved money to buy his drying-paper and a copy of Sir J. E. Smith's 'Compendium.' In 1820 he came to London, and on the recommendation of the younger Aiton was appointed to a place in the Royal Garden at Kensington.

In 1822 he was removed to the propagating-pits at Kew. The Botanic Garden then was the private property of the Crown, and consisted of nine acres, enclosed by a wall, the wages of the young gardeners being twelve shillings a week. At the age of twenty-five he was promoted to be foreman of the hothouses and propagating department, and soon began to take a special interest in ferns. At that time there were about forty hardy and as many tender exotic ferns in the Kew collection. Between 1823 and 1840 Kew was at its lowest ebb, and when, at the death of William IV., Lindley, Bentham, and Paxton were appointed a commission to investigate its condition, they reported that whatever names were attached to the plants " have been furnished by Mr. Smith, the foreman, and that the Director does not hold himself answerable for them."

With 1841 came the transfer of the gardens to the Commissioners of Woods and Forests. Sir W. J. Hooker was appointed Director, and John Smith was continued as Curator, with results which I need not recapitulate here. In 1846 the collection of ferns had increased from 80 to 400 species, in 1857 to 600, and in 1866, when Sir Wm. Hooker died and John Smith resigned, to about 1000 species and well-marked varieties. A considerable number of the additions were raised from spores taken from dried specimens. During the Aitonian period Smith had contributed two papers to the ' Transactions of the Linnean Society,' one on Ergot in 1838, and his well-known paper on *Cœlobogyne* in 1839. In 1841 he contributed to Hooker's Journal an enumeration of the magnificent collection of ferns made by Cuming in the Philippine Islands. His scheme for a new classification of ferns was laid before the Linnean Society in 1841, and published in Hooker's Journal in 1841–2. His primary divisions, *Desmobrya* and *Eremobrya*, were original, but in his idea of founding genera on venation he was anticipated by Presl, whose ' Tentamen Pteridographia' appeared in 1836. His ideas on fern-classification were further explained in his contributions to Hooker & Bauer's ' Genera Filicum' in 1842, and in his enumeration of the ferns gathered by Seemann during the exploring expedition of the ' Herald,' published in 1856. In 1861 his sight began to fail, and in 1863 he retired upon a pension, having been in the service of the Garden forty-four years. His collection of dried ferns, consisting of 2000 species on 6000 large folio sheets, was purchased in 1866 for the British Museum. His wife died in 1838, and he lost his six children one after the other by consumption, the last in 1871. His son Alexander held posts in the Museum, and afterwards in the Herbarium at Kew.

In spite of his blindness, Mr. Smith still continued to take a keen interest in botanical and horticultural matters, and his memory and energy were wonderful up to the very last. He lived in lodgings at Kew, and had a young lady secretary, whom he kept employed for about six hours a day, reading to him and writing for him. The principal books which he produced under these circumstances are his 'Ferns, British and Foreign,' 1866 ; ' Domestic Botany,' 1871; ' Historia Filicum.' 1875 ; ' Bible Plants,' 1878 ; Records of the Botanic Garden, Kew,' 1880 ; and ' Dictionary of Economic Plants,' 1882. He died suddenly on the 12th of February, 1888, and was buried in the churchyard on Kew Green, beside his wife and children. His funeral was attended by Sir J. D. Hooker and nearly the whole of the present staff of the Kew establishment.

J. G. BAKER.

# A SYNOPSIS OF *TILLANDSIEÆ*.

## By J. G. BAKER, F.R.S., F.L.S.

(Continued from p. 82.)

177. TILLANDSIA WARMINGII Baker   *Vriesea Warmingii* E. Morren in Belg. Hort. 1884, 260, t. 12–13.—Leaves about 30, ensiform from an ovate base, 3 ft. long, 2 in. broad at the middle, firm in texture, subglabrous, tinged with red beneath, sometimes maculate with anastomosing transverse green lines. Peduncle about 2 ft. long; bract-leaves many, small, adpressed, imbricated. Spike simple, erect, 2 ft. long, dense upwards, lax in the lower half; flowers 30 or more; flower-bracts broad, ovate, greenish yellow, 2 in. long Calyx reaching to the tip of the bract; sepals oblong, obtuse, bright yellow. Petal-limb small, obovate, falcate, bright yellow. Stamens a little longer than the petals.

Hab.   South Brazil, sent alive by Dr. Glaziou to Prof. Morren in 1880.

178. T. AMETHYSTINA Baker.   *Vriesea amethystina* E. Morren in Belg. Hort. 1884, 330, t. 15–16. — Leaves 20–30 in a rosette, lanceolate from an ovate base, 1–1½ ft. long, ¾–1 in. broad at the middle, thin, flexible, subglabrous, narrowed gradually to the point, plain bright green on the face, claret-purple all over beneath. Peduncle as long as the leaves; bract-leaves many, small, lanceolate, adpressed, not imbricated. Inflorescence a simple lax erect spike 6–8 in. long; flowers 8–10, all distant, erecto-patent; flower-bracts ovate, acute, much shorter than the calyx. Calyx bright yellow, 1¼ in. long; sepals oblong. Petals twice as long as the calyx; limb lingulate, bright yellow. Stamens a little longer than the petals.

Hab.   South Brazil. Sent alive by Dr. Glaziou to Prof. Morren in 1881.

179. T. PLATZMANNI Baker.   *Vriesea Platzmanni* E. Morren in Belg. Hort. 1875, 349, t. 23. — Leaves about a dozen in a rosette, lorate, a foot long, marbled with red-brown, obtuse with a cusp. Peduncle erect, more than twice as long as the leaves; bract-leaves all small, scariose and adpressed. Spike simple, secund, laxly 10–12-flowered; flower-bracts ovate, brownish, an inch long. Calyx yellowish, above an inch long. Corolla yellow, half as long again as the calyx. Stamens shorter than the petals.

Hab.   South Brazil; island on the coast of Parana, *Platzmann*. Introduced into cultivation about 1875.

180. **T. Selloana**, n. sp. — Leaves lorate from a dilated oblong base 3 in. long, 2 in. broad, flexible, subglabrous, 1½ ft. long, 1¼ in. broad at the middle, deltoid-cuspidate at the apex. Peduncle 1½ ft. long; lower bract-leaves with a produced lanceolate blade; upper scariose, entirely adpressed. Inflorescence a simple lax erect distichous spike; lower flowers erecto-patent; flower-bracts oblong, 1½ in. long, an inch round. Calyx just reaching the tip of the bract; sepals very convolute. Petal-blade ¼ in. long.

Hab.   South Brazil, *Sello* 192! (Herb. Mus. Brit.).

181. T. Wawranea Baker. *Vriesea Wawranea* Antoine, Brom. 1, t. 1–2.—Leaves about 30 in a rosette, lorate, 1½–2 ft. long, 2½–3 in. broad, obtuse, cuspidate. Peduncle shorter than the leaves; bract-leaves small, adpressed, imbricated. Inflorescence a lax simple spike of 8–10 erecto-patent flowers; flower-bracts ovate, acute, 1¼ in. long, dark green. Calyx yellowish green, 2 in. long; sepals oblong, acute. Petal-blade oblong, spathulate, yellowish, ¾–1 in. long. Stamens shorter than the petals.

Hab. Locality unknown, probably Brazil. Described by Antoine in 1884 from living plants in the Imperial Garden at Vienna.

182. **T. orizabensis**, n. sp. — Leaves 20 or more in a rosette, lanceolate from a dilated ovate base 2 in. broad, flexible, sub-glabrous, a foot long, 1½ in. broad at the middle, narrowed gradually to the point. Peduncle longer than the leaves; lower bract-leaves with large lanceolate free points. Inflorescence a lax simple dis-tichons spike 6–8 in. long, 2½–3 in. broad; flowers 12–20, erecto-patent; flower-bracts oblong, obtuse, 1–1½ in. long. Calyx ½ in. longer than the bract; sepals obtuse. Corolla not seen. Capsule-valves 2 in. long, ½ in. broad.

Hab. Central Mexico, in the Province of Orizaba, *Bourgeau* 3055! *Hahn!* Differs from *T. ensiformis* by bracts shorter than calyx.

183. T. haplostachya Sauvalle, Fl. Cub. 169.—Leaves lanceo-late from a dilated ovate base. Peduncle twice as long as the leaves; bract-leaves adpressed. Inflorescence a lax simple spike, with a flexuose rachis and flowers twice as long as the space between them; flower-bracts broad-ovate, subobtuse, longer than the calyx. Sepals oblong. Petals greenish, twice as long as the calyx; blade spathulate. Stamens shorter than the petals.

Hab. Eastern Cuba, near Monte Verde, *Wright*.

184. **T. gradata**, n. sp. — Leaves lorate from an ovate base 3–4 in. long, 3 in. broad, flexible, subglabrous, 1½ ft. long, 1¼–1½ in. broad at the middle, deltoid-cuspidate at the apex. Peduncle much shorter than the leaves; bract-leaves imbricated, adpressed. In-florescence a simple erect spike 8–9 in. long, with 20–25 flowers all erecto-patent, contiguous in its upper half, a little spaced out in its lower half; bracts bright red, ovate, the lower 1½ in. long, above an inch broad, the upper suborbicular, about an inch long and broad. Calyx ½ in. longer than the bract; sepals acute. Petals ½ in. longer than the calyx; blade oblong. Stamens a little longer than the petals.

Hab. South Brazil, *Glaziou* 15473! Near *T. ensiformis* Vellozo.

185. **T. unilateralis**, n. sp.—Leaves above a dozen in a rosette, lorate from an oblong base 3 in. long, 1½–2 in. broad, thin, flexible, subglabrous, above a foot long, an inch broad at the middle, plain pale green on both surfaces, rounded at the apex to a minute cusp. Peduncle rather longer than the leaves; bract-leaves imbricated, adpressed. Inflorescence a simple unilateral spike 6–8 in. long; flowers 9–12, erecto-patent; flower-bracts ovate, an inch long. Calyx an inch long, with a short stout pedicel. Petals not seen. Valves of the capsule 1½ in. long, ½ in. broad.

Hab.  South Brazil; Woods near Santos, *Burchell* 3347 !  Near
*T. Platzmanni.*

186. **T. heterostachys,** n. sp. — Leaves lorate from an oblong
dilated base 3-4 in. long, 2½ in. broad, flexible, subglabrous, a foot
long, above an inch broad at the middle, deltoid-cuspidate at the
apex.  Peduncle as long as the leaves ; bract-leaves imbricated,
adpressed.  Spike simple, erect, 6-8 in. long ; flowers of upper
half erect, crowded ; of lower half spaced out, erecto-patent ; flower-
bracts oblong, obtuse, 1½-1¾ in. long, an inch round.  Calyx
reaching to the tip of the bract.  Petals not seen.
Hab.  South Brazil, *Glaziou* 13260 !  Near *T. ensiformis* and
*Warmingii.*

187. **T. ENSIFORMIS** Vell. Fl. Flum. iii. t. 129.  *Vriesea ensiformis*
Beer, Brom. 92.  *V. conferta* Gaudich, Atlas Bonité t. 65 ; Wawra
in Oester. Bot. Zeit. 1880, 184 ; Itin. Prin. Cob. 162 ; Antoine,
Brom. 3, t. 3. — Leaves lanceolate from a dilated ovate base 3 in.
long, 2 in. broad, flexible, subglabrous, 1½-2 ft. long, 1½ in. broad
at the middle, narrowed gradually to the point.  Peduncle stout,
erect, 2 ft. long ; bract-leaves scariose, imbricated, adpressed.
Inflorescence a simple erect distichous spike 1-1½ ft. long, 3 in.
broad ; flowers 20-30, all erecto-patent ; flower-bracts oblong,
acute, dark red, 1½-2 in. long, 1 in. round.  Calyx reaching to the
tip of the bract ; sepals acute.  Petal-blade oblong, ½ in. long,
bright yellow.  Capsule 2 in. long.
Hab.  Southern provinces of Brazil, *Gaudichaud, Glaziou* 13263 !
Entre Rios, *Wawra & Maly* 126.  *T. imbricata* Vell. Fl. Flum. iii.
t. 131 (*Vriesea imbricata* Beer, Brom. 94), known only from a rough
figure, appears to be closely allied to this species, but to differ by
its larger flower-bracts, more gradually narrowed to an acute point.

188. **T. RECURVATA** Baker.  *Vriesea recurvata* Gaudich. Atlas
Bonité, t. 69.— Leaves lorate from an ovate base 2 in. broad, thin,
flexible, 2 ft. long, an inch broad at the middle, deltoid-cuspidate
at the tip.  Peduncle short ; bract-leaves adpressed, imbricated.
Inflorescence a simple 10-12-flowered spike, lax in the lower half ;
bract-leaves oblong-navicular, with a recurved cusp, 2 in. long, an
inch round low down.  Calyx reaching the tip of the bract ; sepals
acute.  Petals unknown.
Hab.  South Brazil, *Gaudichaud.*

189. **T. PLATYNEMA** Griseb. in Gott. Nacht. 1864, 19.  *Vriesea
platynema* Gaudich. Atlas Bonité, t. 66.  *V. bituminosa* Wawra in
Oester. Bot. Zeit. xii. 372 ; xxx. 221 ; French transl. 71 ; Reise,
Kais. Max. 157, t. 86 ; Itin. Prin. Cob. 168, t. 38A. — Leaves
numerous, lorate, 2-2½ ft. long, 3-3½ in. broad at the middle, thin,
flexible, subglabrous, tinged with red-brown beneath, rounded to a
cusp at the apex.  Peduncle stout, stiffly erect, 1½-2 ft. long, an
inch thick at the base ; bract-leaves small, adpressed, imbricated.
Inflorescence a simple distichous spike a foot long, 3 in. broad ;
flowers 20-30, spreading ; flower-bracts ovate, 1½-2 in. long.
Calyx ½-¾ in. longer than the bract, very glutinose ; sepals obtuse.
Petal-blade small.  Valves of the capsule 1½ in. long, ½ in. broad.
Hab.  Forests of South Brazil, *Burchell* 2321 ! (Feb. 1826),

*Gaudichaud, Wawra & Maly*, ii. 25. Jamaica, near Manchester, *Purdie*! Eastern Cuba, near Monte Verde, *Wright* 1523! Venezuela; mountains of Tovar, alt. 6000 ft., *Fendler* 2540, teste *Grisebach*.

190. T. FENESTRALIS Hook. fil., in Bot. Mag. t. 6898. *Vriesea fenestralis*, Linden and André, Ill. Hort., n. s. t. 215 ; E. Morren in Belg. Hort. 1884, 65, t. 4–5.—Leaves 20–30, lorate from an ovate base, 1½ ft. long, 3 in. broad at the middle, thin, flexible, subglabrous, copiously decorated with vertical and horizontal green veins on a pale groundwork ; base reddish brown ; back copiously maculate with reddish brown ; apex deltoid-cuspidate, Peduncle a foot long ; bract-leaves small, ovate, imbricated, entirely scariose. Spike simple, distichous, a foot long, 4 in. broad ; flowers 20–30, many lower spreading ; flower-bracts ovate, green, above an inch long. Calyx glossy green, half an inch longer than the bract ; sepals oblong, obtuse. Petal-limb yellowish white, lingulate, ½ in. long. Stamens reaching nearly to the tip of the petals.

Hab. Parana, introduced into cultivation by Linden in 1878. Description made from a specimen that flowered at Kew in June, 1886. Glaziou's Nos. 14342 and 15466 are either this or very closely allied, but I have only seen them in the dried state.

191. T. JONGHEI K. Koch in Wochen, 1868, 91 ; E. Morren in Belg. Hort. 1874, 291, t. 12–13. *Encholirion Jonghei*, Libon ; R. Koch in Berl. Allgem., Gart. 1857, 22. *Vriesea Jonghei* E. Morren in Belg. Hort. 1878, 257 ; Antoine Brom. 24, t. 16.—Leaves 30–50 in a rosette, lorate, bright green, much recurved, glabrous, 1½–2 in. long, 1½ in. broad at the middle, 3 in. at the dilated base, which is tinged with violet on the back. Peduncle about a foot long ; bract-leaves green, much imbricated ; lower with large lanceolate free points. Inflorescence a simple distichous spike half a foot long, 2½–3 in. broad ; flowers 12–20, spreading ; flower-bracts ovate, 1–1½ in. long, green with a dark purple margin. Calyx equalling or slightly exceeding the bract, greenish. Petal-blade yellow, tinged with red-brown, very short, orbicular. Stamens shorter than the petals.

Hab. Minas Geraes, *Libon*. Sent alive to De Jonghe in 1856. I have not seen Platzmann's Parana plant, which has been referred to the species, or Wawra and Maly's doubtful plant from Teresopolis, described Itin. Prin. Cob. 167.

192. T. CORALLINA K. Koch. App. Ind. Sem. Berol. 1873, 5. *Encholirion corallinum*, Linden Cat. 1865, 27 ; Ill. Hort. n. s., t. 70 ; Floral Mag., n.s., t. 116. *Vriesea corallina*, Regel Gartenfl. 1870 ; 354, t. 671 ; Antoine Brom. 26, t. 17.—Leaves 20–30 in a rosette, lorate from a dilated ovate base 3–4 in. broad, thin, subglabrous, faintly mottled with green, tinged with purple on the back, above a foot long, 2 in. broad at the middle, deltoid-cuspidate at the apex. Peduncle as long as the leaves ; bract-leaves imbricated, adpressed, red-brown. Inflorescence a simple distichous spike 6–9 in. long, 3–3½ in. broad ; flowers 20–30, spreading ; flower-bracts ovate, dark red, 1–1¼ in. long. Calyx pale red or yellow, very glutinose, ½ in. longer than the bract ; sepals obtuse. Petals yellow, scarcely longer than the calyx. Stamens as long as the petals; anthers loosely cohering.

Hab. Minas Geraes, *Libon*. Introduced into cultivation by Linden in 1875. *E. roseum* Hort. Linden ; Morren Cat. 1873, 8, is a variety with green lower leaf-bracts, leaves glaucous green beneath, shorter flower-bracts and a yellow calyx.

193. **T. amazonica**, n. sp.—Leaves lorate from an ovate base 3-4 in. diam., flexible, subglabrous, 3 in. broad at the middle, plain pale green on both sides, deltoid at the apex, with a very large cusp. Peduncle stout, stiffly erect, twice as long as the leaves ; lower bract-leaves with large lanceolate erect free points. Inflorescence a moderately dense erect unilateral spike above a foot long ; flowers patent or rather deflexed ; flower-bracts broad ovate, an inch long. Calyx an inch long ; sepals much imbricated, oblong, obtuse, half an inch broad. Petals not seen. Capsule-valves 1½ in. long, ⅓ in. broad.

Hab. Forests of the Amazon Valley, near Para. *Burchell*, 9440. Allied to *T. Platzmanni* and *Jonghei*.

194. T. GUTTATA Baker. *Vriesea guttata* André and Linden in Ill. Hort. n. s. t. 200 ; E. Morren in Belg. Hort. 1880, 1, t. 1–3.— Leaves lorate from an ovate base 2 in. diam., a foot long, 1-1½ in. broad at the middle, firm in texture, suberect, little arcuate, copiously decorated with irregular transverse bands of large claret-purple spots, deltoid-cuspidate at the apex. Peduncle much longer than the leaves ; bract-leaves small and adpressed. Spike simple sublax, drooping, above a foot long, 3 in. diam. ; flowers all erecto-patent ; flower-bracts ovate, acute, reddish white, 1½ in. long, above an inch broad low down. Calyx yellow, a little larger than the bract. Petal-blade lingulate, yellow, under an inch long. Stamens longer than the petals.

Hab. South Brazil *Glaziou*, 15474 ! Introduced into cultivation in 1870 by seed sent from the province Sta. Catherina by M. Gautier. Flowered at the Luxembourg and by Dr. Le Bele at Mans in 1878.

195. T. SCALARIS Baker. *Vriesea scalaris* E. Morren in Belg. Hort. 1880, 309, t. 15.—Leaves about 16 in a rosette, lorate from an ovate dilated base 2 in. broad, thin, flexible, subglabrous, 1-1¼ ft. long, ¾-1 in. broad at the middle. Peduncle short, slender, cernuous ; bract-leaves small, adpressed, imbricated. Inflorescence a lax simple pendulous spike a foot long, with a very flexuose rachis and 10-12 spreading flowers ; flower-bracts oblong, acute, bright red, 1½ in. long. Calyx bright yellow, ½-¾ in. longer than the bract ; sepals obtuse. Petals greenish-yellow, ¼ in. longer than the calyx. Stamens rather longer than the petals.

Hab. South Brazil ; forests of St. Paulo. *Burchell* 3197 ! (year 1826). Introduced into cultivation by Binot in 1877.

196. T. PSITTACINO × CARINATA. *Vriesea Morreniana* Hort. ; E. Morren in Belg. Hort. 1882, 287, t. 10–12, fig. 2. Recedes from typical *psittacina* by its more numerous closer flowers, 12–15 in a spike, 4-6 in. long, with the red bracts and yellow calyx of its parents.

197. T. PSITTACINO × SCALARIS. *Vriesea retroflexa* E. Morren in Belg. Hort. 1884, 185, t. 10.—Leaves short, lorate, plain green,

thin. Raceme cernous, with a red rachis and 6–8 flowers, the lower patent; flower-bracts oblong, acute, bright red, 1½ in. long. Calyx yellow, rather exceeding the bract. Petal-limb greenish yellow, ½ in. long. Stamens longer than the petals. A cross made by Professor Morren in 1879, which flowered in 1884.

198. T. RINGENS, Griseb. Pl. Cub. 255.—Leaf lanceolate from an ovate base, 3 in. long, 2 in. broad, flexible, subglabrous, narrowed gradually to the point, 1½ ft. long, 1¼–1½ in. broad at the middle. Peduncle 1½ ft. long : lower bract-leaves with large lanceolate free points. Inflorescence a forked lax spike ; flowers few, ascending ; flower-bracts ovate, acute, 1½–2 in. long. Calyx ½–¾ in. shorter than the bract ; sepals acute. Petals recurved, twice as long as the calyx ; blade oblanceolate-unguiculate. Stamens and style reaching to the tip of the petals.

Hab. Eastern Cuba, near Monte Verde, *Wright*, 1518.

199. **T. Chagresiana,** n. sp.—Leaves lorate from a slightly dilated base, thin, flexible glabrous, 1½–2 ft. long, above 2 in. broad at the middle, deltoid at the apex. Peduncle as long as the leaves ; bract-leaves adpressed, much imbricated. Panicle a foot long, consisting of about four laxly few-flowered arcuate ascending branches with a rather flexuose rachis ; flowers 3–4 to a branch, erect ; flower-bracts ovate, acute, 1¼–1½ in. long, 1 in. broad at the base. Calyx an inch long, falling a little short of the bract ; sepals acute. Petals not seen. Capsule as long as the calyx.

Hab. Chagres, isthmus of Panama, *Fendler*, 448 !

200. T. STENOSTACHYA, Baker. *T. glutinosa* Griséb. Fl. Brit., West Ind., 597, non Mart.—Leaves lorate, thin, flexible, subglabrous, above 2 ft. long, 2 in. broad at the middle, deltoid-cuspidate at the apex. Inflorescence a forked spike, the ascending branches of which are 1–1½ ft. long ; flowers close, very ascending ; flower-bracts oblong-navicular, 2 in. long, ¾–1 in. round. Calyx ½ in. shorter than the bract. Petal-blade oblanceolate-oblong, ¾ in. long, protruding half an inch beyond the tip of the bract.

Hab. Trinidad ; Maraccas Waterfall, *Dr. Crueger*, year 1845 !

201. T. DISSITIFLORA Sauvalle Fl. Cub. 168. *T. excelsa* Griseb. Cat. Cub. 254, non Fl. Brit. West Ind. 597.—Leaves lanceolate, flexible, subglabrous, 3 ft. long, 5 in. broad low down, narrowed gradually to an acute point. Panicle ample ; lower branches peduncled, above a foot long ; flowers ascending, contiguous ; flower-bracts broad ovate, almost orbicular, an inch long. Calyx an inch long. Petal-limb oblong, ½ in. long. Stamens not exserted.

Hab. Cuba, *Wright* 3276 !

202. T. DEPPEANA Steud. ; Mart. et Gal. Enum. 8 ; Schlecht. in Linnæa xviii. 424. *T. paniculata* Cham. et Schlecht in Linnæa vi. 54, non Linn. *T. excelsa* var. *latifolia* Griseb. in Gott. Nachtrage 1864, 17. *T. incurvata* Sauv. Fl. Cub. 169.—Leaves lanceolate, thin, flexible, subglabrous, 2–3 ft. long, 4–5 in. broad low down, narrowed gradually to an acute point. Panicle ample ; lower branches a foot long, peduncled ; flowers all contiguous, erecto-patent ; flower-bracts oblong, obtuse, 1¾–2 in. long, an inch broad.

Calyx equalling the bract.   Petals not seen.   Capsule-valves
1½-2 in. long, ⅓-½ in. broad.

Hab. Central Mexico at Xalapa, &c.   *Schiede* and *Deppe*! (type
specimen at British Museum).   *Galeotti* 4915.   Venezuela, mountains of Tovar, *Fendler* 1516!   Eastern Cuba, near Monte Verde,
alt. 6000 ft., *Wright* 1522!

203. T. GIGANTEA Mart.; Roem. et Schultes Syst. vi. 1224.—
Leaves lorate, flexible, subglabrous, 1½ ft. long, 3 in. broad, deltoid-
cuspidate at the apex.   Peduncle with panicle 3 ft. long, the latter
a foot broad; branches spreading, 8–10 in. long, 7–8 flowered;
flowers secund, subcernuous; flower-bracts ovate, ¾-1 in. long.
calyx under an inch long; sepals obtuse.   Valves of the capsule
1¼ in. long.

Hab. Forests of the Rio Negro, *Martius*.

204. T. GLUTINOSA Mart; Roem. et Schultes, Syst. vii. 1225.—
Leaves lanceolate from a dilated ovate base, 4–5 in. long, 3 in.
broad, thin, flexible subglabrous, 1½-2 ft. long, 1½-2 in. broad at
the middle, narrowed gradually to an acute point.   Panicle 1-1½ ft.
long; branches numerous, erecto-patent, the lower ½-1 ft. long;
flowers erecto-patent, not contiguous; flower-bracts oblong, acute,
1¼ in. long, ⅜-¾ in. broad.   Calyx equalling the bract.   Petal-
blade narrow, ½ in. long.   Stamens a little longer than the petals.

Hab. South Brazil; forests of Rio Janeiro and St. Paulo,
*Martius, Burchell* 4367!   *Glaziou* 12225! 14338! 16469! 16471!

205. T. ITATIAIÆ Baker.   *Vriesea Itatiaiæ* Wawra in Œster.
Bot. Zeitsch. xxx. 221; French trans. 70; Itin. Prin. Sax. Cob.
169, t. 31 and 34 C.—Leaves 30–40 in a dense rosette, thin,
flexible, glabrous, plain green, 1¼ ft. long, 4 in. broad, rounded to
a brown cusp at the apex.   Peduncle a foot long, reddish; bract-
leaves small, scariose, adpressed.   Panicle 1-1½ ft. long; branches
few, short, erecto-patent; flowers 8–10 to a branch, moderately
close, secund; flower-bracts ovate, brownish, an inch long.   Calyx
about as long as the bract.   Petal-blade small, obovate, greenish
white.   Stamens shorter than the petals.

Hab. Central Brazil; plateau of the Serra Itatiaia, alt. 9000 ft.,
*Wawra* and *Maly*.

206. T. HIEROGLYPHICA Hort. Bull.   *Vriesea hieroglyphica* E.
Morren in Ill. Hort. 1884, t. 514; Belg. Hort. 1885, 57, t. 10–12.
*Massangea hieroglyphica* Carriere in Rev. Hort. 1878, 175, with
figure.—Leaves 30–40 in a dense rosette, 4–5 ft. diam., lorate from
a dilated base, 4–5 in. broad, thin, flexible, subglabrous, 2½-3 ft.
long, 2½-3 in. broad at the middle, rounded at the apex to a small
cusp, bright green, with conspicuous cross-bands of black blotches
up to the tip.   Peduncle much shorter than the leaves; lower
bract-leaves with large lanceolate free points.   Inflorescence an
ample panicle with 12–20 erecto-patent branches; branch-bracts
ovate acuminate; flowers 12–20 to the lower branches, spaced out,
erecto-patent; flower-bracts ovate, green, 1-1¼ in. long.   Calyx an
inch long, reaching to the tip of the bract.   Corolla yellowish, a
little longer than the calyx.   Stamens shorter than the petals.

Hab. South Brazil; forests of Rio Janeiro and St. Paulo,

*Glaziou* 11684! 11694! 13261! Introduced into cultivation in 1878. First flowered by M. Lubbers at the National Botanic Garden at Brussels in 1885. The conspicuous black hieroglyphic-like marks quite disappear in drying.

(To be continued.)

## BIOGRAPHICAL INDEX OF BRITISH AND IRISH BOTANISTS.

By James Britten, F.L.S., and G. S. Boulger, F.L.S.

(Continued from p. 89).

**Berkenhout, John** (1730–1791): b. Leeds, Yorkshire, 1730; d. Besselsleigh, Oxford, 3rd April, 1791. M.D., Leyden, 1765. 'Clavis Anglica Linguæ Botanicæ,' 1762. 'Outlines of Nat. Hist. of Great Britain,' 1770–1. Pritz. 24 ; Jacks. 521 ; 'European Magazine,' 1788, 156; Gent. Mag. lxi. 388, 485 ; Hutchinson, Biog. ; Dict. Nat. Biog. iv. 369.

**Berry, Andrew** (fl. 1819): M.D. Of Madras. Contributed to Calcutta Garden. Friend of Roxburgh. Roxburgh, 'Coromandel Plants,' 1819, iii. 60 ; R. S. C. i. 307. *Berria* Roxb. = *Berrya* DC., 1824.

**Bicheno, James Ebenezer** (1785–1851): b. Newbury, Berks, 1785; d. Hobart Town, Tasmania, 25th February, 1851. F.L.S,, 1812. Sec. L.S., 1824–1832. Colon. Sec., Van Diemen's Land, 1842. Pritz. 27; Jacks. 17; Proc. Linn. Soc. ii. 181; R. S. C. i. 358; Dict. Nat. Biog. v. 1; Gent. Mag. xxxvi. n. s. ; 'Annual Register,' 1851. Oil-portr. by Eddis at Linn. Soc. *Bichenia* D. Don.

**Biddulph, Susanna** (fl. 1807). E. B. 1762. *Biddulphia* Gray = *Conferva Biddulphiana* Sm.

**Bidwill, John Carne** (1815–1853): b. Exeter, Devon, 1815; d. Tinana, Maryborough Marsh, Wide Bay, New South Wales, 1853. 'Rambles in New Zealand,' 1841. Pritz. 27; Journ. Bot. 1853, 252; R. S. C. i. 360; Gard. Chron. 1853, 438; 1856, 20; Ann. & Mag. viii. 1842, 438; Gent. Mag. 1853; Dict. Nat. Biog. v. 18. *Araucaria Bidwilli.*

**Bigsby, John Jeremiah** (1793–1881): b. Nottingham, 14th August, 1793; d. Gloucester Place, London, 10th February, 1881. M.D. Edin., 1814. F.L.S., 1823. F.G.S., 1823. F.R.S., 1869. 'Flora and Fauna of the Silurian,' 1868. 'Flora and Fauna of the Devonian and Carboniferous,' 1878. Jacks. 522; Proc. Geol. Soc. 1880–81, 39 ; Dict. Nat. Biog. v. 27; Journ. Bot. 1881, 96.

**Bingley, William** (1774–1823): b. Doncaster, Yorkshire, 1774; d. Charlotte Street, Fitzroy Square, London, 11th March, 1823; bur. Bloomsbury Church. Clerk. B.A., Camb., 1799. M.A., 1803. F.L.S., 1800. 'Flora of the Snowdonian Mountains,' 1798–1801, in Appendix to Jones' 'Illustrations of the scenery,' 1829. 'Practical Introduction to Botany,' 1817. Pritz. 27;

Jacks. 522 ; Mag. Zool. Bot. ii. 170 ; Gent. Mag. 1823 ; Diet.· Nat. Biog. v. 55.

**Binney, Edward William** (1812–1881) : b. Morton,¦Notts., 1812 ; d. Manchester, 19th December, 1881.  F.G.S., 1853.  F.R.S., 1856.  ' Sigillaria,' 1858, &c.  Jacks. 181 ; Dict. Nat. Biog. v. 56.

**Bird, Frederick John** (d. 1874) : d. 28th April, 1874.  A.L.S., 1840.  F.L.S., 1862.  M.D.  Brother of Golding Bird.  ' The artificial arrangement of British Plants.'  Mag. Nat. Hist. ii. 1838, 604.

**Bird, Golding** (1814–1854) : b. Downham, Norfolk, 9th December, 1814 ; d. Tunbridge Wells, Kent, 27th October, 1854.  M.D. St. Andrews, 1838.  M.A., 1840.  F.R.C.P., 1845.  F.L.S., 1836.  F.R.S.  Dict. Nat. Biog. v. 74 ; Mag. Nat. Hist. ·ii. 1838.

**Bird, Richard** (fl. 1862).  Of Tame Valley, Dukinfield, Cheshire. Working-man.  Journ. Hort. iii. 767.

**Bishop, David** (1788–1849) : b. Scone, Perth ? c. 1788 ; d. Malone, Belfast, 4th August, 1849.  Curator, Belfast Botanical Garden, circ. 1830.  ' Causal Botany,' 1829.  Pritz. 28 ; Jacks. 68 ; Cott. Gard. ii. 306.

**Black, Allan A.** (1832–1865) : b. Forres, Morayshire, 1832 ; d. in Bay of Bengal, 4th December, 1865 ; buried on Table Island, in Cocos group.  A.L.S., 1858.  Apprenticed at Dunkeld.  Kew gardener.  Curator of Kew Herbaria.  Superintendent of Bot. Garden, Bangalore.  Contributed list of Japan plants to ' Bonplandia,' and to Hodgson's ' Japan ' and ' Treasury of Botany ;'  Pritz. 28 ; Gard. Chron. 1866, 102 ; Journ. Bot. 1866, 64.  *Allanblackia*, Oliver.

**Blackburn or Blackburne, John** (1690–1786).  Of Orford, Lancashire.  Built first hot-house in north of England and first ripened pine-apples.  Catalogue of his garden by Adam Neal, published 1779.  Loudon ' Arboretum,' 56.  Rich. Corr. xxx. 324 ; Gent. Mag. lvii. 204 ; Dict. Nat. Biog. v. 123. *Blackburnia*, Forst. to him and his daughter Anna.

**Blackstone, John** (d. 1753) : Apothecary, of Fleet Street, London.  ' Fasciculus pl. circa Harefield . . . . . ,'  1737. ' Specimen Botanicum,' 1746.  Pritz. 28 ; Jacks. 523 ; Fl. Midd. 389–91 ; Rich. Corr. 351–5 ; Dict. Nat. Biog. v. 132.  Print in Hope Collection.  *Blackstonia*, Hudson = *Chlora*, L.

**Blackwell, Alexander** (c. 1709–1747) : b. Aberdeen ?, circ. 1709 ; beheaded, Stockholm, 9th August, 1747.  Printer.  Husband of Elizabeth Blackwell.  Abridged Miller's ' Botanicum Officinale ' for her ' Herbal.'  Gent. Mag. 1747, 424-6.  Dict. Nat. Biog. v. 142.

**Blackwell, Elizabeth** (c. 1700–c. 1747) : b. Aberdeen, c. 1700 ; d. London (?) after 1747 ; m. Dr. Alexander Blackwell of Chelsea.  ' A Curious Herbal,' cuts drawn, engr. and col. by herself, 1737-9.  Pult. ii. 251–6 ; Pritz. 28 ; Jacks. 31 ; Gent. Mag. xvii.  ' Lives of Eminent Men of Aberdeen,' by James Bruce, 307–18.  Dict. Nat. Biog. v. 144.  *Blackwellia*, Comm.

**Blackie, Thomas.** Sent to Switzerland in 1775, to collect, by Fothergill and Pitcairn. Loudon, Encycl. Gardening, 277.

**Blair, Patrick** (d. 1728): b. Dundee, Scotland; d. Boston, Lincoln ?, 1728. M.D. F.R.S. Practised at Dundee, London and Boston. Out in the '15; but pardoned through Sloane. 'Botanick Essays,' 1720. 'Pharmaco-botanologia,' 1723–8. Pult. ii. 134-140; Rees; Pritz. 28; Jacks. 523; Chalmers; Dict. Nat. Biog. v. 163; Sloane MSS. 4038; Biog. Univ. *Blæria*, L.

**Blake, John Bradby** (1745–1773): b. Great Marlborough Street, London, 4th November, 1745; d. C nton, China, 16th November, 1773. Supercargo to H.E.I.C. Sent many plants and seeds to Europe. Pritz. 28; Dict. Nat. Biog. v. 170.

**Blandford, George, Marquis of,** afterwards 4th Duke of Marlborough [*See* SPENCER-CHURCHILL, GEORGE].

**Bligh, William** (1753 or 1754–1817): b. Tinten, Cornwall ?, 1753 or 1754; d. Bond Street, London, 7th December, 1817; bur. Lambeth Churchyard. Admiral, 1811. F.R.S. 1801. Governor of New South Wales, 1805. Dict. Nat. Biog. v. 219; Ann. Bot. ii. 570. *Blighia*, Koenig.

**Blinkworth, Richard.** Collected at Kumaon. Correspondent of Wallich. *Blinkworthia*, Choisy.

**Bloxam, Andrew** (1801–1878): b. Rugby, Warwick, 22nd September, 1801; d. Great Harborough, Leicester, 2nd February, 1878. Clerk. Rector of Twycross, Leicestershire, and afterwards of Harborough. Visited S. America as naturalist to the 'Blonde,' 1824–5 (Mag. Nat. Hist. 1831, 145). 'Botany of Twycross," Phyt. 'Botany of Charnwood Forest," w. Dr. Churchill Babington, in Potter's History. Critical in *Rubi* and *Fungi*. Fungi and MSS. in Herb. Mus. Brit. Jacks. 523; Journ. Bot. 1878, 96; Dict. Nat. Biog. v. 264; Gard. Chron. 1878, i. 311. Water-colour by Turner of him and five brothers at funeral of their uncle, Sir Thomas Lawrence, in National Gallery.

**Bobart, or Bobert, Jacob** (1597, 1598, or 1599–1680): b. Brunswick, 1597, 1598, or 1599; d. "in his garden-house," Oxford, 4th February, 1679; bur. St. Peter's-in-the-East, Oxford. Made Superintendent, Oxford Botanic Garden, by Danby, 1632. 'Catalogus . . . . . Oxoniensis,' 1648, 2nd. ed., with Stephens and Browne, 1658. Pult. i. 312; Biog. Sketch by H. T. Bobart, 1884; Dict. Nat. Biog. v. 285; Jacks. 415; Pritz. 30; Felton, 108; Journ. Hort. 1876, 364, with portr. Wood, Fasti, ii. 189. Oil portr. by D. Loggan, 1675, at Linn. Soc.; engr. by Burghers; engr. at Kew. *Bobartia*, L.

**Bobart, Jacob** (1640 or 1641–1719): b. Oxford, 2nd August, 1641; d. Oxford, 28th December, 1719; bur. St. Peter's-in-the-East, Oxford. Succeeded his father in 1679 as Superintendent of Bot. Gard., and in 1683, after Morison, Prof. Bot. Oxford. Published Morison's 'Historia,' vol. iii. 1699. Herbarium in 12 vols. at the Bot. Garden, Oxford. Pult. i. 312. Biog. Sketch by H. T. Bobart, 1884. Grey's Hudibras. Rich. Corr. 10, 152. Nich. Illustr. i. 342, 357, 361. Dict. Nat. Biog. v. 286. Oil portr. at Oxford Bot. Gard. Engr. in 'Oxford Almanack,' 1719. *Bobartia*, L.

**Bohler, John** (1796–1872) : b. South Wingfield, Derbyshire, 31st December, 1797; d. Sheffield, 24th September, 1872. ' Lichenes Britannici,' 1835–7. ' Flora of Roche Abbey ' in Aveling's ' Roche Abbey,' 1870. Pritz. 32 ; Jacks. 243 ; Journ. Bot. 1872, 384 ; Dict. Nat. Biog. v. 304.

**Bohn, Henry George** (1796–1884) : b. London, 4th January, 1796; d. Twickenham, Middlesex, 22nd August, 1884. Bookseller and Publisher. Edited Gordon's ' Pinetum,' 1880. Jacks. 524 ; Dict. Nat. Biog. v. 304 ; Gard. Chron. 1884, ii. 283.

**Bollar, Nicholas, or Bollard** (fl. 1500 ?). Of Oxford. " A Tretee of Nicholas Bollard departed in three parties : 1. Of gendryng of Trees, 2 of graffynge, the third forsoth of altracions." MSS. Cotton MS., Jul. D. viii. 11 ; Addit. MS. 5467 ; Pult. i. 23–24 ; Haller i. 232 ; Tanner Bibl. Brit. 110 ; Dict. Nat. Biog. v. 324.

**Bolton, James** (fl. 1775–1795). Of Halifax. ' Filices Britannicæ', 1785–90. ' History of Funguses about Halifax,' 1788-91. Plants in Watson's History of Halifax, 1775. Pritz. 33 ; Jacks. 524 ; Dict. Nat. Biog. v. 327 ; Monthly Review, vols. lxxvi., lxxix. and 2nd ser. viii.

**Booth, William Beattie** (c. 1804–1874) : b. Perthshire, c. 1804 ; d. London (?), 18th June, 1874. A.L.S. 1825. At Chiswick, 1824–1830. Assistant Secretary, R. Hort. Soc. from 1858–1874. ' Illustrations of Camellieæ,' 1831. Contributed to ' Bot. Register,' &c. Gard. Chron. 1874, i. 838 ; Proc. Linn. Soc. 1874, 5, xxxvii.

**Boott, Francis** (1792–1863) : b. Boston, Mass., 26th September, 1792 ; d. 24, Gower Street, London, 25th December, 1863. M.D. Edinb. 1824. F.L.S. 1819. Sec. L.S. 1832–1839. Treas. 1856. V.P. 1861. Lecturer on Bot., Webb St. School of Medicine. ' Illustrations of the Genus Carex,' 1858–67. Proc. Linn. Soc. 1864, xxiii. Massachusetts pl. in Hookerian Herb., Kew. Pritz. 35 ; Jacks. 524 ; Gard. Chron. 1864, 51 ; Dict. Nat. Biog. v. 393. Oil copy of portr. by Gambardella and photo. at Linn. Soc. Portr. at Kew.

**Borlase, William** (1695–1772) : b. Pendeen, Cornwall, 2nd February, 1695 ; d. Ludgvan, 31st August, 1772. Clerk. M.A. Oxon. Ordained 1719. D.C.L., Oxon., 1766. F.R.S., 1750. Rector of Ludgvan, 1720. ' Natural History of Cornwall,' 1758. Pult. i. 355–6 ; Dict. Nat. Biog. v. 398 ; Nich. Anecd. iii. 78, 689 ; v. 291–3 ; Nich. Illustr. iv. 227, 445, 460, 468 ; Gent. Mag. lxxiii., II., 1114–7.

**Borrer, William** (1781–1862) : b. Henfield, Sussex, 13th June, 1781 ; d. Henfield, 10th January, 1862. F.L.S., 1805. F.R.S. Had a salicetum. ' English Botany Supplement.' ' Lichenographia.' Herb. at Kew. Pritz. 36 ; Jacks. 525 ; Journ. Bot. 1863–81 ; Proc. Linn. Soc. 1862, lxxxv. ; R.S.C. i. 499 ; Dict. Nat. Biog. v. 406. Portr. at Kew. *Borreria*, G. W. Meyer.

(To be continued.)

## SHORT NOTES.

THE NOMENCLATURE OF SPARGANIUM. — With reference to the nomenclature of the Spargania used by me in my previous paper, I may say it was no "adjustment" of mine, as I simply followed the last edition of the 'Students' Flora.' Nor did I see any necessity in a paper on Scotch Botany to refer to a plant which (although an object of my northern search) was unknown as British, *i.e.*, the *S. natans* Fries in Diar. Bot. Not. a. 1849, non alior.—the *S. Friesii* Beurl.; and I must demur to the suggestion that I was unaware of its existence, for I had four or five years ago examined specimens collected by F. Ahlberg, which showed it differed considerably from *S. minimum* Fr. or *S. affine* Schnizl., although Syme was unable to say "if it and *S. affine* were really distinct." Which of these plants should bear the name of *S. natans* L. (for it has been ascribed to each of the three segregates mentioned) I leave Mr. Daydon Jackson to decide. If it be the *S. natans* Fries (and of Linn. Fl. Lapp.), the *S. Friesii* Beurl., then it will be borne by a plant of limited distribution, confined as it is, according to Nyman, to Ross. Fenn. and Suec., and apparently against the views of the majority of European authorities. If bestowed upon *S. affine* Schnizl. (*S. longifolium* Don MS.), then it will be given to a widely distributed plant, the *S. natans* L. p.p. (*fide* Nyman), of Gren. et Godr., of Hooker and Arnott, of Babington, of Hook. fil. et auct. var., but not of Fries *fide* Hartm. In either case the difficulties pointed out when these names are used in a restricted sense would be experienced, and which I felt when I wrote *S. ramosum* Huds., since I had to add *S. neglectum* not seen in order to convey the fact that it was the segregate I referred to. I have never stated, suggested, nor thought that *S. affine* Schnizl. was identical with *S. Friesii* Beurl.; it is advisable "to avoid making a man say what he has not said." In ed. i. Fl. Lapp. (an *ante binomial* work) the plant is described as *Sparganium foliis natantibus plano convexis*, and is so quoted in ed. ii. of Spec. Pl.; but Linnæus also quotes as synonyms under *S. natans* Ray's *Sparganium minimum* and Dillenius' (Cat. Giss.) *Sparganium non ramosum minus*, which are certainly *not* Fries' *natans*.—G. C. DRUCE.

With Mr. Druce's permission, the Editor has allowed me to read the above article, and I will add a note respecting *Sparganium natans*. The lengthened citation given by Nyman—"*S. natans* (L.) Fr. in Diar. Bot. Not. a. 1849 (non alior.)"—is rendered necessary, at all events temporarily, by reason of the repeated misapplications of the name *natans* by botanists writing on the flora of countries in which the true plant does not occur. I think that *S. natans* L. and *S. natans* Fries are synonymous. The work of Fries (who, however, did not understand *S. affine*) was to expound the original plant of Linnæus, rather than to add greatly to the description; and he continually quotes the description of Linnæus with admiration. *S. natans* Fries was quoted by me as an example of this form of name sometimes actually used, and as being preferable to *S. Friesii* Beurling. In the Supplement to the 'Flora Danica' Dr. Lange is satisfied to cite the name simply as *S. natans* L., without any reference to Fries as the authority. I wish to lay particular stress on the following statement, which contains the justification of what I have previously written—the original description of Linnæus in 'Flora Lapponica' (referred to in Spec. Plant.) is distinctly *not applicable* to *S. affine* Schnizlein, and it is not surprising therefore that the latter author, in his monograph of the '*Typhaceæ*' (where the Fl. Lapp. description is duly discussed), should recognise in his *S. affine* a species not previously described.—W. H. BEEBY.

NOTES ON THE FLORA OF EASTERNESS, BANFF, ELGIN, AND WEST ROSS. — *Cerastium arcticum* Lange var.? *fide* Lange ex Arthur Bennett. Corrie Leacainn, *93. Very similar to *C. alpinum*, which grew with it. — *C. triviale* Link., var. *alpestre* "Lindb." (? var. *alpinum* Koch). Glen Ennich, *96. — *C. alpinum* L., var. *pubescens* Syme. I see that Syme considers the *piloso-pubescens* of Bentham to be equal to *Smithii* Syme.

*Sagina saxatilis* L. Glen Ennich, 96, *fide* Lange.

*Galium verum* L. Over flower, but I believe this species, on shingle by Loch Torridon, *105.

*Hieracium gracilentum* Backh. Glen Ennich, *96.—*H. anglicum* Fr., var. *acutifolium* Backh. Glen Ennich, *96.— *H. strictum* Fr. Kingussie, *96. — *H. globosum* Backh. Corrie Sneachda, 96; and the same species, I believe, in Glen A'an, near the Loch, *94.— *H. pallidum* Biv., var. *crinigerum* Fr. Glen Ennich, east side, *96. The determination of these Hieracia I owe to the kindness of Mr. Hanbury.

*Agrostis canina* L., f. *grandiflora* Hack. Near Kinchurdy, *96. Kenlochewe, *105. Moore, and between Dunphail and Forres, *95. A much coarser plant than our heathland *canina*, and probably often overlooked. — *A. canina* L., var. *mutica* Gaud. By the Findhorn, *95. A slender shade-grown form. — *A. alba* L., var. *coarctata* Hoffm. A maritime form occurring sparingly by Loch Torridon, *105.

*Deschampsia cæspitosa* Beauv., var. *alpina* (Gaud.). The Cairngorms, at high elevation, *94, *96. Does it differ from *brevifolia* (Parn.)? — *D. flexuosa* Trin., var. *montana* Huds. Beautiful specimens on the Cairngorms, Glen Ennich, 94, 96; and on Ben Eay, 105. The above grasses were kindly determined by Prof. Hackel. —G. C. DRUCE.

---

## NEW PHANEROGAMS PUBLISHED IN BRITAIN IN 1887.

THE periodicals cited in this list are: 'Annals of Botany,' 'Botanical Magazine,' 'Gardener's Chronicle,' 'Icones Plantarum,' 'Journal of Botany,' 'Journal' and 'Transactions' of the Linnean Society of London.

We have included one species published in Dr. H. B. Guppy's 'Solomon Islands,' as there is some danger of plants published in works of this kind being overlooked. The volume in question contains names of other novelties, but these are not described, and thus have no claim to recognition.

The novelties in Mr. im Thurn's Roraima paper (Linn. Trans. (Bot.) ii.) in July last were previously published in ' Timehri ' for December, 1886 ; these of course date from their first publication, and are thus excluded from our list.

New genera are indicated by an affixed asterisk. We have added in square brackets the publishers of certain names which are cited from the MS. description or notes of those who stand as the authority for them.

ACACIA XIPHOCLADA *Baker*. Madagascar. J. L. Soc. xxii. 468.

ACANTHOPANAX DIVERSIFOLIUM *Hemsl*. China. J. L. Soc. xxiii. 340.

ACIDANTHERA LAXIFLORA *Baker*. Trop. Africa. Trans. L. Soc. (Bot.) ii. 350.

ACRANTHERA MUTABILIS *Hemsl*. Perak. J. Bot. 204.

ADINA RUBESCENS *Hemsl*. Perak. J. Bot. 204.

ÆCHMEA FLEXUOSA *Baker*. Gard. Chron. i. 8. — Æ. MYRIOPHYLLA *Baker*. Brazil. Bot. Mag. t. 6939.

AGAVE HENRIQUESII *Baker*. Mexico. Gard. Chron. i. 732. — A. MORRISII *Baker*. Gard. Chron. i. 543, fig. 105.

AJUGA OOCEPHALA *Baker*. Madagascar. J. L. Soc. xxii. 574.

ALBIZZIA TRICHOPETALA *Baker*. Madagascar. J. L. Soc. xxii. 468.

ALOCASIA EMINENS *N. E. Br*. E. Indies. Gard. Chron. i. 105. —A. MARGINATA *N. E. Br*. E. Indies. Id. ii. 712.—A. PERAKENSIS *Hemsl*. Perak. J. Bot. 205.

ALOE HAWORTHIOIDES *Baker*. Madagascar. J. L. Soc. xxii. 529.— A. JOHNSTONI *Baker*. Trop. Africa. Trans. L. Soc. (Bot.) ii. 351, t. 63.

ALPINIA FRASERIANA *Oliv*. Borneo. Ic. Pl. 1567.—A. ZINGIBERINA *Hook. f*. Siam. Bot. Mag. t. 6944.

ALYXIA LUCIDA *Baker*. Madagascar. J. L. Soc. xxii. 503.

AMASONIA CALYCINA *Hook. f*. Brit. Guiana. Bot. Mag. t. 6915.

AMOMUM ALBOVIOLACEUM *Ridl*. and A. ERYTHROCARPUM *Ridl*. Angola. J. Bot. 130.

*AMPHOROCALYX (Melastomaceæ Oxysporeæ) MULTIFLORUS *Baker*. Madagascar. J. L. Soc. xxii. 476.

ANEMONE HENRYI *Oliv*. China. Ic. Pl. 1570.

ANEILEMA TENERA *Baker*. Madagascar. J. L. Soc. xxii. 530.

ANGRÆCUM AVICULARIUM *Rchb. f*. Gard. Chron. i. 40.—A. CALLIGERUM *Rchb. f*. Id. ii. 552.

ANTHERICUM DIANELLÆFOLIUM *Baker*. J. L. Soc. xxii. 529. — A. RUBELLUM and A. VENULOSUM *Baker*. Trop. Africa. Trans. L. Soc. ii. 352.

ANTHOCLEISTA AMPLEXICAULIS and A. RHIZOPHOROIDES *Baker*. Madagascar. J. L. Soc. xxii. 506.

ANTHURIUM ACUTUM *N. E. Br*. Brazil. Gard. Chron. ii. 776.—A. BREVILOBUM *N. E. Br*. Id. i. 380. — A. PURPUREUM *N. E. Br*. Brazil. Id. 575.

ANISOTES PARVIFOLIUS *Oliv*. Trop. Africa. Trans. L. Soc. (Bot.) ii. 346.

ANTIDESMA ALNIFOLIA, A. ARBUTIFOLIA, A. BRACHYSCYPHA, all of *Baker*. Madagascar. J. L. Soc. xxii. 518, 519.

APHELEXIS FLEXUOSA, A. STENOCLADA, A. SULPHUREA, all of *Baker.* Madagascar. J. L. Soc. xxii. 492, 493.

APHLOIA MINIMA *Baker.* Madagascar. J. L. Soc. xxii. 444.

APODYTES EMIRNENSIS *Baker.* Madagascar. J. L. Soc. xxii. 458.

APOROSA BENTHAMIANA *Hook. f.* Malacca. Ic. Pl. 1583.

ARDISIA DISSITIFLORA and A. LEPTOCLADA *Baker.* Madagascar. J. L. Soc. xxii. 500, 501.

ARGOSTEMMA INVOLUCRATUM *Hemsl.* Perak. Ic. Pl. 1556.

ARGYROLOBIUM MEGARHIZUM *Bohn.* S. Africa. J. L. Soc. xxiv. 175.

ARISÆMA ANOMALUM and A. WRAYI *Hemsl.* Perak. J. Bot. 205.

ARISTEA PLATYCAULIS *Baker.* S. Africa. Gard. Chron. i. 732.

ARISTIDA MULTICAULIS *Baker.* Madagascar. J. L. Soc. xxii. 533.

ARTHROSELEN LATIFOLIUS *Oliv.* Trop. Africa. Trans. L. Soc. (Bot.) ii. 348.

ASPHODELUS COMOSUS *Baker.* E. Indies. Gard. Chron. i. 799.

*ASTEPHANOCARPA (Compositæ Inuloideæ) ARBUTIFOLIA *Baker.* Madagasear. J. L. Soc. xxii. 493.

ASTER PERFOLIATUS *Oliv.* S. Africa. Ic. Pl. 179.

ASTEROPEIA SPHÆROCARPA *Baker.* Madagascar. J. L. Soc. xxii. 479.

ASTILBE POLYANDRA *Hemsl.* China. J. L. Soc. xxiii. 265.

BARLERIA KITCHINGII and B. PHILLYREÆFOLIA *Baker.* Madagascar. J. L. Soc. xxii. 510.

BEAUMONTIA BREVITUBA *Oliv.* China. Ic. Pl. 1582.

BEGONIA BOWENI and B. FRAGILIS *Baker.* Madagascar. J. L. Soc. xxii. 479, 480. — B. CYCLOPHYLLA *Hook. f.* China. Bot. Mag. t. 6926.—B. EGREGIA *N. E. Br.* Brazil. Gard. Chron. i. 346. —B. HENRYI *Hemsl.* China. J. L. Soc. xxiii. 322.—B. JOHNSTONI *Oliv.* Trop. Africa. Trans. L. Soc. (Bot.) ii. 334.

BELMONTIA EMIRNENSIS *Baker.* Madagascar. J. L. Soc. xxii. 507.

BENNETTIA LONGIPES *Oliv.* India. Ic. Pl. 1596.

BOEA LAWESII *H. O. Forbes.* New Guinea. J. Bot. 348.

BOMBAX JENMANI *Oliv.* British Guiana. Ic. Pl. 1720.

*BRACHYLOPHORA (Malpighiaceæ Banisterieæ) CURTISII *Oliv.* Penang. Ic. Pl. 1566.

BREWERIA TILIÆFOLIA *Baker.* Madagascar. J. L. Soc. xxii. 508.

BUDDLEA SPHÆROCALYX *Baker.* Madagascar. J. L. Soc. xxii. 505.

BYRSOCARPUS BOWENI *Baker.* Madagascar. J. L. Soc. xxii. 462.

BYTTNERIA BAUHINIOIDES and B. MELLERI *Baker.* Madagascar. J. L. Soc. xxii. 451.

CÆSIA SUBULATA *Baker.* Madagascar. J. L. Soc. xxii. 530.

CALOPYXIS MALIFOLIA *Baker.* Madagascar. J. L. Soc. xxii. 474.

CAPPARIS HAINANENSIS *Oliv.* China. Ic. Pl. 1588.

CARAGANA DECORTICANS *Hemsl.* Afghanistan. Ic. Pl. 1725.

CARDAMINE JOHNSTONI *Oliv.* Trop. Africa. Trans. L. Soc. (Bot.) ii. 328.

CAREX SCAPOSA *C. B. Clarke.* China. Bot. Mag. t. 6940.

CATASETUM COSTATUM *Rchb. f.* Gard. Chron. i. 72.

CATOPSIS FENDLERI, Venezuela; C. FLEXUOSA, Bolivia; C. HAHNII, Mexico; C. STENOPETALA, Guatemala: all of *Baker.* J. Bot. 175, 176.

CELASTRUS MARITIMUS *Bolus.* S. Africa. J. L. Soc. xxiv. 173.

CELOSIA MICRANTHA *Baker.* Madagascar. J. L. Soc. xxii. 514.
CELTIS GOMPHOPHYLLA *Baker.* Madagascar. J. L. Soc. xxiii. 521.
CEPHALOCROTON CORDIFOLIUS *Baker.* Madagascar. J. L. Soc. xxii. 520.
CEROPEGIA MONTEIROÆ *Hook. f.* S. Africa. Bot. Mag. t. 6927.
CHEILOTHECA MALAYANA *Scort. MS.* [*Hook. f.*]. Perak. Ic. Pl. 1564.
CHIMONANTHUS MITENS *Oliv.* China. Ic. Pl. 1600.
CHLORANTHUS ANGUSTIFOLIUS *Oliv.* China. Ic. Pl. 1580.
*CHLOROCYATHUS (Asclepiadeæ Periploceæ) MONTEIROÆ *Oliv.* Delagoa Bay. Ic. Pl. 1557.
CHLOROPHYTUM CHLORANTHUM *Baker.* Madagascar. J. L. Soc. xxii. 529.
CIRRHOPETALUM LENDYANUM *Rchb. f.* Gard. Chron. ii. 70. — C. STRAGULARIUM *Rchb. f.* Id. ii. 186, 214.
CLADIUM FIMBRISTYLOIDES *Baker.* Madagascar. J. L. Soc. xxii. 531.
CLAVIJA ERNSTII *Hook. f.* S. America. Bot. Mag. t. 6928.
CLERODENDRON CEPHALANTHUM *Oliv.* Zanzibar. Ic. Pl. 1559. — C. JOHNSTONI *Oliv.* Trop. Africa. Trans. L. Soc. (Bot.) ii. 346.— C. MIRABILE *Baker.* Madagascar. J. L. Soc. xxii. 513.
CLITORIA HANCEANA *Hemsl.* China. J. L. Soc. xxiii. 187.
COELOGYNE FOERSTERMANNI *Rchb. f.* Sonda. Gard. Chron. i. 798.
—C. SANDERIANA *Rchb. f.* Sonda. Gard. Chron. i. 764.
COLEOTRYPE BARONI *Baker.* Madagascar. J. L. Soc. xxii. 530.
COMMIPHORA FRAXINIFOLIA and C. LAXIFLORA *Baker.* Madagascar. J. L. Soc. xxii. 459.
CONYZA AMPLEXICAULIS, C. ELLISII, and C. SERRATIFOLIA, all of *Baker.* Madagascar. J. L. Soc. xxii. 488, 489.
CORCHORUS HAMATUS *Baker.* Madagascar. J. L. Soc. xxii. 452.
CORNUS HONGKONGENSIS *Hemsl.* China. J. L. Soc. xxiii. 345.
COSTUS GIGANTEUS *Welw.* (*Ridl.*). Angola. J. Bot. 131.
CRASSULA FRAGILIS *Baker.* Madagascar. J. L. Soc. xxii. 469.
CRINUM CRASSIPES *Baker.* Africa. Gard. Chron. ii. 126. — C. MODESTUM *Baker.* Madagascar. J. L. Soc. xxii. 528.
CROTALARIA GRIQUENSIS *Bolus.* S. Africa. J. L. Soc. xxiv. 174.— C. LUTEORUBELLA and C. MACROPODA *Baker.* Madagascar. J. L. Soc. xxii. 462, 463.
CROTON VERNICOSUS *Baker.* Madagascar. J. L. Soc. xxii. 519.
CRYPTOCARYA PAUCIFLORA *Baker.* Madagascar. J. L. Soc. xxii. 515.
CRYPTOLEPIS MONTEIROÆ *Oliv.* Delagoa Bay. Ic. Pl. 1591.
CUSCUTA KILIMANJARI *Oliv.* Trop. Africa. Trans. L. Soc. (Bot.) ii. 343.
CYCLEA MADAGASCARIENSIS *Baker.* Madagascar. J. L. Soc. xxii. 443.
CYPERUS CUSPIDATUS, C. DEBILISSIMUS, C. MONOCEPHALUS, C. PLATYCAULIS, and C. SUBÆQUALIS, all of *Baker.* Madagascar. J. L. Soc. xxii. 531, 532.
CYRTANDROMÆA MEGAPHYLLA *Hemsl.* Ic. Pl. 1555.
DALBERGIA POOLII and D. SCORPIOIDES *Baker.* Madagascar. J. L. Soc. xxii. 466.
DANAIS LYALLII and D. NUMMULARIFOLIA *Baker.* Madagascar. J. L. Soc. xxii. 481.
DENDROBIUM AURANTIACUM *Rchb. f.* E. Indies. Gard. Chron. ii. 98.

DEUTZIA DISCOLOR *Hemsl.* China. J. L. Soc. xxiii. 275. ·
DEYEUXIA EMIRNENSIS *Baker.* Madagascar. J. L. Soc. xxii. 533.
DICHÆTANTHERA CRASSINODIS *Baker.* Madagascar. J. L. Soc. xxii. 476.
DICORYPHE GUATTERIÆFOLIA, D. LAURIFOLIA, and D. RETUSA, all of
 *Baker.* Madagascar. J. L. Soc. xix. 473, 474.
DICHROSTACHYS UNIJUGA *Baker.* Madagascar. J. L. Soc. xxii. 467.
DIDYMOCARPUS ALBOMARGINATUS *Hemsl.* Perak. J. Bot. 204.
DIDYMOCAPSA PUSILLUS *Baker.* Madagascar. J. L. Soc. xxii. 508.
DIOSCOREA CRYPTANTHA *Baker.* Madagascar. J. L. Soc. xxii. 528.
DIPLACHNE ARISTATA *Baker.* Madagascar. J. L. Soc. xxii. 534.
DIRICHLÆTIA INVOLUCRATA, D. TERNIFOLIA, and D. TRICHOPHLEBIA, all
 of *Baker.* Madagascar. J. L. Soc. xxii. 482, 483.
DOMBEYA ACERIFOLIA, D. BARONI, D. BIUMBELLATA, D. INSIGNIS, and
 D. MEGAPHYLLA, all of *Baker.* Madagascar. J. L. Soc. xxii.
 449, 450.
DORSTENIA ZANZIBARICA *Oliv.* Zanzibar. Ic. Pl. 1581.
DYPSIS CONCINNA, D. CURTISII, D. HETEROPHYLLA, D. POLYSTACHYA,
 and D. RHODOTRICHA, all of *Baker.* Madagascar. J. L. Soc.
 xxii. 525, 526.
ELÆOCARPUS DALECHAMPIOIDES *Baker.* Madagascar. ·J. L. Soc.
 xxii. 452.
ELÆODENDRON GYMNOSPOROIDES *Baker.* Madagascar. J. L. Soc.
 xxii. 460.
ELATOSTEMA HEXADONTUM *Baker.* Madagascar. J. L. Soc. xxii. 524.
ELEUTHEROCOCCUS HENRYI *Oliv.* China. Ic. Pl. 1711. — E. LEU-
CORRHIZUS *Oliv.* Id.
EPALLAGE DISSITIFOLIA *Baker.* Madagascar, J. L. Soc. xxii. 494.
—— EPIDENDRUM KIENASTII *Rchb. f.* Mexico. Gard. Chron. ii. 126.
ERICA ADENOPHYLLA, E. ASPALATHIFOLIA, E. BAURII, E. BROWNLEEÆ,
 E. CAFFRORUM, E. COOPERI, E. ERIOCODON, E. HÆMANTHA, E. INOPS,
 E. LEROUXII, E. MISSIONIS, E. NATALITIA, E. TETRASTIGMATA, E.
 TRACHYSANTHA, E. TRICHADENIA, E. TYSONI, and E. URNA-VIRIDIS,
 all of *Bolus.* S. Africa. J. L. Soc. xxiv. 178–187.
ERISMANTHUS SINENSIS *Oliv.* China. Ic. Pl. 1568.
ERYTHROXYLUM AMPULLACEUM and E. SPARSIFLORUM *Baker.* Mada-
 gasear. J. L. Soc. xxii. 455.
EUCHRESTA TENUIFOLIA *Hemsl.* China. J. L. Soc. xxiii. 200.
EUCOMIS PALLIDIFLORA *Baker.* S. Africa. Gard. Chron. ii. 154.
EUGENIA AGGREGATA and E. OLIGANTHA *Baker.* Madagascar. J. L.
 Soc. xxii. 474, 475.—E. FLUVIATILIS *Hemsl.* China. Id. xxiii.
 296.
EUPHORBIA ALCICORNIS and E. ORTHOCLADA *Baker.* Madagascar.
 J. L. Soc. xxii. 517.
EVODIA DISCOLOR and E. FLORIBUNDA *Baker.* Madagascar. J. L. Soc.
 xxii. 456, 457. ·
FICUS ALBIDULA, F. BOTRYOIDES, F. COCCULIFOLIA, F. PACHYCLADA, F.
 PHANEROPHLEBIA, F. PULVINIFERA, F. SAKALAVARUM, and F. TRICHO-
 CLADA, all of *Baker.* Madagascar. J. L. Soc. xxii. 521–524.
—— GALEANDRA FLAVEOLA *Rchb. f.* Gard. Chron. i. 512.
*GAMOPODA (Menispermaceæ) LEPTOPODA *Baker.* Madagascar. J. L.
 Soc. xxii. 443.

GARCINIA CAULIFLORA, G. CERNUA, G. ORTHOCLADA, and G. POLY-
PHLEBIA, all of *Baker*. Madagascar. J. L. Soc. xxii. 446, 447.
GARDENIA SUCCOSA *Baker*. Madagascar. J. L. Soc. xxii. 483.
GAZANIA DIFFUSA *Oliv*. Trop. Africa. Trans. L. Soc. (Bot.) ii.
340, t. 61.
GERBERA EMIRNENSIS *Baker*. Madagascar. J. L. Soc. xxii. 498.
GLADIOLUS PAUCIFLORUS and G. SULPHUREUS *Baker*. Trop. Africa.
Trans. L. Soc. (Bot.) ii. 350.
GLEDITCHSIA AUSTRALIS *Hemsl*. China. J. L. Soc. xxiii. 208, t. 5.
GNAPHALIUM DIFFUSUM *Baker*. Madagascar. J. L. Soc. xxii. 490.
*GOMPHOCALYX (Rubiaceæ Spermacoceæ) HERNIARIOIDES *Baker*.
Madagascar. J. L. Soc. xxii. 485.
GOMPHOCARPUS BISACCULATUS *Oliv*. Trop. Africa. Trans. L. Soc.
(Bot.) ii. 341.
GRAVESIA PORPHYROVALVIS *Baker*. Madagascar. J. L. Soc. xxii. 477.
GUZMANNIA CRISPA *Baker*. Angola. J. Bot. 173.
GYMNEMA PARVIFOLIUM *Oliv*. Trop. Africa. Trans. L. Soc. (Bot.)
ii. 342.
GYMNOSPORIA CUNEIFOLIA *Baker*. Madagascar. J. L. Soc. xxii. 460.
GYNURA SONCHIFOLIA *Baker*. Madagascar. J. L. Soc. xxii. 495.
HEDYOTIS JOHNSTONI *Oliv*. Trop. Africa. Trans. L. Soc. (Bot.)
ii. 335.
HELICHRYSUM AMPLEXICAULE, H. ARANEOSUM, H. FARINOSUM, and H.
PLATYCEPHALUM, all of *Baker*. Madagascar. J. L. Soc. xxii.
491, 492.—H. KILIMANJARI *Oliv*. Trop. Africa. Trans.L. Soc.
(Bot.) ii. 338.
HIBISCUS CYTISIFOLIUS, H. NUMMULARIFOLIUS, H. OBLATUS, and H.
XIPHOCUSPIS, all of *Baker*. Madagascar. J. L. Soc. xxii. 447, 448.
HOYA GUPPYI *Oliv*. Solomon Islands. Guppy, p. 299.
HUERNIA ASPERA *N. E. Br*. Zanzibar. Gard. Chron. ii. 364.
HUTCHINSIA PERPUSILLA *Hemsl*. Tibet. Ic. Pl. 1599.
HYDRANGEA LONGIPES *Hemsl*. China. J. L. Soc. xxiii. 273.
*HYDROTHRIX (Pontederiaceæ) GARDNERI *Hook. f*. Brazil. Ann.
Bot. 90, t. vii.
HYPERICUM KIBOENSE *Oliv*. Trop. Africa. Trans. L. Soc. (Bot.)
ii. 329.
HYPOESTES ACUMINATA, H. CHLOROCLADA, H. CONGESTIFLORA, H. MICRO-
PHYLLA, H. OBTUSIFOLIA, H. PHYLLOSTACHYA, and H. SESSILIFOLIA,
all of *Baker*. Madagascar. J. L. Soc. xxii. 511–513.

(To be continued.)  p 186

---

## NOTICES OF BOOKS.

*The Flora of Howth. With map and an introduction on the Geology
and other features of the promontory.* By H. C. HUNT, B.A.,
F.L.S. Dublin: Hodges, Higgis & Co. 1887. 8vo, pp. 138.

THE "Hill of Howth" and "Ireland's Eye" are familiar
objects to the passenger by the steamer from Holyhead to Kings-
town, and Howth itself is well known as a sea-bathing resort for

the citizens of Dublin.   It has also, Mr. Hart tells us, been at all times favourite ground for botanists, from the time of Threlkeld, whose ' Synopsis Stirpium Hibernicarum ' appeared in 1727, down to the present.   Some of the writers on its botany have possessed a zeal not according to knowledge, for Mr. Hart mentions two pamphlets ·which " unfortunately contain many erroneous statements."

The present Flora—the value of which is the more apparent when we consider how comparatively few districts of Ireland have been thoroughly examined—is the outcome of the author's personal researches during most of the last twenty years.   It is " of particular interest in two special ways : (1) from the variety of several of the species found ; and (2) on account of the large number of forms assembled in so small a space."   Howth itself comprises an area of 2670 acres, and Ireland's Eye is about a mile in circumference.   Mr. Hart's list contains 545 species of phanerogams and ferns, of which 25 are introductions—a total " probably above the average—certainly as regards Ireland—for a district of about four square miles in the British Islands."   Mr. Hart's introduction occupies only ten pages, but is singularly full of information and comparative statistics, and may well be taken as a model of what such essays should be.

Among the more interesting plants of the flora of Howth, many of them now first recorded, may be noted *Lavatera arborea*, in two localities " difficult or impossible to reach except from a boat ; " *Erodium maritimum*, a very local species in Ireland ; *Ornithopus perpusillus*, very rare as an Irish plant ; *Ligustrum vulgare*, " native on steep grassy cliffs in almost inaccessible places," growing " in a fringe at the juncture of the sea rocks with the steep grassy slopes, prostrate and stunted, having stems often an inch in diameter." Mr. Hart considered this and the Waterford coast to be the only indigenous stations for the Privet yet discovered ; and he can bear testimony to the nativity of the plant at Tramore, in habitat exactly similar to that above quoted.   There are several appendices devoted to plants excluded from the Flora on various grounds, with a list of the species found in Dublin County but not occurring in Howth.   An excellent map of the island completes the work, which is appropriately dedicated to Mr. A. G. More.

---

*Flora of the Hawaiian Islands: a description of their Phanerogams and Vascular Cryptogams.*   By WILLIAM HILLEBRAND, M.D. Annotated and published after the author's death by W. F. HILLEBRAND.   London : Williams & Norgate.   8vo, pp. xcvi. 673.   4 maps.

THIS is an extremely interesting and valuable work, and in many ways a remarkable addition to our list of Floras.   The lamented author, who died on the 13th of July, 1886, had only corrected a few pages of proof when his long and trying illness was terminated by death ; and all will regret that he was thus prevented from seeing the outcome of the twenty years of unremitting study which he devoted to the Hawaiian flora.   His son has, however, carried out the work in a way which leaves little, if anything, to be desired.

One special characteristic of the book is the feeling which every page conveys that the author is not working with herbarium specimens, but is recording observations which he has made in the field. It is evident at a glance that Dr. Hillebrand knew intimately in a living state most of the material on which his Flora is based; and this gives an interest to his observations and conclusions which is usually wanting in books of this kind.   The author has prefixed to the work the outlines of Botany which Mr. Bentham prepared for our series of Colonial Floras; and is to some extent responsible for the interesting introduction.

We regret that space will not permit us to notice the work as fully as we could wish; but on glancing through the pages the following points strike us as noteworthy.   *Lepidium* is the only indigenous genus of *Cruciferæ*, and of this there are three species, one (*L. arbuscula*) new; the other representatives of the order—*Senebiera didyma, Cardamine hirsuta, Nasturtium officinale*, and *Brassica nigra*— are all introductions.   There are ten Pittosporums, five of them new; the exclusively Hawaiian Caryophyllaceous genus *Schiedea* is increased to seventeen species, five now first described, and an amended description of Mann's endemic and monotypic *Alsinidendron* is given.   Of Seemann's *Gossypium drynarioides*, originally described from a specimen in the British Museum collected by David Nelson, three trees have been found, as well as of two of a variety; but these are disappearing, if they have not already disappeared.   *Pelea*, a Rutaceous endemic genus, has now twenty species, eight now first described, and *Platydesma*, also endemic, has its number of species increased from two to four.   In *Sapindaceæ* a doubtful new genus is described under its native name, *Mahoe;* the tree is as yet imperfectly known.   Among the not very numerous *Leguminosæ* the novelties are comparatively few, but two of the three indigenous Acacias are new.   There are only seven *Umbelliferæ*—two of them introduced and two (species of *Peucedanum*) new.   In *Araliaceæ*, a new genus, *Pterotropia*, is established for the plant published by Seemann in this Journal (1868, 130) as *Dipanax Manni*, with two others, one of them new : *Pterotropia* had been placed by Horace Mann as a section of *Heptapleurum*, and it seems likely that Seemann's name will have to stand.

The endemic Rubiaceous genus *Kadua* now numbers sixteen species, five of them new.   In *Lobeliaceæ*, the most characteristic order of the Flora, the novelties are numerous, twenty of the fifty-eight species being here first described; five of the six genera— *Brighamia, Clermontia, Rollandia, Delissea*, and *Cyanea*—are endemic, as are three out of the five Lobelias : the description of this interesting order is elaborated with especial care.   *Labordea*, an endemic Loganiaceous genus, has three new species out of the nine enumerated.   In *Cyrtandra* many new forms are described—eleven out of twenty-eight; the polymorphism of the species is said by Dr. Hillebrand to be extraordinary—no single form extends over the whole group, and not many are common to more than one island.

In *Labiatæ*, seventeen species of *Stenogyne*, a genus peculiar to the islands, are enumerated, five of them new. Gray's section *Noto-trichium* of *Ptilotus* is raised to generic rank; it is based on *Ptilotus sandwicensis*, and two more species are added. *Euphorbiaceæ* are not numerous, seventeen species being described, one only being new and five introduced.

The Monocotyledons are not very numerous, and the novelties are comparatively few; there are only three Orchids, one, *Habenaria holochila*, being new. Most of the new species are among the *Cyperaceæ* and Grasses; among the latter, the genus *Eragrostis* is notable as having five novelties and three introductions out of a total of eleven species. Among the ferns, *Asplenium*, with thirty-nine species, holds the first place, but the proportion of new species is greater in *Lindsaya*, four out of eight being first described. There is a new genus, *Schizostege*, founded on the plant described by Baker as *Cheilanthes Lidgatii*.

A word of praise is due to the Darmstadt printer for the admirable manner in which he has executed his task.

----

*A School Flora for the use of Elementary Botanical Classes.* By W. MARSHALL WATTS, D. Sc. (Lond.), Physical Science Master in the Giggleswick Grammar School. Revised and enlarged edition. Rivingtons, 1887. 8vo, pp. viii. 199.

THIS is an extremely useful little volume, intended "to provide the student who has mastered the elements of botanical science with a Flora of such small size as to be easily carried on country rambles, which shall enable him easily to identify the common plants with which he will meet." It is arranged throughout on the principle of determining a plant by deciding which of two opposite characters it possesses, and the plan is well executed. The typography and arrangement are excellent, and the book might well be adopted as a text-book in school natural history societies.

The first edition, which we have not seen, was compiled for the young botanists of Giggleswick School, and confined to the plants of that district; but it is now enlarged so as to include the species marked with a higher number than 50 in the 8th edition of the ' London Catalogue,' and the rarer plants growing within reach of certain schools have been included. These have been ascertained by lists furnished by those connected with the schools enumerated. Among those absent from the list we note Eton, Harrow, and Stonyhurst: Eton has never done much in the way of botany, but the Flora of Harrow and the Stonyhurst list are easily accessible.

Of course in a book of this kind critical distinctions would be out of place; it is pleasant to meet once more our old friends *Ranunculus aquatilis*, *Rubus fruticosus*, and *Rosa canina*. One or two omissions surprise us : there is only one *Drosera*, for example, and one *Erythræa*. The authors' names are not attached to the genera or species, which we think an undesirable omission, as it is desirable to accustom young botanists to the correct method of citing names. A useful little glossary is appended.

*The Characeæ of America.* Part I. By Dr. T. F. Allen. New York. 1888. 8vo. Price 4 dollars.

Dr. Allen has commenced another monograph of the American *Characeæ*, and the present work promises to be a much more practicable undertaking than the large quarto book commenced some years ago. The first part, now issued, is devoted to an account of the structure and classification of the order, it being proposed in a second part to give a description of the American species. As stated by the author in the introduction, the part relating to structure contains but little original matter, but a very fair *résumé* is given of most of the points of interest. Several woodcuts from Sachs, De Bary, and Nordstedt are reproduced, and there are a number of others from original drawings. The latter are rather diagrammatic than artistic, and some of them, notably those of the young nucules, are unnecessarily large and clumsily executed. A key to all the known species is added by Dr. Nordstedt, closely following that by the same author in Braun's 'Fragmente,' but including the plants since described and several previously undescribed species. Dr. Allen has adopted an excellent plan in giving a drawing of the peculiar characteristics of each group in the key.

The author's views on the subject of nomenclature appear to be extremely vague and unsatisfactory. His opinion that all pre-Braunian names "must be discarded" as representing a number of different species will scarcely be adopted by any botanist of the present day. In the key itself fortunately this view is not supported. To get rid of all the old and uncertain names in this way would of course much simplify the work of a monographer, but such a course would be neither just nor conducive to finality.

The book altogether is a very useful addition to the literature of the order. ──────── H. & J. Groves.

The last (March) number of Hooker's 'Icones Plantarum' contains a large number of interesting novelties, chiefly from Tibet, China, and South Africa, and among them a new genus, *Actinotinus* Oliv. (Caprifoliaceæ).

────────

## ARTICLES IN JOURNALS.

*Annals of Botany* ("Feb.").— W. M. Woodworth, 'The Apical Cell of Fucus' (1 plate).—T. Johnson, 'The Procarpium and Fruit in *Gracilaria confervoides* (1 plate).—J. R. Green, 'The germination of the tuber of *Helianthus tuberosus*.'—F. W. Oliver, 'On the sensitive labellum of *Masdevallia muscosa*' (1 plate).—Miss A. Bateson, 'The Effect of Cross-fertilisation on inconspicuous flowers.'— E. Sanford, 'Microscopical Anatomy of *Gymnosporangium macropus*' (1 plate).—F. O. Bower, 'Normal and abnormal developments of the oophyte in Trichogynes' (3 plates).—D. H. Scott & H. Wager, 'Floating-roots of *Sesbania aculeata*.' — W. C. Williamson, 'Anomalous Cells within tissues of fossil plants of coal-measures'

(1 plate). — H. M. Ward, ‘ Recent publications bearing on the sources of nitrogen in plants.’ — C. B. Clarke, ‘ *Acalypha indica.*’— W. Gardiner, ‘ Power of contractibility exhibited by protoplasm of certain plant-cells.’ —I. B. Balfour, ‘ The replum in Cruciferæ.’ — Botanical Necrology for 1887.—J. H. Hart, ‘ Calcareous deposits in *Hieronyma alchorneoides.*’

*Botanical Gazette* (Feb.). — J. D. Smith, ‘ Undescribed plants from Guatemala ’ (*Chrysochlamys Guatemaltecana, Harpalyce rupicola, Bauhinia Rubeleruziana, B. Quetzal, Triolena palmata*).—S. M. Tracy & B. T. Galloway, ‘ *Uncinula polychæta.*’ — S. B. Parish, *Phacelia heterosperma,* n. sp.

*Bot. Centralblatt* (Nos. 10—12).—C. Dünnenberger, ‘ Bacteriologisch-chemische Untersuchung über die beim Aufgeben .des Brotteiges wirkenden Ursachen.’ — G. R. v. Beck (Nos. 10, 12), ‘ Geschichte des Wiener Herbariums.’ — (No. 10). —. Sotereder, ‘ Ueber den systematischen und phylogenetischen Werth der Gefässdurchbrechungen auf Grund früherer Untersuchungen und einiger neuer Beobachtungen.’—(No. 11). C. von Tubeuf, ‘ Ueber die Wurzelbildung einiger Loranthaceen.’—Id., ‘ Eine neue Krankheit der Douglastanne.’—K. Starback, ‘ Beitrage zur Ascomyceten-Flora Schwedens.’ — (No. 12). H. F. G. Strömfelt, ‘ Untersuchungen über die Haftorgane der Algen.’

*Bot. Zeitung* (Mar. 2, 9). — A. F. W. Schimper, ‘ Ueber Kalkoxalatbildung in den Laubblättern.’ —(Mar. 16, 23). F. Schütt, ‘ Ueber die Diatomeengattung Chætoceros ’ (1 plate).

*Bull. Soc. Bot. Belgique* (xxxv. : Comptes Rendus i ; Mar. 1).— G. Rouy, ‘ Notes sur la Géographie Botanique de l’Europe.’— —. Hue, ‘ Lichens de Miquelon.’ — H. de Vilmorin, ‘ Expériences de croisement entre des blés différents.’ — L. Flot, ‘ Sur les tiges aériennes.’--A. Daquillon, ‘ Sur la structure des feuilles de conifères.’ —F. Gray, ‘ Sur les *Ulothrix* aériens.’—L. du Sallon, ‘ Sur les poils radicaux des Rhinanthés.’ — K. J. Foncaud, ‘ Variété nouvelle du *Ceratophyllum demersum.*’ — P. Duchartre, ‘ Organisation de la fleur du *Delphinium elatum.*’

*Bull. Torrey Bot. Club* (March).—W. Deane, Memoir of Asa Gray. — T. Morong, ‘ Sparganium ’ (1 plate : *S. Greeni, S. subglobosum,* spp. nn.).—B. D. Halsted, ‘ Trigger-hairs of the Thistle flower.’—E. L. Greene, ‘ *Castalia* and *Nymphæa.*’

*Flora* (Jan. 1). — C. Müller, ‘ Musci cleistocarpici novi.’—F. Arnold, *Muellerella thallophila,* n. sp.—(Jan. 11, 21 ; Mar. 21). J. Müller, ‘ Lichenologische Beiträge.’—(Jan. 21). K. Schliephacke, ‘ Das Mikromillimeter.’ — (Feb. 1, 11). H. Karsten, ‘ Ueber Pilzbeschreibung und Pilzsystematik.’ — G. Lagerheim, ‘ Ueber eine durch die Einwirkung von Pilzhyphen enstandene Varietät von *Stichococcus bacillaris.*’ — (Feb. 21 ; Mar. 1). — F. Arnold, ‘ Lichenologische Fragmente.’ —(Mar. 1, 11). O. Schultz, ‘ Vergleichende physiologische Anatomie der Nebenblattgebilde.’

*Gardeners’ Chronicle* (Mar. 3).—*Rodriguezia Bungerothii* Rchb. f., *Aeranthus trichoplectron* Rchb. f., *Ponthieva grandiflora* Ridl., spp. nn. — Fruiting of *Brugmansia lutea* (fig. 42). — ‘ Adventitious bulbs in Scilla' (figs. 45, 46).—(Mar. 17). *Cypripedium dilectum* Rchb. f. "n. sp. (hyb.-nat.)."—(Mar. 24). W. Watson, 'Prolifera-tion in *Utricularia*' (fig. 54).

*Journal de Botanique* (Mar. 1).—A. Franchet, 'Les Mutisiacées du Yun-nan' (*Nouelia* (gen. nov., 1 plate) *insignis, Gerbera raphani-folia, G. ruficoma, G. Delavayi, Ainsliæa yunnanensis, A. pertyoides,* 1 plate, spp. nn.).—H. Douliot, 'Sur le périderme des Légumi-nensis.'—(Mar. 16). E. Strasburger, 'Sur la division des noyaux cellulaires, la division des cellules, et la fécondation.'—J. Costantin, 'Note sur un *Papulaspora*' (1 plate).

*Journ. Linn. Soc.* (Bot.) xxiv. 162 (Mar. 12).—J. R. Vaizey, 'On Anatomy and Development of Sporogonium of Mosses' (4 plates).—G. Henslow, 'Transpiration as a Function of living Protoplasm. Transpiration and Evaporation in a saturated atmo-sphere.'—H. N. Ridley, 'Revision of *Microstylis* and *Malaxis*' (many new species).

*Midland Naturalist* (March).—W. Mathews, 'History of County Botany of Worcester.—J. E. Bagnall, 'The Warwickshire Stour Valley and its Flora.'

*Oesterr. Bot. Zeitschrift.* (Mar.).—K. Fritsch, 'Zur Nomenclatur unserer *Cephalanthera*-Arten.'—L. v. Virkotinovic, 'Neue Eichen-formen.'—L. Celakovsky, 'Orientalische Pflanzenarten' (con-cluded).—A. Hausgirg, 'Zur Algenflora Böhmens' (cont.).—P. Conrath, 'Zur Flora von Bosnien.'—P. G. Strobl, 'Flora des Etna' (cont.).—E. Formánek, 'Flora von Nord-Mähren' (con-cluded).—Zukal, 'Wahring der Priorität.'

---

## LINNEAN SOCIETY OF LONDON.

*March* 1st, 1888.—William Carruthers, F.R.S., President in the chair.—The following were elected Fellows of the Society: Messrs. J. T. Baker, J. B. Farmer, H. P. Greenwood, J. F. Maiden, A. G. Renshaw, A. E. Shipley, and J. A. Voelcker.— An interesting collection of Ferns from the Yosemite Valley was exhibited by Mr. W. Ransom, who also showed some admirable photographs of rare plants, many of them of the natural size.— The first paper of the evening was then read by Mr. E. G. Baker, 'On a New Genus of *Cytinaceæ* from Madagascar.' This curious plant, to which the author has given the name of *Botryo-cytinus*, grows parasitically on the trunks of a tree of the natural order *Hamamelideæ*. Its nearest ally is *Cytinus*, of which the best known species grows on the roots of the Cistuses of the Mediter-ranean basin. The Madagascar plant is without any stem, and the sessile flowers grow in clusters, surrounded by an involucre. Each cluster is universal, and the ovary is unicellular, with about a dozen parietal placentæ, and innumerable minute ovules. It was dis-covered during a recent exploration of the Sakalava-country by the Rev. R. Baron, of the London Missionary Society.—The next paper, by Mr. J. F. Cheeseman (communicated by Sir Joseph

Hooker), was entitled 'Notes on the Fauna and Flora of the Kermadec Islands,' and, as regards the flora, might be considered as supplementary to a paper on the flora of these islands published by Sir Joseph Hooker more than twenty years ago (Journ. Linn. Soc. 1856). These islands, situated about 450 miles N.E. of New Zealand, between that country and Fiji, were shown to be of volcanic origin, with a fauna and flora resembling to a great extent those of New Zealand. A few land birds were noted as common to New Zealand, and to the list of plants drawn up by Sir Joseph Hooker from collections made by Macgillivray several new species were added by Mr. Cheeseman, chiefly ferns. A discussion followed, and in illustration of Mr. Cheeseman's remarks Mr. J. G. Baker exhibited specimens of a new endemic *Davallia* closely allied to the well-known *D. canariensis* of the Canary Islands and Madeira.

*March 15th.*—W. Carruthers, F.R.S., President, in the chair.— The following were elected Fellows of the Society:—Messrs. J. W. Taylor, W. Gardiner, and David Sharp. The following were admitted Fellows of the Society:— Messrs. A. G. Renshaw and A. E. Shipley.—The first paper of the evening was then read by Mr. George Massee entitled, "A Monograph of the *Thelophoreæ*," and drawings of several of these Fungi were exhibited. Remarks were made by Mr. A. W. Bennett and Prof. Marshall Ward.— In the absence of the author, a paper by Mr. E. A. Batters, describing three new Marine Algæ, was then read by the Botanical Secretary, Mr. B. Daydon Jackson, who exhibited the drawings made to illustrate the paper. After some critical remarks from the President, Mr. Harting pointed out the indirect influence of the Gulf Stream in causing a deposition of northern sea-weeds upon the north-east portion of the English coast, where some of the species described had been found.

---

## OBITUARY.

On February 24th, 1887, WILLIAM CURNOW, of Penzance, died at the age of seventy-eight, and, as I have not yet seen the occurrence noted in this Journal, I thought I would contribute a line in memory of him. There are certainly some readers of the Journal who were personally acquainted with him; my friendship with him was limited to the exchange of long letters and numerous specimens, the latter chiefly consisting in later years of *Hepaticæ* and Mosses. I always found him to be a charming, generous, and earnest correspondent, his letters being delightful. He did much towards the investigation of the plants of his district, and wrote several papers, among which the following appeared in the 'Transactions of the Penzance Natural History and Antiquarian Society,' of which he was an honorary member:—' The *Hepaticæ* of West Cornwall '; ' The Sphagnums or Bog Mosses of West Cornwall '; and, in combination with John Ralfs, ' The Mosses of West Cornwall.' This short notice may induce someone with better knowledge to write a more extended notice.—WM. WEST.

Tab. 282.

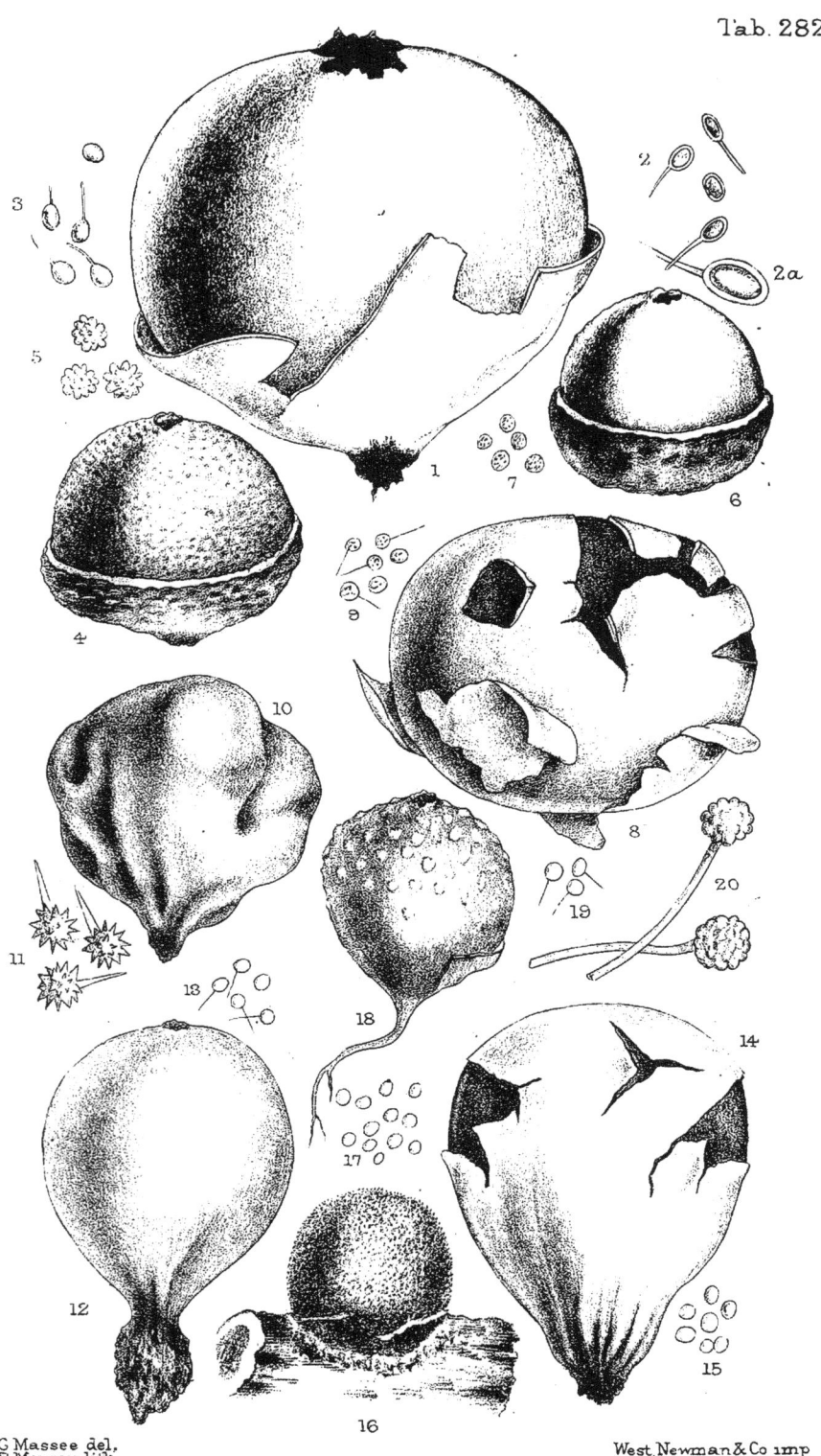

G Massee del,
R.Morgan lith.

West, Newman & Co imp

Bovista

# A REVISION OF THE GENUS *BOVISTA* (DILL.) FR.

## BY GEORGE MASSEE.

### (PLATE 282).

*BOVISTA* (Dill.) Fr.—Peridium double; outer (cortex) distinct, fragile, deciduous; inner (peridium) persistent, dehiscing by a definite or irregularly torn apical orifice; capillitium springing from every part of inner surface of peridium; columella-like sterile base absent. Pers. disp. p. 6; Link diss. i. p. 32; Fr. Syst. Orb. i. p. 138; Fr. Syst. Myc. iii. p. 21. *Lycoperdon* Vitt., Mon. Lycoper. (in part). *Globaria* Quel. Champ. Jur. et Vosg., p. 361, (in part).

Allied to several genera, of which it may be considered the nucleus, and having, perhaps, most affinity with *Lycoperdon*, the points of difference between typical forms of the two genera being as follows :—*Bovista*, cortex free, falling away in patches, sterile base absent, capillitium springing from every portion of inner wall of peridium. *Lycoperdon*, cortex becoming broken up into warts or spines, sterile base present from which the capillitium originates.

In *Bovista* the threads of the capillitium are much more branched and darker coloured than in most Lycoperdons, but in one section are almost colourless and simple, whereas in some species of *Lycoperdon* the threads are as irregular and highly coloured as in *Bovista*. A few species of *Bovista* have the cortex becoming sub-gelatinous, and, as it ceases to grow sooner than the peridium, is torn by the expansion of the latter into scale-like minute patches, which remain permanently attached to the peridium, but in such instances the definite pyramidal structure of the warts of *Lycoperdon* is absent. Pedicellate spores are common to both genera, and of no value, specific or otherwise, in either, their presence or absence depending on age when collected, conditions during drying, and with time drop off in every species. The pedicel of the spore is in reality a sterigma, four of which, as in typical *Basidiomycetes* are produced at the apex of a basidium, and when the spores are mature, instead of breaking away from the tips of the sterigmata as is usual, remain firmly attached, the sterigmata breaking off close to the basidium and thus constitute the pedicels of the spores. *Mycenastrum* is readily distinguished from *Bovista* by the spinulose capillitium threads and corky peridium, *Hippoperdon* differs in the dense elastic capillitium having permanent, small, irregular cavities scattered throughout its substance, somewhat resembling in appearance the sterile basal-stratum of a *Lycoperdon*, and inclines towards such genera as *Scleroderma* and *Polysaccum*.

In the specific character the expression " mass of spores and capillitium " means the colour as seen with the naked eye, whereas the colour of the *threads* and *spores* is as seen by transmitted light. Spores are not, as some people appear to imagine, turned in a lathe, and afterwards all dipped in the same staining solution, but, like other vegetable cells, are liable to vary in size, form, and colour; hence statements respecting spores are those expressing the most frequent condition, and further, are considered as *one* factor only in a specific character; to me it appears quite as unsatisfactory to frame so-called new species on the strength of real or supposed minute differences in spore size, form, colour, or ornamentation, as is the custom with some at the present day, as it was on the part of the older mycologists to depend altogether on naked eye or pocket-lens features. Nevertheless it is much to be desired that in every specific description, spore characteristics, as given above, should be stated, as after the extremists of the present day have passed away, future fungologists, whom it is to be hoped will recognise the importance of *all parts* of the structure, may be glad of the information.

A. *Spores globose, warted or spinulose.*

1. BOVISTA JUGLANDIFORMIS Berk. in Herb. no. 4584.— Subglobose, sessile, cortex thick, persistent and cupulate at the base, peridium polished, rigid, dark brown, mouth small, apical; mass of spores and capillitium olive tinged rufous; threads flaccid, pale, rarely branched, much curled and interlaced; spores brown, globose, minutely warted, about 16 $\mu$. diam., pedicels long, (60-70 $\mu$.) thick, reddish olive. (Type in Herb. Berk. Kew, no. 4584).* Remarkable for the size of the spores and the very long, stout, coloured pedicels. On the ground. S. Africa. From 1-1½ in diameter.

2. B. ZEYHERI Berk. in Herb., no. 4588.—Subglobose, often with a small rounded rooting base; cortex thick, ochraceous, cupulate and persistent below, peridium cinnamon, minutely granulose or powdery, mouth small; mass of spores and capillitium umber; threads pale, simple, much curled; spores brown, coarsely spinulose, 10-12 $\mu$. including spines. (Type in Herb. Bk. Kew, no. 4588.) On the ground. S. Africa. About 1½ in. diam.

3. B. HYALOTHRIX Cke. & Mass. (Grev., March, 1888). — Subglobose; cortex very thick and fibrous, forming a persistent cupulate base; peridium minutely rugulose, dehiscing by a small apical pore; mass of spores umber; threads of capillitium colourless, simple, much curled and interwoven, about 5-6 $\mu$. diam.; spores globose, pale brown, coarsely spinulose, 10-12 $\mu$. diam. (Type in Herb. Kew). On the ground. N.W. of Lake Allacutya, Victoria. (C. French). Allied to *B. circumscissa* B. and *B. juglandiformis* B. A little more than an inch in diameter.

4. B. IRREGULARIS Berk. in Herb. no. 4585. — Subglobose, irregularly nodulose or lobed; cortex pale, fibrous, soon disappearing; peridium thick, elastic, ochraceous, becoming purple-brown; mass of spores and capillitium brown with tinge of purple; threads simple, pale, much curled; spores globose, pale purple-brown,

---

* The bracketted reference in all instances indicates the value and locality of the specimen from which the specific diagnosis, spore measurements, &c., have been taken.

coarsely spinulose, 12 $\mu$. diam. (Type in Herb. Berk. 4585). Remarkable for the varied irregularity of form. From 1–1½ in. across. No locality.

5. B. MUELLERI Berk. Linn. Journ., v. 13, p. 171.— Subglobose, with a short stout rooting basè; cortex soon broken up into minute pale subpersistent warts; peridium firm, rather thick, brown; mass of spores and dense capillitium reddish brown; threads flaccid, pale, but little branched; spores globose, reddish umber, coarsely spinulose, 10–12 $\mu$. including spines. (Type in Herb. Berk. no. 4599.) Remarkable for the very large spinulose spores. From ½ to 1 in. diam. Australia.

6. B. PANNOSA Cooke.—Subglobose; cortex thin, papery, becoming torn and adhering in patches, brown; peridium tough, smooth like leather, cinnamon; mass of spores and capillitium umber with purple tinge; threads thick-walled, brown, branched, thinner than diameter of spores, which are globose, warted, umber with purple tinge, 7 $\mu$. diam. Cooke in Grevillea. (Type in Herb. Kew.) Rio Janeiro. About 2 in. diameter.

7. B. LATERITIA Berk in Herb. no. 4593.—Subglobose; cortex evanescent; peridium pale, thin; mass of spores and exceedingly dense capillitium bright rust-colour; threads thick-walled, brown, branched, tapering; spores spherical, coarsely warted, pale brown, 8 $\mu$. diam. (Type in Herb. Berk. 4593. Sent as a queried specimen by Montagne). About 1¼ in. diam. Locality not known.

8. B. DEALBATA Berk. in Herb. no. 4597.—Subglobose; cortex evanescent, peridium very thin, fragile, silvery; mass of spores and dense capillitium brown; threads thick-walled, brown, branched, tapering towards the tips; spores pale brown, globose, coarsely warted, 8 $\mu$. diam. From 1½–2 in. across. N. Zealand. E. Nepaul.

9. B. ARGENTEA Berk.—Oval, depressed; cortex evanescent, peridium very thin, papery, shining, silvery outside; mass of spores and capillitium dingy red-brown; threads pale, sparingly branched; spores globose, minutely warted, pale brown, 8–9 $\mu$. diam. "Larger diameter 2⅔ inch., smaller 2⅛ inch., depth about 1 inch. Inner peridium resembling very thin paper which has been washed with silver. The sporidia and flocci resemble in colour coffee and cream." (M. J. B.) Berk., Ann. Nat. Hist. iii. p. 400. (Type in Herb. Kew). Madras.

10. B. CIRCUMSCISSA Berk. and Curt. Grev. ii. p. 50.—Subglobose; cortex thick, subcoriaceous, breaking in a circumscissile manner, the lower portion persistent and cupulate, peridium thickish grey, furfuraceous, mouth small, rather silky; mass of spores and capillitium umber with olive shade; threads flexuous, pale, rarely branched; spores globose, minutely verruculose, olive, about 5 $\mu$. diam. *Bovista subterranea* Peck, Ellis N. Amer. Fung. no. 522. (Type in Herb. Berk. 45·98). From ⅔–1 in. diam. The species of Peck is simply a form in which owing to the locality, "in hard trodden paths by the roadside," the cortex is persistent; in every other respect the two are identical. Spores sometimes almost or quite smooth. United States.

11. B. tosta B. & C. — Globose, with a thin cord-like root; cortex evanescent, peridium thick, rigid, dark brown, cracking and falling away in patches; mass of spores and dense capillitium bright cinnamon; threads simple, pale; spores globose, indistinctly granulated, ochraceous, sometimes pedicellate, 5-6 $\mu$. diam. Berk. in Herb. (Type in Herb. Berk. no. 4591). Cuba. The spores are in many instances smooth. Three inches or more in diameter.

12. B. glauco-cinerea Speg. Fung. Argent. Pug. iv. p. 101.— Globose, base rooting; cortex absent, peridium thin, cartilaginous, rather fragile, even, smooth, greyish lead-colour; threads of capillitium hyaline (4 $\mu$. thick) smooth, sparingly septate; spores globose (3-8 $\mu$. diam.), minutely reticulato-papillose, glaucous, sub-hyaline. Sandy ground. Tuyú, Argentina. (25-30''' diam.) Resembling *B. plumbea* in general appearance.

13. B. pampeana Speg. Fung. Argent. Pug. iv. p. 103.—Sub-globose, dehiscing by a more or less regular opening; cortex fugacious, peridium thickish, flexible, then parchment-like, dark chestnut, smooth, even; mass of spores brownish olive; threads of capillitium slender (3 $\mu$ thick), smooth, sparingly septate; walls thick, branched, spores globose, (10-11 $\mu$ diam.) coarsely and densely papillate, fuliginous; pedicels flexuous. Tuyú, Argentina. (25-30'' diam.)

B. *Spores globose, smooth.*

14. B. pila Berk. & Curt. Grev. ii. p. 49. — Subglobose, sessile; cortex fibrillose, evanescent; peridium rather thick, tough, dirty ochraceous, minutely tomentose; mass of spores and capillitium brownish umber, with a slight purple tinge; threads thick-walled, brown, stout, much branched, tips tapering; spores globose, pale umber, 5 $\mu$. diam. (Type in Herb. Berk. Kew. no. 4600). About 1½ in. across. United States.

15. B. nigrescens Pers., Syn. p. 136.—Subglobose; cortex papery, whitish, soon breaking away; peridium thin, tough, shining, blackish umber, dehiscing by an irregular apical orifice; mass of spores and capillitium umber with a purple tinge; threads 12-18 $\mu$. at thickest part, thick-walled, bright brown, much branched, tapering towards the tips; spores same colour, globose, smooth, pedicellate, 5-6 $\mu$ diam. Fr. Syst. Myc. iii. p. 23; Karst, Myc. Fenn. (Basidiomycetes) p. 359; Berk. Engl. Fl. v. p. 302; Outl. Fung. p. 301, pl. 20, f. 5, Cke. Hdbk. p. 371. *Lycoperdon nigrescens*, Vitt. Mon. p. 176. *L. globosum* Bolt. t. 118; Withering iv. p. 350; *L. bovista* Sow. t. 331; *L. ardosiacum* Sowerby Herb. specimen now in Herb. Berk. (specimen in Herb. Berk.) Exs.:— Desm. Cyrpt. Fr. ser. i. 527; Fuckel Fung. Rhen., Karsten Fung. Fenn. 117; Cke. Fung. Brit. 521. Dry pastures and heathy places. From 1-2 in. across. Europe, North America, New Zealand.

16. B. plumbea Pers. Syn. p. 137, t. 3, f. 1.—Globose; cortex thin, whitish, persistent towards the base; peridium tough, thin, lead-coloured, dehiscing by a small irregular mouth; mass of spores and capillitium umber-brown; threads thick-walled, stout,

12–16 $\mu$. at thickest part, brown, much branched, tapering at the tips ; spores paler, subglobose, smooth, pedicellate, 5–6 $\mu$ diam. Fr. Syst. Myc. iii. p. 24; Berk. Engl. Fl. v. p. 302; Berk. Outl. p. 301, pl. 20, f. 6 ; Cke. Hdbk. p. 372 ; Karst. Myc. Fenn. (Basid.) p. 360, Corda Ic. v. f. 47 ; Mich. t. 97, f. 6. *Lycoperdon ardosiacum* Bull. t. 192, A. B. ; With iv. p. 351. *L. plumbeum* Vitt. Mon. p. 174. Exs. :—Klotzsch Fung. Germ. 57 ; Oudemans Fung. Neerl. 117 ; Kx. Rech. Fl. Crypt. Fland. 1261 ; Fuckel Fung. Rhen. 1262 ; Klotzsch Herb. Myc. 143. (Specimen in Herb. Berk.) About 1 in. diam. Dry grassy or heathy places. Europe. St. Michael's, Azores. California. The thin papery cortex and absence of hyaline border to the spores separates this species from *B. ovalispora* Cke. & Mass., which it much resembles externally.

17. **B. olivacea** Cke. & Mass. n. sp.—Globose ; cortex very thin, fugacious ; peridium thick, soft, becoming brittle and breaking away in patches, pale ochraceous, at length nearly white ; mass of spores and very dense capillitium citrin then olive ; threads thin, pale, flaccid, simple ; spores globose, smooth, pale yellow, sometimes pedicellate, 5 $\mu$. diam. (Type in Herb. Kew.) From 1½–2 in. across. Allied to *B. pannosa* Cke. Wimmera, Australia. (F. Reader). England. (Specimen in Herb. Broome, Brit. Mus.)

18. B. BRUNNEA Berk. Fl. N. Z. ii. p. 189.—Globose, with a minute apiculate rooting base ; cortex thin, evanescent; peridium brownish umber, smooth, shining, dehiscing by a small irregular slit ; mass of spores and capillitium brown ; threads thick-walled, brown, branched, tapering ; spores smooth, globose, brown, 4–5 $\mu$ diam., usually furnished with a long slender pedicel. Handbk. Fl. N. Z. 618. (Type in Herb. Berk. Kew, no. 4587. About 1 inch across. New Zealand.

19. B. VELUTINA B. & Br., Journ. Linn. Soc. xiv. p. 78.— Globose ; cortex thick, evanescent above, cupulate and persistent at the base where it becomes effused and whitish ; peridium bright brown, velvety ; mass of spores and capillitium umber ; threads rarely branching, tapering, pale ; spores globose, smooth, pale, 3–4 $\mu$ diam. (Type in Herb. Berk. no. 4603). From ½–⅔ in. diam., growing on small branches. Ceylon.

20. B. AMMOPHILA Lév. Ann. Sci. Nat. ser. 3, ix. p. 129, pl. 9, f. 5.—Broadly obovate, plicate below and passing into a long, stout, tapering root ; cortex whitish, broken up into tomentose warts, peridium thin, pallid, dehiscing by a small irregularly torn apical mouth ; mass of spores and capillitium olive ; threads branched, thick-walled, olive ; spores globose, smooth, pale, pedicellate, 5–6 $\mu$. diam. B. and Br. Ann. Nat. Hist. no. 1033 ; Cooke, Hdbk. p. 372. (Specimen determined by Rev. M. J. Berkeley, in Herb. Berk. no. 4594). On the ground in sandy places. 1–1½ in. high. France. Britain.

21. B. (?) PALUDOSA Lév. Ann. Sci. Nat. 3 ser., v. p. 163.—Gregarious. Subglobose or depressed, plicate below and passing abruptly into a stout stem-like base becoming attenuated downwards ; cortex white, tomentose, evanescent, peridium thin, firm, ochraceous, becoming dark brown ; mass of spores and capillitium

olive; threads rufous-olive, thick-walled, branched, tips tapering; spores same colour, globose, smooth, pedicellate, 4–5 $\mu$. (Specimen from Léveille in Herb. Berk. no. 4595). Agrees with *Lycoperdon* in having a sterile base, but is a true *Borista* in other respects. Peridium about 1½ in. diam., 1 in. high; stem 1 in. long, ¾ in. thick at apex. Amongst sphagnum. Malesherbes.

22. B. ABYSSINICA Mont. Syll. Crypt. p. 287.—Obovate, tapering into a stout rooting base; cortex broken up into minute white persistent warts; peridium papery, smoky-brown, or lead-colour, mouth small, determinate; mass of spores and capillitium dingy olive; threads thick-walled, much branched, tapering, pale brownish olive; spores same colour, globose, smooth, sometimes with a long slender pedical, 5–6 $\mu$. diam. (Specimen from Montagne in Herb. Berk. no. 4596). Abyssinia. (11,000 ft.). About 1½ in. high by 1 in. broad.

23. **B. radicata** Mass. n. sp.—Globose or broadly obovate, slightly plicate below, and contracted into a stout rooting base; cortex breaking up into minute whitish persistent warts; peridium thin, tough, silvery, becoming ferruginous towards the base; mouth small, subrotund; capillitium dense and with mass of spores umber-brown, tinged olive; threads thick-walled, brown, much branched and tapering, often flexuous towards the tips; spores globose, smooth, pale brown. sometimes shortly pedicellate, 4–5 $\mu$. diam. (Type in Herb. Kew). From ⅔–1½ in. across. Cameroon Mountains, 8–10,000 ft. (Mann).

24. B. CERVINA Berk. Ann. Nat. Hist., ix. p. 447 (1842).— Subglobose, with an abrupt stout rooting base; cortex rigid, evanescent above or remaining as scale-like patches, usually persistent below; peridium thin, pale cinnamon; mouth small, apical, subrotund; mass of capillitium and spores brown; threads flaccid, unbranched, pale; spores globose, smooth, very pale brown, minutely pedicellate, 5–6 $\mu$. diam. (Type in Herb. Berk. Kew, no. 4586). *Borista aspera* Lév. Ann. Sci. Nat. ser. 3, v. p. 162, as proved by examination of type-specimen in Herb. Mus. Paris. "Driest part of plains, Rio Negro, Patogonia, Darwin." Chili. Ceylon. New Zealand. From ½–1 in. across.

25. B. URUGUAYENSIS Speg. Fung. Argent. Pug. iv. p. 102.— Globoso-depressed; cortex rather thick, persistent and cupulate at the base, dirty brown, fragile; peridium thickish, grey or brownish grey, coriaceous, rigid, smooth above, tomentose below, dehiscing by a small, regular, subfimbriate mouth; mass of spores and capillitium pale olive; threads slender (3 $\mu$. thick), branched, here and there furnished with warts, brownish olive; spores globose, (3–5 $\mu$. diam.) smooth, pale olive. Related to *B. cervina* Berk. (10–15''' wide by 8–10''' high). Sandy places near Concordia, Uruguay.

26. **B. obovata** Mass. n. sp.—Broadly obovate, plicate below, with a short, thick, rooting base; cortex whitish, evanescent; peridium fragile, papery, silvery; capillitium dense and with mass of spores umber; threads thick-walled, branched, pale umber; spores same colour, globose, smooth, 4 $\mu$. diam. (Type in Herb.

Kew.) Wingate to Zuni, S.E. New Mexico; 6–8000 ft. (Prof. Moseley). About 1½ in. high by 1 in. across.

27. B. STUPPEA Berk., Grev. ii. 2, p. 50.—Subglobose, or ellipsoid, sessile ; cortex whitish, thin, evanescent ; peridium bright brown becoming silvery, shining ; mass of spores and dense capillitium brown ; threads branched, thick-walled, brown ; spores globose, smooth, pale umber, sometimes with a short pedicel, 5 $\mu$. diam. (Type in Herb. Berk. 4602). About 2 in. by 1¼ in. On the ground. Texas.

28. B. CINEREA Ellis, Bull. Washburn Coll. i., no. 2, p. 40 (1885).—Globose, 5–6 cm. in diameter with a short, subfusiform root-like base ; peridium coriaceous, about 1 mm. thick, smooth or nearly so, entire ; capillitium cinereous-grey, abundant, the threads 3–4 $\mu$. diam., more or less branched, and attached on all sides to the inner surface of the peridium ; spores globose, clay-coloured or cinereous, echinulate, 4–5 $\mu$. in diameter. Prairie Ford, Co. Kansas, August.

29. B. AMETHYSTINA Cke. & Mass., Grev. (March, 1888). — Globose, or rather attenuated and plicate below ; cortex evanescent, peridium shining, papery, fragile, breaking away in patches ; mass of spores and capillitium bright amethyst ; threads about as thick as diameter of spores, branched, equal, almost colourless ; spores pale lilac, globose, smooth, very copious, 5–6 $\mu$. diam. (Type in Herb. Berk. Kew). Niger Expedition. (Barter.) About 1½ in. diam.

30. B. CASTANEA, Lév. Ann. Sci. Nat., ser 3, v. p. 162.— Globose ; cortex . . . . . ; peridium parchment-like, black, byssoid below ; mass of spores and dense capillitium bright olive ; spores globose, smooth, with long pedicels. Size of a chestnut. Cape of Good Hope.

31. B. CISNEORI* Speg. Fung. Argent. Pug. 4, p. 100.—Globoso-depressed or globoso-turbinate ; cortex absent ; peridium very smooth, membranaceo-cartilaginous, above brownish orange, brighter below ; threads of capillitium fulvous slender (3–5 $\mu$. thick) simple, smooth, sparingly septate ; spores globose, (3–5 $\mu$. diam.) smooth (or very minutely rugulose) tawny orange. In dry sandy places. Uruguay. Related to B. bicolor Lev. (40·70''' diam. =30–40''' alt.)

32. B. DUBIOSA Speg. Fung. Argent. Pug. iv. p. 101.—Sub-globose, more or less rooting ; cortex white, almost smooth or sparingly verruculose, thin, evanescent, densely interspersed with particles of sand ; peridium very thin, at first white then pale olivaceo-fulvous, more or less covered with persistent fragments of the cortex ; threads of capillitium slender (3–5 $\mu$. thick), branched,

---

* The extreme length and amount of detail introduced into Spegazzini's specific characters suggest the idea that they are detailed descriptions of *individuals* rather than otherwise. The constant repetition of general features is objectionable, and the statement that the peridium is at first closed, then becoming open, is simply absurd. Full details as to the changes taking place in the gleba are highly instructive and interesting, but out of place in a specific diagnosis.

smooth, olive; spores globose, (3-5 $\mu$. diam.) epispore thick, smooth, with pedicels from once to twice as long as diameter of spores. Common. (18-24''' diam.) Argentina.

### c. *Spores elliptical.*

33. B. BICOLOR Lév. Ann. Sci. Nat. 3 ser. v. p. 162.—Globose; cortex evanescent; peridium thin, glabrous, papery, brownish or cinnamon, dehiscing by a small determinate mouth; capillitium dense and with mass of spores, bright cinnamon; threads pale, flexuous, rarely branched; spores elliptic-oblong, pallid, smooth, 6 × 3-4 $\mu$. (Specimen from Léveille in Herb. Berk. no. 4590). The peridium is often irregularly torn above. There is sometimes a stout rooting base. From 2-4 in. across. Bombay. Ceylon.

34. B. OVALISPORA Cke. & Mass. Grev. xvi. p. 46. — Subglobose, sessile; cortex whitish or ochraceous, breaking away in patches above, subpersistent towards the base; peridium thin, flaccid, smooth, dull lead-colour, dehiscing by an irregular apical rupture; capillitium and spores umber in the mass; threads 12-16 $\mu$. at thickest part, much and vaguely branched, tapering to long slender tips, walls thick, dirty umber; spores oval, brownish umber, with a narrow hyaline border; pedicels long, stout, hyaline, 6 × 4-5 $\mu$. (Type in Herb. Kew). On the ground; Kew Gardens. Nelson (New Zealand). Carolina (U.S.A.). Differing from *B. plumbea* in being larger (2 in. or more), in the oval spores, with a hyaline border, and much thicker cortex; and from *B. nigrescens* in the oval spores and absence of purple tinge in the capillitium and spores.

35. **B. fulva** Mass. n. sp.—Globose; cortex persisting for some time in form of minute whitish warts; peridium rather thick, lead-colour with brown tints, dehiscing by a small apical mouth; mass of spores and capillitium bright fulvous; threads thick, branched, brown, tapering; spores very pale yellow, broadly elliptical, pedicellate, 5-6 × 3-4 $\mu$. (Type in Herb. Kew). On the ground. Simla. About 1 in. across.

D. *Species that cannot be arranged under either of the sections given, owing to absence of information respecting the spores.*

36. B. TUNICATA Fr. Syst. Myc. iii. p. 25.—Globose; cortex very thin, free, tunicate; peridium papery, pliant, lead-colour; mouth torn; mass of spores and capillitium smoky olive. About the size of a walnut. Grassy cliffs. Sweden.

37. B. FUSCA Lév. Ann. Sci. Nat. v. 5, p. 303.—Globose; peridium brown, dehiscing vertically; mass of spores and capillitium blackish-purple; threads dichotomously branched, attenuated; spores globose. On the ground. New Grenada.

38. B. CRANIIFORMIS Schwein.—Stipitate, stem obconic, thick, three inches diameter, short, scarcely rooting; peridium resembling in form a human skull; cortex at first minutely furfuraceous, membranaceous and irregularly torn, glabrous inside; peridium densely floccose; mass of spores and capillitium yellowish-ochre at length greyish; spores pedicellate. Schweinitz, Syn. Fung.

Amer. Bor. in Amer. Phil. Soc. 1831, p. 256. Peridium some-times a foot in diameter. Gregarious, in orchards. Bethlehem, United States.

The above description includes the leading characters and is taken from the lengthened description by Schweinitz.

39. B. spumosa Lév. Ann. Sci. Nat. 1844, xi. p. 219.—Globose, small, covered with a dense white frothy cortex which disappears ; mass of capillitium and spores brown. On the ground. Sumatra.

SPECIES EXCLUDED.

*Bovista lilacina* Berk. and Mont., transferred to *Lycoperdon*.

*B. delicata* Berk.=*Lycoperdon delicatum* B. and C., which again is synonymous with *Lycoperdon Berkeleyi* Mass. Mon. Lyc. n. 33.

DESCRIPTION OF FIGURES.—Fig. 1, *Bovista ovalispora*, nat. size ; 2, spores of same × 400; 2 *a*, spore of same × 750. 3, spores of *B. nigrescens* × 400. 4, *B. hyalothrix*, nat. size ; 5, spores of same × 400. 6, *B. circumscissa*, nat. size ; 7, spores of same × 400. 8, *B. olivacea*, nat. size ; 9, spores of same × 400. 10, *B. irregularis*, nat. size ; 11, spores of same × 400. 12. *B. radicata*, nat. size ; 13, spores of same × 400. 14, *B. amethystina*, nat. size ; 15, spores of same × 400. 16, *B. velutina*, nat. size ; 17, spores of same × 400. 18, *B. cervina*, nat. size ; 19, spores of same × 400. 20, spores of *B. juglandiformis* × 400.

A SYNOPSIS OF *TILLANDSIEÆ*.

By J. G. BAKER, F.R.S., F.L.S.

(Continued from p. 111.)

207. Tillandsia gracilis Griseb. in Gott. Nacht. 1864, 17. *Vriesea gracilis* Gaudich, Atlas Bonité t. 67 ; Wawra in Œster. Bot. Zeitsch. 1880, 218 ; French trans. 68 ; Itin. Prin. Sax. Cob. 164 ; Regel Gartenfl. 1886, 161, fig. 11 ; Descr. Plant, Nov. fasc. x. 30.—Leaves 30-40 in a dense rosette, lanceolate from an ovate base, 2-3 in. broad, thin, flexible, subglabrous, light green spotted with purple, 1-1½ ft. long, 1½ in. broad at the middle, narrowed gradually to the point. Peduncle slender, as long as the leaves ; bract-leaves small, adpressed. Panicle lax, diffuse ; lower branches 6-9 in. long, 6-8 flowered, with a slender flexuose rachis ; branch-bracts small, ovate ; lower flowers patent ; flower-bracts oblong, under an inch long, red with a green tip. Calyx greenish, an inch long. Petal-blade yellow, oblong, ⅓ in. long. Capsule-valves lanceolate, 1½ in. long.

Hab. Forests of Southern Brazil, *Burchell* 2354 ! *Gaudichaud* ; Demerara, *Jenman* 3828 ! *Wawra* and *Maly* ii. 202, 515. Intro-duced lately into cultivation by Dr. Glaziou. A plant gathered by Salzmann at Bahia differs only by its lorate leaves with a deltoid cuspidate tip. Grisebach refers here Kegel's 234 from Dutch Guiana, and K. Koch *T. patens* Herb. Willd., gathered by Humboldt at Caripe, briefly described by Schultes Syst. Veg. vii. 1229. *Platystachys patens* K. Koch. Ind. Sem. Hort. Berol. 1873.

App. iv. p. 5.) In Burchell's specimen the flowers are copiously viviparous.

208. T. RODIGASIANA Baker.    *Vriesea Rodigasiana* E. Morren in Ill. Hort. n.s. t. 467 ; Antoine Brom. 16, t. xi.—Leaves 30-40 in a dense rosette, lorate from a dilated base 2 in. diam., thin, flexible, subglabrous, plain green, 1-1½ ft. long, an inch broad at the middle, obtuse with a cusp at the tip.  Peduncle shorter than the leaves ; bracts small, adpressed, not imbricated.  Panicle a foot long, with a red axis, several laxly 4-5 flowered short ascending branches and small red branch-bracts ; flower-bracts oblong, yellow, ½ in. long.  Calyx bright yellow, an inch long.  Petals half as long again as the calyx, bright yellow.  Stamens as long as the petals.

Hab. South Brazil, introduced into cultivation by Linden about 1882.  Closely allied to *T. gracilis*.  I have not seen the allied *Vriesea billbergioides* E. Morren, founded upon *Glaziou's* 3630, mentioned by name under the figure above cited.

209. T. PROCERA Mart. ; Roem. et Schultes Syst. Veg. vii. 1224. —Leaves lorate, thin, flexible, glabrous, 1½ ft. long, 1½-2 in. broad at the middle.  Peduncle stout, erect, 2 ft. long ; lower bract leaves ½ ft. long.  Panicle 2 ft. long, a foot broad ; branches erecto-patent, 6-9 in. long, 6-10 flowered ; flowers spaced out, spreading ; flower-bracts oblong, under an inch long.  Calyx 13-14 lines long ; sepals obtuse.  Petal-blade ligulate.  Stamens rather shorter than the petals.  Capsule-valves 1½ in. long, ¼ in. broad.

Hab. Brazil, near Itahype, *Martius*.  Allied apparently to *T. gracilis* and *Rodigasiana*.

210. **T. Tweedieana,** n. sp.—Leaves a foot or more long, lorate from an ovate base, thin, flexible, subglabrous, 1½ in. broad at the middle, above 2 in. at the dilated base.  Peduncle about a foot long ; bract-leaves all small and adpressed.  Panicle ½-1 ft. long ; branches 8-10, short, with a flexuouse rachis, few-flowered ; branch-bracts small, ovate ; flower-bracts ovate, ½-⅝ in. long.  Calyx ¾-⅞ in. long ; sepals obtuse, much imbricated.  Petal-blade oblong, ¼ in. long.  Stamens a little longer than the petals.

Hab.  South Brazil ; Rio Janeiro, *Tweedie* 1342 !  Collected in the year 1837.  *Glaziou* 16467 !

211.  T. PHILIPPO-COBURGI Baker.  *Vriesea Philippo-Coburgi* Wawra in Œster. Bot. Zeitschrift 1880, 219 ; French trans. 68 ; Itin. Prin. Sax. Cob. 165, t. 29 and 37 t. A.—Leaves 20 in a rosette, lorate thin, flexible, subglabrous, 1½-2 ft. long, 2-2½ in. broad at the middle, brown at the tip and rounded to a small cusp, much recurved, spotted with purple on the back towards the base.  Peduncle stout, a foot long ; upper bract-leaves bright red. Panicle 1½-2 ft. long ; branches many, ascending, lower forked, sometimes a foot long ; flowers erecto-patent, laxly disposed, 6-10 to a branch ; branch-bracts bright red, lower lorate, 4-5 in. long ; flower-bracts ovate, acute, bright red, ¾-1 in. long.  Calyx yellowish, 1-1¼ in. long ; sepals oblong-lanceolate.  Petals greenish, half as long again as the calyx.  Stamens a little longer than the petals.

Hab. South Brazil; forests of the neighbourhood of Rio Janeiro and St. Paulo, *Wawra* and *Maly*, *Longman*! *Glaziou* 8017! 16473! *Mosen* 3248!

212. T. RETICULATA Baker in Gard. Chron. 1887 i. 140.—Leaves 30–40 in a dense rosette, lorate from a dilated ovate base 4 in. broad, thin, flexible, subglabrous, 1½-2 ft. long, 3 in. broad at the middle, deltoid-cuspidate at the apex, grey green reticulated with copious maculations of pale green. Peduncle nearly as long as the leaves; bract-leaves imbricated. Panicle a foot or more long; branches many, erecto-patent; flowers spaced out, erecto-patent; flower-bracts ovate, greenish, under an inch long. Calyx 1¼-1½ in. long; sepals obtuse. Petal-blade small, orbicular, whitish, falcate. Stamens just protruded beyond the petals.

Hab. South Brazil; Rio Grande de Sul. Described from a plant that flowered with Mr. Wm. Bull, in March, 1873.

213. T. MORRENI Baker. *Vriesea Morreni* Wawra in Œster. Bot. Zeit. xxx. 219; French trans. 69; Itin. Prin. Sax. Cob. 166, t. 30 and 37 B.—Leaves 30 or more in a dense rosette, lorate from a dilated ovate base, 4 in. broad, 1½ ft. long, 2½-3 in. broad at the middle, thin, flexible, subglabrous, rounded to a cusp at the apex, green, transversely banded with brown flexuose lines. Peduncle under a foot long. Panicle rhomboid, 2 ft. long; branches few, long, erecto-patent, many flowered; branch-bracts small, ovate; flowers secund, erecto-patent; flower-bracts oblong, brownish, an inch long. Calyx half as long as the bract; sepals oblong-lanceolate. Petals greenish yellow, 2 in. long; blade oblanceolate-oblong. Stamens not larger than the petals. Capsule twice as long as the calyx.

Hab. South Brazil; waterfall of Itamarati, near Petropolis, *Wawra* and *Maly* ii. 72; and a doubtful variety with distichous flowers and much smaller capsule at Teresopolis, *Wawra* and *Maly* ii. 350.

214. T. TESSELLATA Linden and André in Ill. Hort. 1873, 78; 1874, 123, t. 179; E. Morren in Belg. Hort. 1882, 381, t. 14–16. Leaves 30–40 in a rosette, lorate from a dilated ovate base, 1½-2 ft. long, 3–4 in. broad at the middle, rounded to a cusp at the apex, thin, flexible, subglabrous, tessellated all over with fine green cross-lines on a pale ground. Peduncle as long as the leaves; lower bract-leaves with spreading tips, also tessellated. Panicle 3–5 ft. long; branches many, ascending, lower a foot long; flowers secund, ascending; flower-bracts oblong, 1¼-1½ in. long. Calyx green, 2 in. long. Petals yellowish white, half an inch longer than the calyx. Stamens as long as the petals.

Hab. South Brazil; province of Santa Catherina, introduced by Linden about 1872.

215. T. REGINA Vell. Fl. Flum. iii. t. 142. *Vriesea regina* Beer, Brom. 97; E. Morren in Belg. Hort. xiv. 325; Wawra in Oester Bot. Zeit. xxx. 185; French trans. 66; Itin. Prin. Sax. Cob. 163, t. 36 A; Gard. Chron. 1875, fig. 41; Antoine, Brom. t. 9–10. *V. gigantea* Gaudieh. Atlas Bonité t. 70; Lemaire in Ill. Hort. t. 516. *V. Glazioviana* Lemaire Ill. Hort. xiv. Misc. 43. *V. geniculata*

Wawra Reise Kais Max. 159, t. 25.—Leaves 30-50 in a dense rosette, lorate from an ovate base 6-9 in. diam., flexible, glabrous, plain green with a glaucous bloom, 2-3 ft. long, 3-6 in. broad at the middle, narrowed gradually to the point. Peduncle about as long as the leaves; bract-leaves with large free lanceolate points. Panicle very large; branches erecto-patent in the type; branch-bracts large, green, ovate-cuspidate; flowers spaced out, usually erecto-patent; rachis stout, flexuose; flower-bracts oblong, obtuse, 1-1¼ in. long. Calyx 1½-2 in. long; sepals oblong, obtuse. Petals oblanceolate, obtuse, whitish, an inch longer than the calyx. Stamens as long as the petals. Style twice as long as the calyx. Capsule 1½-2 in. long.

Hab. South Brazil; forests of Rio Janeiro and St. Paulo, *Burchell* 2916! (year 1826). *Glaziou* 15468! 15469! 15470! *Mosen* 3247! *V. gigantea* Lemaire (*Glaziou* 11685, 1685!) is a form with narrower leaves than the type, shorter deflexed spikes and crowded flowers. Introduced into cultivation by Linden about 1868.

216. T. GRANDIS Schlecht in Linnæa xviii. 424.—Leaves lorate from a dilated base 3 in. diam., thin, flexible, 1½-2 ft. long, 2 in. broad at the middle, deltoid cuspidate at the apex. Inflorescence an ample panicle; lower branches 1½ ft. long, with a spike a foot long; flowers ½ in. apart; flower-bracts ovate, obtuse, shorter than the calyx. Calyx 1½ in. long; sepals oblong, obtuse, above ½ in. broad. Petals unknown. Capsule 2 in. long.

Hab. Mexico; Hacienda de la Laguna, *Schiede*.

217 T. PANICULATA Linn. Sp. 410 (Plum. Ic. t. 237), Roem. et Schultes Veg. vii. 1220.—This is founded entirely upon two figures of Plumier. The published one cited represents a plant with a dense rosette of lanceolate leaves and a huge chandelier-like panicle with spreading main branches and short secund laxly-flowered ascending branchlets. The unpublished figure represents a flower 4 in. long, with an oblong acute flower-bract shorter than the calyx, a calyx half as long as the corolla, with acute sepals, a lanceolate petal-blade as long as the calyx and stamens longer than the petals. Nothing at all like it is known in St. Domingo or the other West Indian islands at the present day.

*Imperfectly known and undescribed species.*

218. VRIESEA ? SANGUINOLENTA Cogn. and March. in Album Dall. ii. 1874, t. 52; André in Ill. Hort. n. s. t. 200.—Leaves lorate, a foot or more long, an inch broad at the middle, deltoid-cuspidate at the apex, green, copiously irregularly spotted with claret-red as in *V. guttata*, but not utriculate and dilated at the base. Inflorescence unknown. New Granada at Choco, discovered by Roezl in 1872.

219. V. KRAMERI, mentioned Gard. Chron. 1883, 510.

220. V. PRODIGIOSA Lemaire in Ill. Hort. xvi. App. 92.— Peduncle above 4 ft. long. Inflorescence composed of about 30 dense 10-12 flowered spikes 3-4 in. long, subtended by large

ovate-lanceolate bracts. Mexico, *Gheisbreght.* Is very likely one of the large Mexican species of *Platystachys* (Sp. 75-78).

Subgenus X, CYATHOPHORA K. Koch.—Leaves densely rosulate, lanceolate, acuminate. Inflorescence more or less decidedly multifarious. Petal-blade lingulate; claw not scaled. Stamens and style long. Differs from *Allardtia* and *Platystachys* by its non-distichous inflorescence.

Inflorescence a dense capitulum with
single flowers in the axils of large
bracts . . . . . . Sp. 221-226.
Habit small; panicle dense . . Sp. 227-233.
Large, with a lax panicle . . Sp. 234.

221. T. CAPITATA Griesb. Cat. Cub. 255.—Leaves about a dozen in a dense rosette, lanceolate, acuminate from a dilated ovate base an inch broad, 5-6 in. long, $\frac{1}{2}$ in. broad at the middle, recurved, sub-coriaceous, narrowed gradually from the base to a long point, thinly persistently lepidote on both surfaces. Peduncle 4 in. long; bract-leaves crowded, with long lanceolate recurved points. Flowers in a dense oblong head, $1\frac{1}{2}$ in. long; bracts large, ascending, ovate-acuminate, firm in texture, lepidote, the lower 2-3 in. long. Calyx $\frac{1}{2}$ in. long; sepals oblong, obtuse. Petals not seen. Capsule above an inch long.

Hab. Cuba, *Wright* 3274!

222. **T. sphærocephala,** Baker n. sp.—Root-leaves few, lanceolate, acuminate from an ovate dilated base $1\frac{1}{2}$ in. long: blade $\frac{1}{2}$ ft. long, $\frac{1}{2}$ in broad at the middle, firm in texture, thinly lepidote on both surfaces, with convolute edges. Peduncle 3 in. long, quite hidden by the large lanceolate acuminate bract-leaves. Inflorescence a dense oblong multifarious capitulum, with single flowers in the axils of large ovate bracts of firm texture with a long convolute cusp, the lower 2-3 in. long. Calyx flattened, an inch long, enclosed in a pair of boat-shaped bracteoles nearly as long as itself; sepals acute. Petals not seen. Capsule-valves lanceolate, an inch long.

Hab. Andes of Bolivia near Sorata, alt. 9000-10,000 ft. *Mandon* 1188! Near *T. capitata* Griseb.

223. **T. oxysepala,** n. sp.—Leaves about a dozen in a rosette, lanceolate-acuminate from a dilated ovate base an inch broad, 8-9 in. long, $\frac{1}{2}$ in. broad at the middle, tapering gradually to a long point, firm in texture, persistently finely lepidote on both surfaces. Peduncle stiffly erect, as long as the leaves; bract-leaves imbricated, with ovate bases and squarrose linear-subulate points. Inflorescence a few-flowered capitulum, each flower in the axil of a lepidote ovate-acuminate bract of firm texture, 2-2$\frac{1}{2}$ in. long. Calyx $\frac{3}{4}$-$\frac{3}{4}$ in. long; sepals lanceolate, acute; pedicel erecto-patent, $\frac{1}{8}$ in. long. Petals not seen.

Hab. Andes of Peru; Quebrada of Guerreras, *Bridges!* Described from a specimen given by Lord Colchester to Mr. Bentham in 1832.

224. **T. cryptantha,** n. sp. — Tufts crowded. Basal leaves few, lanceolate-acuminate from a dilated ovate base an inch broad, 8-9 in. long, recurved, moderately firm in texture, $\frac{1}{2}$ in. broad at the middle, finely persistently lepidote on both surfaces, tapering very gradually to a long point. Peduncle about 3 in. long; bract-leaves very large, crowded, lanceolate-acuminate, recurved. · Inflorescence a dense globose capitulum, much overtopped by the ovate-acuminate bracts, which are 4 in. or more long and quite similar to the leaves in texture. Flowers single in the axils of the bracts on short stout pedicels. Calyx above $\frac{1}{2}$ in. long; sepals oblong. Capsule-valves lanceolate, $1\frac{1}{4}$ in. long.

Hab. Mexico; Cuernavaca, *Bourgeau* 1423!

225. **T. macrochlamys,** n. sp. — Basal leaves 1-1$\frac{1}{2}$ ft. long, lanceolate-acuminate from a dilated oblong base 1$\frac{1}{2}$ in. broad, $\frac{1}{2}$-$\frac{3}{4}$ in. broad at the m'd lle, tapering gradually to a long point, firm in texture, obscurely lepidote. Peduncle a foot or more long; bract-leaves crowded, large, lanceolate from a loosely-clasping ovate base $\frac{3}{4}$-1 in. broad. Inflorescence a large oblong capitulum, with single flowers in the axils of thin glabrous ovate-acuminate bracts 3-4 in. long, 1-1$\frac{1}{2}$ in. broad. Calyx $\frac{3}{4}$ in. long, enclosed in a pair of subcoriaceous oblong-navicular bracteoles; sepals obtuse, much imbricated. Petals and capsule not seen.

Hab. Central Mexico, *Hahn*! Gathered during the French Expedition of 1865-6.

226. **T. longipetala,** n. sp.—Leaves short, green above, tinged with red beneath. Inflorescence a lax multifarious spike 3-4 in. long, each flower subtended by a chartaceous rose-tinted spreading or falcate lanceolate bract 3-4 in. long. Calyx 1 in. long. Petal-blade white, oblanceolate-unguiculate, protruding an inch beyond the calyx. Stamens nearly as long as the petals.

Hab. Andes of Columbia, alt. 3000-4000 ft. I know this very distinct species only from a dried spike sent to Kew for determination by Messrs. Sander & Co., of St. Alban's, in March, 1881.

227. T. BIFLORA Ruiz et Pavon, Fl. Peruv. t. 268 b; Roem. et Schultes, Syst. Veg. vii. 1228. *Diaphoranthema biflora* Beer, Brom. 156. — Acaulescent. Leaves 20-30 in a dense rosette, lanceolate from an ovate base an inch broad, thin, subglabrous, pale green, half a foot long, $\frac{1}{2}$-$\frac{3}{4}$ in. broad at the middle, narrowed gradually to the point. Peduncle as long as the leaves; bracts large, lanceolate. Flowers in a small moderately dense non-distichous panicle 3-4 in. long; lower branch-bracts thin, ovate-acuminate, about as long as the congested few-flowered spikes; flower-bracts small, ovate, obtuse. Calyx glabrous, $\frac{1}{3}$-$\frac{1}{2}$ in. long; sepals obtuse. Petals not seen. Capsule-valves lanceolate, more than twice as long as the calyx.

Hab. Peru; Andes of Muna, *Ruiz & Pavon*. Eastern Cordillera at Tabina, *Lechler* 2132! Andes of Bolivia, alt. 8000-9000 ft., *Mandon* 1174!

228. T. GRISEBACHIANA Baker. *T. tetrantha* Griseb. in Gott. Nachtrag. 1864, 18, non R. & P. — Acaulescent. Leaves about a dozen in a dense rosette, oblanceolate from a short dilated ovate base, ½ ft. long, an inch broad above the middle, thin, pale green, subglabrous, narrowed from above the middle to an acute point. Peduncle rather shorter than the leaves ; bracts ovate-lanceolate, thin, erect, imbricated. Inflorescence a narrow moderately dense decidedly dense multifarious panicle 4–5 in. long ; branch-bracts thin, ovate-cuspidate, very ascending, 1–1½ in. long, equalling the congested few-flowered spikes ; flower-bracts small, thin, ovate. Calyx glabrous, ½ in. long ; sepals obtuse. Petal-blade oblong, ⅙ in. long. Stamens shorter than the petals.

Hab. Venezuela ; mountains of Tovar, *Fendler* 1509 !

229. T. XIPHOPHYLLA Baker. *Anoplophytum T. foliosa* E. Morren in Bourg. Pl. Mexic. Exsic. No. 1906, non Mart. et Gal. — Leaves about 20 in a dense rosette, lanceolate from a large ovate dilated base 1–1½ in. broad, thin, subglabrous, 8–9 in. long, ½ in. broad at the middle, narrowed gradually to an acute point. Peduncle as long as the leaves ; bract-leaves large, lanceolate, imbricated, erect. Panicle non-distichous, 4–6 in. long, dense in the upper, lax in the lower half ; branch-bracts ovate or ovate-lanceolate, lower 2–3 in. long ; spikes dense, short, erecto-patent ; flower-bracts oblong-lanceolate, ¾ in. long. Calyx ½ in. long, falling short of the bract ; sepals oblong, obtuse. Petals not seen. Capsule-valves lanceolate, an inch long.

Hab. Mexico ; Valley of Cordova, *Bourgeau* 1906 ! Nearly allied to *T. biflora* R. & P.

230. **T. phyllostachya**, n. sp. — Leaves about 20 in a rosette, lanceolate-acuminate from a dilated ovate base above an inch broad, a foot long, ½–¾ in. broad at the middle, moderately firm in texture, subglabrous. Peduncle rather shorter than the leaves quite hidden by the large lanceolate bract-leaves. Inflorescence a moderately dense panicle 3–4 in. long ; branches short, few-flowered, much shorter than the recurved lanceolate bracts, the lower about 3 in. long, the upper an inch long, similar to the leaves in texture ; flower-bracts ovate, acute, ⅓ in. long. Calyx ½ in. long ; sepals oblong. Petals twice as long as the calyx ; blade oblong.

Hab. Central Mexico, *Hahn* ! Gathered during the French Expedition of 1865–6.

231. T. ACORIFOLIA Griseb. in Gott. Nacht. 1864, 19. — Leaves lanceolate-acuminate from a short ovate base, 2 ft. or more long, under an inch broad, thin in texture, subglabrous. Bract-leaves of the stem lanceolate. Inflorescence a dense panicle, the spikes about as long as the lanceolate-acuminate bracts. Flower-bracts crowded, multifarious, roundish, one third the length of the capsule.

Hab. Mountains of Venezuela, alt. 6500 ft., *Fendler* 1771. Not seen.

232. **T. rhodocincta**, n. sp. — Basal leaves lanceolate, rigidly coriaceous, a foot long, an inch broad at the middle, narrowed gradually to an acute point, plain green, obscurely lepidote, with a

distinct narrow scariose red-brown margin.  Peduncle as long as
the leaves, stiffly erect; bract-leaves many, large, lanceolate, persistently lepidote on both sides, the uppermost 3-4 in. long.
Inflorescence a congested oblong multifarious panicle 2-3 in. long;
branch-bracts ovate-acuminate, lower 1¼ in. long; spikes dis-
tichons, lower as long as its bract, 5-6-flowered; flower-bract
oblong-navicular, acute, glabrous, ½-⅔ in. long.  Calyx reaching
the tip of the bract; sepals much imbricated.  Petal-blade minute,
oblong.

  Hab.  British Guiana; Humirida Mountains, *Appun* 1416!
Summit of Mount Roraima, *Im Thurn* 316!

  233. **T. Turneri**, n. sp. — Basal leaves not seen.  Peduncle
3-4 in. or more long, quite hidden by the large erect lanceolate-
acuminate persistently lepidote bract-leaves of firm texture, which
are 3-4 in. long.   Inflorescence a dense oblong multifarious
capitulum 3-4 in. long, 2-3 in. diam.; branch-bracts ovate, with a
lepidote cusp, the lowest with a blade an inch long and a cusp half
as long; spikes flattened, distichous, oblong, as long as the bracts,
the lowest 3-4-flowered; flower-bracts oblong-navicular, ¾ in. long.
Calyx reaching to the tip of the bract.  Petal-blade oblong, ¼ in.
long.  Stamens not exserted beyond the tip of the petals.

  Hab.  Andes of Bogota, *Turner*!  General habit of *Chevalliera*.

  234. T. ᴜᴛʀɪᴄᴜʟᴀᴛᴀ Linn. Sp. Pl. 286 (Sloane, Jam. 76; Hist.
i. 188); Willd. Sp. Pl. ii. 11; Lam. Encyc. i. 617; Leconte in
Ann. Lyc. Nat. Hist. New York, 1826, ii. 129; Roem. et Schultes,
Syst. Veg. vii. 1220; Chapm. Fl. S. U. States, 471.  *T. Nuttalliana*
R. & S. *l. c.*  *T. Bartrami* Nutt. in Sillim. Journ. v. 2, 292, non
Ell.  *Platystachys utriculata*·Beer. Brom. 266.  *Allardtia Potockii*
Antoine in Oester. Bot. Zeit. 1878, 56, with figure.—Acaulescent.
Leaves densely rosulate, lanceolate-acuminate from a large ovate
base 1½-2 in. broad. 1-1½ ft. long, very narrow and acuminate in
the upper half, firm in texture, thinly persistently lepidote on both
surfaces.  Peduncle a foot long; bract-leaves many, small, adpressed, with a small subulate free point.  Panicle 1-1½ ft. long,
very lax, with many slender long ascending branches; flowers
spaced out, multifarious, adpressed to the rachis; flower-bracts
ovate or oblong, obtuse, ½-¾ in. long.  Calyx ¾ in. long; sepals
oblong, obtuse.  Petals narrow, whitish, convolute, ½-¾ in. longer
than the calyx.  Stamens a little longer than the petals.  Capsule-
valves lanceolate. above 1½ in. long.

  Hab.  Florida, *Chapman*!  *Curtiss* 2843!  Bahamas, *Brace* 139!
Cuba, *Wright* 3748! Jamaica, *Houston*! *Hort. Bull* 1881!  Trinidad,
*Fendler* 815!  Venezuela, alt. 6500 ft., *Fendler* 2160.  Most like
*T. flexuosa*, from which it differs by its non-distichous spikes.
Flowered at Kew, July, 1887, from a plant received from the
Bahamas from Mr. Taylor.

(To be continued.)

# BIOGRAPHICAL INDEX OF BRITISH AND IRISH BOTANISTS.

## By JAMES BRITTEN, F.L.S., AND G. S. BOULGER, F.L.S.

(Continued from p. 114).

**Boswell, John Thomas Irvine,** *né* **Syme,** and afterwards **Boswell-Syme** (1822–1888): b. Queen Street, Edinburgh, 1st December, 1822; d. Balmuto, Fife, 29th January, 1888; bur. Kinghorn, Fife. LL.D., St. Andrew's, 1875. F.L.S., 1854. Curator, Bot. Soc. Edin., 1850; Bot. Soc. London, 1851 Lecturer at Charing Cross and Westminster Hospitals. 'English Botany,' ed. 3, 1863–1872. Jacks. 525; Journ. Bot. 1888, 82. Portr. Illustr. Lond. News, 11th February, 1888. *Symea* Baker = *Solaria* Philippi.

**Bowdich, Thomas Edward** (1790–1824): b. Bristol, 20th June, 1791; d. Bathurst, Gambia, 10th January, 1824. African traveller. Cape Coast Castle, 1814. 'Mission to Ashantee,' 1819. 'Excursion to Madeira,' 1825. Pritz. 37; Jacks. 525; Baillon, i. 481; Vallot, 'Flore du Sénégal,' 13; Dict. Nat. Biog. vi. 41. *Bowdichia* H. B. K.

**Bowdich, Mrs. T. E.** [*see* LEE, Mrs. R.].

**Bowerbank, James Scott** (1797–1877): b. Bishopsgate, London, 14th July, 1797; d. 8th March, 1877. LL.D. F.L.S., 1845. F.R.S., 1842. 'Fossil Fruits of the London Clay,' 1840. Had 100,000 fossil plants from Sheppey, now in Brit. Mus. Pritz. 37; Jacks. 182; R. S. C. i. 553; 'Portraits of Men of Eminence,' 1864; Trans. Bot. Soc. Edin. xiii. 123; Proc. Geol. Soc. 1877–8, 36; Geol. Mag. 1877, 91; Dict. Nat. Biog. vi. 53. Photo. at Linn. Soc.

**Bowie, James** (d. 1853): b. Oxford Street, London; d. Cape of Good Hope ?, 1853. Son of seedsman in West Oxford Street. At Kew, 1810. Sent to Brazil, 1814–1817, and to the Cape, 1817–1823 and 1827–1853. Superintendent of Baron Ludwig's garden at Cape. Plants in Herb. Brit. Mus. Drawings at Kew. Pritz. (*sub* Pritchard) 253; Journ. Bot. 1830, 20; Gard. Chron. 1881, ii. 568; Dict. Nat. Biog. vi. 65. *Bowiea* Haworth = *Aloë* in part. *Bowiea* Harvey. *Bowiesia* Grev. = *Delissea* Lamx.

**Bowles, George,** or **Bowle** (fl. 1650). Of Chiselhurst, Kent. M.D., Leyden, 1640. F.R.C.P., 1664. First found *Impatiens Noli-me-tangere;* Johnson's Gerard, 446; Pult. i. 136; Parkinson, 'Theatrum,' 297, 954; How, 'Phytologia.' Johnson's Gerard, Pref. and *passim.* Munk. i. 332.

**Bowles, William** (1705–1780): b. near Cork, 1705; d. Madrid, 25th August, 1780. 'Introducion a la historia natural . . . de España,' 1775. Dict. Nat. Biog. vi. 69. *Bowlesia* Ruiz & Pavon.

**Bowman, David** (1838–1868): b. Arniston, Edinburgh, 3rd September, 1838; d. Bogotá, New Granada, 25th June, 1868;

bur. British Cemetery, Bogotâ. Foreman at Chiswick. Introduced *Cyanophyllum Bowmani*, &c. Journ. Hort. xv. 1868, 172;
Gard. Chron. 1868, 924, 942.

**Bowman, Edward Macarthur** (d. 1872 or 1873 ?). Of Queensland. Discovered *Ricinocarpus Bowmanii, Pimelea Bowmanii*, &c.
Gard. Chron. 1873, 177.

**Bowman, John Eddowes** (1785–1841): b. Nantwich, Cheshire,
30th October, 1785; d. Manchester, 4th December, 1841.
Banker, of Wrexham. F.L.S., 1828. 'Fossil Trees discovered
near Manchester,' Trans. Geol. Soc. 1841. Pritz. 37; Proc.
Linn. Soc. i. 135; R. S. C. i. 553; Mem. Lit. Phil. Soc. Manchester, 1846; Dict. Nat. Biog. vi. 73. *Bowmania* Gardn.

**Bowman, R. B.** Of Newcastle, and Richmond, Yorkshire.
Correspondent of Hewett Watson. Herb. at Newcastle-on-Tyne.
Mag. Zool. Bot. i. (1837) 205; Top. Bot. ed. i. 519.

**Bowring, Sir John** (1792–1872); b. Exeter, 17th October, 1792;
d. Exeter, 23rd November, 1872. Linguist. Editor ' Westminster Review.' M.P. LL.D., Groningen, 1829. Knighted,
1854. F.R.S. F.L.S., 1820. 'Autobiographical Recollections,'
1877. Dict. Nat. Biog. vi. 76. Journ. Bot. 1853, 237. *Bow-
ringia* Hook. = *Brainea* J. Sm., to him and his son, John C.
Bowring.

**Bradbury, Henry** (1831–1860): b. 1831; d. 2nd September,
1860. Printer. Published Moore and Lindley's ' Ferns,' 1855.
' British Sea-weeds.' Jacks. 240. Dict. Nat. Biog. vi. 150.

**Bradley, Richard** (d. 1732): d. Cambridge, 5th November, 1732.
F.R.S., 1720. Prof. Bot. Cambridge, 1724. ' Hist. Pl. Sucen-
lentarum,' 1716–1727. ' Dictionarium Botanicum,' 1728, &c.
Pult. ii. 129–134; Rees; Pritz. 38; Jacks. 525; Phil. Trans.
xxix. 486, 490; Nich. Anec. i. 446–451; Cott. Gard. v. 93;
Baillon, i. 487; Haller, Bibl. Bot. ii. 133; Dict. Nat. Biog. vi.
172. *Bradlea* Adanson = *Apios* Boerh. *Bradlæia* Necker =
*Siler* Scop. *Bradleia* Banks = *Glochidion*.

**Bradshaugh, —.** Of Lancashire. Sent *Rubus Chamæmorus* to
Parkinson. ' Theatr.' 1015.

**Brand, William** (1807–1869): b. Blackhouse, Peterhead, 1807;
d. 15th October, 1869. W.S., 1834. Pupil of Graham at
Edinburgh, 1830–1. Orig. Memb. Bot. Soc. Edin. ·1836;
Trans. 1836. Found *Astragalus alpinus*, 1831. Trans. Bot. Soc.
Edin. x. 284.

**Bray, John** (fl. 1400). Physician. Pensioned by Richard II.
' Synonyma de nominibus Herbarum,' Sloane MS. 282, 24.
Pult. i. 22; Jacks. 501; Haller i. 219; Tanner, Bibl. Brit.
122; Dict. Nat. Biog. vi. 237.

**Bree, Robert Francis.** Of Camberwell, and afterwards of Chichester. Clerk. F.L.S., 1815. A.L.S., 1827. "Qui primus
animum illustrissimi R. Brown capitulis dioicis Serratulæ
tinctoriæ adtendit." *Breea* Lessing = *Cephalonoplos* Neck.
= *Serratula arvensis* L.

**Bree, William Thomas** (1787–1863): b. Coleshill, Warwick,
1787; d. Allesley, Warwick, 1863. Clerk. B.A., Oxon, 1808.

M.A., 1816. Rector of Allesley. 'Nugæ Helveticæ,' 1856. Contributed to 'Phytologist,' 'Mag. Nat. Hist.,' 'Ann. & Mag.,' &c. F. L. Colvile, 'Worthies of Warwickshire.' R. S. C. i. 593; Journ. Bot. 1863, 160; Mag. Nat. Hist. 1831, 29, 162; Ann. & Mag. vi. (1841), 401; Preface to Haworth's 'Saxifragarum,' xiii.

**Brewer, Samuel** (before 1700 ?–c. 1742): b. Trowbridge, Wilts; d. Bierley, Bradford, Yorkshire, after 1742; bur. Cleck-Heaton, Yorkshire. Lived at Bangor, and from 1727 in Yorkshire. Became gardener to Duke of Beaufort at Badminton. Collected for Dillenius. Wrote MS. 'Botanical Journey through Wales' (in Bot. Dept., Brit. Mus.). Herb. bought for £20 by Richard Richardson, jun. Pult. ii. 188–9; Pritz. 40; Rich. Corr.; Nich. Illust. i. 261, 288; Cash, 5; Baker, 'Fathers of Yorkshire Botany'; Dict. Nat. Biog. vi. 295. *Helianthemum Breweri. Breweria* R. Br.

**Bridges, Thomas** (1807–1865): b. Lilly, Herts, May 22nd, 1807; d. in the Pacific, 9th November, 1865; bur. Lone Mt. Cemetery, San Francisco. Son-in-law of Hugh Cuming. Kew Collector in Chili, Peru, Bolivia, and California, 1827–1865. F.L.S. Introduced *Victoria regia.* Pritz. 40; Proc. California Acad. Nat. Sci. iii. 1866, 236; R. S. C. vii. 259; Gard. Chron. 1865, 1226; Journ. Bot. 1834, 177; 1837, 64; 1845, 571; 1866, 64; Trans. Bot. Soc. Edin. viii. *Bridgesia* Hook. = *Polyachyrus* Lag. *Bridgesia* Hook. & Arn. = *Ercilla* A. Juss. *Bridgesia* Cambess.

**Brightwell, Thomas** (1787–1868): b. Ipswich, 18th March, 1787; d. Norwich, 17th November, 1868. Solicitor and Mayor of Norwich. F.L.S., 1821. 'Fauna Infusoria for East Norfolk,' 1848. 'Triceratium,' 1853. 'Memorials,' 1869. Dict. Nat. Biog. vi. 340.

**Brittain, Thomas** (1806–1884): b. Sheffield, 2nd January, 1806; d. Urmston, Lancashire, 23rd January, 1884. President, Manchester Microscopical Society, 1882. 'Micro-fungi,' 1882. Gard. Chron. 1884, i. 155; Dict. Nat. Biog. vi. 359.

**Brodie, James** (fl. 1800). Of that ilk. Discovered *Moneses.* F.L.S., 1795. E. B. 146. *Brodiœa* Sm. = *Hookera* Salisb.

**Brodigan, Thomas** (fl. 1830). Of Pilltown, Meath. 'Botanical, Historical . . . Treatise on the Tobacco Plant,' 1830.

**Bromfield, William Arnold** (1800 or 1801–1851); b. Boldre, Hants, 1800 or 1801; d. Damascus, 9th October, 1851. M.D., Glasgow, 1823. F.L.S., 1836. 'Flora Vectensis,' 1856. Travelled on Continent, Ireland, West Indies, North America, and Egypt. Herbarium at Isle of Wight Philos. Soc., Ryde. MSS. at Kew. Pritz. 41; Jacks. 254, 350; Proc. Linn. Soc. ii. 182; Phyt. iv. pref.; Dict. Nat. Biog. vi. 398; R. S. C. i. 644; Journ. Bot. 1851, 373. Engr. portr. at Linn. Soc. and at Kew, and in 'Flora Vectensis.' *Bromfeldia* Necker = *Curcas* Adans.

**Bromhead, Sir Edward Thomas Ffrench,** Bart. (1789–1855): b. Dublin, 26th March, 1789; d. Thurlby Hall, Newark, 14th March, 1855. F.R.S. F.L.S., 1844. 'Botanical

Alliances,' Edin. Philos. Journ., April, 1838. 'Affinities of *Lythraceæ* ' . . . , Mag. Nat. Hist. ii. 1838, 210; iv. 1840, 329. Pritz. 41; Proc. Linn. Soc. ii. 405; R. S. C. i. 644. *Bromheadia* Lindl.

**Brookshaw, George** (fl. 1812). Artist. 'Pomona Britannica,' 1812. Some plates are stated to be by Richard Brookshaw (fl. 1804). Pritz. 42; Dict. Nat. Biog. vi. 440.

**Brookshaw, Richard** (fl. 1804). Artist. 'Pomona Britannica,' 1804. Redgrave's Dict. of Artists, 1878; Dict. Nat. Biog. vi. 440.

**Broome, Christopher Edmund** (1812–1886): b. Berkhampstead, Herts, 24th July, 1812; d. Furnival's Inn, Holborn, 15th November, 1886. M.A., Camb., 1836. F.L.S., 1866. Mycologist. Pritz. 42; Journ. Bot. 1887, 148; R. S. C. i. 655; vii. 274; Proc. Linn. Soc. 1886–7, 34. Herbarium in Mus. Brit. *Broomeia* Berk.

**Broughton, Arthur** (d. 1796 or 1803?); d. Jamaica, 1796 or 1803? Of Bristol. M.D., Edin., 1779. 'Hortus Eastensis,' 1792. 'Enchiridion Botanicum,' 1782. 'Cat. of the . . . Bot. Garden of Liguanea,' 1794. Herb. in City Library, King Street, Bristol. Pritz. 42; Jacks. 527; Baillon, i. 499; Dict. Nat. Biog. vi. 459. *Broughtonia* R. Br.

**Brown (e ?), Alexander.** Surgeon in E. Indies. Correspondent of Plukenet. Discovered new plants in East Indies and at the Cape. Pult. ii. 62–3; Pritz. 44. *Eriocephalos Bruniades* Pluk. = *Brunia* L.

**Brown, George Dransfield** (c. 1828–1885): b. *circa* 1828; d. Ealing, Middlesex, 17th July, 1885. M.R.C.S. F.L.S., 1876. Practised at Ealing. Microscopist and cryptogamist. Proc. Linn. Soc. 1883–6, 138.

**Brown, John** (d. 1851): d. Boston, 30th January, 1851. M.D. Of Boston, Lincolnshire. F.L.S., 1826. Proc. Linn. Soc. ii. 132.

**Brown, John** (d. 1873): d. Edinburgh, July, 1873. Sec. Edinb. Nat. Field Club. Trans. Bot. Soc. Edin. xi. 427, 470; xii. 19; R. S. C. vii. 278.

**Brown, John Wright** (1836–1863): b. Edinburgh, 19th January, 1836; d. Edinburgh, 23rd March, 1863. Assistant in Edinburgh Herbarium. Associate Bot. Soc. Edin. 'List of Plants in Elie, Fife,' in Trans. Bot. Soc. Edin. vii. 519. Dict. Nat. Biog. vii. 21.

**Brown, Littleton** (b. 1699): b. Bishop's Castle, Salop, 1699. Clerk. M.A., Oxon, 1722. F.R.S. Of Shropshire. "An ingenious young clergyman," Dillenius, 'Historia Muscorum.' Rich. Corr. 233.; Linn. Letters, ii. 145. Letters to Dillenius in the Sherard Room at Bot. Gard., Oxford.

**Brown, P. J.** (d. 1842): d. Thun, Switzerland, 1842. Colonel. Lived at Eichenbühl, Thun, Switzerland, from 1822–1842. Memb. Bot. Soc. Edinb. February, 1838. Mag. Nat. Hist. 1833, 469; Catalogue des pl. de Thoune, 1843; Pritz. 43; Jacks. 344; R. S. C. i. 659.

**Brown, Philip** (d. 1779): d. Manchester?, 1779. M.D. Of

Manchester. ' Cat. of very curious pl., collected by the late . . . lately deceased, to be sold at his garden near Manchester,' 1779. Pritz. 43 ; Dict. Nat. Biog. vii. 23.

**Brown, Robert** (1767?–1845) : b. Perth?, 1767? ; d. near Philadelphia, Penn., 20th September, 1845. Nurseryman, of Perth. Found *Menziesia cærulea.* Introduced Swede. Visited America with James McNab in 1834. Gard. Chron. 1845, 755 ; Lond. Gard. Mag. xii. ; Loudon, 'Arboretum,' 182 ; Trans. Hort. Soc. iv. 285.

**Brown, Robert** (1773–1858) : b. Montrose, 21st December, 1773 ; d. Gerard Street, Soho, London, 10th June, 1858 ; bur. Kensal Green. A.L.S., 1798. F.L.S., 1822. President, 1849–1853. F.R.S., 1811. D.C.L., Oxon, 1832. LL.D. Knight " pour le mérite." Naturalist to Flinders' Expedition, 1801–5. Librarian to Banks, 1810, and to the Linn. Soc. First Keeper of Bot. Dep. Brit. Mus., 1827. 'Prodromus Floræ Novæ Hollandiæ,' 1810–1830. "Botanicorum facile princeps." Pritz. 43 ; Jacks. 527 ; Dict. Nat. Biog. vii. 25 ; Proc. Linn. Soc. 1859, xxv. ; Baillon, i. 501 ; R. S. C. i. 660 ; vii. 279 ; 'Times' quoted in Cott. Gard. xx. 176. Oil-portr. by Pickersgill and bust by Slater at Linn. Soc., bust and portr. at Brit. Mus. Nat. Hist., and portr. at Kew. *Brunonia* Sm.

**Brown, Thomas.** Captain, Forfarshire Militia. F.L.S., 1816. Added notes, &c., to White's ' Selborne,' 1833.

(To be continued.)

---

## NOTES ON HIGHLAND PLANTS.

### By the Rev. E. S. Marshall, M.A., F.L.S.

I spent a month at the end of last summer (Aug. 19 to Sept. 19) in various Highland districts, and append such botanical results of my visit as seem to be of any interest. The season was too far advanced for some groups of plants, and the superabundance of rainy days prevented or greatly interfered with several intended expeditions.

The places visited were Lawers, Mid-Perth (88) ; Loch Awe, Argyll (98)—merely a stroll of an hour or so, near the station ; Fort William, Westerness (97) ; Altnaharra, Durness, and Inchnadamph, W. Sutherland (108), with a few hours on the E. side of Ben Klibreck, E. Sutherland (107) ; and Blair Athole, Mid- and E. Perth (88, 89).

In my opinion the hills around Glen Nevis (97), and the neighbourhood of Inchnadamph (108), with its varied soil and elevation, will well repay careful search. Ben Klibreck (107, 108) is a rather disappointing mountain ; but the corrie on the S.E. (below the "Craig-an-lochan"), which the mist prevented me from working, looked much better than the rest. There are some good Hieracia of the Accipitrine group about Altnaharra, of which I was too late to gather more than scraps.

I am extremely indebted to Messrs. Arthur Bennett and Beeby, who have given me the benefit of their knowledge of many critical forms, besides sending doubtful forms for determination to Prof. Hæckel, Dr. Buchenau, and Dr. Lange. My thanks are also due to Mr. F. J. Hanbury, who kindly looked through the Hieracia, Prof. C. Haussknecht, Dr. Buchanan White, Prof. A. W. Bennett, and Mr. H. N. Ridley.

The nomenclature is usually that of the 'London Catalogue,' ed. 8. * denotes a "new county record"; † a form apparently not known before as British.

*Ranunculus Flammula* L., var. *radicans* Nolte (*pseudo-fluitans* Syme) occurred in Lochan-na-Chait, Ben Lawers (88). The extreme specimens resemble a plant from Littlesetter Loch, Shetland, but differed greatly in appearance, when fresh, from Swiss *R. reptans* L., and are considerably coarser than my Loch Leven example. It shaded off gradually towards the type.

*Caltha palustris* L., c. *minor* Syme, was still in good flower on Carn Liath (89) on Sept. 19th.

*Arabis petræa* Lam. Stob Ban, Mamore Forest (97), at about 2500 ft. The glabrous form; leaves broader and less deeply cut than in the Aberdeenshire plant.

*Cochlearia officinalis* L. Fruiting plants from Ben Lawers, Ben Nevis, and Stob Ban show scarcely any venation of the pods, even when dry, and seem to me different from the Teesdale "*alpina*," which is distinctly veined.

*Viola sylvatica* Fr. A very dwarf and compact form occurs on the limestone hillocks about Durness (108); something like it grows on Cronkley Fell, N.W. Yorks.

*Polygala serpyllacea* Weihe. Abundant near Altnaharra and Inchnadamph (108).

*Cerastium arcticum* Lange. The "*C. latifolium* var. *Smithii*" of Ben Lawers (88) is, I think, clearly this, after comparison with the Cairntoul plant. I failed to get it in fruit. There are specimens of it in the S. Kensington herbarium from Snowdon, Ben Lawers, Ben Hope, and Ben More of Assynt. Is the specimen figured in Smith's E. B., and preserved there, the same plant? Seeds from Aberdeenshire do not appear to differ appreciably from those of the Shetland var. *Edmonstonii*. — *C. alpinum* L., b. *pubescens* Syme. What I suppose to be this was gathered on Ben Nevis (97). I have not yet seen type-specimens.

*Arenaria sulcata* Schlecht. On the summit of Ben Lawers this was just like the Shetland "*forma condensata*"; lower down it was more luxuriant, but the variation was not marked. — *A. sedoides* Schultz is plentiful on both sides of Ben Klibreck, and should therefore be recorded for 107 as well as 108.

*Sagina nivalis* Fr. In advanced fruit the pedicels are certainly not "always erect," as stated in the books. On Aug. 24th nearly all were decumbent, and the plant greatly resembled small states of *S. maritima* in appearance.—*S. nodosa* E. Mey. Near Durness (108). It ascends to fully 2000 ft. above Lochan-na-Lairige (88).

*Ilex aquifolium* L. is clearly native about Inchnadamph, growing on the limestone cliffs, far from houses. I did not notice it in cultivation.

*Trifolium pratense* L. A beautiful form, with roundish, compact heads of very dark purple flowers, grows beside the Traligill at Inchnadamph; probably it comes under b. *sylvestre* Syme. — *T. hybridum* L., b. *elegans* (Savi), was plentiful as an escape near Lawers, and occurred between Blair Athole and Killiecrankie (88).

*Anthyllis Vulneraria* L., b. *Dillenii* Schultz, should have been recorded from the cliffs west of Melvich (108) in last year's paper.

*Rosa mollis* Sm. Lawers and Blair Athole (88); Inchnadamph (*108).— *R. tomentosa* Sm., b. *subglobosa* (Sm.). Common near Lawers; a very handsome bush when in ripe fruit.— *R. canina* L., e. *dumalis* Bechst. Lawers. — w. *subcristata* Baker. Near Blair Athole (88).

*Saxifraga stellaris* L. Descends to sea-level at the Smoo Cave, Durness.—*S. sponhemica* Gmel. Stob Ban, 2500 ft.

*Sedum Rhodiola* DC. Central corrie of Ben Klibreck (107).

*Callitriche hamulata* Kuetz. (*fide* Ar. Bennett). Lochan-na-Chait, Ben Lawers. Lochan below Stob Ban, at about 2200 ft.; a very 'slender state.

*Epilobium obscurum* Schreb. Roadside near Skaig Bridge, Loch Assynt (108). Named by Dr. Lange, "*E. virgatum* Fr."; but Mr. Beeby pointed out that this is "nomen erroneum," and his opinion is confirmed by Prof. Haussknecht. — † A very curious plant was found growing in some quantity by a rill above the path from Inchnadamph to Ben More of Assynt, at about 1600 ft. Dr. Lange suggested: " Forsan hybridum ex *E. alpino* et *E. montano*, vel *E. collino* Gmel." I did not fancy it to be a hybrid, not having observed either of the supposed parents ; but Dr. Haussknecht now writes: "It greatly interested me ; it is certainly a hybrid, namely, *E. anagallidifolium* × *obscurum* = *E. Marshallianum* mihi, ined. On examining the locality more closely you will be sure to discover the parents." As several good specimens were sent, there can be little doubt that this determination is correct. Stem 4–8 in. high, with two pubescent lines, round (or very faintly 4-angled when dry). Stolons much as in *obscurum*. Buds erect ; capsules very slender. Leaves light green, rather shining, sessile or shortly stalked, mostly rather crowded, alternate or opposite. Stigmas ?

*Circæa alpina* L. Near Loch Awe Station. A curious plant, which I at first thought might be *intermedia* Lond. Cat. (non Ehrh.), but some of the flowers have bracteoles. Queried in Top. Bot.

*Carum verticillatum* Koch. Glen Nevis, by the road-side, in no great quantity.

*Viburnum Opulus* L. By the Traligill, Inchnadamph ; native.

*Galium saxatile* L. Ascends to within 100 ft. of the observatory on Ben Nevis.

*Solidago Virgaurea* L., c. *cambrica* Huds. Meall Garbh (88); Stob Ban ; abundant on the coast around Durness, here passing gradually into the type.

*Gnaphalium supinum* L. E. side of Ben Klibreck (107).

*Matricaria inodora* L., var. *phæocephala* Ruprecht (*fide* Lange).

Durness, in oat-fields.  I think it also grew in an oat-field at Alt-naharra.  Leaves fleshy, subrigid ; segments longer and less closely set than in *salina* Bab.  Height 6–18 in.; larger and branching more than the Shetland specimens that I have seen.

*Cnicus arvensis* Hoffm.  Two forms were collected ; one at Lawers, which Mr. Ar. Bennett places "near var. *mitis* Koch"; the other at Durness ; this has the leaves woolly underneath, and he names it "var. *vestitus* Koch ?"

*\*Hieracium calenduliflorum* Backh.  A single specimen (the other plants being past flower) was gathered on Stob Ban, which Mr. Hanbury agrees with me, after careful examination, in placing here. As at Dhu Loch, it was associated with *H. lingulatum*.  I hope to verify this record on a future visit. — *H. senescens* Backh.  Ben Lawers ; Meall Garbh.  Ben Nevis (*97), at about 3000 ft. — *H. chrysanthum* Backh. b. *microcephalum* Backh.  Ben Lawers, above 3000 ft.  Clearly identical with the Lochnagar plant.  I only saw two specimens. — *\*H. nitidum* Backh.  Stob Ban, 2500 ft.  Quite typical.—* *H. cæsium* Fr.  Ben Nevis, 3000 ft.— *\*H. vulgatum* Fr. Near Loch Awe Station ; about \*Inchnadamph.—*\*H. Dewari* Bosw. Lawers (88), by several streamlets.  By a streamlet running down from Stob Ban (*97) into the Nevis Water at Achriach.  I have seen authentic specimens from Glen Devon, E. Perth.  Though not mentioned in Nyman, this is a very marked plant ; possibly known to Scandinavian botanists under another name. — *H. Eupatorium* Griseb.  Lawers.  Lower part of Glen Nevis.

*Leontodon autumnalis* L., b. *pratensis* (Koch) is common in the Breadalbanes.  A tall form of it occurs near Inchnadamph.  A plant between the type and var. *sordida* Bab. was gathered in Glen Nevis.

*Campanula rotundifolia* L., b. *lancifolia* Koch.  Meall Garbh, Lawers, at fully 2500 ft.

*Armeria maritima* Willd.  A specimen from Durness, with very pubescent scape, and channelled, slender leaves, is referred by Mr. Beeby to var. *duriuscula* Bab.—b. *planifolia* Syme was met with on Meall Garbh, on Stob Ban (*97), and on Coniveall (*108) ; in no case below 2500 ft.  This is the only form I have yet seen in alpine situations.

*Lysimachia nemorum* L. ascends to nearly 3000 ft. on Meall Garbh.

*\*Gentiana Amarella* L., *forma multicaulis* Lange.  Abundant on limestone about Inchnadamph and Durness.  I have not seen the type in the Highlands.

*Myosotis palustris* With., b. *strigulosa* (Reichb.).  By the road-side close to the inn at Lawers.

*Mimulus luteus* L. is well established near Inchnadamph.

*Euphrasia officinalis* L.  Several forms were noticed ; a large and handsome one by Loch Tay, Lawers ; a very minute one, with highly-coloured flowers, at 3500 ft. on the west side of Ben Lawers ; a stout one, with crowded and somewhat fleshy leaves, on the Durness coast.—b. *gracilis* (Fr.) grew lower down on Ben Lawers, and abounded on the heaths round Altnaharra.

*Melampyrum pratense* L., d. *montanum* Johnst.  Near Altnaharra and Loch Assynt, in fruit only.  Capsules erect-patent.

*Thymus serpyllum* Fr., var. *prostrata* Hornem. Ben Klibreck (108), up to 1500 ft. Apparently the usual form in N. Scotland, and the only one I have come across.

*\*Stachys palustris* L. Cultivated ground, Durness; also *S. arvensis* L.

*Plantago lanceolata* L. Some of the Durness coast-plants were between type and var. *eriophylla;* others, towards var. *capitata* Presl. — *P. maritima* L. Plentiful inland about Altnaharra, in grassy places; apparently the Shetland *"forma procerior"* Lange. I have seen a similar form in Upper Teesdale.

*\*Oxyria digyna* Hill. Plentiful in the central corrie of Ben Klibreck (107).

†*Betula glutinosa* Fr., var. *parvifolia* Wimm. Herb. (*fide* Lange, sub *B. odorata* Bechst.); Flora Danica, 2917. — *B. carpathica, β. sudetica* Reichb. — *B. odorata, β. microphylla* Hartm. "Ad *B. al-pestrem* Fr. accedens" (Lange). Close to the ferry at Cashil Dhu, below Ben Hope (108); a small tree, 8–10 ft. high, in good fruit. "Leaves small ($\frac{1}{2}$–$\frac{3}{4}$ in. long, about $\frac{1}{2}$ in. broad), rhomboid-elliptic or roundish-ovate, pointed" ('Danske Flora,' ed. 4, p. 241).

' \**Salix Caprea* L. Inchnadamph; native. Alpine exstipulate form, varying in size from a bush to a tree 25 ft. high.—*S. nigricans* Sm., a. *genuina* auct. and h. *hirta* (Sm.). Cliffs near Lochan-na-Lairige.—\**S. Myrsinites* L., b. *procumbens* (Forbes). Stob Ban (97). By the Traligill, Inchnadamph (confirms the record of *Myrsinites* for 108), at 400–600 ft. A very curious plant. Mr. Ar. Bennett writes: "I think correct, but it may be a foreign form." I have only once seen specimens of *S. Grahami*, but hardly think it to be that. — \**S. herbacea* L. E. side of Ben Klibreck (107); plentiful. It descends to about 1500 ft. on the W. side.

\**Juniperus nana* Willd. Below Stob Ban (97); barren. It was fruiting beautifully at Durness and on the lower slopes of Quinag (108).

*Epipactis atrorubens* Schultz. was seen sparingly on limestone near Inchnadamph, as well as abundantly about Durness.

\**Juncus trifidus* L. Top of Ben Klibreck, on the E. side (107) as well as the W.—*J. supinus* Moench.—b. *Kochii* Syme. E. side of Ben Klibreck, at about 2000 ft.; stamens 6.—c. *fluitans* Fr. Lochan below Stob Ban. Slow stream below Quinag, Assynt.— d. *uliginosus* (Roth.) Bealach Pass, below Ben Klibreck (E. side). Lochan Feoir, Assynt. These were passed by Mr. Arthur Bennett as correctly named.—*J. lamprocarpus* Ehrh. A tall, slender, and leafy plant, from a birch-wood by the Garry, a little above Blair Athole (88), was named by Dr. Buchenau: "forma virescens, probabiliter in umbrosis errata."—*J. alpinus* Vill., a. *genuinus* Buchenau. I kept a good look-out for this, and was fortunate enough to find it near Inchnadamph (\*108), growing with *J. lam-procarpus* at about 400 ft. A new station for it in 89 is the moorland below Carn Liath, about a mile from Monzie Farm (alt. 1600 ft.); here not exceeding six inches in height. These were also determined by Dr. Buchenau.

\**Luzula spicata* DC. must be credited to 107 (Ben Klibreck).

\*<i>Potamogeton natans</i> L.    Lochan below Stob Ban (97).—<i>P. polygonifolius</i> Pour., var. <i>angustifolius</i> Fr. (<i>fide</i> Bennett).   Lochan Feoir, Assynt.

Rhyncospora alba Vahl.   The form gathered in Glen Nevis, near Loch Awe, and near Inchnadamph, differs from that of Surrey and the " landes " of W. France in the colour and shape of the heads, and answers to the description of var. sordida in Babington.

\*<i>Carex curta</i> Good., b. <i>alpicola</i> (Wahl.).    W. side of Coniveall (108), at over 2500 ft.—<i>C. stricta</i> Good.   Specimens from Lochan Feoir, Assynt, are referred to this by Dr. Lange.   The fruit, how-ever, appeared veinless, when fresh, nor were the reticulations of the lower sheaths present; and the plants, though densely tufted, were not " tussocky."   If correct, this would be a great northern extension of its range, from Aberdeenshire.   There are some Irish specimens in Herb. Brit. Mus., S. Kensington, which are perhaps the same, under the name of <i>stricta</i>.   Mr. H. N. Ridley considers it <i>Goodenowii</i>; but nothing in the large series of that species at S. Kensington appears to me identical with it; and I should at present be somewhat disposed to doubt its belonging to either.—<i>C. rigida</i> Good.   Abundant on the E. side of Ben Klibreck (107). Obs. Among the Carices gathered by Mr. Hanbury and myself in a bog above the head of Canlochan Glen, Forfar, on July 23rd, 1886, was a form of this, which Dr. Lange now names " var. <i>infuscata</i> Drej., vel <i>C. hyperborea</i> Drej., statu juniori."   Mr. Arthur Bennett tells me that it is not true <i>hyperborea</i>, though coming near it.   He has kindly sent me the following description.—" <i>Carex rigida</i> Good. β. <i>infuscata</i>.   Culmo erectiore, minus crasso, foliis angustioribus, marginibus revolutis, acutioribus erectiusculis, spicis tenuoribus laxioribus auriculisque bractearum infuscatis.   <i>C. saxatilis</i> Wahlenb. fl. lapp. p. 247, ex parte.   Eng. Bot. 29, t. 2047! Faêrö! Island (specimina ambigua).   Grönland!   Ex hac ut mihi videtur, deri-vatur forma spicis multo elongatioribus basi magis rarifloris, peri-gyniis stramineo-viridibus, squamis obtusissimis perigynio minor-ibus, fuscis, in Grönlandia haud rara et ad speciem sequentem [<i>i. e.</i> No. 33 β., p. 460] accedens."   Drejer, in ' Revisio critica Caricium borealium.'—\*<i>C. aquatilis</i> Wahl., b. <i>Watsoni</i> Syme.   By Mudal Water, Altnaharra (108).   Spikelets more slender than in my Wick and Thurso specimens, and more like one from the river Isla, E. Perth, labelled " var. <i>virescens</i> Anderss.' — <i>C. glauca</i> Murr., c. <i>stictocarpa</i> (Sm.).   Cliffs above Lochan-na-Lairige, at over 2000 ft.— †<i>C. pelia</i> F. O. Lang.   In a heathy bog, not far from the head of Loch Naver, Altnaharra (108).   Quite a different-looking plant from Mr. Nicholson's Perthshire specimen (which he kindly lent me to look at) so named by Dr. Christ; the beak of which does not accord with the following description, for which I have again to thank Mr. Bennett, who also determined my plants, gathered in very ripe fruit :—" 99. <i>Carex pelia</i> (πέλιος, i. e., <i>lividus</i>).   Rhizomate stolonifero, culmo acutangulo scabro, foliis carinatis planis latis rigidis margine scaberrimis, spica mascula solitaria, fœmineis 1–2 elongato-lanceolatis laxifloris pedunculatis erectis, fructibus oblongo-ovatis enerviis glabris rostro brevi integerrimo.   Hab. in pal.

Norvegiæ prope Christianiam, Blytt! Ab habitu *C. vaginatæ* colore livido-glauco totius plantæ longe differt. Bracteæ vaginantes. Vaginæ foliorum integerrimæ."—F. O. Lang, in ' Caricineæ Germanicæ et Scandinavicæ.' Linnea (1851) vol. 24. Nyman places it as a subspecies of *C. panicea* L. My specimens appear much nearer to that than to *C. vaginata;* but I do not yet clearly understand the differences, owing to their shabby condition. They were gathered in torrents of rain.—*C. capillaris* L. Stob Ban (97). No doubt simply overlooked before, being one of the commonest " alpines."—*C. binervis* Sm. Exceedingly variable. With glaucous leaves, in very wet ground, Mamore Forest. Nearly 4 ft. high, on the heaths above Loch Assynt.—*C. flava* L. Ascends to 3000 ft. on Meall Garbh, Ben Lawers. — *C. Œderi* Ehrh. Sea-coast, Durness.

*Phleum pratense* L. A very slender state of this (*fide* Haeckel), with small oval heads, grew at over 2000 ft. above Lochan-na-Lairige.

*Agrostis alba* L., c. *maritima* Mey. Coast near Durness; frequent.—*A. vulgaris* With. (*fide* Haeckel). A very dark and compact-flowered form grows on Ben Lawers, at fully 3500 ft.—b. *pumila* (L.). Meall Garbh (ascending to 3000 ft.) Cliffs above Lochan-na-Lairige. Roadside, Altnaharra. — c. *nigra* (With.). Plentiful in oat-fields, Lawers.

*Deschampsia cæspitosa* Beauv., b. *brevifolia* Parn. In several localities about Lawers. Specimens from ledges on the W. side of Ben Lawers are named by Prof. Haeckel " var. *alpina* Gaud."—c. *pseudo-alpina* Syme. Ben Nevis. Ben Klibreck. Coniveall.—Var. *pallida* Koch, which I reported doubtfully from Betty Hill, last year, was found in beautiful condition, in the neighbourhood of Durness.—*D. alpina* Roem. & Schult. Ben Lawers; flowering. Ben Nevis (*fide* Haeckel); viviparous only.—*D. flexuosa* Trin. Plentiful, E. side of Ben Klibreck (107).

*Arrhenatherum avenaceum* Beauv., b. *nodosum* Reichb. Abundant and well-marked in oatfields at Lawers.

*Phragmites communis* Trin., b. *nigricans* Gren. & Godr. Bogs about Altnaharra.

*Molinia cærulea* Moench, b. *depauperata* (Lindl.). Mamore Forest. Heath near Loch Eriboll (108). — Var. *minima* Rab. Below Stob Ban, at 2000 ft. I rather doubt the value of these " varieties."

*Poa annua* L., var. *supina* (Schrad.). Wet shale, Ben Lawers, above 3000 ft. A form scarcely separable from this, but with green instead of variegated spikelets, grows in mossy rills on Coniveall. Prostrate. I have seen specimens from the tableland above Corrie Ceannder, S. Aberdeen.—*P. glauca* Sm. A tall plant from Meall Garbh, which appeared to tone down on exposed rocks into the ordinary form of this, was sent to Prof. Haeckel, who writes:— " This is, I think, the true *glauca* Sm.; which, however, I cannot separate from the *P. cæsia* of the same author." I do not yet know how to distinguish *P. Balfourii* from it, though there clearly are at least two forms in the Ben Lawers district. — *P. nemoralis*

*L., b. glaucantha* Reichb. Stob Ban (97).—*P. trivialis* L. (*fide* Haeckel). Ascends to over 2000 ft., above Lochan-na-Lairige ; and to 2500 ft., on Carn Liath.

*Asplenium viride* Huds. Grows near Inchnadamph as low down as 400 ft., with *Polystichum Lonchitis* Roth.

\*Athyrium alpestre* Milde. E. side of Ben Klibreck (107).

\*Equisetum sylvaticum* L., var. *capillare* Hoffm. Lawers (88) ; much less frequent than the type. Nyman apparently places this as a mere synonym of *sylvaticum* ; but Fries (Mantissæ) and Lange (' Danske Flora,' ed. iv. Appendix) treat it as a very well-marked variety. I believe that I saw it near Marburg, Hesse, some years ago.—\*E. variegatum* Scheich., a. *arenarium* Newman. By the Traligill, near Inchnadamph (108).

---

## SHORT NOTES.

SOME NEW RUBI RECORDS FOR 1887.—Prof. Babington has just named or confirmed my naming of the following ; all, I believe, new county records. For Berks :—*R. nitidus* W. & N. Snelsmore Common, in plenty.—*R. incurvatus* Bab., *R. carpinifolius* W. & N,, and *R. saxicolus* P. J. Müll. Cold Ash Common.—*R. villicaulis* var. *pampinosus* Lees, and *R. corylifolius* var. *fasciculatus* P. J. Müll. Beedon Wood. For South Hants :—*R. nitidus* var. *hamulosus* P. J. Müll., and *R. thyrsiflorus* Wirtg. (? & Genev., but probably not Focke). Lyndhurst Road, New Forest ; both in great quantity.— *R. cordifolius* Genev. (non Focke). Wood east of Hern Railway Station. For Dorset : — *R. nitidus* W. & N. and *R. præruptorum* Boul. Wareham and Bere Road. — *R. hemistemon* P. J. Müll. Heath west of Poole Junction.— *R. thyrsoideus* var. *fragrans* Focke. Branksome Avenue.—*R. mutabilis* Genev. Sutton Holms. Of these last *fragrans* was found by Mr. T. R. Archer Briggs ; *mutabilis* by Mr. J. C. Mansel-Pleydell and me ; *nitidus* and *præruptorum* by us three. The S. Hants plants were all found by Mr. Briggs and me, while the Berks ones and the Dorset *hemistemon* were of my collecting. Prof. Babington has also now named *præruptorum*, a bramble found some years ago by Mr. Briggs between Holsworthy and Thornbury, in North Devon.—W. MOYLE ROGERS.

CAREX LAGOPINA Wahlenberg. — In his recent interesting paper (p. 23) Mr. Druce substitutes for this name that of *C. approximata* Hoppe (1800). But I see from Nyman that there is another *C. approximata* Hoppe (1795), a variety of *C. ericetorum* Poll. There-fore the proposed change involves the adoption of an ambiguous name, and can hardly be pressed.—EDWARD S. MARSHALL.

THE NOMENCLATURE OF SPARGANIUM (p. 115). — In the haste which preceded my leaving England at the end of March for some weeks, I did not observe that Mr. Druce had added a sentence to the proof of his note on this matter which materially affects Mr. Beeby's note immediately following it. Mr. Beeby is made to say that he had read Mr. Druce's article, whereas he had only read part of it, and it is of course only to that part that his answer applies : while his note, as it stands, is stultified by the addition made by Mr. Druce. I am in no way blaming Mr. Druce, who was of course at liberty to correct his proof; but it is obvious that Mr. Beeby is entitled to this explanation, and to an expression of regret for the oversight on my part.—ED. JOURN. BOT.

DR. BOSWELL'S HERBARIUM. — Our readers will be glad to know that Dr. Boswell's Herbarium has been purchased by Mr. F. J. Hanbury, who generously proposes to arrange the plants and to place it at the disposal of botanists who may wish to consult it. The collection is a very large one, containing many critical notes and some unique specimens. The collection will be kept intact, and entirely distinct from Mr. Hanbury's own herbarium.

---

## NOTICES OF BOOKS.

*The Native Flowers of New Zealand illustrated in Colours.* By Mrs. CHARLES HETLEY. Part. I. London : Sampson Low. Imp. 4to. 12 plates. £1 1s.

FOR this handsome work we are indebted to the pen and pencil of a lady who, " believing that few people are aware of the number and great beauty of the flowers indigenous to New Zealand," has determined to do her part towards enlightening their ignorance. She has succeeded in producing extremely good plates of *Clematis indivisa, Olearia semidentata, Epacris microphylla, Senecio perdicioides, Celmisia Monroi, Metrosideros lucida, Pimelea longifolia, Areca sapida, Dysoxylon spectabile, Geranium Traversii, Ranunculus Lyallii,* and *Loranthus Adamsii.* This last is said to be a new species, but there is no diagnosis, nor does the plate contain any dissections.

It always seems to us unfortunate that, when so much pains has been bestowed upon a book as is the case in the present instance, the author should not secure that botanical aid which is necessary to establish the scientific value of the work. A few dissections might have been so arranged as to interfere very little with the artistic appearance of Mrs. Hetley's plates ; and the descriptions, which are very meagre, could easily have been amplified. As it is, the scientific importance of the work is comparatively slight, and this is the more to be regretted on account of the manifest care and accuracy with which the plates have been prepared.

---

### ARTICLES IN JOURNALS.

*Bot. Centralblatt.* (No. 14). — V. F. Brotherus, 'Musci novi transcaspici' (*Tortula desertorum, T. transcaspica, T. Raddei, Barbula excurrens,* spp. nn.). — —. Hartig, *Herpotrichia nigra,* n. sp.— (Nos. 15, 18). E. Godlewski, 'Einige Bemerkungen zur Auffassung der Reizerscheimungen an den wachsenden Pflanzentheilen.'—

C. O. Harz, 'Ueber eine Entstehungsart des Dopplerites.'—K. Wilhelm, 'Memoir of A. de Bary.

*Bot. Gazette* (March).—W. G. Farlow, Memoir of Asa Gray.— B. D. Halsted, 'Iowa *Peronosporeæ.*'—(April). Asa Gray, 'New or rare plants' (*Blepharipappus lævis, Hieracium Howellii*, spp. nn.). —J. D. Smith, 'Undescribed plants from Guatemala' (*Mimosa sesquijugata, Melampodium brachyglossum, Ardisia Tuerckheimii, Cobæa triflora, Beloperone Pausamalana, Thyrsacanthus geminatus, Scutellaria lutea, Asplenium Vera-pax*).—J. M. Coulter & J. N. Rose, 'Notes on Western Umbelliferæ' (*Peucedanum Canbyi, P. Sandbergii, Angelica Hendersoni, Sanicula Howellii*, spp. nn.).—L. H. Bailey, 'Notes on Carex' (*C. ablata, C. pansa*, spp. nn.).—L. M. Underwood, 'Distribution of Isoetes' (*I. mexicana, I. maritima*, spp. nn.).

*Botaniska Notiser* (heft 2).—G. Lagerheim, 'Ueber eine neue *Peronospora*-Art aus Schwedisch-Lappland' (*P. lapponica*, sp. n.). — L. M. Neuman, 'Om tvenne Rubi från mellersta Halland' (*Rubus hallandicus, R. eluxatus*, n. sp.). — K. Starbäck, 'Kritisk utredning af *Leptosphæria modesta.*' — A. M. Lundström, 'Om fär glösa oljeplastider och oljedropparnes biologiska betydelse hos vissa Potamogetonarter.' — C. J. Johanson, 'Jakttageber rörande några torfmossar i södra Småland och Halland.' — K. F. Dusén, 'Om några Sphagnumprof från djupet af sydsvenska torfmossar.' —O. F. Andersson, 'Om *Palmella uvæformis* och hvilsporerna hos *Draparnaldia glomerata.*'—A. S. Trolander, 'Växt lokalen i Nerike.'

*Bull. Soc. Bot. France* (xxxv. 2: May 1).— P. Duchartre, 'Organisation de la fleur du *Delphinium elatum.*'— —. Colomb, 'Classification des Fougères de France.' — P. van Tieghem & H. Douliot, 'Des tubercles radicaux des Légumineuses.'—L. du Sablon, 'Réviviscence du *Selaginella lepidophylla.*' — G. Rouy, 'Excursions botaniques en Espagne' (*Microlonchus spinulosus,|Crepis scorzoneroides, Thymus Webbianus, T. micromerioides*, spp. nn.). — J. de Seynes, '*Ceriomyces & Fibrillaria.*'—P. A. Dangeard, 'Observations sur les Cryptomonadinées.' — E. J. Camus, '*Potentilla procumbens* Sibth.' A. Legrand, 'Essai de réhabilitation des genres de Tournefort.'— L. Dufour, 'Développement et fructification du *Trichocladium asperum.*' — E. Wasserzug, 'Sur les spores chez les levûres.'— P. A. Dangeard, 'Sur la gaine foliaire des *Salicorniea.*'

*Bull. Torrey Bot. Club* (April). — N. L. Britton, 'New or noteworthy American Phanerogams' (*Cerastium Texanum, Cyperus Martindalei, Dichronema Watsoni, Scirpus Pringlei, Scleria graminifolia*, spp. nn.: 1 plate).—E. L. Sturtevant, '*Capsicum umbilicatum.*' — E. L. Greene, 'Bibliographical Notes' (*Gleditschia inermis, Hesperochiron nanus*).

*Gardeners' Chronicle* (Mar. 31).—*Agave Baxteri* Baker, *Oncidium detortum* Rchb. f., spp. nn.— (Ap. 7). *Cynosorchis elegans* Rchb. f., *C. Lowiana* Rchb. f., spp. nn. — A. M. Jones, 'The Crossing of Ferns.' —(Ap. 14). *Cypripedium Rothschildianum* Rchb. f., *Anthurium Chamberlainii* Mast. (figs. 66, 67), spp. nn.—(Ap. 28). *Coelogyne lactea* Rchb. f., n. sp.

*Journal de Botanique* (Ap. 1).—C. Flahault, 'Les herborisations aux environs de Montpellier.' — A. G. Garcin, 'Sur le fruit des Solanées.' — (Ap. 16). — Boulay, 'Sur les plantes fossiles des grés tertiaires de Saint-Saturnin.'—P. A. Dangeard, 'Les Péridiniens et leurs parasites' (1 plate). — P. Duchartre, Memoir of Asa Gray.

*Journ. Royal Microscopical Soc.* (April). — G. Massee, 'On the type of a new Order of Fungi' (*Matuleæ: Matula poroniæforme* Mass.).

*Magyar Növénytani Lapok* (Nos. 128-130). — Memoirs of A. de Bary and J. Panchich (1814–Feb. 10, 1888).

*Notarisia* (April).—G. Lagerheim, 'Sopra una nuova specie del *Pleurocapsa*' (*P. fluviatilis*).'—G. B. De-Toni, 'Manipolo di Alghe Portoghesi raccolte del S. F. Moller.' — A. Piccone, 'Nuove Spigolature per la Ficologia della Liguria.'—G. B. De-Toni, 'Conspectus generum Chlorophycearum.'

*Nuovo Giornale Bot. Ital.* (April). — A. N. Berlese, 'Monografia dei generi *Pleospora, Clathrospora*, e *Pyrenophora*' (10 plates).— C. Massalongo, 'Contribuzione alla teratologia vegetale (2 plates). —G. B. De-Toni, 'Sopra un curioso *Flos-aquæ* osservato a Parma.' — A. Bottini, 'Appunte di Briologia Toscana.' — G. Arcangeli, 'Sul *Saccharomyces minor*.' — E. Tanfani, 'Sul frutto e sul seme delle Apiacee.' — P. Pichi & A. Bottini, 'Prime Muscinee dell' Appennino casentinese.' — R. Ricci, 'Nota sulla *Festuca alpina*.'— G. Arcangeli, 'Sull' influenza della luce nell' accrescimento delle foglie.'

*Oesterr. Bot. Zeitschrift* (April).—F. Sauter, 'Zwei neue Formen von *Potentilla*' (*P. porphyracea, P. Botzanensiformis*, spp. nn.).— A. Hausgirg, 'Neue Beiträge zur Kenntniss der halophilen der thermophilen und der Berg-Algenflora, Sowie der thermophilen Spaltpilzflora Böhmens.'—B. Blocki, '*Rosa Liechtensteinii*, sp. n.'— A. v. Degen, 'Pressburger Flora.' — E. Woloszczak, '*Heracleum simplicifolium* Herb.' — P. Conrath, 'Zur Flora von Bosnien.'— J. Bornmüller, 'Einiges über *Vaccaria parviflora* & *V. grandiflora*.' —C. Jetter, 'Ausflug nach Dalmatien.' — P. G. Strobl, 'Flora des Etna.'—A. Tomaschek, 'Symbiose von Bacterien.'

*Scottish Naturalist* (April). — A. Bennett, 'Additional records of Scottish Plants for 1887.'

*Trans. Linnean Soc. London* (Botany, iii. pt. i.: April).— J. E. T. Aitchison, 'The Botany of the Afghan Delimitation Commission' (*Ranunculus leptorhynchus* Aitch. & Hemsl., *Delphinium Zalil* A. & H., *Isatis bullata* A. & H., *Ruta affinis* A. & H., *R. rotundifolia* A. & H., *Astragalus Nawabianus, A. Stephenianus, A. Barrowianus, A. Cottonianus, A. Rawlinsianus, A. Grisebachianus, A. Merkianus, A. Lumsdenianus, A. Durandianus, A. Weirianus, A. Hollichianus, A. Goreanus, A. Talbotianus* (all of Aitch. & Baker), *Hedysarum Maitlandianum* Aitch. & Baker, *H. Wrightianum* A. & B., *Onobrychis megalobotrys* A. & B., *O. caloptera* A. & B., *Prunus calycosus* Aitch. & Hemsl., *Bryonia monoica* A. & H., *Carum leptocladum* A. & H., *Ferula suaveolens* A. & H., *Dorema serratum* A. & H., *Johnsonia platypoda* A. & H., *Gaillonia dubia* A. & H., *Codonocephalum Peacockianum* A. & H., *Anthemis caulescens* A. & H., *Cousinia*

*Winkleriana* A. & H., *Jurinea variabilis* A. & H., *Centaurea plumosa* A. & H., *Lactuca longirostra* A. & H., *Acantholimon Ecæ* A. & H., *A. speciosissimum* A. & H., *Cistanche Ridgewayana* A. & H., *C. laxiflora* A. & H., *Chamæsphacos afghanicus* A. & H., *Stachys trinervis* A. & H., *Eremostachys persimilis* A. & H., *E. Regeliana* A. & H., *Habenaria Aitchisoni* Rchb. f., *H. Josephi* Rchb. f., *Iris Fosteriana* Aitch. & Baker, *I. drepanophylla* A. & B., *Allium leucosphærum* A. & B., *A. Yatei* A. & B., *A. xiphopetalum* A. & B. : 48 plates).

---

## LINNEAN SOCIETY OF LONDON.

*April 5th*, 1888.—William Carruthers, F.R.S., President, in the chair. — The following were admitted Fellows of the Society: Messrs. D. Sharpe, J. B. Farmer, and J. A. Voelcker. Mr. G. B. Sowerby was balloted for, and elected a Fellow.—Amongst the exhibitions of the evening Mr. D. Morris showed a curious native bracelet from Martinique. Although formed apparently of seeds, or beads of wood, or bone, its real composition had puzzled both botanists and zoologists, and until microscopically examined could not be determined. — Mr. J. G. Baker, F.R.S., exhibited a series of specimens of *Adiantum Fergusoni* and *A. Capillus Veneris*, and offered some remarks upon their specific and varietal characters. —Mr. Clement Reid exhibited a series of fruits and seeds obtained by Mr. J. Bennie from interglacial deposits near Edinburgh, affording evidence of a colder climate formerly than that now prevailing in the lowlands of Scotland.

*April 19th.* — William Carruthers, F.R.S., President, in the chair.—Messrs. Alexander Whyte and G. B. Sowerby were admitted Fellows of the Society, and the following were balloted for and elected: F. E. Weiss, Rev. W. Johnston, and R. G. Alexander. The following were elected Auditors: for the Council, Mr. A. D. Michael and Dr. John Anderson; for the Fellows, Messrs. D. Morris and G. Murray.—Mr. George Murray exhibited some specimens of *Spongocladia*, with explanatory coloured diagrams, and made some interesting remarks on the presence of sponge-spicules on Algæ at present unaccounted for. — Mr. D. Morris, of Kew, exhibited and made remarks upon the Bird-catching Sedge, *Uncinia jamaicensis.* — Mr. John R. Jackson, of Kew, exhibited some table-mats from Canada, made of the highly-scented grass, *Hierochloe borealis*, and a sample of the so-called pine wool prepared from the leaves of the American long-leaved or turpentine-yielding pine, *Pinus australis*, with a mat made from the wool, an industry which has recently been started on a large scale at Wilmington, North Carolina.—The first paper of the evening was by the Rev. George Post (communicated by Mr. Thiselton Dyer), and contained descriptions of new plants from Palestine. In the absence of the author, the salient points in the paper were admirably demonstrated by Mr. J. G. Baker, F.R.S., who exhibited specimens of the plants alluded to. — A paper was then read by the Botanical Secretary, Mr. B. Daydon Jackson, on behalf of Prof. Fream, " On the Flora of Water Meadows."

# ASA GRAY.

## (WITH PORTRAIT.)

THE removal from among us of another leader of English-speaking botanists is an event that can hardly be passed over with the short and formal notice which is all that space will usually allow us to devote to obituary notices.  No one among our teachers has more generally received, as not one has more thoroughly merited, the respectful admiration of workers in all branches of science.  Specialist as he was, the tributes of esteem and respect which flowed in upon him three years since, on his seventy-fifth birthday, largely sent, as they were, by American botanists, were furnished also by leaders in every other branch of science; and it may be doubted whether any botanist has ever more naturally attracted to himself the affection, as well as the admiration, of his fellow-workers.  In this respect, indeed, Asa Gray formed a remarkable contrast to the great English systematist whose loss we deplored some years since.  No one could fail to respect and admire the marvellous power of steady work and indomitable perseverance which Mr. Bentham brought to bear upon the science to which he devoted his life; but in Asa Gray we have lost one who, in addition to an equal power of work and a wider range of thought, had that personal charm of manner which is sometimes denied to the greatest among the leaders of men.

To Dr. Asa Gray, perhaps more than any other foreign botanist, the great collections of this country were for years familiar; and, since his first visit to our shores in 1839, they have constantly been consulted by him in connection with his works upon the Flora of the western world.  Before the Kew Herbarium existed, the Hookerian collection on which it has been so largely based was examined by him while yet in the hands of its then owner at Glasgow; while the old collections contained in the British Museum were appreciated by him at their true value as affording material for the history of American botany.  At every visit to Europe, a certain period was set apart for work at these collections; while at other times frequent communications crossed the Atlantic containing questions which could only be finally answered by a reference to the actual specimens of Gronovius or other early collectors.  It is, indeed, hardly too much to say that the National Herbarium was known more thoroughly by Dr. Gray than by any botanist not officially connected with it; while so far as the American specimens were concerned, his knowledge was as exhaustive as it was critical.

Asa Gray was born at Sanquoit, in the township of Paris, in Oneida County, on November 18th, 1810.  At the age of seven, his father moved to Paris Furnace, and established a tannery, and Asa was employed, in the intervals of school, in feeding the bark-mill, and driving the horse that turned it,  His education had begun at the age of three, and by this time he was a champion speller in the numerous matches—the prototypes of the " spelling-bee " which had a brief popularity on these shores—which took place in the school. In 1826 he entered the Medical College at Fairfield, where he subsequently took his degree of M.D. in 1831.

It was while at Fairfield that young Gray's attention was turned especially towards Botany. In the winter of 1827-8, he read the article on that science in the 'Edinburgh Encyclopædia,' and his interest was so excited, that he bought Eaton's 'Manual of Botany,' longing for spring to come that he might observe for himself the plants described in it. "He sallied forth early," says Prof. Barnes,* "discovered a plant in bloom, brought it home, and found its name in the 'Manual' to be *Claytonia virginica*, the species *Caroliniana* to which the plant really belonged, not being distinguished then. . . . In the frequent rides about the country to visit patients he had abundant facilities for observing and collecting plants, and, besides studying out their names, he began a herbarium. In the autumn, when he returned to the medical school, he took with him a bundle of specimens which had puzzled him;" and shortly afterwards opened correspondence with Dr. Lewis C. Beck, of Albany. In 1830 he visited New York, taking with him a packet of undetermined specimens, and a letter of introduction to Dr. Torrey. The latter, however, was absent, but shortly returned the plants named. The correspondence thus begun lasted until Dr. Torrey's death in 1873, and soon led to a closer companionship, for in 1833 young Gray became assistant in Dr. Torrey's chemical laboratory. Having by this time made up his mind to devote himself to botany rather than to medicine, Gray was anxious to obtain the help which he knew Torrey would readily give.

During this winter Dr. Gray spent his spare time in herbarium work, and in December, 1834, he read before the Lyceum of Natural History in New York, the first of that long series of papers with which his name has long since been associated. This was a monograph of the North American species of *Rhynchospora;* it exhibits that careful elaboration of material and anxiety to exhaust all possible sources of information which always characterised Dr. Gray's work, as well as that acknowledgment of what had been done by others which he never omitted to make.† He had previously issued the first of two volumes, each containing a century of North American *Gramineæ* and *Cyperaceæ*, which he dedicated to his former instructor at Fairfield, Dr. James Hadley, and which contains a new grass, *Panicum xanthophysum*, from the Oneida Lake district, the first of the many hundreds of species to which his name is attached as author. This collection, in the opinion of Sir W. J. Hooker, "may fairly be classed among the most beautiful and useful works of the kind that we are acquainted with. The specimens are remarkably well selected, skilfully prepared, critically studied, and carefully compared with those in the extensive and very authentic Herbarium of Dr. Torrey." ‡

---

* 'Botanical Gazette,' Jan., 1886.
† Reprinted in 'Companion to Bot. Magazine,' ii. 26—38 (1836).
‡ 'Companion to Bot. Magazine,' i. 14 (1835).

In 1835, Dr. Gray obtained, through Torrey's influence, the post of curator and librarian of the Lyceum of New York, and at once set to work upon the "task of preparing an original work, expressly adapted to the use of the student of North American Botany," under the title, 'Elements of Botany'; this was published at New York in 1836. It is interesting to note that his first and last work bear the same title, although the 1887 volume (noticed at p. 92 of this Journal) is rather intended to supply the place of his 'Lessons in Botany.' In the same year Gray began his communications to the 'American Journal of Science and Arts,' of which he became an assistant editor in 1853, and an associate editor in 1871; his valuable 'Bibliographical Notes' were a feature of the periodical.

It was originally arranged that Gray should have occupied the position of botanist to the Wilkes Exploring Expedition, but owing to delays and uncertainties he resigned the appointment, and devoted himself to working with Torrey at the 'Flora of North America,' the first part of which appeared in 1838. In the summer of this year he was appointed to the botanical chair (which, however, he never filled) in the newly-founded University of Michigan, "a position," says Prof. Sargent,* "he accepted only upon condition of being allowed to pass a year in Europe for the purpose of consulting there the herbaria which contained the old collections of American plants,—types upon which the early species had been founded,—and which his work upon the Flora made it necessary to critically examine. He left New York in November, 1838, and returned the following November. The year for him was one of great scientific activity and interest. No American botanist had ever been so cordially received by his associates in the Old World."

The summary given by Prof. Sargent of Gray's first European visit is so interesting, that we reproduce it in full :—

"In Glasgow he made the acquaintance of William Jackson Hooker, the founder of the greatest of all herbaria, the author of many important works upon botany, and then about to publish his 'Flora Boreali-Americana,' in which were described the plants of British North America, a work just then of special interest to the young American, because it first systematically displayed the discoveries of David Douglas, of Drummond, Richardson, and other English travellers in North America. At Glasgow, too, was laid the foundation for his lifelong friendship with the younger Hooker, then a medical student seven years his junior, but destined to become the explorer of New Zealand and Antarctic floras, the intrepid Himalaya traveller, the associate of George Bentham in the authorship of the 'Genera Plantarum,' a president of the Royal Society, and, like his father, the director of the Royal Gardens at Kew. At Edinburgh he saw Greville, the famous cryptogamist; while in London, Francis Boott, an American long resident in England, the author of the classical history of the

* In a pamphlet reprinted from the 'Sun' of January 3rd, 1886, to which we are indebted for much of the above information.

genus *Carex*, and at that time secretary of the Linnæan Society, opened to him every botanical door. Here he saw Robert Brown, then the chief botanical figure in Europe, with the exception, perhaps, of De Candolle; and Menzies, who fifty years before had sailed as naturalist with Vancouver on his great voyage of discovery; and Lambert, the author of the sumptuous history of the genus *Pinus*, in whose hospitable dining-room were stored the plants upon which Pursh had based his North-American Flora. Here, too, he met Bentham and Lindley and Bauer, and all the other workers in his scientific field.

" A visit to Paris brought him the acquaintance of the group of distinguished botanists then living at the French capital: P. Barker Webb, a writer upon the botany of Spain; the Baron Delessert, Achille Richard, whose father had written the Flora of Michaux; Mirbel, already old, still actively engaged in investigations upon vegetable anatomy; Spach; Decaisne, then a young *aide naturaliste* at the Jardin des Plantes, of which he was afterwards to become the distinguished director; Auguste St. Hilaire, the naturalist of the Duke of Luxembourg's expedition to Brazil, and at that time in the full enjoyment of a great reputation earned by his works upon the Brazilian flora; Jacques Gay; Gaudichaud, the naturalist of the voyage of L'Uranie and La Physicienne; the young Swiss botanist, Edmond Boissier, the Spanish traveller, and, later, one of the most important contributors to systematic botany in his classical ' Flora Orientalis;' Adrien de Jussieu, grand-nephew of Bernard, and son of Laurent de Jussieu, himself a worthy and distinguished representative of a family unequalled in botanical fame and accomplishment.

"At Montpellier, Mr. Gray passed several days with the botanists Delile and Dunal, and then hurried on to Italy, where at Padua, in the most ancient botanical garden in Europe, he made the acquaintance of Visiani, at that time one of the principal botanists in Italy; at Vienna he saw the learned Endlicher, the author of a classical ' Genera Plantarum ;' and at Munich, Von Martius, the renowned Brazilian traveller, the historian of the palms, and the earliest contributor to that stupendous work, the ' Flora Brasiliensis,' which bears his name ; and here, too, was Zuccarini, the collaborator with Von Siebold in the ' Flora Japonica.' Geneva then, as at the present time, was a centre of scientific activity ; and there he made the personal acquaintance of the De Candolles, father and son, and worked in their unrivalled herbarium and library. He saw Schlechtendal at Halle ; and at Berlin, Klotzsch, Kunth, and Ehrenberg,—familiar names in the annals of botanical science. Alphonse De Candolle and Sir Joseph Hooker alone are left of the brilliant group of distinguished naturalists who cordially welcomed the young American botanist in 1839.

" The scientific results of this journey were important. The identity of many doubtful American species was settled, and confused synonymy made clear, by critical examinations of the plants in the Linnæan Herbarium, then in London, as well as those in the herbaria of Clayton, Catesby, and Plukenet; the plants of Michaux, Pursh, Douglas, Drummond, Mitchell, Bradbury, and Richardson; the herbaria of Schkuhr, Willdenow, and Lehmann, and many others which it is not necessary to mention in order to demonstrate the intellectual industry and botanical zeal of the younger of the two authors of the new American Flora."

In 1840, an excellent paper from Dr. Gray's pen, entitled, ' Notices of European Herbaria, particularly those most interesting to the North-American Botanist,' appeared in the ' American Journal of Science and Arts.' His account of the Linnæan herbarium is very full, and his evident appreciation of the work of the older men—Menzies, Aiton, Dryander, and Banks—is very pleasant. " The collections at the British Museum," he says, " are scarcely inferior in importance to the Linnæan herbarium itself, in aiding the determination of the species of Linnæus and other early authors."

When Grey returned, arrangements at Michigan were not completed, so he obtained a renewal of leave, and settled down to work with Torrey in New York, at the ' Flora of North-America.' The first volume was completed in 1840, the first part of vol. ii. appeared in 1841, and the second in 1842. The elaboration of the *Compositæ*, an order in which he was always specially interested, was entirely from his pen. In this year, on the occasion of a visit to Boston, he was offered the Fisher Professorship of Natural History in Harvard University. Accepting this, he went to Cambridge in July of the same year, and remained there till his death.

From this period onward, Asa Gray's life was unremittingly devoted to his favourite science. The herbarium, the lecture-room, and the study, engrossed the whole of his time; and it was thus that, besides the systematic and descriptive publications which constantly issued from his pen, he found time to write popular introductions to botany and text-books of the science, as well as reviews of the work of others, biographical memoirs, papers on Darwinism, and miscellaneous contributions to a long list of journals, among which we are glad to be able to include our own. Whatever he did, he did well.

It would be impossible here even to enumerate Dr. Gray's contributions to botanical science, nor is it necessary to do so. One or two of the most important may be referred to in passing. What student of North-American botany does not take as his text-book the ' Manual of the Botany of the North United States,' the first edition of which appeared in 1848, and the eighth issue of the 5th edition in 1878? The ' Botanical Text-book,' which was first issued in 1842, had attained a 5th edition in 1857, each edition being, as the author tells us, "in good part re-written." The growth of the science had been such, however, that in 1885 it was considered desirable to divide what had been one book into four; the first, on ' Structural Botany,' written by Dr. Gray, appeared in 1879; the last— a sketch of the Natural Orders, which he " hoped rather than expected himself to draw up," must be undertaken now by other—though hardly by abler—hands.* His paper on the ' Relations of the Japanese with the North-American Flora,' was published in 1859, and is styled by Prof. Sargent his " most remarkable contribution to science." His two little books of ' Botany for Young People,' entitled, ' How Plants Grow,' and ' How Plants Behave,' show that Dr. Gray possessed the by no means common gift of presenting scientific truths in simple language with perfect accuracy. In 1878 he took up his long-interrupted work on the Flora of North America, publishing then the *Gamopetalæ* after *Compositæ*, and later (in 1884) a revision, which, of course, was really a new book, of the *Compositæ* and few preceding Gamopetalous orders. He had previously elaborated the *Gamopetalæ* for the handsome volumes of the ' Botany of California.'

His visits to Europe for the purpose of consulting various herbaria were continued as opportunity served, and were always shared by Mrs. Gray. It was during one of them (in 1869) that the Thames boat-race between crews of Oxford and Harvard Universities took place, and Dr. Gray, with his fellow-professor, the poet Longfellow, was among the crowd of sympathising spectators. During his visits to England, he lived at Kew, and was thus enabled to interchange opinions with Mr. Bentham and Sir Joseph Hooker, and to be consulted by them on certain particulars regarding the ' Genera Plantarum,' when that great work was in progress. In 1877 he accompanied Sir Joseph on a visit to the Rocky Mountain region, the results of which are recorded in a suggestive paper published in the names of both botanists.

Of the long and intimate connection which existed between Gray and Darwin it is unnecessary to speak ; it is sufficiently manifested in the ' Life and Letters' of the latter, from which we learn, *inter alia*, that Darwin was led to take up the subject of climbing plants by reading a paper ' On the Coiling of Tendrils,' published by Gray in 1858. One of the first to welcome the ' Origin of Species,' and throughout a warm supporter of Darwin and his views, Asa Gray was never to be found in the ranks of those who oppose the doctrine of development to the teachings of revealed religion. While admitting, for instance, the influence of " natural selection," he is careful to say—" if this term is to stand for sufficient cause and rational explanation, it must denote or include that inscrutable something which produces, as well as that which results in the survival of, ' the fittest.' " [†] " Natural law," he says elsewhere,[‡] " is the human conception of continued and orderly Divine action ; " and this, says Prof. Dana,[§] was "his firm faith to the end." Gray himself thus summed up his creed:

---

* It must always be a matter of regret that Sir Joseph Hooker has never fulfilled the hope he expressed, in the preface to the first edition of the ' Student's Flora,' of preparing a companion to that work which should contain " a record of those physiological and morphological observations on British plants which have given so great an impulse and zest to botanical pursuits."

† ' Darwiniana,' p. 388.

‡ Amer. Journ. Science and Arts, 1860, p. 183.     § Id., 1888, p. 199.

" I am scientifically, and in my own fashion, a Darwinian ; philosophically, a convinced theist ; and religiously, an accepter of the ' creed, commonly called the Nicene,' as the exponent of the Christian faith." *

Asa Gray's seventy-fifth birthday in 1885 was made an occasion of congratulation by his fellow-botanists, whose tribute of affection took the form of a handsome silver vase, embossed with representations of North-American plants, among which those with which his name is connected hold a conspicuous place. The greetings of one hundred and eighty American botanists were thus represented ; while others, not botanists, also offered their meed of respect. Mr. James Russell Lowell's lines may be quoted:

" Just Fate ! prolong his life, well spent,
Whose indefatigable hours
Have been as gaily innocent
And fragrant as his flowers."

His last visit to this country took place last year. His step was almost as elastic and his manner as bright and kindly as ever, and yet there were signs which made us feel that we should probably not see him again. He received the degree of Doctor from the Universities of Oxford, Cambridge, and Edinburgh; and attended the meeting of the British Association in Manchester. The likeness which, by the courtesy of Mr. Guttenberg, the photographer, we are enabled to reproduce here, is from a group consisting mainly of the botanists who were present at the Association, taken while the meetings were in progress. Two gaps in this group have already been made by death, and two which it will be by no means easy to fill; De Bary and Asa Gray are no longer with us.

Dr. Gray returned home in October, and at once resumed work upon the ' Flora.' But his labours were nearly done. At the end of last November, a paralytic stroke closed his long botanical career ; and although he lingered on until the 30th of January, his powers of speech never returned, and on that day he passed quietly from among men. He was buried in the Mount Auburn Cemetery ; but his name will live for ever in the annals of science, and will long stand pre-eminent in the history of the botany of his country.

JAMES BRITTEN.

---

# A SYNOPSIS OF *TILLANDSIEÆ*.
## By J. G. BAKER, F.R.S., F.L.S.
(Concluded from p. 144.)

Subgenus XI. CONOSTACHYS Griseb. — Leaves densely rosulate, thin, subglabrous, lorate or lanceolate. Inflorescence more or less decidedly multifarious. Petal-blade lingulate ; claw with a pair of scales at the base. Differs only from *Vriesea* by its non-distichous inflorescence.

Spikes simple . . Sp. 235–237.
Spikes panicled . . Sp. 238–241.

* ' Darwiniana,' Preface.

235. T. MUCRONATA Griseb. in Gott. Nacht. 1864, 20. — Leaves lorate from a dilated oblong base 3 in. broad, 2 ft. long, 2 in. broad at the middle, flexible, subglabrous, narrowed gradually to an acute point. Peduncle shorter than the leaves; bract-leaves crowded, erect, lanceolate, imbricated, with large free points, the upper 5–6 in. long. Inflorescence a dense oblong multifarious spike 3–4 in. long, 2 in. diam.; bracts broad-ovate, acute, much imbricated, 1½–2 in. long. Calyx about an inch long; sepals acute. Corolla reaching to the tip of the bract.

Hab. Venezuela; mountains of Tovar, alt. 6500 ft., *Fendler* 2159!

236. **T. strobilantha, n. sp.** — Leaves about a dozen in a rosette, lanceolate from an oblong utricular base 4–5 in. long, 2–2½ in. broad, 1½ ft. long, under an inch broad at the middle, moderately firm in texture, subglabrous, narrowed gradually to an acute point. Peduncle ½ ft. long; bract-leaves crowded, with large erect lanceolate free points. Inflorescence a simple multifarious spike 6–9 in. long, 3 in. diam.; bracts broad-ovate, glossy, coriaceous, tinged with bright red, the lower ½ ft. long, the upper 3 in. Calyx flattened, 1½ in. long, enclosed in a pair of bracteoles. Petals ½ in. longer than the calyx. Stamens longer than the petals; anthers ¼ in. long.

Hab. Mexico; Province of Orizaba, *Bourgeau* 2389! There are two fine specimens of the same species from Pavon at the British Museum. A very fine plant, allied to *T. mucronata* Griseb.

237. T. MALZINEI Baker in Bot. Mag. t. 6495; Hemsl. in Biol. Cent. Amer. iii. 321. *Vriesea Malzinei* E. Morren in Belg. Hort. 1874, 313, t. 14.—Acaulescent. Leaves 15–20 in a dense rosette, lorate from a dilated ovate base 3 in. broad, about a foot long, 1½–2 in. broad at the middle, subglabrous, plain green on the face, tinged with red-brown on the back. Peduncle rather shorter than the leaves; bract-leaves many, small, adpressed, ovate-lanceolate, imbricated. Spike lanceolate, dense, not distichous, 6–8 in. long; flowers erecto-patent, bracts ovate, red or yellow, glabrous, above an inch long. Calyx protruded beyond the bract; sepals oblong, obtuse. Petals white, twice as long as the calyx; blade lingulate. Stamens and pistil shorter than the petals.

Hab. Mexico; Province of Cordova, discovered by M. de Malzine. Introduced into cultivation and flowered by M. Jacob-Makoy, of Liege, in 1872.

238. T. SAUNDERSII K. Koch in Ind. Sem. Berol. 1873, App. 6. *Encholirion Saundersii* André in Ill. Hort. n.s. t. 132; E. Morren, Cat. 1873, 8. — Leaves about 20 in a dense rosette, lorate from an ovate dilated base 1½–2 in. broad, very flexible, recurved from below the middle, a foot long, above an inch broad, dull glaucous green on the face, with copious small claret-red spots towards the base, spotted with claret-red all over the back, especially in the lower half. Peduncle with panicle 1½–2 ft. long; bract-leaves small, lanceolate, with scarcely any free points; panicle about a foot long, with 3–4 short erecto-patent laxly-flowered branches;

flowers ascending; flower-bracts ovate, bright yellow, about an inch long. Calyx bright yellow, glabrous, 1½ in. long; sepals oblong, obtuse, much imbricated. Petal-blade lingulate, bright yellow, ½ in. long. Stamens and pistil shorter than the petals.

Hab. Brazil, imported by Mr. Wilson Saunders about 1870. Introduced into commerce by De Smet, of Ghent.

239. T. CAPITULIGERA Griseb. Cat. Cub. 254.—Leaves lanceolate, 2–3 ft. long, 2 in. broad at the middle, moderately firm in texture, subglabrous, narrowed gradually to an acute point. Peduncle 1½ ft. long; bract-leaves crowded, erect, with large lanceolate free points. Inflorescence a panicle nearly a foot long, consisting of 8–9 spaced-out subglobose nearly sessile multifarious capitula; upper branch-bracts ovate; lowest lanceolate, 2–3 in. long; flower-bracts oblong-navicular, rigid, cuspidate, an inch long. Calyx ¾ in. long; sepals oblong, obtuse. Capsule scarcely longer than the calyx.

Hab. Cuba, *Wright* 3275!

240. T. PLEIOSTACHYA Griseb. in Gott. Nacht. 1864, 19.—Leaves lorate from a dilated ovate base 2½–3 in. broad, 1½ ft. long, 1½ in. broad at the middle, thin in texture, subglabrous, deltoid-acuminate at the apex. Peduncle with panicle 2½–3 ft. long; bract-leaves imbricated, with large lanceolate free points. Inflorescence a panicle 1½ ft. long, of several erecto-patent oblong multifarious spikes; flower-bracts ovate, ¾–1 in. long. Calyx ⅓–½ in. longer than the bract; sepals obtuse, much imbricated. Petals not seen. Capsule-valves an inch long, ⅓ in. broad.

Hab. Venezuela; mountains of Tovar, alt. 7000–8000 ft., *Fendler* 1514!

241. T. VENTRICOSA Griseb. in Gott. Nacht. 1864, 19. — Leaf lanceolate from a dilated ovate base above 3 in. broad, 2½–3 ft. long, 2–2½ in. broad at the middle, thin in texture, glabrous, narrowed gradually to the point. Peduncle shorter than the leaves; upper bract-leaves ovate-lanceolate, entirely scariose, imbricated. Inflorescence a lax panicle of a few oblong very multifarious spikes, the upper the longest, half a foot long, the side ones erecto-patent; branch-bracts ovate-oblong, scariose; flower-bracts ovate, about an inch long, wrapped loosely round the calyx. Calyx ¾ in. long; sepals oblong. Petals greenish yellow; blade small, oblong. Capsule-valves lanceolate, an inch long.

Hab. Venezuela; mountains of Tovar, alt. 7000 ft., *Fendler* 1517!

INDEX OF SPECIES.

[The following is a complete list of the species and of their synonyms under *Tillandsia*, enumerated in this paper. Species published for the first time have a * affixed; synonyms are in italics. After each name is given in brackets the number prefixed to the species by Mr. Baker in his enumeration, followed by the year and page of the Journal in which it will be found].

acorifolia (231) 1888, 143
*aloifolia* (109) 1887, 346
amazonica* (193) 1888, 108
amethystina (178) 1888, 104
anceps (42) 1887, 239
andicola (14) 1887, 213
angustifolia (63) 1887, 245
argentea (81) 1887, 280　　　　[16
*argentea* (53) 1887, 242; (125) 1888,
aurantiaca (94) 1887, 304
axillaris (134) 1888, 41
azurea (27) 1887, 235

Bakeriana* (22) 1887, 234 †
Balbisiana (65) 1887, 245
bandensis (21) 1887, 234
Barclayana* (41) 1887, 239
Barilleti (167) 1888, 79
*Bartramii* (46) 1887, 241; (234) 1888
Benthamiana* (123) 1888, 15
*bicolor* (121) 1888, 14
biflora (227) 1888, 142.
billbergiæ (159) 1888, 48
Bourgæi* (76) 1887, 278
brachycephala* (130) 1888, 40
brachycaulos (129) 1888, 40
brachyphylla* (127) 1888, 16
*brachystachys* (174) 1888, 81
brachypoda* (32) 1887, 237
brassicoides (115) 1888, 12
*bracteata* (69) 1887, 277
brevibracteata (111) 1887, 346
brevifolia* (39) 1887, 239
*breviscapa* (33) 1887, 237
bryoides (2) 1887, 213
bulbosa (43) 1887, 240

cærulea (99) 1887, 305
*cærulea* (100) 1887, 305
cæspitosa (13) 1887, 213
*cæspitosa* (163) 1888, 49
canescens (45) 1887, 240
capillaris (7) 1887, 213
capitata (221) 1888, 141
capituligera (239) 1888, 169
Caput-Medusæ (44) 1887, 240
caracasana* (144) 1888, 44
carinata (165) 1888, 49
Chagresiana* (199) 1888, 109
chontalensis (31) 1887, 237
chrysostachys (166) 1888, 79
*cinerascens* (52) 1887, 242
*circinalis* (26) 1887, 235
*circinnata* (59) 1887, 244
*coarctata* (50) 1887, 242
Commelyna (156) 1888, 47
compacta (139) 1888, 42

complanata (133) 1888, 41
*complanata* (38) 1887, 238
*compressa* (42) 1887, 239
conspersa (85) 1887, 280
corallina (92) 1888, 107
corcovadensis (110) 1887, 346 ‡
Cossoni* (78) 1887, 279
*crinita* (1) 1887, 212
crocata (17) 1887, 214
cryptantha* (224) 1888, 142
cyanea (140) 1888, 43

dasyliriifolia* (93) 1887, 304
Deppeana (202) 1888, 109
dianthoidea (120) 1887, 14
didisticha (126) 1888, 16
dissitiflora (201) 1888, 109
distachya (54) 1887, 243
disticha (52) 1888, 242
*disticha* (161) 1887, 48
divaricata (55) 1887, 243
drepanocarpa* (132) 1888, 41
Dugesii* (75) 1887, 278
Duratii (26) 1887, 235
Duvaliana (160) 1888, 48

elata* (153) 1888, 46
elongata (74) 1887, 278
*eminens* (43) 1887, 240
ensiformis (187) 1888, 106
erecta (9) 1887, 213
erectiflora* (112) 1887, 346
erubescens (37) 1887, 238
*erubescens* (128) 1888, 40
*erythræa* (43) 1887, 240
excelsa (152) 1888, 45
*excelsa* (201) 1888, 109
*excelsa* var. (202), 1888, 109

fasciculata (69) 1887, 277
Fendleri (146) 1888, 44
fenestralis (190) 1888, 107
filifolia (106) 1887, 345
flabellata* (51) 1887, 242
*flavescens* (35) 1887, 238
flexuosa (109) 1887, 346.
floribunda (50) 1887, 242
*floribunda* (49) 1887, 242
foliosa (68) 1887, 277
*foliosa* (154) 1888, 46
fusca (16) 1887, 213

Gardneri (125) 1888, 16
geminiflora (124) 1888, 15
gigantea (203) 1888, 110
Gilliesii (10) 1887, 213
gladioliflora (169) 1888, 80

---

† See note to *tricholepis*.　　　　‡ See note to *ventricosa*.

glaucophylla (56) 1887, 243
globosa (119) 1888, 13
glutinosa (204) 1888, 110
*glutinosa* (200) 1888, 109
goniorachis* (89) 1887, 303
gracilis (137) 1888, 137
gradata* (184) 1888, 105
graminifolia* (88) 1887, 281
grandis (140) 1888, 140
grisea (62) 1887, 245
Grisebachii* (100) 1887, 335
Grisebachiana (228) 1888, 143
guttata (194) 1888, 108
gymnobotrya* (57) 1887, 243
*gymnophylla* (131) 1888, 41

Hamaleana (156) 1888, 46
haplostachya (183) 1888, 105
*havanensis* (69) 1887, 277
heliconioides (161) 1888, 48
*heliconioides* (131) 1888, 41
heptantha (101) 1887, 306
heterophylla (135) 1888, 41
heterostachya* (186) 1888, 166
hieroglyphica (206) 1888, 110
humilis (79) 1887, 279

*imbricata* (187) 1888, 106
*inanis* (43) 1887, 240
*incana* (125) 1888, 16
incurva (82) 1887, 280
incurvata (164) 1888, 49
*incurvata* (202) 1888, 109
*inflata* (164) 1888, 49
ionantha (128) 1888, 40
Itatiaiæ (205) 1888, 110
ixioides (18) 1887, 214

Jenmani* (107) 1887, 345
Jonghei (191) 1888, 107
*juncea* (46) 1887, 241
*juncifolia* (46) 1887, 241

Kalbreyeri* (142) 1888, 45
Karwinskiana (103) 1887, 344
Krameri (219) 1888, 140
Kunthiana (66) 1887, 246

latifolia (71) 1887, 277
laxa (175) 1888, 81
*laxa* (97) 1887, 305
Lescaillei (67) 1887, 246
Lieboldiana (58) 1887, 244
limbata (108) 1887, 345
Lindeni (155) 1888, 46
*Lindeniana* (155) 1888, 46
linearis (20) 1887, 234
loliacea (102) 1887, 344
longibracteata* (173) 1888, 81

longicaulis* (171) 1888, 80
longipetala* (226) 1888, 142
Lorentziana (60) 1887, 244

macrochlamys* (225) 1888, 142
macrocnemis (83) 1887, 280
maculata (142) 1888, 43
Malzinei (237) 1888, 168
martinicensis (150) 1888, 45
Mathewsii* (29) 1887, 236
megastachya* (154) 1888, 46
meridionalis (122) 1888, 15
micrantha (90) 1887, 303
monadelpha (87) 1887, 281
Morreni (213) 1888, 139
*Morreniana* (155) 1888, 46
mucronata (235) 1888, 168
multiflora (92) 1887, 304
myosura (15) 1887, 213
myriantha* (49) 1887, 242

narthecioides (104) 1887, 344
*Nuttalliana* (234) 1888, 144

*odorata* (19) 1887, 214
oligantha* (105) 1887, 345
orizabensis* (182) 1888, 105
oxysepala* (223) 1888, 141

pachycarpa* (36) 1887, 238
pachychlamys* (162) 1888, 48
paleacea (80) 1887, 279
paniculata (217) 1888, 140
*paniculata* (114, 202) 1888, 12, 109
parabaica (176) 1888, 82
Parkeri* (137) 1888, 42
Parryi* (70) 1887, 277
parvifolia (91) 1887, 303
parvispica* (61) 1887, 244
*patens* (207) 1888, 137
paucifolia (43) 1887, 240
penduliflora (151) 1888, 45
Philippo-Coburgi (211) 1888, 138
phyllostachya* (230) 1888, 143
*picta* (168) 1888, 79
*pityphylla* (118), 1888, 13
platynema (189) 1888, 106
platypetala* (157) 1888, 47
Platzmanni (179) 1888, 104
pleiostachya (240) 1888, 169
plumosa* (116) 1888, 13
polystachya (64) 1887, 245
*polytrichoides* (2) 1887, 213
procera (209) 1888, 138
prodigiosa (220) 1888, 140
propinqua (5) 1887, 213
pruinosa (33) 1887, 237
psittacina (174) 1888, 81
psittacino × carinata (196) 1888, 108

psittacino × scalaris (197) 1888, 108
*pulchella* (118) 1888, 13
pulchra (118) 1888, 13
pumila (30) 1888, 237
*pumila* (43) 1887, 240
*punctulata* (46) 1887, 241
purpurea (28) 1887, 236
pusilla (4) 1887, 213

*quadrangularis* (46) 1887, 241
*Quesneliana* (28) 1888, 40

rectangula (6) 1887, 213
recurvata (12) 1887, 213
*recurvata* (188) 1888, 106‡
*recurvifolia* (120) 1888, 214
regina (215) 1888, 139
reticulata (212) 1888, 139
retorta (8) 1887, 213
*revoluta* (26) 1887, 235
rhodocincta* (232) 1888, 143
rigidula* (148), 1888, 44
ringens (198) 1888, 109
robusta (72) 1887, 278
Rodigasiana (208) 1888, 128
Roezlii (147) 1888, 44
*rosea* (120) 1888, 14
rubella* (145) 1888, 44
rubida (124) 1888, 15
rubra (143) 1888, 43
rupicola* (117) 1888, 13

sanguinolenta (218) 1888, 140
Saundersii (238) 1888, 168
scalaris (195) 1888, 108
scalarifolia (24) 1887, 235
*Schiedeana* (35) 1887, 238
Schlectendahlii (163) 1888, 49
*scoparia* (95) 1887, 304
Scopus (128), 1888, 40
secunda (73) 1887, 278
*Selloa* (46) 1887, 241
Selloana* (180) 1888, 104
*sericea* (19) 1887, 214
setacea (46) 1887, 241
*setacea* (42) 1887, 239
*simplex* (174) 1888, 81
Sintensii* (113) 1888, 12
soratensis* (25) 1887, 235
sphærocephala (222) 1888, 141
spiculosa (138) 1888, 42
splendens (168) 1888, 79
*squamulosa* (99) 1887, 305
staticiflora (106) 1888, 345
stenostachya (200) 1888, 109
straminea (95) 1887, 304

streptocarpa* (47) 1887, 241
streptophylla (59) 1887, 244
stricta (121) 1888, 14
*stricta* (120) 1888, 14
strobilantha* (236) 1888, 168
*suaveolens* (19) 1887, 214
subimbricata* (96), 1887, 304
sublaxa* (86) 1887, 280
*subulata* (121) 1888, 15
Swartzii* (114) 1888, 12

tectorum (53) 1887, 242
*tenuifolia* (46, 109) 1887, 241, 346
tessellata (214) 1888, 139
tetrantha (141) 1888, 43
*tetrantha* (228) 1888, 143
tortilis (34) 1887, 237
*tortilis* (59) 1887, 244
*trichoides* (1) 1887, 213
tricholepis (3) 1887, 213
*tricholepis* (22) 1887, 234 †
*tricolor* (42) 1887, 239
triglochinoides (40) 1887, 239
triticea* (136) 1888, 42
Turneri* (233) 1888, 144
Tweediana* (210) 1888, 138

umbellata (158) 1888, 47
unca.(23) 1887, 234
undulata (11) 1887, 213
unilateralis (185) 1888, 105
usneoides (1) 1887, 212
utriculata (234) 1888, 144

valenzuelana (97) 1887, 305
variabilis (98) 1887, 305
*variegata* (43) 1887, 240
ventricosa (241) 1888, 169
*ventricosa* (110) 1887, 346 ‡
vernicosa* (48) 1887, 241
vestita (35) 1887, 238
*vestita* (123) 1888, 15
viminalis (170) 1888, 80
violacea* (77) 1887, 279
virginalis (135) 1888, 41
viridiflora (172) 1888, 81
*viridiflora* (170) 1888, 80

Warmingii (177) 1888, 104
Wawreana (181) 1888, 105

xiphioides (19) 1887, 214
xiphophylla (229) 1888, 143
xiphostachys (38) 1887, 238

yucatana* (84) 1887, 280

zebrina (168) 1888, 97

---

† Mr. Baker overlooked the fact that he had already bestowed this name upon another species (No. 3 of this enumeration) : we propose therefore to call the present species *T. Bakeriana*.

‡ This name appears twice in this enumeration (Nos. 110 and 241) ; it must be retained by the latter. The former may take the name *corcovadensis*.

# NOTES ON THE BOTANY OF NORTHERN PORTUGAL.

## By the Rev. R. P. Murray, M.A., F.L.S.

On the morning of June 6th, 1887, I arrived at Oporto by sea, meaning to spend five or six weeks in Portugal, and to devote as much of this time as possible to a study of the flora of the country. I hope that a slight sketch of the results may prove interesting to some of the readers of this Journal.

During the greater part of the time I spent in Portugal I was the guest of my friend Mr. A. W. Tait, partly at Oporto and partly at his bungalow in the Gerez; and I owe not only the pleasure, but to a very large extent the botanical success of my trip to his kindness and knowledge of the country. Certainly, but for his guidance, I should have missed some of the finest plants of the country. To Prof. Henriquez, and to Signor Moller, of the University of Coimbra, and to many others, I beg also to tender the expression of my heartfelt thanks for their unvarying kindness to me.

My explorations were chiefly confined to three centres: Oporto, the Gerez Mountains in the extreme north of the kingdom, and four or five days spent with Mr. Moller on the Serra da Estrella, or great central range of Portugal. Two or three short walks near Coimbra also produced some good plants. It may be convenient to take these localities in their order.

And first, Oporto.—Here my work was chiefly confined to an examination of the sandy coast-line running north from the mouth of the Douro as far as Matasinhos, a distance of five or six miles; but only about half this distance is available for botanical purposes, the suburbs of Oporto stretching far in this direction. In all this part of the country the underlying rock is granite, large tracts being bare rock, sometimes clothed with a scanty covering of barren soil. Trees are almost wanting, except the two species of pine (*Pinus Pinea* L. and *P. Pinaster* Soland.), which are common throughout Portugal, either as natives or sown. My first walk in this direction (June 7th) was taken in the company of Mr. E. Johnston, a gentleman who has devoted much time to the botany of the Oporto district. It is of course impossible to give anything like a complete list of the plants observed, so that I must confine myself to a mention of those most likely to be interesting to an English botanist. *Lobelia urens* L. and *Trixago viscosa* Stev. were common, and served to remind one of the botany of the South-west of England, as also *Erica ciliaris* L. *Bellis perennis* L. was scarce, but was probably common earlier in the year. It was curious to find *Spiranthes autumnalis* Rich. in full flower early in June, the more so as Messrs. Wilkomm and Lange give August and September as the flowering months for this plant in Spain. *Spiræa Ulmaria* L. was pointed out to me in one spot; it is one of the rarest plants in Portugal. *Cyperaceæ* seemed scarce about Oporto: some fifteen miles further south, near Esmoriz, they were more numerous. About Oporto I remember only *Cyperus longus* L. and *? C. badius* Dsf. (if this be really distinct); *Carex trinervis* Degl., *C. divulsa*

Good., and *C. paniculata* L.   Clovers were represented by *T. fragi-ferum* L. and *T. resupinatum* L.   Other noticeable plants were *Malcomia litorea* Br., abundant; *Anagallis linifolia* L., also common. This is one of the most beautiful plants I have ever seen, and might well be introduced into gardens; the flowers, which are much larger than those of our English pimpernel and of an intense blue, being produced freely.   *Sedum arenarium* Br. was plentiful in pure sand in one pine wood : it is a tiny annual, very slightly or not at all branched, and it is hard to imagine it to be only a form of our familiar *S. anglicum* Huds.   *Chrysanthemum Myconis* L. here replaces our "corn marigold," and is accompanied by another plant, *Lepido-phorum repandum* DC., from which it can hardly be distinguished, except by the absence of scales on the receptacle and by the different shape of the seeds.   *Erica umbellata* L. is common quite close to the coast, but is much dwarfed in comparison with its appearance even a few miles inland.   *Cistineæ* were less numerous than I had expected to find them, *Cistus salvifolius* being the sole representative of its genus ; while of *Helianthemum* I could only find *H. guttatum* Mill. and *H. Tuberaria* Mill.   Other coast-plants were *Eudiantha læta* Fzl., *Silene portensis* L. and *S. litorea* Brot., *Linum angustifolium* Huds., *Lavatera sylvestris* Brot., *Lythrum acutangulum* Lej., *Paronychia argentea* Lam., *Ormenis mixta* DC. (very common), *Diotis candidissima* Dsf., *Soliva lusitanica* Lees, *Cirsium filipendulum* Lge., *Tolpis barbata* G., *Erythræa maritima* P., *Verbascum sinuatum* L., *Scrophularia frutescens* L., *Crucianella maritima* L., *Genista tri-acanthos* Brot., and many others, including the lovely *Pancratium maritimum* L., which was just coming into flower on July 1st.   The ovaries of this plant are much infested by a large brightly-coloured lepidopterous larva.   I was unaware of this when I collected the plants, but on my return to England towards the end of the month I found a considerable number among my papers.   I endeavoured to rear some in order to determine the species, but they all escaped, having very quickly gnawed through the box in which I had placed them.   In one spot I gathered a few small specimens of *Laurentia tenella* DC., looking like a miniature *Lobelia urens*, to which it is closely allied.

Six weeks later I spent a few hours at Esmorig, which produced a few interesting plants, but the ground was difficult to work, much of it consisting of drifting sand, exposed to the full heat of the sun. Some marshy fields near the station produced some interesting *Cyperaceæ*, viz., *Scirpus Tabernæmontani* G., *S. pungens* Vahl, *S. mucronatus* L., and *Eleocharis acicularis* Br. ; while *Holoschænus vulgaris* Lk. and *Schœnus mucronatus* L. grew in patches on the sand. Under the pines *Corema album* D. Don grew in profusion, conspicuous by its white berries.   Other plants of this neighbourhood were *Drosera intermedia* Hay, *Potamogeton natans* L., *Alisma Plantago* L., and *A. ranunculoides* L.; while hedgerows were full of myrtle and of *Rosa sempervirens* L., with occasional clumps of *Euphorbia pubescens* Vahl.   Myrtle I found to be very plentiful in all kinds of situations in the warmer parts of Northern Portugal, such as hedge-rows, river-banks, and waste places ; it has all the appearance of a

true native, nevertheless it appears to have been introduced at some
early period from Western Asia. Now it has thoroughly taken its
place as a dominant plant. Returning to the station I found I had
a few minutes to wait for the train, so strolled down to the shore of
the neighbouring lagoon, where I was delighted to find *Eryngium
corniculatum* L., well characterised by the central bract of each head
being prolonged into a sharp spine, projecting from a quarter to
half-an-inch beyond the flowers. Brotero, however, says that the
plant is very variable, and that this spine is often wanting. I would
willingly have given more time to this lagoon, but I could hear the
train in the distance, and it was already so dark that I could
botanise only with my fingers; so I returned to the station.

The banks of the River Douro afford a thoroughly interesting
study to the botanist, and I know few things more enjoyable than
to be rowed slowly along, landing from time to time to examine any
spot that may look promising. The harbour-walls produced *Koniga
maritima* Br. and *Trachelium cæruleum* L., the latter possessing
exactly the habit of a *Centranthus;* it is possible, however, that
these may have been introduced. A marshy stretch just above the
town was covered by a profusion of *Veronica anagalloïdes* Guss. The
stalked glands with which the inflorescence is covered give a very
distinct look to this plant, but I can see no other character by
which to separate it from *V. Anagallis* L. Soon after, a sand-bank
afforded several large tufts of *Eragrostis pilosa* Ptz. and two or three
plants of *Pharnaceum Cerviana* L. (*Mollugo Cerviana* Ser.), an
addition, as I believe, to the Portuguese flora. Here also *Chenopodium Botrys* L. occurred. A little higher up the banks became
steeper and more wooded, and several interesting plants began to
occur, among which I may mention *Tunica Saxifraga* Scop., *Dianthus
monspessulanus* L. (in great quantity), *Psorelia bituminosa* L., *Inula
salicina* L., *Clematis Vitalba* L., *Linaria Tournefortii* Lge., and
*Allium vineale* L., this last abundant and always capsuliferous.
Here also I collected *Scrophularia canina* L., var. *pinnatifida.*

Another expedition was to the dry and barren granite hills some
six miles N.E. of Oporto. These are for the most part treeless, but
a certain proportion of the surface is covered with pine and cork-
oak. The flora is apparently very limited as regards the number
of species, but some of these are of extreme interest, especially
*Drosophyllum lusitanicum* Lk., which was plentiful, growing in the
driest and most exposed places, and seeding abundantly. Unfortunately it was nearly out of flower at the time of my visit
(the beginning of July). So far as I could see, the number of
insects captured at that time of year was not very great; probably the leaves were more or less exhausted. Tiny seedlings
about an inch in height are now (March) quite busy in my greenhouse. Associated with the *Drosophyllum* were an abundance of
*Erica umbellata* L., and occasional plants of *Polygala microphylla*
L. and of *Odontites tenuifolia* G. Don. Here also was *Helianthemum ocymoides* P. in abundance. In a small wooded glen
were a few plants of the pretty little *Leucojum autumnale* L., and,
if I remember right, of *Centaurea uliginosa* Brot. I saw it plenti-

fully in another spot some two or three miles to the south of Oporto. *Thapsia villosa* L. and *Margotia laserpitioides* Boiss. occurred under pine-trees.

Some very common plants in lanes, &c., close to the city were *Digitalis purpurea* L., *Arenaria montana* L., *Galactites tomentosa* Mch., *Reseda media* Lej., and *Echium plantagineum* L.; while *Anarrhinum hirsutum* Lk. occurred more rarely. Ferns were plentiful, including *Adiantum Capillus-Veneris* L., *Cheilanthes odora* Sw., *Asplenium Adiantum-nigrum* L. (always in the form or subspecies *acutum*), *Cystopteris fragilis* Bernh., and *Gymnogramma leptophylla* Desv.

THE SERRA DO GEREZ.—This lies about sixty miles due north of Oporto, and reaches to the Spanish frontier. The village of Caldas do Gerez, celebrated for its hot springs which attract large numbers of Portuguese and Brazilians during the season, forms a convenient centre for exploring the district. It is easily reached from Oporto by taking the train to Brage, where a carriage can be procured for the remaining thirty miles or so of the journey. The road is an excellent one, but I should imagine the hotels (which are apt to be crowded) would be very noisy. Fortunately I had no occasion to try the experiment.

Like all the other serras of Northern Portugal, the Gerez is a mass of granite. The mountains rise to a considerable height; the loftiest, Borrageiro, reaches 4750 ft. These mountains, which in their higher portions consist chiefly of enormous boulders and bare sheets of granite, are divided by deep and narrow valleys, well supplied with most delicious water. Much of the old forest still remains.

The flora of the district is a very interesting one, but not so numerous in species as might have been expected. In 1884 Prof. Henriquez published a list in which he enumerates 353 species of flowering plants and ferns. To this number I was able to make about 70 additions.

The principal trees and shrubs are *Quercus pedunculata* Ehrh. and *Q. Tozzi* Bose., and *Castanea vulgaris* Lamk., which all reach a large size; *Acer Pseudo-platanus* L., *Arbutus Unedo* L., *Prunus lusitanica* L., and *Pyrus communis* L.

Ferns are numerous and luxuriant: many of the tiny glens in the immediate neighbourhood of Caldas being almost choked with *Woodwardia radicans* Cav. The fronds are often from ten to fourteen feet in length. It is generally accompanied by *Osmunda regalis* L. On a wall at Bouro (half-way from Braga) I saw *Cheilanthes odora* Sw.; and on oak-trees, a little nearer Braga, some beautiful specimens of *Davallia canariensis* Sw.

Two plants peculiar to this serra are *Iris Boissieri* Henriq. and *Armeria Willkommi* Henriq. I had the pleasure of collecting both. The *Iris* is a singularly beautiful plant, and will probably before long become a favourite in gardens. It grows in grassy places between 1900 and 2800 ft. The *Armeria* seems to be confined to the summit of Borrageiro.

Other noticeable plants, *Agrostis truncatula* Parl., a very remarkable *Carex*, possibly new, which I found in one place on Borrageiro;

a *Gladiolus*, apparently identical with our New Forest plant ; *Potamogeton microcarpus* Boiss. & Reut., for the determination of which I am indebted to Mr. A. Bennett ; it seems, however, to be merely a slight var. of *P. polygonifolius*. So far as the *name* goes, it is an addition to the flora of Portugal. *Gymnadenia conopsea* R. Br., a small patch near the top of Borrageiro—new to Portugal ; *Luzula lactea* E. Mey., a close ally of *L. nivea* DC., is abundant in the woods ; *Ruscus aculeatus* L., *Paradisia Liliastrum* Bertol., *Simethis bicolor* Kth., *Asphodelus cerasiferus* Gay, and several other species of *Liliaceæ*.

The curious *Cytinus Hypocistis* L. was not uncommon on the roots of *Cistineæ*. *Daphne Guidium* L. was the only representative of its genus : it is a very common plant in N. Portugal. Among the higher rocks of Borrageiro *Thymelæa coridifolia* Endl. was not uncommon ; here also I found *Valeriana montana* L., new to Portugal. Among composites the most notable species are *Pulicaria odora* Rchb., *Phalacrocarpum oppositifolium* Wk., and *Crepis lampsanoides* Fröl. All about Calnas the exquisite *Campanula Loeflingii* Brot. is abundant ; in general appearance it reminds one of *C. patula*, but is much handsomer. *Ericaceæ* are abundant and characteristic ; besides our common *Calluna* and *Dabeocia polifolia* Don, there are no less than six species of *Erica*, most of them exceedingly common. Then we have the beautiful *Thymus cæspititius* Hffgg. Lk., common all the way from Bouro to Caldas ; two or three species of *Echium*, *Lithospermum prostratum* Lois., *Omphalodes lusitanica* Pourr., *Scrophularia Herminia* Hffgg. Lk., *Linaria triornithophora* Willd., &c. Umbelliferæ are not very numerous, but include the strange *Eryngium Duriænum* Gay ; while we are reminded of Cornwall by finding *Physospermum* common in the woods near Caldas, and of Ireland by *Saxifraga umbrosa* L. (this does not seem to vary at all in Portugal).

*Crassulaceæ* were represented by seven species, of which the most remarkable were *Sedum amplexicaule* DC. and *S. brevifolium* DC. *Lythrum acutangulum* Lej. was the only loosestrife observed. *Rosaceæ* seem to be very poorly represented in Portugal ; I only saw one rose in the Gerez district, a very beautiful *canina*-form with aciculate peduncles, queried by Prof. Henriquez as var. *fusiformis* Wlk. Mr. Baker, to whom I showed specimens, remarked that it approached *collina*. The *Rubi* were more numerous, particularly near Caldas, and by the track leading to Leonte. I should place them as follows :—*R. discolor* Weihe (may I be pardoned for not following the latest synonymic split ?) ; locally common in the least elevated parts of the district ; apparently quite the same as the plant so common in the South of England.—*R. Borreri* Bell-Salt. Caldas, rare. — *R. tomentosus* Bork. Sparingly on a bare stony hillside exposed to the sun, not far from the Spanish frontier.—*R. fusco-ater* Weihe (or very near to it). Caldas do Gerez ; apparently rare. I gathered this also in one spot in the Serra da Estrella. Besides these, I found a bramble very commonly in the woods above Caldas, which seems to differ from any hitherto recorded, I venture, therefore, to give the following description :

**Rubus lusitanicus,** n. sp.—R. caule arcuato-prostrato angulato subglabro parce glanduloso, aculeis e basi dilatata declinatis tenuibus, foliis quinatis, foliolis subduplicato-patenti-dentatis, vel dentato-serratis, supra subglabris, subtus pallide viridibus, tomentosis hirto-velutinis vel in venis tantum pilosis, foliolo terminali elliptico acuminato basi subcordato, paniculæ hirtæ tomentosæ setosæ pyramidalis ramis patentibus corymbosis, inferioribus axillaribus, aculeis parvis declinatis, sepalis hirtis tomentosis ovato-attenuatis, petalis albis. — In silvis prope "Caldas do Gerez" (Lusitanic), abundat. Junio.

I much regret that I have been unable to bring this very handsome bramble under any previously named form. Its nearest relations seem to be with *R. villicaulis* W. & N. and with *R. macrophyllus* Weihe. Yet it seems too distinct from either not to require a distinguishing name. The number of aciculi and setæ on the barren stems seems to vary much; in some cases they are present in fair number; in others they are almost wanting. The stamens (green?) exceed the styles. The plant is abundant and very luxuriant in the woods a short distance above the village of Caldas do Gerez. The highest point at which I noticed it was at Leonte (900 m.); here it grew less luxuriantly, and the panicle was in many cases much depauperated. I have not seen fruit.

For Leguminosæ I was a little late, and so probably missed several species. I was pleased to meet with *Vicia Gerardi* Vill. in one spot. *Genista lusitanica* L. grew sparingly on the very top of Borrageiro, while in many places the slopes were thickly covered by a dense growth of *G. tridentata* L. This plant is collected and brought to Oporto in large quantities to serve as "kindling" for fires. It is so dry that I have seen a large patch in full flower spring into flame within a few seconds of a lighted match being applied to it. Other characteristic plants of the lower regions are *Cytisus albus* Lk., *Sarothamnus eriocarpus* Bss. Reul., and *Adenocarpus intermedius* DC. Passing rapidly over several orders, I cannot omit to mention *Hypericum linarifolium* Vahl., *Silene colorata* Poir., *Cistus hirsutus* Lamk., *Helianthemum occidentale* Wlk., and *H. globulariæfolium* P. *Resedaceæ* are represented by *Reseda media* Lej. and *Astrocarpus Clusii* J. Gay, both fairly common; while among *Ranunculaceæ*, the fine *Ranunculus bupleuroides* Brot., *Thalictrum glaucum* Desf., and *Aquilegia dichroa* Freyn. deserve mention.

I must pass rapidly over the remaining localities. In the early part of July I spent a week in exploring the Serra da Estrella with Mr. Moller, of Coimbra. This serra forms a huge granite mass, and may be considered as the back-bone of Portugal. It may be about 120 miles south of the Gerez Mountains, and reaches a height of 7500 ft. Unfortunately we were unable to reach the . highest levels. Many rare plants grow here, but for the most part at long intervals, almost every accessible spot being kept closely cropped by the innumerable sheep which form the wealth of the country. Near S. Romao I found *Digitalis Thapsi* L. and *Lavandula pedunculata* Cav., both in plenty; and a very pretty pink, probably *Dianthus lusitanicus* Brot. In a neighbouring wood I again saw

*Prunus lusitanicus* L., and a little further on *Vincetoxicum nigrum* Mch. At N. S. do Desterro a fine *Verbascum* grows abundantly; I believe it to be *V. Henriquezii* Lange. About Subagueiro (a most filthy village) such plants as *Linaria sapphirina* Hffgg. Ck., *Periballia hispanica* Trim., *Cynosurus echinatus* L., and *Festuca Henriquezii* Hack. were frequent; and here also I first saw the curious *Hispidella hispanica* Lam.

The next day we reached the higher ground in the neighbourhood of the "lakes"—small mountain tarns. Here the countless pools were full of a Batrachian *Ranunculus* very closely allied to *R. hololeucos* F. Sz. It has been described as a distinct species by Freyn, under the name of *R. lusitanicus*. More rarely we came on a pool choked with the rare and pretty grass, *Antinoria agrostidea* Parl. Rocks produced *Saxifraga umbrosa* L., and very sparingly *Campanula Herminii* Hffgg. Lk. This should have been abundant, but the sheep had been before us. Fortunately they had spared a tiny Umbellifer which I detected in one spot. I saw only a few plants, and supposed it at the time to be a *Conopodium*. It proves, however, to be *Butinia bunioides* Bss., a very rare plant hitherto known only from one or two localities in Spain. Then, while we rested, one of the men was sent to explore the rocks above us. He returned with a supply of *Senecio cæspitosus* Brot.

Next day we tramped over miles of *Plantago subulata* L. var. *granatensis*, but saw little else, except a *Hieracium* near *murorum*, which, I understand, may be *H. cinerascens* Jord. The only other hawkweed I saw on the serra was *H. castellanum* B. R., a species allied to our *H. Pilosella* L. Three species of *Genista* brought in by one of our men (*G. lusitanica* L., *G. cinerascens* Lge., and *G. polygalæfolia* DC.) may complete the list of our Estrella spoils.

Lastly, a few words about Coimbra. Here I found myself on limestone, and of course the difference in the vegetation was very marked. But I had little opportunity for investigating it, nearly all my time being spent in the herbarium of the University. I cannot speak too highly of this, nor of the kindness of every one connected with it. I was most generously assisted both by information and by the gift of specimens, and I beg to tender my most grateful thanks to Prof. Henriquez and the members of his staff.

During the short walks for which I was able to find time I gathered, among other plants, *Plantago Lagopus* L., *Jasminum fruticans* L., *Campanula primulæfolia* Brot., *Smilax mauritanica* Dsf., *Pimpinella villosa* Schomb., *Bupleurum paniculatum* Brot., two or three species of *Centaurea*, the allied *Microlonchus salmanticus* DC., *Bourgæa humilis* Coss., and *Acanthus mollis* L. A little earlier in the year the old walls about the city would have afforded good botanising ground; as it was, I was glad to collect *Antirrhinum latifolium* DC., and a *Micromeria*, probably *M. græca* Bth.

# BIOGRAPHICAL INDEX OF BRITISH AND IRISH BOTANISTS.

## By James Britten, F.L.S., and G. S. Boulger, F.L.S.

(Continued from p. 149).

**Browne, Patrick** (c. 1720–1790) : b. Woodstock, Co. Mayo, c. 1720; d. Rushbrook, Co. Mayo, 29th August, 1790; bur. Crossboyne. M.D., Leyden, 1743. Curator, Oxford Botanic Garden. In Antigua, 1737. In Jamaica, 1746–1755. 'Natural History of Jamaica,' 1756. MS., 'Fasciculus Plantarum Hiberniæ' at Linn. Soc. Herbarium purchased by Linnæus for 8 guineas. Pult. ii. 349; Pritz. 44; Jacks. 370; Linn. Letters, i. 39, 42; ii. 480; Trans. Linn. Soc. iv. 31; Baillon, i. 502; Dict. Nat. Biog. vii. 53. *Brownæa* Jacq.

**Browne, Samuel** (d. before 1703). M.D. Surgeon to H. E. I. C. at Madras. Sent plants to Petiver, Phil. Trans. xx. 313; xxiii. Plants in Sloane Herbarium. Pult. ii. 38, 39, 62; Pritz. 44; Rich. Corr. 76; Linn. Letters, ii. 165; 'Museum Petiverianum,' 43.

**Browne, Sir Thomas** (1605–1682): b. St. Michael's, Cheapside, London, 19th October, 1605; d. Norwich, 19th October, 1682; bur. St. Peter Mancroft, Norwich. B.A., Oxon, 1627. M.A., 1629. M.D., Leyden, 1633. M.D., Oxon, 1637. F.R.C.P., 1664. Knighted, 1671. 'The Garden of Cyrus,' 1658. 'Plants . . . in Scripture,' &c., in posthumous works. Pritz. 44; Life and Works, ed. Wilkin, 1836; Nich. Illustr. vi. 830; Munk, i. 321; Felton, 94; Cott. Gard. v. 15; Linn. Trans. vii. 296; Dict. Nat. Biog. vii. 64. Portr. in vestry of St. Peter Mancroft, Norwich; one at R. C. P.; one, with family, by Dobson, at Devonshire House; engr. by R. White in 'Works,' ed. 1686; and engr. at Linn. Soc., from Norwich portr.

**Browne, William** (c. 1628–1678): b. Oxford, c. 1628; d. Oxford, 25th March, 1678; bur. in outer chapel, Magdalen Coll., Oxon. Clerk. B.A., Oxon, 1647. B.D., 1665. Senior Fellow of Magdalen. " Chief hand in the composition of the Catalogue," Wood, Fasti. ii. 282. 'Catalogus Horti Bot. Oxoniensis,' 1658, with Stephens and Bobart. Pult. i. 167; Pritz. 44; Wood's Athen. Oxon. ed. Bliss.; Dict. Nat. Biog. vii. 75.

**Brownlee, J.** (fl. 1842). Clerk. Missionary at the Cape. Had " a good general knowledge of Botany," and sent pl. to Harvey. Journ. Bot. 1842, 16. *Brownleea*, "Harvey MS.," Lindl.

**Brownlow, Lady** (d. before 1819). Daughter of Lady Amelia Hume. *Brownlowia* Roxb.

**Brownrigg, William** (1711–1800): b. High Close Hall, Cumberland, 24th March, 1711; d. Ormathwaite, Keswick, 6th January, 1800. M.D., Leyden, 1737. F.R.S., 1741. 'Electrifying of Plants,' Phil. Trans. 1747, No. 482. Friend of Sloane, Hales, and Franklin. 'Literary Life,' by Joshua Dixon, 1801; Dict. Nat. Biog. vii. 85.

**Bruce, Arthur** (c. 1725–1805): b. c. 1725; d. 1805.  Sec. N. H. Soc. Edin.  Discovered *Polygonatum verticillatum*:  E. B. 128. Herb. bequeathed to Sir J. E. Smith.  Smith Lett. i. 431–3.

**Bruce, James** (1730–1794): b. Kinnaird, Stirling, 14th December, 1730; d. Kinnaird, 27th April, 1794.  Traveller.  Of Kinnaird. Consul at Algiers, 1763–5.  In Abyssinia, 1769–1771.  Pritz. 44; Biog. by A. Murray in 2nd ed. of 'Travels,' 1805; Dict. Nat. Biog. vii. 98; Baill. i. 503.  Portr. engr. Heath, 1790, in his 'Travels.'  *Brucea* J. S. Miller.

**Brunton, John** (fl. 1777).  'Cat. of pl. botanically arranged, .... most of wh. are cultivated and sold by . . .' at . . . Perryhill, Birmingham, 1777.  Pritz. 46; Jacks. 409.

**Brunton, William** (1775–1806): b. 21st October, 1775: d. Ripon, 23rd June, 1806.  Of Ripon.  F.L.S., 1806.  Found *Hypnum squarrosulum*.  E. B. 1709.

**Bryant, Charles** (d. 1799 ?).  Of Norwich.  'Flora Diætetica,' 1783.  'Dictionary of Ornamental . . . . Plants,' 1790.  Linn. Trans. vii. 299; Pritz. 46; Jacks. 528.

**Bryant, Henry** (1721–1799): b. 1721; d. Colby, Norfolk, 4th June, 1799.  Clerk.  B.A., Camb., 1749.  M.A., 1753.  A.L.S., 1795.  Rector of Colby, Norfolk.  Discovered *Tillæa muscosa* in Britain, 1766.  'Enquiry into the cause of . . . . Brand, 1784. Brother of the above-mentioned.  Pritz. 46; Linn. Trans. vii. 297; Gent. Mag. lxix. 1799, i. 532; Dict. Nat. Biog. vii. 155; Smith Lett. i. 33.

**Bryce, James** (1806–1877): b. Killaig, Coleraine, 22nd October, 1806; d. Inverfarigaig, Loch Ness, 11th July, 1877.  B.A., Glasgow, 1828.  LL.D., 1858.  F.G.S.  'Geology of Arran, . . . with an account of the Botany . . . . .' 1872.  Dict. Nat. Biog. vii. 159.

**Buchanan, Francis,** afterward **Hamilton** [see HAMILTON.]

**Buckland, William** (1784 ?–1856): b. Axminster, Devon, 1784?; d. 15th August, 1856; bur. Islip, Oxon.  Clerk.  D.D., Oxon, 1825.  F.L.S., 1821.  P.G.S., 1824 & 1840.  F.R.S., 1818. Dean of Westminster, 1845.  Prof. Mineralogy, Oxon, 1813; of Geology, 1818.  Jacks. 182; R. S. C. i.; Proc. Roy. Soc. viii. 264; Quart. Journ. Geol. Soc. 1857; Dict. Nat. Biog. vii. 206.  *Bucklandia* Sternb. = *Clathraria* Mant.  *Bucklandia* Brong. (fossil Cycad).  *Bucklandia* R. Br. (*Hamamelidæ*).  Portr. in Ipswich series, 1849.

**Buckman, James** (1814 or 1816–1884): b. Cheltenham, Gloucestershire, 1814 or 1816; d. Bradford Abbas, Dorset, 23rd Nov., 1884.  Druggist.  F.L.S., 1850.  Prof. of Bot., Cirencester, 1848–1863.  'Bot. Guide to . . . Cheltenham,' 1844.  'British . . . . Grasses,' 1858.  Pritz. 47; Jacks. 528; R. S. C. i. 705; vi. 611; vii. 298; Proc. Linn. Soc. 1884–5, 104; Proc. Geol. Soc. 1884–5, 43; Dict. Nat. Biog. vii. 216.

**Buddle, Adam** (d. 1715): b. Deeping St. James, Lincoln; d. Gray's Inn, 15th April, 1715; bur. St. Andrew's Holborn. Clerk.  B.A., 1681.  M.A., Camb., 1685.  Vicar of Great Fambridge, Essex, 1703.  Reader at Gray's Inn.  'Hortus Siccus

Buddleanus sive Methodus Nova stirpium Britann.,' Sloane
MSS. 2970–2980. Herbarium in Sloane's, in Mus. Brit. Fl.
Midd. 386–8; Rich. Corr. 87, 95, 103; Dict. Nat. Biog. vii.
222. Portr. at Hadleigh Hall. *Buddleia* L.

**Bulger, George Ernest** (d. 1884 or 1885). Lieutenant-Colonel.
F.L.S., 1864. 'Flora of Windvogelberg.' Student, iv. 1870,
275. R. S. C. vii. 301; Proc. Linn. Soc. 1883–86, 80.

**Bulkley, Edward** (fl. 1695). Surgeon to H. E. I. C. at Madras.
Succeeded Samuel Browne. Sent plants to Petiver. 'Museum
Petiverianum,' 43, 94.

**Bull, Henry Graves** (c. 1818–1885): b. circ. 1818; d. Hereford,
31st October, 1885. M.D. Of Hereford. Mycologist. 'Herefordshire Pomona.' Journ. Bot. 1886, 62. R. S. C. i. 715;
vii. 302.

**Bull, Martin M.** (d. 1879): d. Jersey, 17th August, 1879. M.D.
Of Jersey. Found *Ranunculus chærophyllus*. List of pl. . . . . of
Sark, Journ. Bot. 1872, 199; R. S. C. vii. 302; Journ. Bot.
1879, 288.

**Bulleyn, William** (1500?–1576): b. Isle of Ely, 1500; d. in
prison, London?, 7th January, 1576; bur. in Cripplegate
Church. Clerk and Physician. Both of Oxford and Cambridge.
F.R.C.P., 1560. Practised in Durham. Travelled in Scotland
and Germany. 'A Book of Simples,' 1562. Pult. i. 77–83;
Pritz. 48; Jacks. 25; Wood, Athen. Oxon. ed. Bliss, i. 538;
Tanner, Bibl. Angl. Hibern.; Cooper, Athen. Cantab. i. 343;
Dict. Nat. Biog. vii. 244; Munk; Biog. Brit.; Felton, 84–5;
Cott. Gard. v. 207; Journ. Hort. 1876, 373, with portr. Engr.
portr. in profile in 'Government of Health,' 1559, and full
length (imaginary) in 'Bulwarke of Defence'; head, engr. W.
Stubeley, 1722.

**Bunbury, Sir Charles James Fox**, Bart. (1809–1886): b. Mes-
sina, 1809; d. London?, 19th June, 1886. F.L.S., 1833.
F.R.S. 'Plants of Brazil,' Proc. Linn. Soc. 1849, 108; 'Fossil
Plants from . . . . Jurassic,' Quart. Journ. Geol. Soc. 1851.
Pritz. 48; Proc. Linn. Soc. 1886, 735; Ann. & Mag. vii. 1841,
439; Journ. Bot. 1842, 549; 1843, 15; 1844, 242; R. S. C. i.
715; vi. 612. *Bunburia* Harv.

**Burchell, William John** (1781 or 1782–1863): b. Fulham, 1781
or 1782; d. Churchfield House, Fulham, 23rd March, 1863.
F.L.S., 1803. D.C.L., Oxon, 1834. "Schoolmaster and
acting botanist" to H. E. I. C. at St. Helena, 1805–1810.
Travelled in S. Africa, 1811–1815; in S. America, 1825–1829.
Collected 15,000 spp. Heir to R. A. Salisbury. Plants from
Africa and Brazil at Kew. Pritz. 48; Jacks. 346; Proc. Linn.
Soc. vii. (1864); Bot. Misc. ii. 128; R. S. C. i.; Dict. Nat.
Biog. vii. 290. Portr. at Kew. *Burchellia* R. Br.

**Burgess, Rev. Dr.** (fl. 1777). Of Kirkmichael, Dumfries. Lightfoot, Fl. Scotica, i. xiii. Lichenologist. *Leptogium Burgessii*
Leighton.

**Burke, Joseph** (fl. 1840). Collector for Lord Derby in S. Africa.
Pl. at Kew and in Herb. Mus. Brit. Journ. Bot. 1843, 163;
1845, 644. *Burkea* Hook.

**Burnett, Gilbert Thomas** (1800–1835): b. 15th April, 1800; d. 27th July, 1835. Prof. Bot., King's College, London, 1831, and at Chelsea, 1835. F.L.S., 1832. 'Outlines of Botany,' 1835. Pritz. 49; Jacks. 38, 495; Semple, 186; R. S. C. i. 735; Mag. Gard. iv. (n.s. ii.) 29; Dict. Nat. Biog. vii. 412. *Burnettia* Lindl.

**Burnett, James** (fl. 1836). M.A. 'Magazine of Botany and Gardening,' 1835–7.

**Burnett, M. A.** (fl. 1840–50). Sister of Gilbert Thomas Burnett. 'Plantæ Utiliores,' 1842–50, drawn and coloured by M. A. B., the text mainly by G. T. B. Pritz. 49; Jacks. 192; Dict. Nat. Biog. vii. 412.

**Burton, David** (d. before 1807): d. Australia, before 1807. Gardener. Sent by Sir J. Banks to Port Jackson. Salisbury, Parad. Lond. f. 73. *Burtonia* Salisb. = *Hibbertia*. *Burtonia* R. Br. in Aiton.

**Busk, George** (1807–1886): b. 1807; d. London, 10th August, 1886. Surgeon, R.N. F.L.S., 1846; Sec., 1857–1868. F.R.S., 1850. Pres. R.C.S., 1871. 'Volvox Globator,' with Williamson, 1853. Zoologist and Anthropologist. Proc. Linn. Soc. 1886–7, 36; R. S. C. i. 742. Oil-portr. at Linn. Soc. by Miss E. M. Busk.

**Bute, John, Earl of** [*see* STUART, JOHN].

**Butt, J. M.** (fl. 1825). Clerk. M.A. Vicar of East Garston, Berks. Relation of the following. Smith Lett. i. 440; 'Botanical Primer,' 1825; Pritz. 50; Jacks. 37.

**Butt, Thomas.** Clerk. Of Areley, Salop. F.L.S., 1797. Sent *Gnaphalium margaritaceum* from Wyre Forest to Smith. Smith Lett. i. 435–441.

**Buxton, Richard** (1786–1865): b. Sedgley Hall Farm, Prestwich, Manchester, 15th January, 1786; d. Ancoats, Manchester, 2nd January, 1865. Shoemaker and newsman. 'Botanical Guide to . . . . Manchester,' 1849. Pritz. 50; Jacks. 256; Autobiog. in 'Guide'; Cash, 94; Journ. Bot. iii. 71; Dict. Nat. Biog. viii. 106.

**Caius, John,** *alias* **Key** (1510–1573): b. Norwich, 6th October, 1510; d. London, 29th July, 1573; bur. in Chapel of Caius College, Camb. M.A., Cantab., 1535. M.D., Padua, 1541; of Cambridge, 1558. Pupil of Vesalius. F.R.C.P., 1547; Pres., 1555–60. Physician to Edward VI., Mary, and Elizabeth. Founder of Caius College, Camb. Communicated histories of rare plants to Gesner. 'De Stirpium,' 1570. Pritz. 50; Jacks. 26; Munk, i. 37; Cooper, Athen. Cantab. i. 312; Dict. Nat. Biog. viii. 221. Life by Aikin. Three portr. at Caius Coll.; mez. portr. by J. Faber; woodcut in his 'Works,' 1556; and one in the 'Heroologia.'

**Calcoensis, Henricus** (fl. 1493). Benedictine Prior. Scotch? 'Synopsis Herbaria,' MS. Pult. i. 24.

**Caldcleugh, Alexander** (fl. 1823). F.L.S., 1823. F.R.S. Collected in Chili. Friend of D. Don. *Caldcluvia* D. Don.

**Caldwell, Andrew** (1733–1808): b. Dublin?, 19th December,

1733 ; d. near Bray, Wicklow, 2nd July, 1808. Irish bar, 1760. F.L.S., 1796. Correspondent of Smith. Smith Lett. ii. ; Nich. Illustr. viii. 24 ; Gent. Mag. lxxviii. 746 ; Dict. Nat. Biog. viii. 247.

**Caley, George** (d. 1829) : b. Yorkshire ; d. Bayswater, 23rd May, 1829 ; bur. St. George's burial-ground, Hyde Park. Began life as stable-boy. Assisted by Banks. In New South Wales, 1799–1810. Superintendent, St. Vincent Gardens, 1811–1823. "Botanicus peritus et accuratus," R. Brown. Trans. Linn. Soc. xv. (1811) ; Gard. Chron. xxiv. 1885, 263 ; Mag. Nat. Hist. 1829, 310 ; 1830, 226 ; Cash, 21. *Caleana* R. Br.

**Callcott, Lady,** *née* **Dundas** (1785–1842) : b. Papcastle, Cockermouth, 19th July, 1785 ; d. Kensington Gravel-Pits, 28th November, 1842 ; bur. Kensal Green Cemetery. m. 1. Captain Thomas Graham, R.N., 1809. m. 2. Augustus Wall Calcott (afterwards knighted), 1827. 'Journal of a Residence in India,' 1812. 'Journal of a Voyage to Brazil,' 1824. ' Scripture Herbal,' 1842, &c. Collected in Brazil. Jacks. 20 ; Journ. Bot. 1842, 26 ; Gent. Mag. 1843, i. 98 ; Dict. Nat. Biog. viii. 258 ; Dodd, 'Annual Biography,' 1842, 285.

**Cameron, David** (1787 ?–1848) : b. circ. 1787 ; d. Birmingham, 25th June, 1848. A.L.S. Curator, Birmingham Garden, to 1847. Contrib. to Phyt. i. Gard. Chron. 1847 and 1848.

**Campbell, Archibald, 3rd Duke of Argyle** (1682–1761) : b. Ham House, Richmond, June, 1682 ; d. London, 15th April, 1761. Planted Whitton. Most of his trees went to Kew. Loud. 'Arboretum,' 57–8 ; Dict. Nat. Biog. viii. 341. *Argylia* D. Don.

**Campbell, William H.** (fl. 1841). LL.D. Sec. Bot. Soc. Edin., and joint author with J. H. Balfour and C. C. Babington of ' Cat. Brit. Pl. Edin.,' 1851. *Campbellia* Wight.

(To be continued.)

---

# SHORT NOTES.

SUFFOLK PLANTS.—During a recent short visit to this county, I found the following additions to ' Topographical Botany,' ed. 2, for districts 25 and 26 (E. & W. Suffolk) :—*Viola hirta* L., near Mellis, and in and about Bergate Wood (25). *V. Reichenbachiana* Jord., more abundant than *V. sylvatica* about Mellis, Wortham, and Burgate (25) ; plentiful about Hawstead and Whepstead (26). *Taraxacum officinale* Wigg., var. *erythrospermum* ? Wortham Ling (25) ; probably correct, but only in flower. *Myosotis sylvatica* Hoffm., abundant along small streams near Hawstead and Whepstead. Without personal authority for 26. *Carex stricta* Good, a few tussocks in a ditch a little north of Burgate Wood (25).—EDWARD S. MARSHALL.

PULMONARIA OFFICINALIS L. AS A NATIVE OF BRITAIN.—This plant appears to have been hitherto looked upon as naturalised only, but I am strongly of opinion that it is indigenous in E. Suffolk. Dr. Hind, who agrees in this, kindly refers me to the first publication of the station (Burgate Wood), by Mr. C. J. Ashfield, son of a former Vicar of Burgate, in the 'Phytologist,' n.s., 1862, p. 351 and adds that it also occurs in a small wood in an adjoining parish. I had no thought of its being native until I actually saw it *in situ*, when the probability of this at once struck me as very great. Growing in an apparent remnant of the old forest, the only introduced plants seen being a few young larches at one end, in the company of *Paris*, *Viola*, *Valeriana officinalis*, *Spiræa Ulmaria*, *Allium Ursinum*, &c., it seemed equally wild with any of them, and equally capricious in its selection of spots to grow in, occurring in patches in various parts of the wood. A woman who was binding faggots told me that it had been known there as long as she could remember, and that all the villagers considered it a native. A very important factor in the evidence is the absence of spots on the leaves, which are also darker and less flaccid than in the form so common as a cottage ornament. Nyman, curiously enough, gives it as a rare inhabitant of England, without mark of suspicion ; and the continental distribution is quite in favour of its natural occurrence here. Mr. Watson's rejection of certain E. Anglian plants will carry less weight if (as I have been informed) he never visited that part of the country.—EDWARD S. MARSHALL.

VITALITY OF SPORES OF GYMNOGRAMMA LEPTOPHYLLA.—In 1880 I spent some time in the Channel Islands, and when in Jersey made enquiry of M. Piquet as to the above fern, and was told that it was then (August) too late in the season to find it. He very kindly gave me exact information of one of the known localities, from which I obtained some of the surface-earth. Plenty of plants came up that autumn, and a succession of them for some years since. Only a small portion of the Jersey earth obtained was used at the time, the remainder being left tied up in the box in which it was brought over. In October, 1886, having prepared a small seed-pan, and poured boiling water over the soil in it to destroy any insects, &c., a small quantity of the Jersey earth of 1880 was scattered on the surface, and a bell-glass put on. It was then placed in the greenhouse, where there were no other plants of the species. Some prothalliæ appeared in the autumn, and some time in June, 1887, small fronds were seen, and since then others have come up, and last autumn the pan was thickly covered with young plants. The spores had thus retained their vitality at all events for *seven* years ; what further length of time they may have remained dormant before 1880 it is, of course, impossible to say. I left some of the 1880 earth with the Curator, Mr. Nicholson, at Kew Gardens, on October 2nd, 1886, who said they would try it as a check experiment. On visiting the gardens, July 11th, 1887, I was told that the plants had come up plentifully.—A. SHARLAND.

*NEW PHANEROGAMS PUBLISHED IN BRITAIN IN 1887.*

(Concluded from p. 121.)

INDIGOFERA DESMODIOIDES and I. ORMOCARPOIDES *Baker.* Madagascar. J. L. Soc. xxii. 463, 464.

IPOMÆA BULLATA *Oliv.* Trop. Africa. Trans. L. Soc. (Bot.) ii. 343, t. 62.—I. ROBERTSII *Hook. f.* Australia. Bot. Mag. t. 6952.— I. RUBRO-VIRIDIS and I. SYRINGÆFOLIA *Baker.* Madagascar. J. L. Soc. xxii. 507.

IRIS BILIOTTI, Trebizond. Gard. Chron. i. 738 : I. DUTHIEII, E. Indies. Id. 611 : I. HOOKERIANA, E. Indies. Id. 611 : I. KINGIANA, E. Indies. Id. 611 : I. LUPINA, E. Indies. Id. 732 : all of *M. Foster.*

ISOGLOSSA LAXA *Oliv.* Trop. Africa. Trans. L. Soc. (Bot.) ii. 345.

IXORA PACHYPHYLLA *Baker.* Madagascar. J. L. Soc. xxii. 484.

KALANCHOE BRACHYCALYX, K. BREVICAULIS, K. GOMPHOPHYLLA, K. INTEGRIFOLIA, K. LAXIFLORA, K. PUBESCENS, K. STREPTANTHA, K. SUBPELTATA, and K. SULPHUREA, all of *Baker.* Madagascar. J. L. Soc. xxii. 470–472.

KNIPHOFIA KIRKII *Baker.* Trop. Africa. Gard. Chron. ii. 712.

*LASIOCOCCA (Euphorbiaceæ Acalypheæ) SYMPHILLIÆFOLIA *Hook. f.* India. Ic. Pl. 1587.

LEBECKIA INFLATA (Ic. Pl. 1576), L. LONGIPES, and L. WRIGHTII, all of *Bolus.* S. Africa. Ic. Pl. 1552.

LEEA CUSPIDIFERA *Baker.* Madagascar. J. L. Soc. xxii. 461.

LONCHOCARPUS PAULLINIOIDES *Baker.* Madagascar. J. L. Soc. xxii. 466.

LONICERA PILEATA *Oliv.* China. Ic. Pl. 1585.

*LOPHOPYXIS (Euphorbiaceæ) MAINGAYI *Hook. f.* Malacca. Ic. Pl. 1714.

LOTONONIS FOLIOSA *Bolus.* S. Africa. J. L. Soc. xxiv. 173.

MACARANGA FERRUGINEA and M. RACEMOSA *Baker.* Madagascar. J. L. Soc. xxii. 520, 521.

MAMMILLARIA CORNIMAMMA *N. E. Br.* Gard. Chron. ii. 186.

MASCARENHAISIA GERARDIANA and M. MACROSIPHON *Baker.* Madagascar. J. L. Soc. xxii. 504.

MASDEVALLIA DEMISSA *Rchb. f.* Costa Rica. Gard. Chron. ii. 9.— M. PUSIOLA *Rchb. f.* Columbia. Id. i. 140. — M. SOROCULA *Rchb. f.* Id. ii. 713.—M. WENDLANDIANA *Rchb. f.* Id. i. 174.

MAXILLARIA MOLITOR *Rchb. f.* Gard. Chron. ii. 242.

MEDINILLA DIVARICATA and M. LINEARIFOLIA *Baker.* Madagascar. J. L. Soc. xxii. 478.

*MEGAPHYLLÆA (Meliaceæ Trichilieæ) PERAKENSIS *Hemsl.* Malaya. Ic. Pl. 1708.

*MEGISTOSTIGMA (Euphorbiaceæ Plukenetieæ) MALACCENSE *Hook. f.* Malacca. Ic. Pl. 1592.

MELHANIA GRIQUENSIS *Bolus.* S. Africa. J. L. Soc. xxiv. 172.

MELOCHIA BETSILIENSIS *Baker.* Madagascar. J. L. Soc. xxii. 451.

MEZZETTIA HERVEYANA *Oliv.* Malacca. Ic. Pl. 1560.

MIMOSA MYRIOCEPHALA *Baker.* Madagascar. J. L. Soc. xxii. 467.

MIMULOPSIS AFFINIS *Baker.* Madagascar. J. L. Soc. xxii. 509.

MITREPHORA MACROPHYLLA *Oliv.* Penang. Ic. Pl. 1562.

Modecca hederifolia *Baker*.  Madagascar.  J. L. Soc. xxiii. 479.
Meconeurum sinense *Hemsl*.  China.  J. L. Soc. xxiii. 204.
—— Mormodes vernixium *Rchb. f.*  Guiana.  Gard. Chrom. ii. 682.
Mucuna axillaris *Baker*.  Madagascar.  J. L. Soc. xxiii. 465.—M. sempervirens *Hemsl*.  China.  Id. xxiii. 190.
Mundulea laxiflora *Baker*.  Madagascar.  J. L. Soc. xxiii. 464.
Musaronia unifoliolata *Oliv*.  China.  Ic. Pl. 1709.
Nasturtium Henryi *Oliv*.  China.  Ic. Pl. 1712.—N. Millefolium *Baker*.  Madagascar.  J. L. Soc. xxiii. 444.
Nepenthes Curtisii *Mast*.  Borneo.  Gard. Chrom. ii. 681, 682.
Nepenthes picturata *N. E. Br.*  Congo.  Gard. Chrom. i. 476.
—— Notylia Bungerothii *Rchb. f.*  S. America.  Gard. Chrom. ii. 38.
Nuxia pachyphylla and N. terminalioides *Baker*.  Madagascar.  J. L. Soc. xxiii. 505, 506.
Ochna macrantha *Baker*.  Madagascar.  J. L. Soc. xxiii. 457.
Ocotea trichantha *Baker*.  Madagascar.  J. L. Soc. xxiii. 515.
—— Odontoglossum Schroederianum *Rchb. f.*  Gard. Chrom. ii. 364.
— —— Oncidium Hookeri *Rolfe*.  Brazil.  Gard. Chrom. ii. 520. — O. lucescens *Rchb. f.*  Id. i. 799.
Oncostemum botryoides, O. platycosum, O. macrospherum, O. ? polytrichum, and O. vacciniifolium, all of *Baker*.  Madagascar.  J. L. Soc. xxiii. 501, 502.
*Oxrstia (Orchideæ) elegans *Ridl*.  W. Africa.  J. L. Soc. xxiv. 197, t. 6.
—— Ornithidium ochraceum *Rchb. f.*  N. Granada.  Gard. Chrom. i. 209.
Oxalis catharinensis *N. E. Br.*  Brazil.  Gard. Chrom. i. 149.
Pachypodium brevicaule and P. densiflorum *Baker*.  Madagascar.  J. L. Soc. xxiii. 503.
Pandanus dyckioides *Baker*.  Madagascar.  J. L. Soc. xxiii. 527.
Panax gomphophylla *Baker*.  Madagascar.  J. L. Soc. xxiii. 480.
Pazinarium emirnense *Baker*.  Madagascar.  J. L. Soc. xxiii. 489.
Phyllanthus ? lycioides *Baker*.  Madagascar.  J. L. Soc. xxiii. 516.
Pelargonium madagascariense *Baker*.  Madagascar.  J. L. Soc. xxiii. 454.
Pentas hirtiflora *Baker*.  Madagascar.  J. L. Soc. xxiii. 462. — P. longifolia *Oliv*.  Trop. Africa.  Trans. L. Soc. (Bot.) 335.
—— Peristeria leta *Rchb. f.*  Gard. Chrom. ii. 616. — P. selligera *Rchb. f.*  Id. 272.
*Petrocosmea (Gesneriaceæ Cyrtandreæ) sinensis *Oliv*.  China.  Ic. Pl. 1716.
Phreatites lucidus *Oliv*.  Penang.  Ic. Pl. 1561.
Phalaenopsis Foerstermanii *Rchb. f.*  Gard. Chrom. i. 244. — P. Esmeraldana *Rchb. f.*  Siam.  Id. ii. 746.
Philippia capitata, P. cryptoclada, P. hispida, P. minutifolia, and P. trichoclada, all of *Baker*.  Madagascar.  J. L. Soc. xxiii. 499. 500.—P. tristis *Baker*.  S. Africa.  J. L. Soc. xxiv. 187.
Phrynium textile *Ridl*.  Angola.  J. Bot. 133.
Phyllobea sinensis *Oliv*.  China.  Ic. Pl. 1721.
Pilea Johnstoni *Oliv*.  Trop. Africa.  Trans. L. Soc. (Bot.) ii. 349.—P. macropoda *Baker*.  Madagascar.  J. L. Soc. xxiii. 524.
Piper emirnense *Baker*.  Madagascar.  J. L. Soc. xxiii. 514.
Piptadenia leptoclada *Baker*.  Madagascar.  J. L. Soc. xxiii. 457.

*NEW PHANEROGAMS PUBLISHED IN BRITAIN IN* 1887.

(Concluded from p. 121.)

INDIGOFERA DESMODIOIDES and I. ORMOCARPOIDES *Baker.* Madagascar. J. L. Soc. xxii. 463, 464.

IPOMÆA BULLATA *Oliv.* Trop. Africa. Trans. L. Soc. (Bot.) ii. 343, t. 62.—I. ROBERTSII *Hook. f.* Australia. Bot. Mag. t. 6952.— I. RUBRO-VIRIDIS and I. SYRINGÆFOLIA *Baker.* Madagascar. J. L. Soc. xxii. 507.

IRIS BILIOTTI, Trebizond. Gard. Chron. i. 738 : I. DUTHIEII, E. Indies. Id. 611 : I. HOOKERIANA, E. Indies. Id. 611 : I. KINGIANA, E. Indies. Id. 611 : I. LUPINA, E. Indies. Id. 732 : all of *M. Foster.*

ISOGLOSSA LAXA *Oliv.* Trop. Africa. Trans. L. Soc. (Bot.) ii. 345.

IXORA PACHYPHYLLA *Baker.* Madagascar. J. L. Soc. xxii. 484.

KALANCHOE BRACHYCALYX, K. BREVICAULIS, K. GOMPHOPHYLLA, K. INTEGRIFOLIA, K. LAXIFLORA, K. PUBESCENS, K. STREPTANTHA, K. SUBPELTATA, and K. SULPHUREA, all of *Baker.* Madagascar. J. L. Soc. xxii. 470–472.

KNIPHOFIA KIRKII *Baker.* Trop. Africa. Gard. Chron. ii. 712.

*LASIOCOCCA (Euphorbiaceæ Acalypheæ) SYMPHILLIÆFOLIA *Hook. f.* India. Ic. Pl. 1587.

LEBECKIA INFLATA (Ic. Pl. 1576), L. LONGIPES, and L. WRIGHTII, all of *Bolus.* S. Africa. Ic. Pl. 1552.

LEEA CUSPIDIFERA *Baker.* Madagascar. J. L. Soc. xxii. 461.

LONCHOCARPUS PAULLINIOIDES *Baker.* Madagascar. J. L. Soc. xxii. 466.

LONICERA PILEATA *Oliv.* China. Ic. Pl. 1585.

*LOPHOPYXIS (Euphorbiaceæ) MAINGAYI *Hook. f.* Malacca. Ic. Pl. 1714.

LOTONONIS FOLIOSA *Bolus.* S. Africa. J. L. Soc. xxiv. 173.

MACARANGA FERRUGINEA and M. RACEMOSA *Baker.* Madagascar. J. L. Soc. xxii. 520, 521.

MAMMILLARIA CORNIMAMMA *N. E. Br.* Gard. Chron. ii. 186.

MASCARENHAISIA GERARDIANA and M. MACROSIPHON *Baker.* Madagasear. J. L. Soc. xxii. 504.

MASDEVALLIA DEMISSA *Rchb. f.* Costa Rica. Gard. Chron. ii. 9.— M. PUSIOLA *Rchb. f.* Columbia. Id. i. 140. — M. SOROCULA *Rchb. f.* Id. ii. 713.—M. WENDLANDIANA *Rchb. f.* Id. i. 174.

MAXILLARIA MOLITOR *Rchb. f.* Gard. Chron. ii. 242.

MEDINILLA DIVARICATA and M. LINEARIFOLIA *Baker.* Madagascar. J. L. Soc. xxii. 478.

*MEGAPHYLLÆA (Meliaceæ Trichilieæ) PERAKENSIS *Hemsl.* Malaya. Ic. Pl. 1708.

*MEGISTOSTIGMA (Euphorbiaceæ Plukenetieæ) MALACCENSE *Hook. f.* Malacca. Ic. Pl. 1592.

MELHANIA GRIQUENSIS *Bolus.* S. Africa. J. L. Soc. xxiv. 172.

MELOCHIA BETSILIENSIS *Baker.* Madagascar. J. L. Soc. xxii. 451.

MEZZETTIA HERVEYANA *Oliv.* Malacca. Ic. Pl. 1560.

MIMOSA MYRIOCEPHALA *Baker.* Madagascar. J. L. Soc. xxii. 467.

MIMULOPSIS AFFINIS *Baker.* Madagascar. J. L. Soc. xxii. 509.

MITREPHORA MACROPHYLLA *Oliv.* Penang. Ic. Pl. 1562.

MODECCA HEDERÆFOLIA *Baker*. Madagascar. J. L. Soc. xxii. 479.
MEZONEURUM SINENSE *Hemsl*. China. J. L. Soc. xxiii. 204.
—— MORMODES VERNIXUM *Rchb. f.* Guiana. Gard. Chron. ii. 682.
MUCUNA AXILLARIS *Baker*. Madagascar. J. L. Soc. xxii. 465.—M.
    SEMPERVIRENS *Hemsl*. China. Id- xxiii. 190.
MUNDULEA LAXIFLORA *Baker*. Madagascar. J. L. Soc. xxii. 464.
MUNRONIA UNIFOLIOLATA *Oliv*. China. Ic. Pl. 1709.
NASTURTIUM HENRYI *Oliv*. China. Ic. Pl. 1719.—N. MILLEFOLIUM
    *Baker*. Madagascar. J. L. Soc. xxii. 444.
NEPENTHES CURTISII *Mast*. Borneo. Gard. Chron. ii. 681, 689.
NEPHTHYTIS PICTURATA *N. E. Br*. Congo. Gard. Chron. i. 476.
—— NOTYLIA BUNGEROTHII *Rchb. f.* S. America. Gard. Chron. ii. 38.
NUXIA PACHYPHYLLA and N. TERMINALIOIDES *Baker*. Madagascar.
    J. L. Soc. xxii. 505, 506.
OCHNA MACRANTHA *Baker*. Madagascar. J. L. Soc. xxii. 457.
OCOTEA TRICHANTHA *Baker*. Madagascar. J. L. Soc. xxii. 515.
—— ODONTOGLOSSUM SCHRŒDERIANUM *Rchb. f.* Gard. Chron. ii. 364.
—— ONCIDIUM HOOKERI *Rolfe*. Brazil. Gard. Chron. ii. 520. — O. LU-
    CESCENS *Rchb. f.* Id. i. 799.
ONCOSTEMUM BOTRYOIDES, O. FLEXUOSUM, O. MICROSPHÆRUM, O. ? POLY-
    TRICHUM, and O. VACCINIIFOLIUM, all of *Baker*. Madagascar.
    J. L. Soc. xxii. 501, 502.
*ORESTIA (Orchideæ) ELEGANS *Ridl*. W. Africa. J. L. Soc. xxiv.
    197, t. 6.
ORNITHIDIUM OCHRACEUM *Rchb. f.* N. Granada. Gard. Chron. i. 209.
OXALIS CATHARINENSIS *N. E. Br*. Brazil. Gard. Chron. i. 140.
PACHYPODIUM BREVICAULE and P. DENSIFLORUM *Baker*. Madagascar.
    J. L. Soc. xxii. 503.
PANDANUS DYCKIOIDES *Baker*. Madagascar. J. L. Soc. xxii. 527.
PANAX GOMPHOPHYLLA *Baker*. Madagascar. J. L. Soc. xxii. 480.
PARINARIUM EMIRNENSE *Baker*. Madagascar. J. L. Soc. xxii. 469.
PEDILANTHUS ? LYCIOIDES *Baker*. Madagascar. J. L. Soc. xxii. 516.
PELARGONIUM MADAGASCARIENSE *Baker*. Madagascar. J. L. Soc.
    xxii. 454.
PENTAS HIRTIFLORA *Baker*. Madagascar. J. L. Soc. xxii. 482. —
    P. LONGIFOLIA *Oliv*. Trop. Africa. Trans. L. Soc. (Bot.) 335.
PERISTERIA LÆTA *Rchb. f.* Gard. Chron. ii. 616. — P. SELLIGERA
    *Rchb. f.* Id. 272.
*PETROCOSMEA (Gesneraceæ Cyrtandreæ) SINENSIS *Oliv*. China.
    Ic. Pl. 1716.
PHÆANTHUS LUCIDUS *Oliv*. Penang. Ic. Pl. 1561.
PHALÆNOPSIS FOERSTERMANII *Rchb. f.* Gard. Chron. i. 244. — P.
    REGNIERIANA *Rchb. f.* Siam. Id. ii. 746.
PHILIPPIA CAPITATA, P. CRYPTOCLADA, P. HISPIDA, P. MINUTIFOLIA, and
    P. TRICHOCLADA, all of *Baker*. Madagascar. J. L. Soc. xxii.
    499, 500.—P. TRISTIS *Bolus*. S. Africa. J. L. Soc. xxiv. 187.
PHRYNIUM TEXTILE *Ridl*. Angola. J. Bot. 133.
PHYLLOBŒA SINENSIS *Oliv*. China. Ic. Pl. 1721.
PILEA JOHNSTONI *Oliv*. Trop. Africa. Trans. L. Soc. (Bot.) ii.
    349.—P. MACROPODA *Baker*. Madagascar. J. L. Soc. xxii. 524.
PIPER EMIRNENSE *Baker*. Madagascar. J. L. Soc. xxii. 514.
PIPTADENIA LEPTOCLADA *Baker*. Madagascar. J. L. Soc. xxii. 467.

PITTOSPORUM PACHYPHYLLUM and P. VERNICOSUM *Baker*. Madagascar. J. L. Soc. xxii. 444, 445.

PLECTRANTHUS PARVUS *Oliv*. Trop. Africa. Trans. L. Soc. (Bot.) ii. 347.

PLECTRONIA MICRANTHA *Baker*. Madagascar. J. L. Soc. xxii. 483.

PLEUROTHALLIS INSIGNIS *Rolfe*. Gard. Chron. i. 477.

*POLYDRAGMA (Euphorbiaceæ Crotoneæ) MALLOTIFORME *Hook. f.* Malaya. Ic. Pl. 1701.

POLYGALA LEPTOCAULIS *Baker*. Madagascar. J. L. Soc. xxii. 445.

POPOWIA MICRANTHA *Baker*. Madagascar. J. L. Soc. xxii. 442.

POTAMOGETON MEXICANUS *A. Bennett*. Mexico. J. Bot. 289. — P. TEPPERI *A. Bennett*. Australia. Id. 178.

PRIMULA BLATTATIFORMIS and P. VINCIFLORA *Franchet*. China. Gard. Chron. i. 575 (fig. 108).

PRUNUS HIRTIPES *Hemsl*. China. J. L. Soc. xxiii. 218.

PSIADIA CUSPIDIFERA, P. MODESTA, and P. STENOPHYLLA, all of *Baker*. Madagascar. J. L. Soc. xxii. 489, 490.

PSILOTRICHUM AFRICANUM *Oliv*. Trop. Africa. Trans. L. Soc. (Bot.) ii. 348.

PSOROSPERMUM EMARGINATUM and P. POPULIFOLIUM *Baker*. Madagasear. J. L. Soc. xxii. 453.

PSYCHOTRIA HIRTELLA *Oliv*. Trop. Africa. Trans. L. Soc. (Bot.) ii. 336. — P. PARKERI, P. RETIPHLEBIA, and P. REDUCTA, all of *Baker*. Madagascar. J. L. Soc. xxii. 484, 485.

PTEROLOBIUM PUNCTATUM *Hemsl*. China. J. L. Soc. xxiii. 207.

RESTREPIA PANDURATA *Rchb. f.* Gard. Chron. i. 244.

RHODODENDRON LOCHÆ *F. Muell*. Australia. Gard. Chron. i. 543.

*RHODOSEPALA (Melastomaceæ Osbeckieæ) PAUCIFLORA *Baker*. Madagasear. J. L. Soc. xxii. 475.

RHYNCHOSIA ? HENRYI *Hemsl*. China. J. L. Soc. xxiii. 196. — R. TRICHOCEPHALA *Baker*. Id. xxii. 465.

ROTALA CORDIFOLIA *Baker*. Madagascar. J. L. Soc. xxii. 478.

ROTBOELLIA CÆSPITOSA and R. GRACILLIMA *Baker*. Madagascar. J. L. Soc. xxii. 533.

RUBUS DICTYOPHYLLUS *Oliv*. Trop. Africa. Trans. L. Soc. (Bot.) ii. 332.—R. HENRYI *Hemsl. & O. Kze.* (Ic. Pl. 1705) ; R. ICHANGENSIS *Hemsl. & O. Kze.*; R. KUNTZEANUS *Hemsl.*; R. PLAYFAIRII *Hemsl.* China. J. L. Soc. xxiii. 231, 235.—R. LINTONI *Focke*. England. J. Bot. 331 (R. LUCENS *Linton, non Focke*, id. 82 ; R. LÆTUS *Linton, non Prog.*, id. 331). — R. NEWBOULDII *Bab.* Britain. J. Bot. 20.

SACCOLABIUM PECHEI *Rchb. f.* Moulmein. Gard. Chron. i. 447.— S. SMEEANUM *Rchb. f.* Id. ii. 214.

SAGERETIA FERRUGINEA *Oliv*. China. Ic. Pl. 1710.

SANTIRIA ? BALSAMIFERA *Oliv*. W. Africa. Ic. Pl. 1573.

SAXIFRAGA TABULARIS *Hemsl*. China. J. L. Soc. xxiii. 269.

SCHISMATOCLADA TRICHOLARYNX *Baker*. Madagascar. J. L. Soc. xxii. 480.

SCHOMBURGKIA THOMSONIANA *Rchb. f.* Gard. Chron. ii. 38.

SCILLA JOHNSTONI *Baker*. Trop. Africa. Trans. L. Soc. (Bot.) ii. 351.

*Scortechinia (Euphorbiaceæ Phyllantheæ?) Kingii. *Hook. f.* Malaya. Ic. Pl. 1706.

Sedum filipes and S. polytrichoides *Hemsl.* China. J. L. Soc. xxiii. 284, 286 (t. 7).

Selago Johnstoni *Rolfe.* Trop. Africa. Trans. L. Soc. (Bot.) ii. 344.

Senecio acetosæfolius, S. cicatricosus, S. cyclocladus, S. Hilde-brandtii, S. melastomæfolius, S. monocephalus, and S. verni-cosus, all of *Baker.* Madagascar. J. L. Soc. xxii. 496–498.— S. albopunctatus, S. namaquanus, S. Rehmanni, and S. sociorum, all of *Bolus.* S. Africa. J. L. Soc. xxiv. 175–177.—S. Baurii *Oliv.* S. Africa. Ic. Pl. 1572. — S. Johnstoni *Oliv.* Trop. Africa.· Trans. L. Soc. (Bot.) ii. 340, t. 60.

Smythea macrocarpa *Hemsl.* Perak. Ic. Pl. 1558.

Solanum Wendlandii *Hook. f.* Costa Rica. Bot. Mag. t. 6914.

Spathoglottis Regnieri *Rchb. f.* Cochin China. Gard. Chron. i. 174.

Speranskia Henryi *Oliv.* China. Ic. Pl. 1577.

*Sphyranthera (Euphorbiaceæ) capitellata *Hook. f.* Andaman Is. Ic. Pl. 1702.

Spiræa Henryi *Hemsl.* China. J. L. Soc. xxiii. 225, t. 6.

Stoebe biotoides and S. cryptophylla *Baker.* Madagascar. J. L. Soc. xxii. 494.

Streptocarpus montanus *Oliv.* Trop. Africa. Trans. L. Soc. (Bot.) ii. 344.

Strobilanthes hispidula *Baker.* Madagascar. J. L. Soc. xxii. 509.

Strychnos Baroni *Baker.* Madagascar. J. L. Soc. xxii. 504.

*Temnolepis (Compositæ Helianthoideæ) scrophulariæfolia *Baker* Madagascar. J. L. Soc. xxii. 495.

Thalia cærulea and T. Welwitschii *Ridl.* Angola. J. Bot. 132.

Thesium cystoseiroides *Baker.* Madagascar. J. L. Soc. xxii. 516.

Thladiantha Henryi and T. nudiflora *Hemsl.* China. J. L. Soc. xxiii. 316 (t. 8).

Thunbergia chrysochlamys *Baker.* Madagascar. J.L. Soc. xxii.508.

Tillandsia Barclayana, J. Bot. 239; brachypoda, 237; brevi-bracteata, 346; brevifolia, 239; Bourgæi, 278; chontalensis, 237; Cossoni, 279; dasyliriifolia, 304; Dugesii, 278; erecti-flora, 346; flabellata, 242; goniorachis, 303; graminifolia, 281; Grisebachii, 305; gymnobotrya, 243; Jenmani, 345; Mathewsii, 236; micrantha, 303; myriantha, 242; oligantha, 345; pachycarpa, 238; Parryi, 277; parvispica, 244; scalari-folia, 235; soratensis, 235; streptocarpa, 241; subimbricata, 304; sublaxa, 280; tricholepis, 234; vernicosa, 241; violacea, 279; yucatana, 280; all of *Baker.* — T. reticulata *Baker.* Brazil. Gard. Chron. i. 140.

Tina velutina *Baker.* Madagascar. J. L. Soc. xxii. 469.

Trachyphrynium violaceum *Welw.* Angola. J. Bot. 133.

*Trapella (Pedalineæ) sinensis *Oliv.* China. Ic. Pl. 1595.

Trifolium Johnstoni *Oliv.* Trop. Africa. Trans. L. Soc. (Bot.) ii. 331.

*Trimorphopetalum (Balsamineæ) Dorstenioides *Baker*. Madagasear. J. L. Soc. xxii. 454.

Tristellateia emarginata and T. stenoptera *Baker*. Madagascar. J. L. Soc. xxii. 456.

Turræa rhombifolia and T. venulosa Baker. Madagascar. J. L. Soc. xxii. 458.

Unona Wrayi *Hemsl.* Perak. Ic. Pl. 1553.

Urginea eriosphermoides *Baker*. S. Africa. Gard. Chron. ii. 126.
—U. macrocentra *Baker*. S. Africa. Gard. Chron. i. 702.

Uvaria leptocladon *Oliv.* Trop. Africa. Trans. L. Soc. (Bot.) ii. 327.

Vanda Americana *Rchb. f.* E. Indies. Gard. Chron. i. 764.

Veprecella biformis *Baker*. Madagascar. J. L. Soc. xxii. 477.

Vernonia betonicæfolia, V. capreæfolia, V. exserta, V. grisea, V. rhodopappa, and V. stenoclinoides, all of *Baker*. Madagascar. J. L. Soc. xxii. 486–488. — V. stenolepis and V. Wakefieldii *Oliv.* Trop. Africa. Trans. L. Soc. (Bot.) ii. 337.

Viscum glomeratum and V. rhipsaloides *Baker*. Madagascar. J. L. Soc. xxii. 515, 516.

Vitis humilis *N. E. Br.* Natal. Ic. Pl. 1565. — V. rhodotricha, V. sphærophylla, and V. Voanonala, all of *Baker*. Madagascar. J. L. Soc. xxii. 460, 461.

Weinmannia leptostachya *Baker*. Madagascar. J. L. Soc. xxii. 469.

Wendlandia Henryi *Oliv.* China. Ic. Pl. 1712.

Xylopia stenopetala *Oliv.* Penang. Ic. Pl. 1563.

Ziziphus pubescens *Oliv.* Trop. Africa. Trans. L. Soc. (Bot.) ii. 330.

The following were accidentally omitted in their proper places:—

Begonia Wrayi *Hemsl.* Perak. J. Bot. 203.

Dendrobium Friedricksianum *Rchb. f.* Siam. Gard. Chron. ii. 648.—D. rutriferum *Rchb. f.* Papua. Id. 746.—D. trigonopus *Rchb. f.* Birma. Id. 682.

---

## NOTICES OF BOOKS.

New Books. — G. J. Filet, 'Plantkundig Woordenboek von Nedlandersch-Indië' (ed. 2, Amsterdam, Bussy: 8vo, pp. x. 348).— G. de Saporta, 'Origine paléontologique des Arbres cultivés ou utilisés par l'homme' (Paris, Baillière: 8vo, pp. xvi. 360, 44 cuts; 3 fr. 50).—A. F. W. Schimper, 'Die Wechselbeziehungen zwischen Pflanzen und Ameisen im tropischen Amerika' (Jena, Fischer: 8vo, pp. 97; 3 plates).—E. Strasburger, 'Ueber Kern-und Zelltheilung im Pflanzenreiche' (Jena, Fischer: 8vo, pp. xviii. 258, 3 plates).—P. Vuillemin, 'La Biologie Végétale' (Paris, Baillière: 8vo, pp. 380, 82 cuts). — P. Wossidlo, 'Leitfaden der Botanik' (Berlin, Weidmann: 8vo, pp. viii. 255, 494 cuts). — F. Brendel, 'Flora Peoriana; the Vegetation in the climate of Middle Illinois' (Peoria, Franks: 8vo, pp. 89). — J. Pelletan, 'Les Diatomées, (Paris: vol. i. 8vo, pp. xiii. 322, 5 plates, 265 cuts; 15 fr.). — G.

HENSLOW, 'The Origin of Floral Structures through insects and other Agencies' (London, Kegan Paul : 8vo, pp. xix. 349 ; 88 cuts ; 5s.).—W. ZOPF, 'Untersuchungen über Parasiten aus der Gruppe der Monadinen' (Halle, Nieueyer : 4to, pp. 39, 3 plates). — J. G. BAKER, 'Handbook of the Amaryllideæ' (London, Bell ; 8vo, pp. xii. 216, 5s.

---

## ARTICLES IN JOURNALS.

*Bot. Centralblatt.* (Nos. 19, 20).—E. Godlewski, 'Zur Auffassung der Reizersaheinungen an den wachsenden Pflanzen.'—C. O. Harz, 'Ueber ägyptische Textilstoffe des 4 bis 7 Christlichen Jahr hunderts.' — (No. 21). K. Schilberzky, '*Aspidium cristatum* in Oberungarn.'—(No. 22). A. Tomaschek, 'Ueber *Bacillus muralis*.' —(No. 23). —. Röll, 'Artentypen' und 'Formenreihen' bei den Torfmoosen.'

*Botanical Gazette* (May).— M. S. Bebb, 'Notes on N. American Willows' (*Salix commutata, S. conjuncta,* spp. nn.).— L. M. Underwood, 'Undescribed Hepaticæ from California' (*Jungermania Danicola, J. rubra, J. Bolanderi, Grimaldia californica,* all of Gottsche MSS. : 4 plates).—S. Coulter, Memoir of Jacob Whiteman Bailey (Ap. 29, 1811–Feb. 27, 1857).—T. Morong, *Castalia Leibergi,* n. sp.

*Bot. Notiser* (Haft. 3). — G. E. Ringius, 'Några floristika anteckningar från Wermland.'—F. E. Ahlfvengren, 'Växtgeografiska bidrag till Gotlands flora.'—A. S. Trolander, 'Växtlokaler i Nerike.' — A. Y. Grevillius, 'Om stammens bygnad hos några lokalformer af *Polygonum aviculare.*' — A. N. Lundström, 'Några iakttagelser öfver *Calypso borealis.*'—T. M. Fries, 'Terminologiska smånotiser.' — N. H. Nilsson, '*Scirpus parvulus* och des narmaste förm vandtskaper i vår flora.' — Id., 'Tvänne nya Rumex-hybrider.' — L. M. Neuman, '*Sparganium neglectum* funnen i Danmark.'

*Bot. Zeitung* (Mar. 30). — F. Hildebrand, 'Ueber die Keimlinge von *Oxalis rubella* und deren Verwandten' (1 plate).—(Ap. 6). F. Krasser, 'Ueber den mikrochemischen Nachweis von Eiweisskörpen in der pflanzlichen Zellhaut.'—(Ap. 13, 20). H. de Vries, 'Ueber den isotonischen Coëfficient des Glycerins.'—(Ap. 27). S. Winogradsky, 'Ueber Eisenbacterien.'—(May 4, 11, 18). A. Koch, 'Ueber Morphologie und Entwickelungsgeschichte einiger endosporer Bacterienformen' (1 plate).

*Bull. Torrey Bot. Club* (May).— M. S. Bebb, 'White Mountain Willows' (1 plate). — E. L. Greene, 'Linnæus and his Genera.'— C. H. Kain, 'Diatoms of Atlantic City.' — E. L. Sturtevant, *Capsicum fasciculatum,* sp. nov.

*Flora* (Ap. 1).—R. Chodat, 'Neue Beiträge zum Diagramm der Cruciferenbluthe' (1 plate). — H. G. Reichenbach, 'Orchideæ' (*Cynosorchis compacta, C. Lowiana, C. elegans, Lockhartia cladoniophora, Oncidium oloricolle, Grammatophyllum leopardinum, Dendrochilum cobolbine, Microstylis librosa, M. Mandonii, M. Javesia, M. brachyrrhynchos, M. linguella, M. major, Pleurothallis scoparum, P. Wendlandiana,* spp. nn.). — (Ap. 11, 21). E. Heinricher, 'Zur

Biologie der Gattung *Impatiens*' (1 plate). — (Ap. 11). K. Schlie-phacke, *Bryum subglobosum*, n. sp.—(Ap. 21). U. Dammer, 'Einige Beobachtungen über die Anfassung der Blüthen von *Eremurus altaicus* an Fremdbestäubung.'—P. F. Reinsch, ' Ueber einige neue Desmarestien' (*Desmarestia pteridoides, D. Willii*, spp. nn.).

*Gardeners' Chronicle* (May 5). — *Erica striolata* Rchb. f., *Phalæ-nopsis gloriosa* Rchb. f., spp. nn. — C. B. Plowright, ' Smut in oats and barley.'—' Knaurs and Burrs' (figs. 76, 77). — (May 12). M. Foster, 'Freesias' (*F. alba, F. Leichtlinii* (fig. 79), spp. nn.). — (May 26). *Cypripedium bellatulum* Rchb. f., sp. n. — W. G. Smith, *Heterosporium Ornithogali* (fig. 88).

*Journal de Botanique* (May 1). — P. A. Dangeard, 'Les Péri-diniens et leurs parasites.' — N. Patouillard, 'Fragments mycolo-giques' (new species of *Hyalodesma, Asterina, Microthyrium, Hender-sonia, Uromyces, Leptosphærium*). — —. Boulay, ' Plantes fossiles des grès tertiaires de Saint-Saturnin.' — H. Douliot, ' Note sur la formation du périderme.'— —. Bornet & Flahault, 'Deux nouveaux genres algues perforantes' (*Hyella, Gomontia*). — E. Mer, 'De l'influence de l'exposition sur le developpement des couches annuelles dans les Sapins.'

*Midland Naturalist* (May). — W. Mathews, 'History of the County Botany of Worcester.' — W. B. Grove & J. E. Bagnall, ' The Fungi of Warwickshire' (contd.).

*Oesterr. Bot. Zeitschrift* (May). — A. Hausgirg, ' Zur Algenflora Böhmens.' — H. Braun, ' Zur Flora von Neiderösterreich.' — B. Blocki, *Hieracium Andrzejowskii*, n. sp.—A. Zimmeter, 'Verwilderung von Pflanzen.'—V. v. Borbás, ' *Geum spurium* und *G. montanum*.'

---

*LINNEAN SOCIETY OF LONDON.*

*May 3rd*, 1888.—Dr. John Anderson, F.R.S., Vice-President, in the chair. — The following were elected Fellows of the Society: A. V. Jennings, L. A. Boodle, W. Cash, and A. Henry.—The fol-lowing were elected Foreign Members : Dr. A. Engler, Prof. T. Fries, Prof. R. Hartig, Dr. E. Warming, and Dr. A. Dohrn.—The Chairman announced a Resolution of the Council to found a Gold Medal, to be called " the Linnean Medal," to be awarded at the forthcoming Anniversary Meeting to a Botanist and Zoologist, and in future years to a Botanist and Zoologist alternately, com-mencing with a Botanist.—On behalf of Mr. Miller Christy, the Botanical Secretary, Mr. B. Daydon Jackson, exhibited some speci-mens of the Bardfield Oxlip, *Primula elatior* Jacquin, gathered near Dunmow.—A communication was made by Mr. C. B. Clarke on "Root-Pressure," which will be found in the next issue of this Journal.

The Anniversary Meeting of the Society, being also its Cen-tenary, was attended with circumstances of special interest. We hope to give a full report in our next issue.

# CATALOGUE OF THE MARINE ALGÆ OF THE WEST INDIAN REGION.

### By George Murray, F.L.S.

The region dealt with in the following paper includes the whole West Indian group of Islands, and the mainland coast from Venezuela round the Caribbean Sea and Gulf of Mexico to Florida. I have not only included the Bahamas, which obviously belong to the same group as the other islands, but also Bermuda, for the reason that it is a coral island,—the most northern one,—and its marine flora, so far as it is known, has a large proportion of West Indian forms mixed with those of the temperate Atlantic. As to the mainland, I have thought it right not to take forms farther north than Florida, since the coral practically ends there, and I am upheld in this opinion by Prof. Farlow and Mr. Cosmo Melvill. Florida is the only spot on the mainland known to us to be rich in marine Algæ. These have been described by Harvey ('Nereis Boreali-Americana'), by Mr. Cosmo Melvill in the 'Journal of Botany' (1875), by Prof. Farlow in his 'List of the Marine Algæ of the United States' (Fish Commission Report, 1876), and in the 'Algæ Exsiccatæ Amer. Bor.' The rest of the mainland appears to be mostly sandy wastes and the like; though the coasts of Honduras, the Mosquito coast, that of Colombia and Venezuela may yet yield good collecting-grounds to enterprising travellers. A few forms from Mexico and from La Guayra (Venezuela), collected by Liebman, are all known to me beyond Florida.

From the islands we know a great variety of forms. First of all comes Guadeloupe, which leaves even the famous Floridan keys far behind. MM. Mazé and Schramm, in their 'Essai de classification des Algues de la Guadeloupe,' 1870–77, enumerate 811 marine Algæ (with 129 more from fresh-water and mineral springs) from that island alone! In naming them (and the large number of new forms among them) these accomplished workers had the help of the brothers Crouan, and the result is a model of what such work should be. Their exploration of Guadeloupe alone has done more for our knowledge of West Indian Algæ than all the other workers put together, as the following list abundantly shows. Fortunately the British Museum possesses a magnificent set of these Algæ complete, but for one or two unimportant forms, and I have been able to use it in the comparison of Algæ from the other localities. Of the Algæ of Barbadoes a good deal is known. Prof. Dickie published a list of them in the Linnean Society's Journal, vol. xiv., founded on specimens collected by the then Governor Sir Rawson Rawson and Miss Watts,—and on others in the Gray Herbarium,—all of which are now also in the British Museum. There are also smaller sets from Martinique (Duperrey), Santa Cruz, Cuba (Ramon de la Sagra, Wright—see Farlow in 'American Naturalist,' vol. v., p. 201, and other collectors), Jamaica (Sir Hans Sloane and Chitty), and Bermuda (Kemp in 'Canadian Naturalist,' vol. ii., p. 145; Rein in 'Bericht ü. d. Senckenberg. Gesellsch.,' 1872–73, p. 151; and

Moseley, who collected the 'Challenger' Algæ described by Prof. Dickie in the Linnean Society's Journal, vol. xiv.). I also include those which I collected at Grenada in 1886, when naturalist to the Solar Eclipse Expedition. There are remarkably few *Floridea* among them, mainly owing, I take it, to the presence of much fresh-water from Orinoco floods (see Governor Rawson, 'Nature,' Dec. 19, 1872, and Jan. 2, 1873; and Dickie on the Barbadoes Algæ, *loc. cit.*). In Guadeloupe, for example, the *Floridea* outnumber the green Algæ, as is to be expected, but in Grenada the proportion was very markedly the other way, so far as a six weeks' examination enables me to say. I may note here that I actually obtained in a tow-net, south of Grenada, and outside, in the stream of the equatorial current, specimens of *Spirogyra tropica*, a fresh-water Alga from the Orinoco probably, but known anyhow from the Amazons.

The localities are cited in the order determined by the direction of the equatorial current and the Gulf Stream. The southern forms come first, and after them the others down-stream, as it were, to Bermuda. At the end I shall deal fully with the relationships of this marine flora. Meanwhile I must remark that this list, large as it is, may be extended. Private and other collections doubtless exist containing forms beyond those I have cited, and the present enumeration will have served its purpose if it lead to its own extension.

## I. FLORIDEÆ.

### CERAMIEÆ.

CALLITHAMNION PALLENS Zan.    Guadeloupe, *Mazé*!
  *Geographical Distribution.*    Mediterranean.
C. BYSSACEUM Kütz.    Guadeloupe, *Mazé*!
  *Geogr. Distr.*    Mediterranean.
C. TURNERI Ag.    Florida, *Harvey.*
  *Geogr. Distr.*    Atlantic, Mediterranean, Tasmania.
C. LUXURIANS J. Ag.    Florida, *Melvill*!    Bermuda, *Kemp.*
  *Geogr. Distr.*    N. Atlantic, North Sea.
C. FLOCCOSUM Müll.    Grenada, *Murray*!    Bermuda, *Kemp.*
  *Geogr. Distr.*    Atlantic and North Sea.    Var. PACIFICUM Harv.
    at Vancouver.
C. PLUMULA Ellis.    Bermuda, *Kemp.*
  *Geogr. Distr.*    Atlantic and Mediterranean, Tasmania.
C. POLYSPERMUM Ag.    Florida, *Harvey.*
  *Geogr. Distr.*    N. Atlantic (Europe and America).    Pacific
    (Vancouver).
  . ELLIPTICUM Mont., var. MAJOR Crn.    Guadeloupe, *Mazé*!
  *Geogr. Distr.*    Canary Islands.
  . PEDUNCULATUM Kütz.    Guadeloupe, *Mazé* l
C   *Geogr. Distr.*    Brazil.
  . GORGONEUM Mont.    Guadeloupe, *Mazé*!
  *Geogr. Distr.*    Cape Verde.
  . APICULATUM Crn.    Guadeloupe, *Mazé*!
  . L'HERMINIERI Crn.    Guadeloupe, *Mazé*!
  . CORYNOSPOROIDES Crn.    Guadeloupe, *Mazé*!

C. HYPNEÆ Crn.   Guadeloupe, *Mazé* !
C. AMENTACEUM Crn.   Guadeloupe, *Mazé* !
C. CORNICULIFRUCTUM Crn.   Guadeloupe, *Mazé* !
C. BEANII Crn.   Guadeloupe, *Mazé* !
C. INVESTIENS Crn.   Guadeloupe, *Mazé* !
C. PELLUCIDUM Farlow.   Florida, *Farlow* !
GRIFFITHSIA SCHOUSBŒI Mont.   Guadeloupe, *Mazé* !
    *Geogr. Distr.*   Mediterranean, Atlantic (S. Europe & N. Africa).
G. CORALLINA Ag.   Florida, *Harvey*.
    *Geogr. Distr.*   Atlantic (Europe) and Mediterranean.
G. GLOBIFERA J. Ag.   Guadeloupe, *Mazé* !   ? Florida, *Harvey*.
    *Geogr. Distr.*   Atlantic (N. America).
G. OPUNTIOIDES J. Ag.   Guadeloupe, *Mazé* !
    *Geogr. Distr.*   Mediterranean.
G. SETACEA Ag., var.   Grenada, *Murray* !   Guadeloupe, *Mazé* !
    *Geogr. Distr.*   Atlantic (Europe) and Mediterranean.
CROUANIA ATTENUATA J. Ag.   Guadeloupe, *Mazé* !   Florida, *Harvey,*
    *Melvill* !
    *Geogr. Distr.*   Atlantic (Europe & America) & Mediterranean.
C. AUSTRALIS J. Ag.   Guadeloupe, *Mazé* !
    *Geogr. Distr.*   Australia and Tasmania.
HALOPLEGMA DUPERREYI Mont.   Grenada, *Murray* !   Barbadoes,
    *Dickie* !   Guadeloupe, *Mazé* !   Martinique, *Duperrey*.
    *Geogr. Distr.*   (Cape of Good Hope ?) Pacific.
CERAMIUM BYSSOIDEUM Harv.   Florida, *Harvey, Melvill* !
C. SUBTILE J. Ag.   Barbadoes, *Dickie* !   Guadeloupe, *Mazé* !   Vera
    Cruz, *Liebman*.
C. TENUISSIMUM Lyngb.   Guadeloupe, *Mazé* !   $=$ C. *arachnoideum*
    var. *patentissimum* Harv. $=$ also C. *nodiferum* Crn. ?
    Florida, *Harvey, Melvill* !
    *Geogr. Distr.*   Atlantic (Europe & N. America) & Mediterranean.
C. GRACILLIMUM J. Ag.   Guadeloupe, *Mazé* !
    *Geogr. Distr.*   Atlantic and Mediterranean, Australia ?
C. FASTIGIATUM Harv.   Florida, *Melvill* !   Bermuda, *Kemp* !
    *Geogr. Distr.*   N. Atlantic and Mediterranean.
C. DESLONGCHAMPII Chauv., var. VIMINARIUM Melv. MS.   Florida,
    *Melvill* !
    *Geogr. Distr.*   Atlantic (Europe).
C. STRICTUM Grev. et Harv.   Guadeloupe, *Mazé* !
    *Geogr. Distr.*   North and South Atlantic, Mediterranean,
    Black Sea.
    . DIAPHANUM Roth.   Florida, *Harvey, Melvill* !
    *Geogr. Distr.*   Atlantic (Europe), Cape of Good Hope, Australia ?
    . CORNICULATUM Mont.   Guadeloupe, *Limminghe, Mazé* !
    . RUBRUM Ag.   Grenada, *Murray* !   Bermuda, *Kemp*.
    *Geogr. Distr.*   Throughout all seas.
C . NITENS J. Ag.   Guadeloupe, *Mazé* !   Florida, *Harvey* !   *Melvill* !
    Bermuda, *Rein*.
    . MINIATUM Suhr.   Guadeloupe ? *Mazé*.
    *Geogr. Distr.*   Pacific, Australia.
    . STRICTOIDES Crn.   Guadeloupe, *Mazé* !

C. CORNIGERUM Crn.  Guadeloupe, *Mazé*!
C. CHILENSE Crn.  Guadeloupe, *Mazé*!
C. LANCIFERUM Kütz.  Danish West Indian Islands, *Hohenack.*!
   Meeralgen, 539.
CENTROCERAS CLAVULATUM Ag.  Grenada, *Murray*!  Guadeloupe,
   *Mazé*!  St. Thomas, *A. R. Young*!  Cuba, *Hb. Montagne*!
   Florida, *Harvey*!  *Melvill*!  Bermuda, *Rein*, '*Challenger*'!
Var. CRYPTACANTHUM Crn.  Guadeloupe, *Mazé*!  Danish West
   Indian Islands, *Hohenack.*!  Meeralgen, 441 & 442.  Bermuda,
   '*Challenger*'!
Var. LEPTACANTHUM Crn.  Guadeloupe, *Mazé*!  Danish West Indian
   Islands, *Hohenack.*!  Meeralgen, 443.
Var. HYALACANTHUM Crn.  Guadeloupe, *Mazé*!  Danish West Indian
   Islands, *Hohenack.*!  Meeralgen, 537.
Var. MICRANTHUM Kütz.  There is a specimen in Herb. Mus. Brit.
   on which is written, "Ipse, Ins. Ind. occid. Dan."  I do not
   recognise the identity of "Ipse" from the handwriting.
Var. OXYACANTHUM Crn.  Guadeloupe, *Mazé*!
Var. BRACHYACANTHUM Crn.  Guadeloupe, *Mazé*!
Var. CRISPULUM Mont.  Florida, *Harvey, Melvill*!
   *Geogr. Distr.*  Throughout all tropical and subtropical seas.

(To be continued.)

---

## SALIX FRAGILIS, S. RUSSELLIANA, AND S. VIRIDIS.

### BY F. BUCHANAN WHITE, M.D., F.L.S.

THE question of the relation which Smith's *Salix Russelliana*
bears to *S. fragilis* L., is one that has much exercised the minds of
salicologists, both British and foreign.

I do not know when British authors first abandoned *S. Russelliana* as a species distinct from *S. fragilis*, since I am not able
to refer to all the editions of the various handbooks; and though
the point is of interest it is not a vital one.  In the 4th edition
(1838) of Hooker's 'British Flora'—of special value as including
the opinion of Borrer—both *Russelliana* and *fragilis* are retained
as distinct.  So also in the 1st edition (1843) of Professor Babington's 'Manual'; in the 6th edition *Russelliana* being reduced to
varietal rank, as it also is by Walker-Arnott in the 8th edition
(1860) of Hooker & Arnott's 'British Flora.'  In the 3rd edition
(1873) of 'English Botany,' Boswell Syme makes a new departure,
and refers *S. Russelliana* to *S. viridis* Fr., saying that "it appears
to be one of the series of hybrids between *S. fragilis* and *S.
alba*, as was pointed out by Dr. Andersson," and that "there can
be no doubt, however, that if this be the case *S. Russelliana* Smith
is a departure from Fries' *S. viridis*, in the direction of *S. fragilis*."
Finally in the 'Student's Flora' (1884), it is stated that *S. Russelliana* "is considered a hybrid between *fragilis* and *alba*, and referred
to *S. viridis* Fries.  Mr. Baker says it is a synonym of *S. fragilis*,
and that Fries' *S. viridis* is not a British plant."

Turning now to some of the chief Continental salicologists, Koch, in both editions of the ' Synopsis' (1838 & 1844), makes *Russelliana* a variety of *fragilis.* Grenier, in Grenier & Godron's ' Flore de France (1856), cites *S. Russelliana* Sm. as a synonym of *S. fragilis* L., *β. pendula* Fr. Wimmer (' Salices Europææ,' 1866), while quoting *Russelliana* under *S. fragilis-alba* Wimm., thinks that the name should be abandoned as dubious. Lastly, N. J. Andersson (' Monographia ' 1863, and DeCandolle's ' Prodromus ' 1868), after mentioning the views of many authors, thinks that the name *Russelliana* should be consigned to oblivion since it is impossible to find out now whether it was to a species or definite variety, or rather to a single individual that Smith applied it, and since it has been so variously used by different writers to denote various ambiguous forms between *S. fragilis* and *S. alba.* He, however, gives the name as a synonym of *S. viridis* Fr. Nyman (' Conspectus ') quotes *S. Russelliana* Sm. as a major variety of *S. viridis* Fr., and *S. Russelliana* Koch as a minor variety.

Amongst all this diversity of opinion, a glance at the descriptions makes one thing clear, namely this, that the *Russelliana* of many authors is not the *Russelliana* of Smith.

To clear up, if possible, the confusion, it is necessary to go to the fountain-head, and ascertain that which I suspect many writers on the subject have neglected to do, namely Smith's own views, not of *Russelliana* only, but of *fragilis.*

For willows, more than almost any other class of plants, it is desirable to examine and compare authentic specimens. Unfortunately in Linné's Herbarium, in the possession of the Linnean Society, the only specimen named by Linné *S. fragilis* is one of *S. alba,* and Linné's description of *S. fragilis,* though probably intended for that species, is too vague to discriminate it with certainty from some allied willows. In Smith's Herbarium, also belonging to the Linnean Society, there are either no specimens of *S. fragilis* and its allies, or what seems more probable, the packet containing them has been mislaid. A comparison, therefore, of specimens is impossible. Fortunately, however, in Smith's descriptions and drawings there is no dubiety, and an examination of these reveals the very curious fact that the *fragilis* of Smith is *not* the *fragilis* of many modern botanists !

In this group of willows the most important characters lie in the structure of the female flowers and fruit. Smith's description— I quote from the ' Compendium Floræ Britannicæ'—of these parts, states for *fragilis,* " germinibus ovatis subsessilibus," and for *Russelliana,* " germinibus pedicellatis subulatis." Koch says of *fragilis,* " capsulis ex ovata basi lanceolatis glabris pedicellatis, *pedicello nectarium* bis terve *superante;*" and distinguishes *Russelliana* as a variety, with more pubescent and usually more finely serrated leaves ; Grenier, that *fragilis* has the " capsule ovoïde-conique, atteuuée au sommet, glabre, à pédicelle court (1 mill.), *deux-trois fois plus long que les glandes* à peine visibles ;" and of *β. pendula* Fr. ( = *S. Russelliana* Sm.), " capsules plus petites ;" Andersson, for *fragilis,* " capsulis elongato-conicis, attenuatis, glaberrimis, pedicello nectarium bis terve superante ;" and for *viridis* Fr., under which he quotes as a synonym *S. Russelliana* Sm., " capsulis breve conicis obtusiusculis glaberrimis, pedicellatis, pedicello nectarium subsuperante," a statement which he afterwards modifies as " nectarium ferc duplo superante;" Wimmer, for *fragilis*, " germina in pedicello brevi aut brevissimo,* conico-subulata," and for *fragilis-alba*, " germina pedicello brevissimo conico-cylindracea." British authors have followed Smith more or less closely, the exception being Boswell Syme, who says of *fragilis*, " capsule conical-subulate, glabrous, on a stalk twice or thrice as long as the nectary ;" and for *Russelliana* (as a synonym of *viridis* Fr.), " capsule conical-subulate, glabrous, on a stalk slightly longer than the nectary." In the ' Student's Flora,' little is said about the capsule of *fragilis*, except that it is pedicelled.

From these quotations it is evident that the plant called *fragilis* by Koch, Grenier, Andersson, Wimmer, and Boswell Syme is almost certainly identical with Smith's *Russelliana*, and distinctly different from his *fragilis*. Which of the two plants is Linné's *fragilis* must remain a little uncertain. For my part I retain the name *fragilis* L. for the plant of the Continental botanists. That these writers failed to notice that their *fragilis* and Smith's were not the same can, I assume, only be accounted for thus, that, trusting in Smith's reputation and in his possession of the Linnæan Herbarium, they neglected to compare his description and drawings (or perhaps had not the opportunity), and taking for granted that his and their *fragilis* was the same, attempted to discover some difference between plants which are identical—their *fragilis* and Smith's *Russelliana*.

What then is Smith's *fragilis ?* I think there can be no doubt but that it is an hybrid between *S. fragilis* and *S. alba ;* in other words, one of the series of which *S. viridis* Fr. is the central form.

Putting aside for the moment other points of distinction, the essential characteristics of what, following Andersson, &c., is the female of the real *S. fragilis* L., lie in the long, ovate, subulate capsule, gradually tapering into the style, with a pedicel 2 to 3 times as long as the nectary. *Alba* has a sessile or almost sessile capsule, ovate-conic in shape, and distinctly obtuse. *Viridis*, like other hybrids, varies according as it approaches or recedes from one or other of its parents, and has a capsule, sometimes more and sometimes less, but always more obtuse than in *fragilis, i.e.* not tapering so gradually into the style, and with a pedicel sometimes scarcely as long as the nectary, sometimes twice (or even a little more) as long. Moreover, though I think none of the descriptions point this out, the length of the capsule in *viridis* is intermediate between *fragilis* and *alba*. The scale of size in Andersson's plate shows this, and my measurements corroborate it. In *alba* the capsule is about from 2½–3½ mm. ; in *fragilis*, about 7 mm. ; and in *viridis*, about 5 to 6 mm.

---

* Wimmer's " brevis " will probably have a wrong value placed on it unless his specimens are examined.—F. B. W.

Now read Smith's definition of *fragilis*—" Germinibus ovatis subsessilibus," and compare his plate (E. B. t. 1807), which shows a *sessile or almost sessile* ovary, distinctly *obtuse*, and with scarcely any style. For the female plant in this plate two drawings were made, both of which are in the Botanical Department of the British Museum, along with the rest of the drawings for E. B. The plant figured came from the Rev. J. Holme, Cambridge, and was preferred to another (which Smith says is also right) from the Rev. C. Abbot, Bedfordshire. In the drawing of the latter the ovary is similar but rather more oblong, and with no style. I may here mention that in the plate (E. B. t. 1808), the ovary of *Russelliana* appears as if it were slightly obtuse, but in the drawing this is scarcely the case. Smith notes (for *Russelliana*), "stalk of germen should be shewn," and " green of every part lighter than in *S. fragilis*."

As illustrating " *S. fragilis* L., E. B. t. 1807," Mr. Leefe published specimens in ' Sal. Brit. Exs.,' No. 52, and ' Sal. Exs.' fasc. ii. No. 32, both from Essex. These agree thoroughly well with Smith's description and plate. Of the former (No. 52) Andersson wrote for Watson, ' Specimina foliifera ad *S. viridem* Fr., amentifera ad *S. fragilem* L. pertinent.' How, if as is probably the case, the female catkins seen by Andersson are similar to those I have seen, that distinguished salicologist could refer them to *fragilis* I am at a loss to conceive, unless he had not time to examine them. The almost sessile, short, abruptly-pointed ovaries are in direct opposition to his own definition of *fragilis*. (Andersson, I suspect, made only a hurried examination of some British willows on another occasion, since in the Kew Herbarium several specimens— including one of *S. viridis*—certified by him, manifestly contradict his own definition of the species they are said to belong to). The specimens published by Mr. Leefe are, I have no doubt, Smith's *fragilis*, and from a comparison of them with authentic specimens of *S. viridis*, I have also no doubt but that they are forms of that hybrid.

As a matter of fact, *S. viridis*—taking that name, as Andersson does, to represent the series of hybrid forms between *S. fragilis* and *S. alba*—is not very rare in England and South Scotland. As already mentioned, the most striking characteristic is in the structure of the ovary, but the form of the catkins is also of importance. These are more slender and proportionately longer, and as regards the female more dense-flowered than in *S. fragilis*. The scales of the flowers are, in this group, too variable in shape, size, and pubescence to afford reliable characters, though (as will be seen presently) some attention must be paid to them. The leaves are also variable, sometimes approaching those of *fragilis*, at other times much resembling *alba*. On the whole they are less obliquely acuminate, much more finely serrated, and of a darker green. (Smith, as mentioned above, points out to his artist that the green of " *Russelliana*" is paler than that of " *fragilis*").

As for the male plant figured in E. B. (t. 1807), while it is not certain that it represents *S. viridis*, it is certainly not the common British form of *S. fragilis*. With Smith's drawings in the British Museum a male catkin is preserved, but it does not throw very much light on the subject, though I am inclined to think that it may be that of *viridis*. The best British specimens of male *viridis* which I have seen, were gathered at Malvern Link by Mr. R. F. Towndrow, who has also sent me from the same place almost typical female *viridis*. In Roxburghshire, Mr. A. Brotherston gets another form (*albescens* And.) with leaves very like those of *S. alba*, but with capsules nearer *S. fragilis*. In Boswell Syme's Herbarium (now in the possession of Mr. F. J. Hanbury), are specimens of this form collected at Duddingston, near Edinburgh, by Boswell Syme, and labelled "*S. alba?*"

One point in connection with the Smithian species remains to be noticed. Of his *fragilis* Smith says, " A tall bushy-headed tree, whose branches are set on obliquely, somewhat crossing each other, not continued in a straight line"; while his *Russelliana* is stated to have branches "slender and straight, not angular at their insertion like *S. fragilis*." What this exactly means I do not know. Another point which Smith considers important is, that in "*fragilis*" the branches are more brittle at their base than in "*Russelliana*." Andersson, following Fries, says that *fragilis* has branches springing nearly at a right angle, while those of *alba* make an angle of 35°, and those of *viridis* one of 60°. The latter may be the case in typical *viridis*, but in forms nearer *fragilis* or nearer *alba*, the angle may reasonably be expected to be different.

For various reasons "*Russelliana*"—the Bedford willow—is, Smith states, of greater commercial value than "*fragilis*." Be this as it may, we can now, from a botanical stand-point, discuss the Smithian species only from the descriptions and drawings of their essential organs. The specimens figured came, in the case of "*fragilis*," from Cambridgeshire and Bedfordshire; and in the case of "*Russelliana*," from Crowe's garden; and hence there is no proof that either one or the other of them was the true Bedford willow. At the same time it is by no means improbable that, as in the case of *S. alba* var. *cærulea* (the Huntingdon willow), where we have a "strain," showing great differences in commercial value, with scarcely any morphological distinctions, there may be two "strains" of *S. fragilis*.

Of the true *S. fragilis* there are in Britain two forms—at least of the male plant. In one the male catkin is rather dense-flowered, with the stamens much longer than the scales; in the other the catkins are lax-flowered, with the stamens not much longer than the scales. In the first the stamens, in the second the scales, form the conspicuous feature of the catkin. The first seems to be very scarce in Britain, but is the only form which I have seen in a large series of continental European specimens, and in any of the continental figures examined by me. The second is the common British form, and I have seen no specimens from outside Britain. It may be distinguished as var. *britannica*. All the British female *fragilis* which I have seen seem to belong to the var. *britannica*. The continental female *fragilis* differs from ours chiefly in having shorter scales and rather thicker styles. Since the male occurs in Britain, it is probable that the female may also be found.

Andersson, who says of *S. viridis*, that it is almost never seen except either as a cultivated plant, or as growing in the company of its parents near cultivated ground, distinguishes three chief modifications, namely—*fragilior*, more like *S. fragilis; viridis*, the most intermediate form ; and *albescens*, approaching *S. alba*. All three occur, I think, in Britain.

The synonymy of *S. fragilis* and *S. viridis*, as regards some of our British books, is as follows :—

SALIX FRAGILIS L.—*S. Russelliana* Sm. E. B. t. 1808 ; Hooker, Br. Flora, 4th ed. 358, 14. *S. fragilis*, var. *β*. Hooker & Arnott, Br. Flora, 8th ed. 401, 10. *S. fragilis*, var. *a. genuina*, Boswell Syme, E. B. 3rd ed. viii. 206 (exclude the plate). *S. viridis*, Boswell Syme (not Fr.), E. B. 3rd ed. viii. t. mcccviii. (exclude the description). *S. fragilis*, var. *S. Russelliana* (Sm.), Babington ' Manual,' 8th ed. 324, 3.

SALIX VIRIDIS Fr., And.—*S. fragilis*, Sm., E. B. t. 1807 ; Hooker, Br. Flora, 4th ed. 358, 13 ; Boswell Syme, E. B., 3rd ed. viii. t. mcccvi. (exclude description). *S. fragilis*, var. *a.* Hooker & Arnott, Br. Flora, 8th ed. 401, 10. *S. viridis*, Boswell Syme, E. B. 3rd ed. viii. 207 (exclude the plate). *S. fragilis*, var. *β. S. fragilis* (L.), Babington ' Manual,' 8th ed. 324, 3.

---

## ROOT-PRESSURE.*

### BY C. B. CLARKE, M.A., F.R.S.

I FIRST read some extracts from ' Sachs' Text-book,' pp. 600—613, to remind the meeting what the accepted doctrine regarding root-pressure is :—

" Another kind of motion of water in the plant, depending not on suction but on pressure from below, is caused by the roots. It is the root-pressure which forces out drops at particular points of the leaves."

" The width of the capillary tubes is much too great to raise water to a height of 100 feet or more."

" The question whether the attraction of the cell-walls for water is sufficiently powerful to sustain the weight of a column of water of the height of 100 or even 300 feet may be answered in the affirmative."

" Root-pressure has no share in the ascent of the water, at the time when transpiration is active."

" With the exception of times when the transpiration is small, or when drops exude from the leaves, no root-pressure exists when the plant is uninjured."

" Water rises in cut leafy branches placed with their upper end in water."

---

* Spoken at the meeting of the Linnean Society, May 3rd.

"The hypothesis finally that the water is forced up into the
s m, and even into the leaves, by root-pressure must be aban-
dtmed."

Dr. Vines, in his 'Lectures on Physiology,' 1886, accepts this
doctrine.   He says (p. 91), "The root-pressure causes a flow of
sap from the cut surfaces of plants ; it also causes in many plants
the exudation of drops of sap at the free surface."

The way I propose to reconcile the manifest difficulties in these
statements is by denying that any root-pressure exists in any case.
The foregoing extracts represent that the vital action of the cells of
the root exert a pressure supporting a column of fluid.   If the tree
were 140 feet high, and the fluid about the density of water, the
root-pressure required would be about 60 lbs. to the inch.   Similarly, if a stem is cut 10 feet above the ground, and the manometer
showed that the fluid exuded with a force of 7 lbs. to the inch, the
root-pressure required would be 7 lb., plus the weight of a column
of fluid 10 feet high, in all 11 or 12 lbs. to the inch.

It is to be noted that on this hypothesis the pressure would be
the same in all the cells at the same altitude, and would (in the tree
140 feet high) diminish gradually from 60 lbs. at the base to zero
at the summit.   All this would be so if the cells and vessels in
vegetables were 6 inches in diam., and it is only the conceiving a
"vital force" in the cells that can account for their not being
blown to atoms.   The believers in root-pressure attempt to diminish this difficulty ; they say (relying on manometer experiments)
that the root-pressure will not support a column of water more
than 30 feet high or thereabout, and that the forcing of water
higher than this must be effected by some other force.   This seems
to me to increase the difficulty ; a column 30 feet high would give
a pressure of 12 lb. to the inch or more, which would burst all the
tender cells (unless they are supposed endowed by vital force) ; and
on the other hand, if in a tree 100 feet high, some " other force"
sustains the pressure due to 70 feet of height, why should we feign
the existence of this troublesome and inadequate root-pressure ?
The advocates of root-pressure refer also to the variety in the form
and complex arrangement of the cells to account partially for
these difficulties.   But if the pressure is transmitted from the root
in the way they imagine the equation $p = g_\rho z$ would hold (as they
suppose).   This is their hypothesis.

My hypothesis is, that the cells being all " capillary" (though
varying considerably in size and form), the equation $p = g_\rho z$ does
not hold at all; that the whole *mechanical* fluid action in plants
must be considered in accordance with the laws of capillarity ; and
that the fluid-pressure in every plant cell is very nearly zero.   I do
not say that all the fluid motion observed in plants can be
explained by capillarity ; but capillary action can do a great deal
more than Sachs imagines. It can only raise water in one inelastic
tube a small height, but in a bundle of tubes capable of motion
it can do much more.   The water raised $\frac{1}{4}$ inch in one tube may
laterally (by pressure or other cause) pass into another, and there rise
by capillary action another $\frac{1}{4}$ inch, and so on nearly *ad infinitum*.

An excellent example of this, known to every one, is the wet-and-dry-bulb thermometer: some cotton wrapped round the bulb has the ends placed in a cup of water, whatever the rate of evaporation, the bulb is thus kept wet. There is certainly no root-pressure and no vital action here. Another experiment is to place a large very dry sponge gently in a shallow vessel of water ; the water rises suddenly about $\frac{1}{4}$ inch in the sponge ; if the sponge be left still, it will yet be found wet to the top in six hours; if it be given any motion it will be wet to the top in six minutes. It is difficult to imagine how fluid can rise in a tree by capillary action if the tree were kept absolutely still ; but it is almost equally difficult to imagine a tree absolutely still, and the most minute vibratory motion would be sufficient.

The contradictory " facts " stated by physiologists have been arrived at mainly from experiments with the manometer. I hold that the manometer records the capillary action at the point (very imperfectly), and nothing about the pressure at the root. We have, in a cut stem, a bundle of tubes of various sizes, besides the wall of cells. In a single inelastic capillary tube, whether fluid exudes from the top or not, depends on several things—the viscosity of the fluid, the roughness of the tube, the state of the atmosphere, &c. When an elastic tube is cut across with an oblique ragged edge, it would be still more difficult to say what the manometer measures. In the case of a stem cut across we have additional complications from the varying size of the tubes, the creeping down of air-bubbles, &c. I cannot discuss all the recorded manometer experiments. The following is selected by Dr. Vines as one of the most decisive :—" The exudation of drops from the leaves depends upon the forcing of water into the cavities of the vessels by the root-pressure. For if the stem be cut off and placed in water, no more drops will appear on the leaves. Again, Mohl has shown that if the root-pressure be replaced in the case of a cut-off branch by the pressure of a column of mercury, an exudation of drops will take place." This proves to me nothing as to what takes place in the uninjured plant. A piece of sponge might fill a glass tube, and you might force water through it by a pressure of so many pounds to the inch ; or the water might get through (without any pressure) by capillary action. In the case of the plant, it seems to me quite possible that fluid might (before the stem is injured) be rising by the cells and the smaller vessels, and at the same time be descending in the larger vessels. Herr Mohl cuts this stem and forces water up the larger vessels by a pressure of so many pounds to the inch. What is proved by this experiment?

But I do not pretend to explain all the facts of fluid-motion in plants. My point is, that the difficulties of the subject are in no way decreased but rather increased by the fiction of root-pressure. I point physiologists to capillary action (rather than to their equation $p = g_\rho z$), as the place where they should look for the mechanical portion of their explanations.

In the subsequent discussion, Mr. A. W. Bennett said he had for some time past regarded " root-pressure " as one of the class of scientific terms useful to conceal our ignorance. He was prepared to accept in the main the views Mr. Clarke had put forward. He said that the old experiment of water rising in cut stems, as when cut flowers revived on being placed in water, showed that water could rise without root-pressure. It was, therefore, not necessary to presume a root-pressure, which involved grave and acknowledged difficulties.

Prof. Ward, on the other hand, said he could not see his way to do without the hypothesis of root-pressure altogether. He thought Mr. Clarke had not sufficiently estimated the complication of structure in a vegetable stem. He referred to the observed fact, that when stimulus (as warmth) was applied to the roots of a stem (which had been cut across and fitted with a manometer), the pressure was observed to increase.

Mr. Clarke, in reply to Mr. Ward, maintained that the action, so far as it was transmitted mechanically in this case, would be transmitted in accordance with the laws of capillary action, not with the equation $p = g_{\rho}z$. He said that, in the case of capillary tubes of any length, or of a system of such tubes of considerable length, a very high pressure would be requisite to force fluid through them.

---

## NOTES ON SOME HIERACIA NEW TO BRITAIN.

### By Frederick J. Hanbury, F.L.S.

I purpose making the following notes as brief as possible, and would say at once that they by no means treat of all the new species and forms found by myself and others during the summers of 1885, 1886, and 1887, chiefly in the extreme north. It seemed desirable, however, not to delay their publication any longer, as another collecting-season has come round, and the knowledge of what has been already done may serve to stimulate others to make further researches in the same direction. Nearly all the plants here treated of have been seen by Professor Babington and Mr. Backhouse, and all by Dr. C. J. Lindeberg, the well-known authority on the Scandinavian Hieracia. These gentlemen have made copious notes and given most valuable opinions on the specimens submitted to them. I was accompanied in 1885 by the Rev. H. E. Fox, and in 1886 by the Rev. E. S. Marshall. Though availing myself freely of the kind help afforded me by others, it must be understood that they are in no way collectively responsible for the statements here made.

The following species, recognised by continental authors have not hitherto been recorded as British:—

Hieracium Schmidtii Tausch.—Though this name is used synonymously with *H. pallidum* Biv., the plant to which Dr. Lindeberg applied it differs so conspicuously from any plants I have hitherto seen, either growing or in our herbaria, that I cannot doubt its being a new form to Great Britain. It occurred abundantly by the Naver, in North Sutherland, and is marked by Dr. Lindeberg "*typicum*!" Other plants, such as we have been accustomed to call *H. pallidum* Biv., he has called *H. Schmidtii*, vars.*

*H. Oreades* Fr.—From rocky places by the sea, on the east coast of Caithness.

*H. bifidum* Kit.—From Glen Caness, in Forfarshire; from Teesdale; and by the Rev. Augustin Ley, from Carnarvonshire.

*H. stenolepis* Lindeb.—From Herefordshire, by the Rev. Augustin Ley; and a less typical form from near Braemar by myself.

*H. Sommerfeltii* Lindeb.—From the east coast of Caithness, and from the Cairngorms in Western Aberdeenshire; also by the Rev. W. R. Linton, from the north coast of Sutherland.

*H. pulchellum* Lindeb. — From Burrafirth, Unst, Shetland, by Mr. W. H. Beeby; and subsequently in greater quantity, by the Rev. W. R. Linton, from one of whose specimens the species was determined.

*H. Friesii* Hartm.—From the north coast of Sutherland; the Cairngorms; and by Dr. F. Arnold Lees, from Ingleboro', Yorks. The broader-leaved forms of this species are identical with Mr. Backhouse's *H. gothicum* var. *latifolium*. *H. Friesii* var. *vestitum* Lindeb., from Caithness.

*H. dovrense* Fr., from Shetland, by Mr. W. H. Beeby.

*H. orarium* Lindeb.—From the north coast of Sutherland and from Caithness. A very marked and striking plant.

H. AURATUM Fr. — From Tongue, Altnaharra, Betty Hill and Melvich, in Sutherland; from Dunbeath, in Caithness; from Braemar; and by the Rev. W. R. Linton, from near Killin. It is a curious fact, that this very beautiful and apparently widely-distributed species should have so long escaped the observation of British botanists. It is a native of North America, and is not only an interesting addition to the flora of Great Britain, but to that of Europe, having previously been recorded as an introduction only, near Upsala. Dr. Lindeberg identified it without any hesitation, and both Professor Babington and Mr. Backhouse concur in the correctness of the name given.

H. ANGUSTUM Lindeb. (= *H. crocatum* var. *angustatum* Fr.).— From near the Spittal of Glen Shee in Forfar, and from Teesdale.

I will now briefly refer, under the names by which I propose calling them, to two or three species which appear to be new. The conclusions at which I have arrived have not been hastily formed. Two of the plants have been collected during three successive seasons, and cultivated in my garden. I should not, with my limited knowledge of foreign species, have ventured to speak of any plant as new, had I not first consulted the best authorities on the subject. The work of detailed description and drawing I purpose leaving for the illustrated monograph that I hope to commence publishing in January next.

---

* Whilst the name "*pallidum*," is doubtful (Fries himself included several species under it—*H. Schmidtii, H. saxifragum* and *H. bifidum*), the name "*Schmidtii*" is definite and certain.

**H. Langwellense.**—Approaching *H. anglicum* Bab., but differing from it in the blunter, shorter, less porrect phyllaries, subglabrous ligules, and in the radical leaves, which are broader at the base and abruptly decurrent. From the east coast of Caithness.

**H. pollinarium.** — Perhaps nearest to *H. murorum* Linn. pt., differing from it in the closely-aggregated heads, the short, straight-based peduncles, which are mealy with floccose down (hence the name), the broader phyllaries, and the scarcely-toothed, round-based, fleshy radical leaves, &c. From the north coast of Sutherland.

**H. scoticum.**—I purpose giving this name to the plant collected in several places from the coasts of Caithness and Sutherland, first in 1885, and which was recorded and distributed by me as *H. norvegicum* Fr. Dr. Lindeberg now considers, that although some of the specimens resemble that species in general appearance, yet that others are so different that they cannot be united to it. It differs from *H. norvegicum* in the leaves being fewer, much broader, and rapidly decreasing upwards in size; the broader, concolorous, adpressed phyllaries, the outer ones of which do not descend into the peduncle, and in other minor points.

There are several other interesting forms that I hope to record later on, but as I expect to revisit Sutherlandshire during July, I think it is better to wait and see whether any fresh light is thrown on them.

Amongst the identifications of British Hieracia I have recently received from Dr. Lindeberg are typical *H. cæsium* Fr., and *H. strictum* Fr. I mention this because the great bulk of specimens found under these names in British herbaria are wrongly so called.

The following are, I believe, new county records:—*H. argenteum* Fr., Caithness, 109; *H. gothicum* Fr., Caithness, 109; *H. nitidum* Backh., W. Sutherland, 107.

---

## PIMINA, NOVUM HYPHOMYCETUM GENUS.

### Descripsit W. B. Grove, B.A.

**Pimina.** — Hyphæ steriles repentes, hyalinæ v. subcoloratæ; fertiles erectæ, fuligineæ, sursum basidiis coronatæ. Conidia simplicia, hyalina, acrogena.

Genus e Stachylidieis Fuckelinæ peraffinis, sed habitu distinctum.

**P. parasitica,** hyphis sterilibus longis, flexuosis, tenerrimis, hinc inde septatis, et inter septa coloratis; fertilibus curtis, e parte colorata oriundis, clavatis, e binis cellulis compactis, inferiore cylindrica, olivacea, superiore subfalcata, clavata, denigrata, apice sterigmatibus oblongo-ovatis, hyalinis plerumque quaternis coronata; conidiis globulosis, solitarie acrogenis, 5 $\mu$. diam.

Hab. Parasitice in hyphis *Polyactidis*, in pagina inferiora foliorum *Passiflora principis* et *P. quadrangularis* aridorum, in horto apud Monkstown, Dublin, Hiberniæ (Mr. Greenwood Pim).

# CENTENARY OF THE LINNEAN SOCIETY OF LONDON.

THE Anniversary Meeting of the Linnean Society this year, being also the Centenary of its existence, was made an occasion for special observances of an interesting kind. The meeting was held in the library, which was beautifully decorated with flowers, and the attendance was exceptionally large. After the financial statement by the Treasurer, Mr. F. Crisp, which showed a total income of £3246 for the past year, beginning with a balance of £321, and leaving one of £302, an abstract of a history of the Linnean collections was read by the senior Secretary, Mr. B. Daydon Jackson, after which the President, Mr. W. Carruthers, delivered his annual address.

After a reference to the losses sustained by the Society during the past year, prominent among which are the names of De Bary and Asa Gray, Mr. Carruthers addressed himself to the special circumstances under which the members were assembled. He said that on that day they had to survey a century rather than a year. The acquisition by Dr. J. E. Smith of everything which Linnæus possessed relating to natural history or medicine, with his entire library, manuscripts, and correspondence, raised him at once to a position of high eminence among the students of natural history in England. The transference of the collections to England created a second centre for naturalists in London. Sir J. Banks had opened his house and given fresh access to his collections and library to scientific inquirers; and he rendered an unselfish and important service to science by exerting his influence to induce Smith to secure a rival and finer collection. The system of Linnæus had then completely displaced all others. The happy invention and careful definition by Linnæus of the words he employed, the precision of his descriptive characters, his terminal nomenclature, and, above all, the clear and certain divisions of his sexual system presented such favourable contrasts to the systematic works of earlier authors that he had secured absolute sway over English naturalists. There existed at the same time a small society in London devoted to the study of natural history. It seemed to have been a kind of mutual improvement society which did not publish memoirs. The Natural History Society continued to hold its meetings for several years after the beginning of this century; and when the meetings could not be kept up and the society was dissolved the books and other property were handed over to the Linnean Society, including the ivory hammer still used by the President. The new impetus given to natural history by the arrival of the Linnean collections showed the urgent need of a society which did not limit its operations to the mutual benefit of its members; and this led to the formation of the Linnean Society, whose first year's income was £65 17s. 6d. For the first fifty years the members were satisfied with annual parts of Transactions, two, three, or four years being required to make up a volume. In 1855 a quarterly journal had become necessary. The distinguished position of the Society was due less to its age than to the remarkable activity of its Fellows, the importance of their work, and the speedy and efficient manner in which the communications were put before the world.   During the past year the Society had published seven parts of Transactions, four devoted to botany and three to zoology, containing 429 pages, 89 plates, and 2 maps.   During the same period there had been issued 20 numbers of the Journal, 9 being botanical and 11 zoological, containing 1151 pages, 56 plates, and 54 woodcuts, together with the Proceedings for the year, requiring 65 pages of letterpress.   These publications contained papers of the highest importance in all departments of science. Not everything submitted to the Society found a place in its publications.   Every communication was reported upon by one or more experts, and was afterwards carefully considered by the council, and only real contributions to knowledge, expressed in fitting language, were published.   Fellowship was not limited to men of science, but it was extended to lovers and patrons of science, who often rendered valuable services.   At one time the reading of papers at the Society's meetings was not followed by discussion, and the proposal to allow discussion was at first opposed as an innovation "that would turn the meeting-room into an arena for gladiatorial combats of rival intellects and lead to the ruin of the society."

In conclusion, Mr. Carruthers gave some account of the Society's collections, and, after the usual vote of thanks, proceeded to read a eulogium on Linnæus, prepared for the occasion by Prof. Thöre Fries, the present occupant of the Chair of Botany at Upsala, who was not able to be present in person.

Prof. Fries began by referring to the profound sleep of the natural sciences through the middle ages, to the hard battles that had to be fought before men of science could liberate themselves from the fetters of a narrow orthodoxy, and to the restraining bands men of science had forged for themselves by attaching infallibility to Greek and Roman authors rather than to the works of Nature.   They worked slowly forward to a truer conception through the 16th and 17th centuries, longing for one who should bring order and quickening life.   At last came Linnæus, to whom, although a poor and unknown youth, the world almost immediately paid homage as a master of the extensive dominion of natural history.   And to-day his name was mentioned with the highest respect in all lands upon which culture had shed its benign rays. Passing over the story of his eventful life the eulogist surveyed the part taken by Linnæus in the development of the sciences to which his penetrating activity extended itself.   Upon botany his systematic mind stamped its impress for all time.   Industrious naturalists had described as well as they could plants brought from all parts of the world; but their descriptions were a shapeless mass of material. There was no lack of system, but none satisfied even the unassuming demands of those times.   The Upsala student, at the age of twenty-two, exhibited to his teacher some outlines of a system which, when published under the name of the sexual system, rapidly supplanted all predecessors.   It was so simple that a child could grasp it. Contemporaries and successors rejoiced at the discovery of the thread of the labyrinth which for centuries had been sought in vain. Linnæus, with clear insight, had openly suggested the weakness of the system and put forward the establishment of a natural system which he laboured to find. Down to our days botanists had tried to raise the edifice of a natural system of plants without getting it complete or even being able to agree on a ground-plan. But all that agreed that Linnæus, over against an artificial system, set forth in a clear light the character and form of the natural one, marked out the way for its development, and secured its supremacy. By successive works Linnæus reconstructed descriptive botany in almost every detail and that in such a manner that the opinions he expressed and the laws he established are even to this day approved of as in all essentials correct. From botanical language he swept away its inrooted barbarism, and gave the proper stability by accurately limiting every botanical idea and furnishing it with definite, appropriate nomenclature. For describing plants and naming them he set up simple practical rules based on a careful analytical examination of the structure of many thousand species, especially their flowers and fruits. In opposition to all his predecessors he drew a sharp line between species and variations. To the then known 8000 species he gave not only new and appropriate names, but also new definitions, and he added critically tested statements of their nomenclature by prior authors, together with an account of their native country, manner of appearing, properties, uses, and so forth, and all this in a way easily apprehended in accordance with the simple laws he himself had established. All his work he endeavoured to arrange on the most natural and easily comprehended plan. In small as well as large things he proved himself a master yet unsurpassed in producing regularity and order where previously ignorance, carelessness, or arbitrariness had generated obscurity and confusion. It was sometimes said he was not qualified for the study of vegetable anatomy, and revealed a one-sided love for descriptive botany; but the reproach usually came from one-sided anatomists. The amount of what he did bordered on the miraculous. He himself admitted that the naming, describing, and classifying of plants was not the only or the highest function of the science, but only a necessary condition for a successful study of the more important parts. It was almost impossible to point to an investigator in botany who had studied the world of plants from so many sides, and who pointed out so many new aspects from which it ought to be examined. Much that had been said about botany applied also in the department of zoology. By establishing new, easily-understood laws, he made scientific, descriptive zoology, and he laid the first groundwork of a real system. In the history of mineralogy he occupied a by no means unimportant position, chiefly through his re-arrangement of the mineral kingdom. More conspicuous was his energetic zeal in the field of medicine. He attempted to arrange scientifically the different forms of diseases. It was easy now, compared with what it was in the time of Linnæus, to bring together collections from

widely distant places.   Untiring was his zeal and unparalleled his
power of stimulating persons of the most varied positions in life—
monarchs and students, lords and poor seamen, bishops and
ignorant tradesmen—all to work to one end.   Devoted scholars,
young and old, surrounded his chair.   His disciples went to
unknown regions to collect for him the treasures of nature, and
many of them perished in foreign lands as the martyrs of natural
science.   Nowhere, next to his own native land, had his name been
so revered as in England.   The botanist Dillenius pressed him to
remain at Oxford " to live and die with him."   He was in active
correspondence with nearly all England's naturalists, several of
whom had enjoyed his instruction in Upsala.   England, unluckily
for Sweden, finally became his heir.   In conclusion, Professor
Fries said :  " Many are consequently the ties by which the memory
of Linnæus is united with England, the strongest, however, is the
Linnean spirit—the genuine spirit of freshness and enterprise
in which scientific research has continued, and still continues, in
England.   Is it not probable that this fact is due, in some measure
at least, to the transfer of the Linnean collections here ?  At any
rate it was that which gave the primary incentive to the formation
of this Society, which has now, for a hundred years, uninter-
ruptedly manifested its vigorous life, extending its useful activity
more and more over the whole globe.   The precious gift of Sir
James Edward Smith was indeed a noble seed, since grown up
into a strong plant, which has borne flowers and fruits from year to
year in abundance.   Its vitality is a guarantee that it will thrive
and flourish, so long as the *Linnæa borealis*, ever green, spreads its
fragrance over young and old, high and low, rich and poor, in the
mighty forests of the north."

Sir Joseph Hooker pronounced the eulogium on Robert Brown,
who was recognized as the greatest botanist of his age.   Passing
over the life, history, and personality of Robert Brown, the eulogist
gave some account of his investigations and discoveries relating to
the morphology, classification, and distribution of plants, and
especially to their reproductive organs, their structure and economy
—investigations which display an untiring industry, an accuracy of
observation and exposition, a keenness of perception, together with
sagacity, caution, and soundness of judgment, in which he has not
been surpassed by any botanical writer.   Where others have
advanced beyond the goal he attained to, it has been by working on
the foundations he laid, by the light and aids of correlative
advances in chemistry and physics, and by the use of optical instru-
ments unknown in his day.   His collection of about 4000 species
of plants belonging to all orders, and three-fourths of them new to
science, in nine years, was a feat unexampled in the history of
botanical science.   In the course of a detailed review of his works,
Sir Joseph gave some personal reminiscences, including these :—
' His appetite for acquiring botanical knowledge amounted, I believe,
to voracity, while his wonderful memory enabled him to retain, and
his methodical faculties to classify all he had acquired.   Of that
memory and of his readiness in utilizing it I had, thanks to

his kindness, much experience. He seemed to me never to forget a plant that presented any feature of interest if he had but once seen it, and he could single out the specimen that he had examined from a sheet full of duplicates. It was the same with books ; those of the old authors especially, as Ray, Linnæus, Rumph, and Rhede, they were all familiar to him, and he could often turn to a volume, and sometimes to a page, for a statement or figure without the aid of a reference. Thus, at the age of twenty-eight, when he sailed for Australia, it was as an accomplished botanist.'

Professor Flower pronounced the eulogy on Charles Darwin, who, he said, had special claims on their consideration, inasmuch as a large and very important portion of his work was first communicated to the world by means of papers read at their meetings and published in their journal. His life and work, however, were so familiar and had been exhaustively treated so recently that the task assigned him could be discharged with a brevity which would be by no means the measure of their appreciation. They were concerned chiefly with those great characteristics of Darwin which dominated all others and made him what he was—the consuming, irrepressible longing to unravel the mysteries of living nature, to penetrate the shroud which conceals the causes and methods by which all the wonders and all the diversity, all the beauty, yea, and all the deformity too, which we see around us in the life of animals and plants, have been brought about. Against our ignorance on those subjects his life was one long battle ; the work of others, by comparison, was irregular guerilla warfare. His main victory was the destruction of the conception of species as being beyond certain narrow limits fixed and unchangeable—a conviction which prevailed almost universally before his time. It might be admitted that others had prepared the way, and that the work was carried on simultaneously by others who might have attained to the same conclusion ; but the fact remained that he was the main agent in the conversion of almost the whole scientific world from one conception to a totally opposite conception of one of the most important operations of nature. Such a revolution, with its momentous consequences on the study of zoology and botany, was without a parallel in the history of science. This rapid conversion was much facilitated by the fascinating nature of the theory of the operation of natural selection in intensifying and fixing variation as originally propounded in the rooms of the Society independently and simultaneously by Darwin and by Wallace. The theory had been subjected to keen criticism, and difficulties had undoubtedly been shown in accepting it as the complete explanation of many of the phenomena of evolution. That other factors had been at work besides natural selection in bringing about the present condition of the organic world probably every one would now admit, as indeed Darwin did himself. That, however, was not the occasion to examine so complex a subject, and indeed the time seemed scarcely yet to have come when it could be done with the necessary calmness and impartiality. But Darwin's work and the controversies that had gathered round it had proved a marvellous stimulus to research.   Though he did not, as it had been too rashly said, tear down the curtain which obscured our gaze and lay bare the birth of life, he had lifted the veil here and there and given us glimpses which would light the path of those who followed in his steps, and, more than this, he showed by his life and by his work the true methods by which alone the secrets of nature may be won.

Mr. Thiselton Dyer delivered the eulogy on George Bentham, whose friendship he had enjoyed.   A nephew of Jeremy Bentham, he was early imbued with a taste for methodizing and analyzing, and through his mother's fondness for plants and the attraction which their classification had for him˗ he was led to study them with marvellous results.   He was President of the Linnean Society from 1863 to 1874, and in his devotion to its interests, which knew no bounds, he shrank from no labour.   He stood in the footsteps of Linnæus ; and, although the descent was oblique, he inherited the mantle of the master whose memory was that day commemorated.

After the proposal of votes of thanks to the writers of these addresses, a very interesting ceremony took place.

The President explained that it had been determined to establish a Linnean Gold Medal to be presented in subsequent years alternately to a botanist and a zoologist ; but on this occasion two were to be presented, and there had not been any question in the council as to who the first recipients were to be.   The medal had on one side a portrait of Linnæus, taken from the bust in the room, and on the reverse the arms of the Society surrounded by the *Linnæa borealis.*   The President first made the presentation to Professor Owen, recounting his distinctions and scientific services, and then, after a similar tribute, presented the second medal to Sir Joseph Hooker.

In the evening the annual dinner was held at the Hôtel Victoria, the President being supported by Sir John Lubbock, Sir Joseph Hooker, Prof. Flower, Prof. P. M. Duncan, and Mr. St. George Mivart.   The toasts of " The Queen" as patron of the Society, " The memory of Linnæus," and " The Linnean Society," having been duly honoured, Prof. Duncan proposed " The health of the Linnean Medallists," Sir Richard Owen and Sir Joseph Hooker. He recalled the time more than forty years ago when, as one of a band of noisy medical students, he sat upon the gallery of the Royal College of Surgeons and saw a wonderful company of men, including judges, bishops, lawyers, and medical men, assembled to hear the marvellous lectures of Prof. Owen in his prime, characterized by fine delivery, wonderful powers of description, and grand generalization.   To him he owed his love of natural history.   For long his name had been synonymous with British science to vast numbers of people wherever science was esteemed.   To him the medical profession owed much of its modern development through its greatly increased interest in physiological science.   As regarded Sir Joseph Hooker he could not refrain from mentioning his exquisite Himalayan journals as among the two or three most charming books of travel and science in the language.   Sir Joseph

Hooker, in responding, said the reception of the medal had given him a gratification of a peculiar kind, which no other Society could have afforded, inasmuch as his father, grandfather, father-in-law, and uncle had been Fellows; he had personally known eight of its Presidents, and many of his own papers had owed their publication to the Society. Moreover, it was Sir J. E. Smith, the founder, who had induced his father to take up the study of botany. He was grateful also to the memory of Linnæus for his own early studies in botany, which were made with a pin and flowers, making out their parts and names according to the Linnean system; and that he believed to be the most valuable way of beginning. Sir John Lubbock proposed "The Health of the President, Officers, and Council of the Society," to which Mr. Crisp, the Treasurer, replied, and the proceedings closed.

On the following evening, Friday, the 25th May, a Conversazione was given by the President and Council, in the rooms of the Society. The Linnean collections were on view, a suitable descriptive catalogue of these having been prepared by Mr. Daydon Jackson; and a large and distinguished company accepted the invitation of the Council. The whole of the arrangements for the Centenary were most efficiently carried out, and reflect great credit upon the Secretaries, and especially upon Mr. J. E. Harting, the Librarian to the Society.

---

# BIOGRAPHICAL INDEX OF BRITISH AND IRISH BOTANISTS.

## By James Britten, F.L.S., and G. S. Boulger, F.L.S.

(Continued from p. 184).

Carey, William (1761–1834): b. Paulerspury, Northants, 17th August, 1761; d. Serampore, 9th June, 1834. Clerk. Baptist Missionary and Orientalist. D.D., 1804. F.L.S., 1823. Founded Bot. Gard. Serampore. Edited Roxburgh's 'Flora Medica.' Pritz. 56; Jacks. 530; Baillon, i. 630. Memoir by Eustace Carey, 1836; Dict. Nat. Biog. ix. 77. *Careya* Roxb.

Cargill, James (fl. 1603). Medical man. Of Aberdeen. Studied at Basle under Caspar Bauhin. Correspondent of Gesner, Lobel, Caspar Bauhin. Described *Fuci*. Pult. ii. 2; Bauhin, 'Prodromus,' 154; Lobel, 'Adversaria'; Dict. Nat. Biog. ix. 80. *Cargillia* R. Br.

Carmichael, Dugald (1772–1827): b. Lismore, Hebrides, 1772; d. Appin, Argyleshire, September, 1827. Captain. F.L.S. Friend of Robert Brown. At the Cape, 1806–1810. At Mauritius and Bourbon, 1810–1814. In India, 1815–1817. Took Tristan d'Acunha, 1817. Linn. Trans. xii. 483. R. S. C. i. 791; Pritz. 56; Bot. Misc. ii. 1, 258; iii. 23; Baill. i. 632. *Carmichaelia* Grev. = *Striaria* Grev. *Carmichaelia* R. Br.

**Carpenter, William Benjamin** (1813-1885): b. Exeter, 29th October, 1813; d. London, 10th November, 1885. M.D., Edin., 1839. LL.D., 1871. C.B., 1872. F.R.S., 1844. F.L.S., 1856. Practised at Bristol till 1844. Fullerian Prof. of Physiology, 1844. Registrar, London University, 1856-1879. 'General Physiology,' 1839. 'Vegetable Physiology,' 1844. Pritz. 56; Jacks. 70, 220; Proc. Linn. Soc. 1885-6, 138; Proc. Geol. Soc. 1885-6, 40; Dict. Nat. Biog. ix. 166. Portr. in Ipswich Museum Series; portr., Men of Eminence, 1864.

**Carroll, Isaac** (1828-1880): b. 1828; d. Aghada, Co. Cork, 17th September, 1880. Visited Lapland, 1864, and Iceland. Studied Cryptogamia. Contributed to 'Cybele Hibernica,' 1866. Herbarium and MS. Flora of Cork at Queen's College, Cork. Journ. Bot. 1881, 128; R. S. C. i. 801; vii. 339.

**Castle, Thomas** (c. 1804-1838): b. Kent, c. 1804; d. Brighton (?), 1838. M.D., Camb. F.L.S., 1827. Practised in Bermondsey. 'Systematic and Physiological Botany,' 1829. 'Medical Botany,' 1829. 'Synopsis of Systematic Botany,' 1833. 'Linnean System, 1836. 'British Flora Medica,' with B. H. Barton, 1837. Pritz. 58; Jacks. 531; Dict. Nat. Biog. ix. 275.

**Castle, R.** (fl. 1840). 'Description of a species of Rose new to the British Flora,' Proc. Sci. Soc. Lond. 1840, ii. 36. R. S. C. i. 821.

**Catesby, Mark** (1679 or 1680-1749): b. Sudbury, Suffolk, 1679 or 1680; d. Old Street, London, 23rd December, 1749. F.R.S., 1733. Of Hoxton and Fulham. In Virginia, 1712-1719; in Carolina, Georgia, Florida, Bahamas, &c., 1722-1726. 'Natural History of Carolina,' 1730-1748. 'Hortus Europæus-Americanus,' 1767. Pult. ii. 219; Rees; Pritz. 58; Jacks. 110, 362; Baillon, i. 655; Rich. Corr. 401; Linn. Letters, ii. 440; Nich. Anec. i. 371; Gent. Mag. 1749, xx. 30; Loudon, 'Arboretum,' 68, 81; Dict. Nat. Biog. ix. 281. *Catesbea* Gronovius.

**Cathcart, J. F. W.** (fl. 1851). Bengal civilian. Collected at Darjeeling, and had drawings made, which are now at Kew. Bot. Mag. t. 4596. *Cathcartia* Hook. fil.

**Catlow, Agnes** (fl. 1847-1855). 'Popular Field Botany,' 1847. 'Popular Garden Botany,' 1855. Pritz. 58; Jacks. 42, 235.

**Chambers, Richard** (1784-1858): b. London, 1784; d. Balderton, Notts, 20th December, 1858. F.L.S., 1822. Of Cecil Court, St. Martin's Lane. Schoolmaster. 'Cat. of Pl. of Tring,' Mag. Nat. Hist. n. s. ii. 1838, 38; R. S. C. i. 868: 'Introduction to the Study of Botany,' 1847; Jacks. 486; Proc. Linn. Soc. 1859, xxx.

**Champion, John George** (1815 ?-1854): b. 1815 ?; d. Scutari, 30th November, 1854. Lieut.-Col. In Ceylon, 1838-1847. Hongkong, 1847-1850. '. . on . . Cape of Good Hope,' Silliman's Journ., 1836, 230. 'Botany in Ceylon,' Journ. Agric. Hort. Soc. India, ii. (1843), 371. 'Ternstrœmiaceæ of Hong Kong,' 1850. Linn. Trans. 1855, xxi. 111. R. S. C. i. 870; Gard. Chron. 1854, 819; Dict. Nat. Biog. x. 33. Plants in Kew Herbarium. *Championia* Gardn.

**Chandler, Alfred** (fl. 1831). Nurseryman and floral artist, of

Vauxhall. "Illustrations . . . of . . . Camelliæ . . . the drawings by Alfred Chandler; the descriptions by William Beattie Booth,' 1831. Pritz. 60; Jacks. 126.

**Chandler, Elizabeth** (1818–1884): b. Hinton-in-the-Hedges, Bucks, 29th April, 1818; d. Isleworth, 29th April, 1884; bur. Isleworth. 'Plants of High Wycombe,' Bot. Chronicle, 1864, 81–84. Bucks plants in Brit. Mus.

**Chanter, Charlotte,** *née* **Kingsley** (fl. 1856). Sister of Charles Kingsley. 'Ferny Combes,' 1856. Pritz. 61; Jacks. 251.

**Charles, James** (fl. 1584). "Collection of Simples most in use, with their names Latin and English," 1584, Sloane MS. Haller, Addend. ii. 675.

**Charlton, Edward** (1814–1874): b. Hesleyside, Northumberland, 1814; d. Newcastle-on-Tyne, 14th May, 1874; bur. Bellingham. Orig. Memb. Bot. Soc. Edin. M.D., Edin., 1836. D.C.L. Trans. Bot. Soc. Ed. xii. 198.

**Charlwood, George** (fl. 1827). Nurseryman. "An assiduous botanist," Sweet, Fl. Australasica, t. 18. *Charlwoodia* Sweet = *Cordyline* Comm.

**Chatterley, William Maddox** (fl. 1839). Hon. Sec. Bot. Soc. Lond. 'Botanical Statistics,' Proc. Bot. Soc. Lond. 1839, 87.

**Chesney, Francis Rawdon** (1789–1872): b. Annalong, Co. Down, 16th March, 1789; d. Mourne, Co. Down, 30th January, 1872. Explorer of Euphrates. F.R.S. General. Pl. descr. in Bertolini, 'Miscellanea Botanica,' i. Journ. Bot. 1872, 96; Dict. Nat. Biog. x. 185. *Chesneya* Bert. = *Pimpinella* L. *Chesneya* Lindl.

**Christy, William** (d. 1839): d. Clapham Road, 24th July, 1839. Of Clapham Road. F.L.S., 1828. Visited Norway and Madeira. Contributed to Gibson's Fl. of Essex, and to Mag. Nat. Hist. 1833, 51; 1837, 25. Gave plants and books to Bot. Soc. Edinb. Jacks. 335; R. S. C. i. 925; Proc. Linn. Soc. i. 67; Gard. Mag. xv. 536. *Christya* Ward & Harv. = *Strophanthus* DC.

**Christison, Sir Robert** (1797–1882): b. Edinburgh, 18th July, 1797; d. Edinburgh, 27th January, 1882. Pupil of Orfila. M.D., Edin., 1819. D.C.L., Oxon, 1865. LL.D., Edin., 1872. P.R.S.E., 1868–1873. Bart., 1871. V.-P. Bot. Soc. Edin., 1839. Prof. of Medical Jurisprudence, Edin., 1822–1832; of Mat. Med., 1832–1877. Lord Rector, 1880. 'Treatise on Poisons,' 1829. Pritz. 62; R. S. C. i. 922; 'Life' (partly autobiog.), 1885–6; Trans. Bot. Soc. Edin. xiv. 266; Dict. Nat. Biog. x. 290. *Christisonia* Gardn.

**Clapperton, Hugh** (1788–1827): b. Annan, Dumfriesshire, 1788; d. Chungary, near Sokota, 13th April, 1827. Captain, R.N. African explorer. In Africa, 1822–1827. Pl. in Herb. Mus. Brit., descr. by Robert Brown in 'Narrative of Travels,' by Denham and Clapperton, 1826. Pritz. 63; Jacks. 346; Dict. Nat. Biog. x. 372. *Clappertonia* Meisn. = *Honckenya* Willd.

**Clarke, Sir Edward Daniel** (1769–1822): b. Willingdon Vicarage, Sussex, 5th June, 1769; d. Pall Mall, London, 9th March, 1822; bur. Jesus Coll. Chapel, Cambridge. Clerk. Traveller

and mineralogist. B.A., Cantab., 1790. M.A., 1794. LL.D.,
1803. Collected in Scandinavia, 1799; Russia, 1800. Ordained,
1805. Vicar of Harlton and Yeldham. Prof. of Mineralogy,
Cambridge, 1808. 'Travels,' 1810–1823. Pritz. 63 ; R. S. C.
i. 935; 'Life,' by Otter, 1825; Nich. Anecd. iv. 389, 721; Nich.
Illust. Dict. Nat. Biog. x. 421. Bust, by Chantrey, at Jesus
Coll. Portr. by Opie, engr. in 'Travels,' vol. i., and in 'Life.'

**Clarke, George** (ᵮ. 1840). Of Mahé, Seychelles. 'Coco de Mer,'
Ann. & Mag. Nat. Hist. v. (1840), 422; and vi. (1841), 408.
Proc. Linn. Soc. i. 1849, 153 ; R. S. C. i. 936.

**Clarke, Robert** (ᵮ. 1843–1863). Surgeon, Sierra Leone. Pritz.
63 ; R. S. C. i. 937.

**Clarke, William Barnard** (ᵮ. 1840). M.D. 'Flora of Ipswich,'
in Mag. Nat. Hist. 1840, 124. R. S. C. i. 937.

**Clarke, William Branwhite** (1798–1878): b. East Bergholt,
Suffolk, 2nd June, 1798 ; d. 17th June, 1878. Clerk. M.A.,
Camb. F.R.S., 1876. Geologist. Discoverer of gold in Aus-
tralia. In Australia, 1839–1878. Various papers on peat bogs,
submerged forests, Carboniferous plants, R. S. C. i. 937. Dict.
Nat. Biog. x. 450.

**Clayton, John** (1686 or 1693 ?–1773): b. Fulham ; d. 15th
December, 1773. Went to Virginia in 1705. 'Flora Virginica,'
1739. Sent pl. to Gronovius. Virginian pl. in Mus. Brit.
Pritz. 63 ; 'Memorials of Bartram,' p. 406 ; Dict. Nat. Biog.
xi. 13; Appleton's Cyclopæd. Americ. Biog. *Claytonia* Gronov.

**Clement, or Clements, John** (d. 1572) : b. Yorkshire? ; d. Bloc-
strate, Mechlin, 1st July, 1572 ; bur. St. Rumbold's Cathedral,
Mechlin. M.D., Oxon. Prof. of Greek. F.R.C.P., 1528 ;
President, 1544. Tutor to Sir Thos. More's children. Jacks.
xxx.; Munk, i. 27 ; Wood, Athen. Oxon. i. 401 ; Dict. Nat.
Biog. xi. 33.

**Clifford, Thomas Hugh** (1762–1825): b. 4th December, 1762 ;
d. Ghent, 25th February, 1825. 'Flora Tixalliana,' 1817.
Pritz. 64 ; Jacks. 260.

**Cobbold, Thomas Spencer** (1828–1886): b. 1828 ; d. Maida
Hill, London, 20th March, 1886. M.D., Edinb., 1847. F.L.S.,
1857. F.R.S. Helminthologist. Prof. Bot. Roy. Vet. Coll.
'Embryogeny of *Orchis mascula*,' Quart. Journ. Micros. Sci.,
1853. 'Embryology of *Achimenes*,' Journ. Quekett Micros. Club,
1879. Proc. Linn. Soc. 1885–6, 140.

**Cockfield, Joseph** (1740 ?–1816): b. 1740 ?; d. March, 1816.
Of Upton, West Ham, Essex. Friend. 'Catalogue of scarce
plants.' 'The Botanist's Guide' (anon.), 1813 ; Letters, 1765–
1771, in Nich. Illust. v. 753–808 ; Pritz. 64 ; Jacks. 256 ;
Friends' Books, i. 438.

**Coel, James.** Of Highgate. Merchant. Lobel's son-in-law.
Had a botanic garden at Highgate. Introduced *Cerasus Lauro-
cerasus*. Pult. i. 125.

**Colden, Cadwallader** (1688–1776): b. Dunse, Scotland, 17th
February, 1688 ; d. Long Island, New York, 28th September,
1776. M.D., Edin., 1705. Practised in Pennsylvania, 1708–

1715. Surveyor-general of New York, 1719. Lieutenant-governor, 1761. Introduced Linnean system into America. ' Plantæ Coldenhamiæ ' in 'Acta Upsaliensia,' 1743. Correspondent of Linnæus. Pritz. 65 ; ' Memorials of Bartram,' p. 19 ; ' Correspondence,' in Silliman's Journ. vol. xliv. ; Dict. Nat. Biog. xi. 260; Linn. Letters, ii. 451–8, 476; Nich. Anecd. v. 484 ; Appleton, ' Cyclop. Americ. Biog.,' with portr. *Coldenia* L.

**Colden, Jane,** afterwards **Farquhar.** [*See* FARQUHAR.]

**Cole, Thomas** (fl. 1725). Of Gloucester. Dissenting Minister. Correspondent of Dillenius. Burnt his herbarium. Pult. ii. 191.

· **Colebrooke, Henry Thomas** (1765–1837): b. London, 15th June, 1765 ; d. London, 10th March, 1837. F.R.S. F.L.S., 1816. Sanskrit scholar. Bengal Civil Service. In India, 1783–1815. Furnished oriental names for Roxburgh's ' Flora Indica.' ' On *Menispermum, Boswellia,* &c.,' Trans. Linn. Soc. 1817–1826. Pritz. 65 ; R. S. C. ii. 12 ; Life by his son, Sir T. Edward, in his ' Miscellaneous Essays,' 1872 ; Dict. Nat. Biog. xi. 282. *Colebrookia* Donn = *Ceratanthera* Horn. *Colebrookea* Sm.

**Coleman, William Higgins** (1816?–1863): b. 1816?; d. Burton-on-Trent, 12th September, 1863. Clerk. B.A., Camb. 1836. M.A., 1838. Ordained, 1840. Master at Christ's Hospital, Hertford; and, from 1847, at Ashby-de-la-Zouch Grammar-school. Described *Œnanthe fluviatilis.* 'Flora Hertfordiensis,' with R. H. Webb, 1849. Pritz. 340 ; Jacks. 253, 255 ; Trans. Bot. Soc. Edin. viii. 13 ; R. S. C. ii. 13 ; ' Flora of Hertfordshire,' 1887, xlii. ; Journ. Bot. 1863, 318 ; Dict. Nat. Biog. xi. 290. *Rubus Colemanni* Bloxam.

**Coles,** or **Cole, William** (1626–1662): b. Adderbury, Oxon, 1626; d. Winchester ? 1662. B.A., Oxon, 1650. B.D. ' Art of Simpling, Introd. to Knowledge of Pl.,' 1656. 'Adam in Eden,' 1657. Rees ; Pritz. 65 ; Wood, Athen. Ox. ed. Bliss, iii. 621 ; Dict. Nat. Biog. xi. 277.

**Collie, Alexander** (d. 1835): d. King George's Sound, December, 1835. Surgeon R.N., F.L.S., 1825. On board H.M.S. ' Blossom,' 1825–1828. Collected about 175 species in California, with Lay, in 1827. Pritz. 66 ; W. J. Hooker, ' Botany of Capt. Beechey's Voyage ' ; Botany of Geol. Surv. Californ. 554.

**Collinson, Michael** (1729 ?–1795) : b. Peckham ? 1729 ? ; d. 11th August, 1795 ; bur. at Sproughton, Suffolk. Of Hendon, Middlesex, and Chantry, Suffolk. Only son of Peter Collinson. Nich. Anec. v. 315.

**Collinson, Peter** (1694–1768): b. St. Clement's Lane, Lombard Street, or near Windermere ? 14th January, 1694 ; d. London, 11th August, 1768. F.R.S., 1728. F.S.A., 1737. Friend. Woollen-draper. Had garden at Peckham till 1749, and then at Mill Hill (afterwards Salisbury's). Friend of Derham, Woodward, Dale, Sloane, Linnæus, Bartram, Franklin, &c. Contributed to Gent. Mag. 1751–1766. Pult. ii. 275 ; Rees ; Pritz. 66. Account by Dr. Fothergill, 1770, with portr. engr. ;

T. Trotter; Nich. Anec. v. 309; ix. 609, with portr. 'Hortus Collinsonianus,' by L. W. Dillwyn, 1843; Friends' Books, i. 443; Linn. Letters, i. 1–77; Linn. Trans. x. 282; Loudon, 'Arboretum,' 54, 81.; Cott. Gard. viii. 143; Dict. Nat. Biog. xi. 382, portr. at Kew. *Collinsonia* L.

**Colquhoun, Sir Robert,** Bart. (fl. 1822). Collected in Kumaon. Pl. in Hort. Bot. Calcutta. Trans. Linn. Soc. xiii. 608. *Colquhounia* Wall.

**Compton, Hon. Henry** (1632–1713): b. Compton Wyniates, Warwick, 1632; d. Fulham, 7th July, 1713; bur. Fulham. Cornet in Guards. M.A., Camb., 1661; and Oxon, 1666. D.D., 1669. Bishop of Oxford, 1674; of London, 1675. Friend of Ray, &c. Introduced many exotics. Loudon, 'Arboretum,' 50; Pult. ii. 105; Pritz. 67; Dict. Nat. Biog. xi. 443; F. L. Colvile, 'Worthies of Warwickshire'; Phil. Trans. xlvii. 243; Cott. Gard. iv. 183; vii. 171. Portr. mezz. by Is. Beckett, after J. Riley; by J. Simon, after Hargrave, 1710; and engr. by D. Loggan, 1679. *Comptonia* Brongn. = *Comptonites* Nilss. ? *Comptonia* Banks MS., Gärtn.

**Cook, James** (1728–1779): b. Marton, Yorkshire, 27th October, 1728; murdered Hawaii, 14th February, 1779. Circumnavigator. Pritz. 68; Jacks. 534; Comp. Bot. Mag. ii. 233; Dict. Nat. Biog. xii. 66. *Cookia* Sonn.

**Cook, Samuel Edward,** afterwards **Widdrington.** [*See* WIDDRINGTON].

(To be continued.)

---

## SHORT NOTES.

EXPERIMENTS WITH GYMNOSPORANGIUM JUNIPERI.—On Saturday, June 4th, 1887, I placed feebly-germinating spores of *Gymnosporangium Juniperi* on two small pieces of moist blotting-paper, and then placed these on two leaflets of a young mountain ash, lacerating them slightly with a lancet at the same time. I likewise placed some spores, *not* on blotting-paper, on the back of a third leaflet. The blotting-paper, &c., was moistened from time to time by a spray-bottle, but during the three days considered necessary by me they were unfortunately allowed to become dry several times. The plant was kept in my room, without the protection of a bell-jar. The experiment seemed to have failed, because I was not able to detect the pustules produced by the first appearance of the spermogones; however, in the autumn, on returning after two months' absence, one of the lacerated leaflets was found to have mature *Ræstelia cornuta*. On Monday, June 13th, with *Gymnosporangium Juniperi* spores which were germinating very freely, producing a great number of pro-mycelium spores, I inoculated a young mountain ash, placing some spores naked on the leaflets, others on blotting-paper, on the upper and under surfaces respectively, without lacerating them. The plant was kept under a bell-jar, and freely moistened by a spray-bottle from time to time. On the morning of the eighth day spermogone pustules made their appearance in considerable number. In the autumn, on my return, the whole lower part of the plant was found badly infected with *Rœstelia cornuta*, but the leaves which had developed above the region of inoculation were scathless. The obvious conclusion is that the down-wash of the spray-bottle had carried the spores to the leaflets below, but that the mycelium could not grow at such a rate as to infect the subsequently produced leaves at a higher level. It is almost needless to add that a third young mountain ash planted out in a garden-plot was entirely free from *Rœstelia*. The results of the experiments have been deposited at the Natural History Museum, South Kensington.—Geo. Brebner.

A Heterodox Onion. — While rambling over the golf-links at Felixstowe a few days ago, I came across a profusion of *Allium vineale*. I am familiar with this plant at Folkestone, but never saw it flower, and only sometimes have I met with the head of sessile bulbils which replaces the inflorescence. The specimens I saw on the ʼlinks attracted attention by the faded yellow colour of their tips and by a slight swelling, indicative, as I thought, of the inflorescence. On examining them, now that I have got home, I find the stem or scape terminates in a long closed extinguisher-shaped scape of a yellowish colour. On slitting this up a sort of cup-shaped membranous perianth is seen, dividing at the margin into irregular laciniæ and lobes, but not presenting any indication of regular perianth-segments. There are no stamens within this cup and no pistils, but a globular head like a free central placenta, covered with straight *ovules* developing centrifugally, those at the top being the oldest. I have never seen anything like this before, nor in the few books I have at present looked into do I find any such condition described. On this account I send you this note, thinking perhaps some of the readers of the Journal may be interested, and find other specimens which will afford some rational explanation of what seems at present a very heterodox onion.— Maxwell T. Masters.

Vicia hybrida L.—Four or five specimens of this vetch have been found by Miss Huc and Mr. A. Steuart, of Ventnor, in a field of saintfoin, in the Undercliff of the Isle of Wight; the field was not mown last year, on account of the drought, and it is hoped that enough may be left this year for the plant to establish itself.— Herbert D. Geldart.

---

## NOTICES OF BOOKS.

*The Flora of West Yorkshire, with a Sketch of the Climatology and Lithology in connection therewith.* By Frederick Arnold Lees, M.R.C.S., L.R.C.P. London: Reeve, 8vo, pp. x. 843.

This Flora has been looked forward to with interest and perhaps somewhat impatient expectation since the issue of ' West Yorkshire ' ten years since. The long delay, Mr. Lees tells us in what he somewhat affectedly styles the " Foreword," is " due to expansion of the scheme originally projected, by the inclusion of the collateral objects, Climatology and Lithology"; and those into whose hands the present volume may come will have no reason to regret the time spent upon its preparation.

In general appearance and arrangement, this latest contribution to our local floras leaves nothing to be desired. The selection and employment of the various types is excellent, and the paper and printing equally so. The actual flora is extremely interesting reading; there is that evidence throughout of personal observation and intimate knowledge of the plants as they occur in the district, which has never been better evidenced than by Mr. Archer Briggs in his ' Flora of Plymouth,' and the absence of which, owing to his lamented death, is a conspicuous defect in Mr. Pryor's ' Flora of Hertfordshire.' Mr. Lees has been fortunate in his helpers, to one of the most important of whom, the Rev. W. W. Newbould, the volume is fittingly dedicated; but in so saying we in no way detract from the claims of Mr. Lees to the credit of the general excellence of the work.

A detailed review of a local flora can only be undertaken satisfactorily by one who has a knowledge of the district to which it is devoted. The present writer can claim no such knowledge, and his remarks must therefore be on the general scope of the work: Mr. J. G. Baker, who is of course *the* man for such a task, considering himself disqualified from undertaking it by the help which he had given to the author. The essays on climatology and lithology, which seem very carefully done, occupy between them 84 pages; they are followed by 13 pages of bibliography (nearly three more are added in the appendix); and after a short " plan of the Flora " (in which we are glad to note a definition of the meaning to be attached to the words " common," " rare," and the like), the Flora proper begins.

The species are numbered throughout—the Flora including 995 phanerogams, undoubted aliens and errors being very properly unnumbered; and the local name, " Hedge Feathers," which is new to us, assigned to the first species, *Clematis Vitalba* (which the author thinks "just possibly native" in West Yorkshire), reminds us to call attention to this too-often neglected feature in a local flora. Mr. Lees has evidently devoted much care to obtaining the plant-names used in the district. He suggests a new " lineal arrangement " for the Batrachian Ranunculi, agreeing in the main with that of the ' Student's Flora.' " O. B. G." seems to us an inconvenient way of referring to the ' Botanists' Guide.' He accepts *Nymphæa (Castalia)* as native, although it is " very rare in a truly native state." The suggestion that the name Fumitory may be a corruption of *fumus terræ*, " from the grey glaucous hue of the curling foliage, resembling masses of cloud or smoke-wreaths," has, we think, been made before; but the old name, as Dr. Prior says, no doubt arises "from the belief that it was produced without seed from vapours rising from the earth." *Hesperis* ranks as a denizen, and is persistent at Bolton. "Hand-flower" is given as a local name for *Cheiranthus*—" formerly (and probably still) the cut blooms were brought to the Leeds market for sale in great bunches, and known and spoken of as "Hand-flower—a penny a bunch"; according to De Théis, the generic name was given by Linnæus for a similar reason. *Barbarea arcuata* Reich., which Prof. Babington and most others "cannot separate" from *B. vulgaris*, has a separate number; "this plant has an elongated raceme in fruit, with slender-arched spreading siliques and long seeds." The date of flowering given for the two violets, *Riviniana* and *Reichenbachiana*, does not accord with general experience; the former is given as "March— July," the latter "May—June"; in the south, and as far north as Cheshire and Lancashire, *Reichenbachiana* is always the earlier. *Oxalis corniculata* appears as a colonist or casual, but *O. stricta* is not given.

We note that *Melilotus parviflora* receives a number and is classed as a colonist; it is said to be "spreading and establishing itself better than either *M. arvensis* or *M. alba*, and to ripen its seed." Twenty-five years ago this species was very frequent about London, but it did not hold its ground. "Fenu-græcum" (p. 193) is one of the very rare misprints in the book. "Proli-ferated" (p. 195) seems to us a new and needless word. *Poten-tilla norvegica* is found in several places, "well-established, spread-ing, and now ineradicable, on the banks and in the masonry of canals and rivers." *Œnothera biennis* is styled a "colonising alien." On *Sonchus arvensis* Mr. Lees notes:—"This is the species well known of farmers, which is called Sow-thistle in West Yorkshire— 'Sow,' because it sows itself so amazingly alike in corn and turnip fields"; is this really the local derivation? A pretty and appro-priate local name—"Bog-bell"—is given for *Andromeda Polifolia*, the first genuine folk-name we have seen recorded for the plant. The Primrose has a very curious name in the north of the county —'Simaruns'—which Mr. Lees says "appears to be an ellipsis of St. Martin's ones, or St. Mary's ones, *i.e.*, St. Martin's or St. Mary's flower, blooming at about the time of the Saint's day (our 'Feast'), as appointed." But we usually associate St. Martin with his "little summer" in late autumn, at which time his feast is kept. The analogies suggested by Mr. Lees are ingenious rather than convincing, as is the case with his suggestion that *Galium Cruciata* might have been used as a substitute for *Polygala*, in what Mr. Lees gracefully terms "the old solemn mockeries of Litany week," and hence called, as it is sometimes called in North Yorkshire, 'Polygally.'" *Daphne Mezereum* is "native or bird-sown."

There is an interesting note on *Elodea canadensis*:—"As bear-ing upon the date of introduction of this plant into Great Britain (usually stated as 1836) in Aveling's 'History of Roche Abbey' (1870) there occurs the curious statement (by J. Bohler probably —he furnished the main part of the plant-list which forms the Addendum): 'We noticed it in several places growing with great freedom, about half-a-century ago, and then it disappeared as mys-teriously as it came.' This could hardly have been penned later than 1865 or 1866, which would carry the first observance back to about the date of Waterloo." *Cypripedium* "still occurs in two places." Mr. Lees still prefers his name *saxumbra* for the interesting variety of *Carex pilulifera*, which Mr. Ridley had named *Leesii* in compliment to its discoverer.

We have 46 ferns and fern-allies; 12 Characeæ; 347 mosses—these and the Hepaticæ, 108 in number, being treated with unusual and gratifying fulness; 234 lichens, raised in the Addenda to 258; nearly a thousand (987) fungi; and 379 fresh-water algæ. It will thus be evident that the flora is the most complete which has yet appeared for any county or vice-county in the kingdom. The only noticeable defect is in the index, which should include species as well as genera.

British botanists will be sincerely grateful to Mr. Lees for this volume. They may be inclined to smile at certain eccentricities of diction, as when Fries is styled (p. 813) "the pater of the specific name" of a plant; or when we are told of a form which is "often by-passed" for something else; but those who know Mr. Lees' other writings will be prepared for these, and they do not detract from the scientific value of the work. The Yorkshire Naturalists' Union, under whose auspices it is issued, is to be congratulated on the publication of this handsome addition to our local floras.

---

*A Manual of Orchidaceous Plants.* Part III. *Dendrobium, Bulbophyllum,* and *Cirrhopetalum.* JAMES VEITCH & SONS, Chelsea. 8vo, pp. 104; plates, woodcuts, and maps. Price 10s. 6d.

THE third part of this valuable work treats of the cultivated species of *Dendrobium*, with a selection of the showier members of the allied genera *Bulbophyllum* and *Cirrhopetalum*. The arrangement is the same as in the two preceding parts, and the descriptions and scientific details maintain the same degree of excellence. Although intended for cultivators, it will be also of great value to botanists, for it is more especially in the showy garden orchids where so much confusion prevails, and the Messrs. Veitch do not scruple to reduce some of the spurious species to their proper rank as varieties, or polymorphisms, of others. About a hundred species of *Dendrobium* are enumerated, so that only a third of the genus may be considered as generally cultivated. A few others are, however, occasionally met with in gardens. A list of fourteen garden hybrids is given, but only one undoubted natural hybrid is admitted, other reported cases being considered doubtful or illusory. It may be pointed out that the *Bulbophyllum Dearei* described on p. 95, with the remark, "we have failed to find any published description," is identical with the plant described and figured in the 'Gardeners' Chronicle,' n.s., xx. p. 108, fig. 17 (1863), as *Sarcopodium Dearei*, a name apparently also of garden origin, as the editor of that paper makes a similar remark respecting it. The work is embellished with a number of woodcuts and two excellent maps, in which the distribution of the cultivated species is approximately given, being printed in colour as nearly as possible where they are known to occur wild: we notice two or three names on the maps which are not to be found in the descriptive portion. Succeeding parts of this valuable work will be awaited with interest.

R. A. ROLFE.

THE most recent (May) issue of the ' Icones Plantarum ' contains, besides a large number of new species, many of them Chinese, the description of a new genus of *Apocyneæ Echitidæ*, from China, which Prof. Oliver names *Sindechites*.

## ARTICLES IN JOURNALS.

*American Naturalist* (April). — J. M. Coulter, ' Evolution in the Plant Kingdom.' — W. J. Beal, ' Rootstocks of *Leersia* and *Muhlenbergia* ' (1 plate).

*Ann. des Sciences Nat.* (May). — G. de Saporta, ' Flore fossile d'Aix en Provence.'

*Bot. Centralblatt.* (Nos. 24–26). — E. Röll, ' Artentypen ' und ' Formenreihen ' bei den Torfmoosen.' — (No. 24). F. Areschoug, ' Ueber *Rubus affinis* & *R. relatus*.' — (No. 27). O. G. Petersen, ' Ueber Quernetze in Gefässen.' — A. N. Lundström, ' Ueber die Salixflora der Jenissej-Ufer.'

*Bot. Zeitung* (May 25, June 1).—A. Koch, ' Ueber Morphologie und Entwickelungsgeschichte einiger endosporer Bacterien-formen.' (June 8, 15). — L. Jost, ' Zur Kenntniss der Blüthen-entwickelung der Mistel ' (1 plate).

*Bull. Soc. Bot. France* (Session Cryptogamique, 1887) (June).— E. Prillieux, ' Les maladies de la Vigne en 1887.' — —. Gomont, ' Note sur le genre *Phormidium*.' — P. A. Dangeard, ' Notes mycologiques.'—E. de Seynes, ' La Moissure de l'Ananas.' — A. Malbranche, ' Plantes rares, litigieuses, ou nouvelles.' — E. Roze, *Geaster Pillotii*, sp. n. — N. Patouillard, *Tubercularia chætospora*, sp. n.—P. Vuillemin, ' Sur une malade des Amygdalées observée en Lorraine en 1887. — —. Boudier, *Ascobolus minutus*, *Ascophanus pallens*, *Ryparobius albidus*, spp. nn. (1 plate).—G. Bernard, *Lepiota echinellus* Quél. & Bern.—C. Richon, *Asterina Scabiosæ*, *Phomatospora Berberidis*, *Anthostomella Berberidis*, *Ramphoria Buxi*, spp. nn. (2 plates).— —. Boudier, ' Sur une forme counidifère curieuse du *Polyporus biennis* ' (1 plate).

*Bull. Torrey Bot. Club* (June). — A. F. Foerste, ' Development of *Symplocarpus fœtidus* ' (1 plate). — F. L. Harvey, ' Fresh-water Algæ of Maine.'—E. E. Stern, ' Peculiarities in seed of *Smilax*.'

*Gardeners' Chronicle* (June 2). — E. Bonavia, ' Slipper of *Cypripedium*.'—(June 16). T. Meehan, ' Knaurs and Burrs.

*Journal de Botanique* (June 1).— —. Masclef, ' Sur la géographie botanique du Nord de la France.' — E. Mer, ' Du développement des conches annuelles dans les Sapins.'—E. Roze, ' Le Jardin des Plantes en 1636.' — (June 16). F. Elfving, ' Sur la courbure des plantes.'

*Journ. Linn. Soc.* (xxiii., No. 155 : June 12). — F. B. Forbes & W. B. Hemsley, ' Index Floræ Sinensis ' (*Caprifoliaceæ*—*Dipsaceæ: Viburnum arborescens, V. brachybotryum, V. Carlesii, V. Henryi, V. propinquum, V. rhytidophyllum, V. utile, Abelia parvifolia, Lonicera Bournei, L. fuchsioides* (pl. ix.), *L. gynochlamydea, L. Henryi, L. similis, L. tragophylla, Hedyotis tenuipes, Myrioneuron Faberii, Diplospora fruticosa, Lasianthus trichophlebius, Leptodermis vestita, Nertera sinensis* (pl. x.), *Patrinia angustifolia, P. saniculæfolia,* spp. nn., all of Hemsley).

*Journ. Royal Microscopical Soc.* (June). — J. Rattray, ' Revision of *Aulacodiscus* ' (3 plates : many new species).

*Midland Naturalist* (June). — W. B. Grove & J. E. Bagnall, ' Fungi of Warwickshire ' (contd.).—W. Mathews, ' County Botany of Worcester ' (contd.).

*Œsterr. Bot. Zeitschrift* (June). — Obituary notice of Hubert Leitgeb (died Ap. 5).—E. Formánek, ' Mährische *Thymus*-Formen.' — B. Blocki, *Hieracium subauriculoides,* sp. n. — F. Krasan, ' Reciproke Culturversuche.'

---

## LINNEAN SOCIETY OF LONDON.

*June* 7, 1888.—The President (Mr. Carruthers) in the chair.— Messrs. G. C. Haité and C. A. Hebbert were elected Fellows of the Society. — The following were nominated Vice-Presidents : Mr. F. Crisp, Dr. Maxwell Masters, Dr. John Anderson, and Mr. C. B. Clarke. — Mr. D. Morris, of Kew, exhibited some drawings of a fungus (*Exobasidium*) causing a singular distortion of the leaves of *Lyonia*, from Jamaica. — A paper was then read by Mr. H. N. Ridley, '' On the Natural History of Fernando Noronha,'' in which he gave the general results of his investigations into the Geology, Botany, and Zoology of this hitherto little-explored island.

*June* 21.—Mr. F. Crisp, Treasurer and V.-P., in the chair, which was subsequently taken by Dr. John Anderson, V.-P.— Messrs. G. C. Haité and R. G. Alexander were admitted Fellows of the Society.—Mr. F. W. Oliver exhibited the aquatic and terrestrial forms of *Trapella sinensis,* of which he gave a detailed account, illustrated by diagrams. — Dr. R. C. A. Prior exhibited a branch of the so-called '' Cornish elm,'' and described its peculiar mode of growth, which suggested its recognition as a distinct species. In the opinion of botanists present, however, it was regarded as merely a well-marked variety of the common elm. — Mr. A. W. Bennett exhibited under the microscope, and made remarks upon, filaments of *Sphæroplea annulina* (from Kew) containing fertilised oospores.— Mr. Thomas Christy exhibited specimens of natural and manufactured Kola nuts, and explained how the latter might always be detected.—The following botanical papers were then read : (1) Mr. H. Bolus, '' On South African *Orchideæ*''; (2) Mr. R. A. Rolfe, '' A Morphological and Systematic Revision of *Apostasiæ*.''

# ON TWO RECENT COLLECTIONS OF FERNS FROM WESTERN CHINA.

## By J. G. BAKER, F.R.S.

IT is most gratifying to see at what a rapid rate our knowledge of the flora of the interior of China is advancing. The two collections of ferns which form the subject of the present paper were made, one by the Rev. Ernst Faber, on Mount Omei, in the province of Szechwan, and the other by Dr. A. Henry, of Ichang, mainly in the Patung district. Both of them were received in England towards the end of 1887. I described the new ferns of a previous collection from Dr. Henry in Journ. Bot. 1887, p. 170. As usual in my fern-papers, the numbers in brackets that precede the names of the novelties indicate their position in the sequence followed in our ' Synopsis Filicum.'

*Gleichenia glauca* Hook. Mount Omei, alt. 3000 ft., *Faber* 1021.

*Hymenophyllum polyanthos* Sw. Summit of Mount Omei, *Faber* 1079.— *H. javanicum* Spreng. Mount Omei, *Faber* 1026. Frequent in the Eastern Himalayas. Gathered lately by Ford in Lofanshan, and Dr. Watt in Manipur.

*Trichomanes radicans* Sw. Mount Omei, 3000 ft., *Faber* 1025.

*Woodsia polystichoides* Eaton. Nanto, *Henry* 3933.

*Davallia marginalis* Baker. Mount Omei, *Faber* 1057. — *D. strigosa* Sw. Mount Omei, *Faber* 1053, 1090. — *D. Clarkei* Baker. Summit of Mount Omei, *Faber* 1089. A rare East Himalayan species, discovered lately by Abbé Delavay in Yunnan. See Hooker's ' Icones,' tab. 1625.

*Lindsaya cultrata* Sw. Mount Omei, 4000 ft., *Faber* 1030.

*Adiantum caudatum* L. Yang-tze-kiang, *Faber* 1049.—*A. Capillus-veneris* L. Ichang, *Henry* 3449. Yang-tze-kiang, *Faber* 1031. Mount Omei, *Faber* 1034.

(51\*). **A. Faberi**, n. sp. Rootstock erect; basal paleæ dense, spreading, linear-acuminate, nearly black. Stipes wiry, naked, castaneous, 4-6 in. long. Lamina oblong-lanceolate, tripinnate, 5-6 in. long, 1¼-1½ in. broad, firm in texture, green, glabrous. Lower pinnæ deltoid. Final segments suborbicular, distinctly petioled, ⅛-¼ in. broad, obscurely crenate on the upper margin when sterile. Veins distinct, free, flabellate. Sori 1-2, small, globose, indented on the upper edge of the frond. Indusium orbicular, glabrous, much smaller than in *A. monochlamys.*—Mount Omei, 3000 ft., *Faber* 1033. Midway between *A. monochlamys* and *A. æthiopicum*, and nearly allied also to the Himalayan *A. venustum.*

*A. pedatum* L. Patung district, *Henry* 3688. Mount Omei, 3000 ft., *Faber* 1032.

*Cheilanthes mysurensis* Wall. Yang-tze-kiang, *Faber* 1001, ex parte.

(28\*). **C. patula**, n. sp. Stipes castaneous, slender, densely tufted, 4-5 in. long, with only a few scattered brown lanceolate acuminate paleæ near the base. Lamina oblong-lanceolate, tripinnate, 4-6 in. long, 2-3 in. broad at the middle, moderately firm in texture, green and glabrous on both surfaces; main rachis castaneous, glabrous, flexuose. Pinnæ deltoid, spreading horizontally, distant, 1–1½ in. long; final segments oblong, obtuse, ⅛–⅙ in. broad. Sori placed all round the free part of the edge of the nearly continuous segments. Indusia narrow, glabrous. — Ichang, *Henry* 3998. Allied to the Himalayan *C. subvillosa* Hook. and *C. Dalhousiæ* Hook. There are two allied undescribed species from Yunnan in the Delavay collection.

*Pellæa geraniæfolia* Fée. Wushang Gorge, Yang-tze-kiang, *Faber* 1011.

*Onychium japonicum* Kunze. Mount Omei and Min River, *Faber* 1024.

*Cryptogramme crispa* R. Br. Summit of Mount Omei, *Faber* 1080. Widely spread in the Himalayas, but new to China.

*Pteris cretica* L. Type; Mount Omei, *Faber* 1014, 1016.— Var. *P. melanocaulon* Fée. Yang-tze-kiang, *Faber* 1013, 1017.— *P. dactylina* Hook. Mount Omei, 3500 ft., *Faber* 1012. A rare East Himalayan species, new to China.

(4*). **P. deltodon**, n. sp. Stipes tufted, slender, naked, stramineous, ½–1 ft. long. Lamina ovate, simply pinnate, membranous, 4–8 in. long, green and glabrous on both surfaces. Pinnæ 3–5, lanceolate, sessile, 2–6 in. long, ½–1 in. broad at the middle, conspicuously inciso-dentate at the sterile tips. Veins lax, simple or forked, ascending, distinct. Sori continuous from the base to within a short distance of the tip of the pinnæ. Indusium narrow, glabrous. —Mount Omei, 3500 ft., *Faber* 1010. Nearest *P. cretica*, but pinnæ few, the lowest simple, the barren part differently toothed, and the veining much laxer.

*P. serrulata* L. fil. Mount Omei, *Faber* 1015. — *P. excelsa* Gaudich. Mount Omei, *Faber* 1018. Agrees well with the Himalayan plant so called, but perhaps not distinct specifically from *P. quadriaurita*.

(6*). **Lomaria deflexa**, n. sp. Caudex erect. Stipes densely tufted, naked, that of the sterile frond 7–8 in. long. Sterile frond oblong-lanceolate, 15–18 in. long, 6–7 in. broad at the middle, narrowed gradually to the base, membranous, green and glabrous on both surfaces, cut down to the rachis into very numerous adnate lanceolate pinnæ, the largest 3–3½ in. long, ⅓ in. broad, tapering gradually to a distinctly-serrated apex, the lower ones much deflexed. Veins lax, distinct, ascending, usually once forked. Pinnæ of the fertile frond remote, linear, the central ones 2–2½ in. long.—Mount Omei, 7000 ft., *Faber* 1023. Nearest to the Norfolk Island *L. acuminata* Baker = *L. norfolkiana* Kunze.

*L. Spicant* Desv. Ichang, *Henry* 3316. Mount Omei, 2500 ft., *Faber* 1028.

*Woodwardia radicans* Sm. Mount Omei, *Faber* 1081.

*Asplenium Trichomanes* L. Patung district, *Henry* 3675. Mount Omei, *Faber* 1087. — *A. normale* Don. Mount Omei, *Faber* 1077. An extreme form, resembling the Neilgherry *A. opacum* Kunze.— *A. crinicaule* Hance. Mount Omei, 2500 ft., *Faber* 1006. — *A. resectum* Smith. Mount Omei, *Faber* 1005, ex parte. An earlier name for this species is *A. unilaterale* Lam. — *A. incisum* Thunb. Patung district, *Henry* 3674. Mount Omei, *Faber* 1088. — *A. Saulii* Hook. Patung district, *Henry* 3789. Mount Omei, *Faber* 1007. Cannot stand as a species distinct from *A. pekinense* Hance. —*A.* (*Darea*) *rutæfolium* Kunze. Ichang, *Henry* 3291. — *A. spinulosum* Baker. Mount Omei, 3500 ft., *Faber* 1050. Found lately in Yunnan by Delavay. — *A. nigripes* Blume, var. Mount Omei, 3500 ft., *Faber* 1048, 1061.

(197*). **A.** (ATHYRIUM) **lastreoides**, n. sp. Lamina very large, membranous, green and glabrous on both surfaces; rachises brownish, without hairs or paleæ. Lower pinnæ oblong-deltoid, $1\frac{1}{2}$–2 ft. long, under a foot broad; pinnules and tertiary segments oblong-lanceolate; final segments oblong, obtuse, deeply pinnatifid, $\frac{1}{8}$–$\frac{1}{8}$ in. broad. Sori subglobose or oblong, contiguous to the midrib of the quaternary divisions. Indusium usually athyrioid, sometimes subreniform.— Mount Omei, 3500 ft., *Faber* 1064. Allied to *A.* (*Athyrium*) *fimbriatum* Wall.

*A. japonicum* Thunb. Mount Omei, *Faber* 1093. — *A. Wichuræ* Mett. Mount Omei, *Faber* 1005, 1068.

' *Allantodia Brunoniana* Wall. Mount Omei, *Faber* 1067. New to China. Frequent in the Eastern Himalayas.

(5*). **Aspidium** (POLYSTICHUM) **xiphophyllum**, n. sp. Caudex stout, erect; basal paleæ linear or lanceolate, bright brown. Stipes tufted, 4–5 in. long, with only a few spreading paleæ near the base. Lamina oblong-lanceolate, rigid, glabrous, simply pinnate, under a foot long, 3–4 in. broad, green on both sides; rachis with a few spreading linear or subulate paleæ. Pinnæ about 20-jugate, the upper sessile, the lower nearly so, lanceolate, $1\frac{1}{2}$–2 in. long, $\frac{1}{3}$ in. broad, cut away on the lower side at the base, with a free or adnate auricle on the upper, finely serrated towards the acuminate tip. Veins immersed, indistinct, pinnate with about 2 veinlets on a side in the centre of the pinnæ. Sori uniserial on each side of the midrib of the pinnæ, medial. Indusium small, glabrous, deciduous. —Mount Omei, 5000 ft., *Faber* 1040. A very distinct species, allied to *A. munitum* and *A. falcinellum*.

*A. auriculatum* Sw. There are two plants in this present collection closely allied to this common and very variable Himalayan species, but which do not quite agree with any of the Indian forms. 1, *submarginale*, nov. var., with the frond and pinnæ of *A. obliquum* Don, but with submarginal sori.—Mount Omei, *Faber* 1038, 1040. Also Kwantung, *Ford* 225. 2, *stenophyllum*, nov. var., approximating to *A. Lonchitis* in habit, with a frond a foot long and scarcely above an inch broad at the middle, with subrhomboid spinulose conspicuously auricled pinnæ much cut away on the lower side of the midrib.—Mount Omei, 3000 ft., *Faber* 1035.

*A. deltodon* Baker. Mount Omei, 3000–3500 ft., *Faber* 1039, 1045. — *A. aculeatum* Sw., var. *A. lobatum* Sw. Mount Omei, 8000 ft., *Faber* 1085. Var. *A. angulare* Sw. Ichang, *Henry* 3289. Patung district, *Henry* 3676. Mount Omei, *Faber* 1055. Var. *A. biaristatum* Blume. Mount Omei, 3500 ft., *Faber* 1091. — *A. amabile* Blume. Mount Omei, *Faber* 1096. Gathered previously in the province of Kiu-kiang by Dr. Shearer.

(26*). **A.** (Polystichum) **capillipes**, n. sp. Stipes densely tufted, thread-like, naked, 3–4 in. long; basal paleæ lanceolate, brown, membranous. Lamina lanceolate, 2–3 in. long, not more than ½ in. broad, deeply bipinnatifid, moderately firm in texture, green, glabrous, with many minute linear dark brown paleæ on the main rachis. Pinnæ lanceolate, ¼ in. long, most of the pinnules lanceolate-aristate and 1-nerved, but the lowest on the upper side much larger than the rest and deeply pinnatifid. Sori one to a pinnule. Indusium large, convex, persistent, membranous.—Mount Omei, *Faber* 1086. Not nearly allied to anything already known. The large bullate membranous indusium recalls *A. craspedosorum*. The habit is most like that of a finely divided form of *Asplenium fontanum*.

*A. aristatum* Sw. Patung district, *Henry* 3678. Mount Omei, 3500 ft., *Faber* 1082, 1083.

(42*). **A.** (Polystichum) **caruifolium**, n. sp. Caudex stout, erect. Stipes densely tufted, slender, stramineous, 2–3 in. long, with a few small lanceolate paleæ. Lamina lanceolate, decompound, 9–12 in. long, 1½ in. broad, moderately firm in texture, bright green, glabrous ; rachis naked, slender, stramineous. Pinnæ lanceolate, multijugate, ¾–1 in. long, ⅓ in. broad, much cut away on the lower side; lower pinnules deltoid, bipinnate; final segments lanceolate-aristate, 1-nerved, 1-16th to 1-12th in. long. Sori not more than one to each final segment, various in position. Indusium small, peltate, glabrous.— Mount Omei, 3000–3500 ft., *Faber* 1027. A very distinct and beautiful species, most like *Asplenium tenuifolium* in habit.

*A. falcatum* Sw. Mount Omei, *Faber* 1060. Patung district, *Henry* 3686, 3687. Yang-tze-kiang, *Faber* 1046 ; and a variety like the Himalayan *A. caryotideum* Wall., with few very large pinnæ, on Mount Omei, at 5000–8000 ft., *Faber* 1058.

*Nephrodium decursivo-pinnatum* Baker. Patung district, *Henry* 3679. — *N. hirtipes* Hook. A form intermediate between the type and the Japanese *N. Dickinsii* Baker, on Mount Ornei, *Faber* 1047. —*N. Dickinsii* Baker. Patung district, *Henry* 3677.—*N. gracilescens* Hook. Mount Omei, *Faber* 1052.—*N. Clarkei* Baker. Mount Omei, *Faber* 1003. An East Himalayan species, new to China.

(70*). **N.** (Lastrea) **unifurcatum**, n. sp. Stipe naked, dull brown, 12–15 in. long. Lamina oblong-lanceolate, 1½–2 ft. long, 8–10 in. broad, bipinnatifid, membranous, green and glabrous on both surfaces ; rachis dull brown, obscurely paleaceous. Pinnæ lanceolate, the lowest not reduced, 5–6 in. long, 1–1¼ in. broad, cut down regularly to a narrow wing into parallel, oblong, obtuse segments ¼ in. broad. Veinlets of the pinnules 7–8-jugate, forked. Sori medial. Indusium small, glabrous, persistent.— Mount Omei, 3500 ft., *Faber* 1051.

*N. lacerum* Baker. Patung district, *Henry* 3680.— *N. odoratum* Baker. Wushan Gorge, *Faber* 1084. — *N. intermedium* Baker. Mount Omei, 3500 ft., *Faber* 1054, 1062. — *N. sophoroides* Desv. Mount Omei, *Faber* 1078. — *N. cicutarium* Baker. Mount Omei, *Faber* 1074. Common in the Eastern Himalayas. New to China.

(1*). **Polypodium** (Phegopteris) **gymnogrammoides**, n. sp. Rootstock creeping below the surface. Stipes spaced out, slender, stramineous, naked, fragile, 4–5 in. long. Lamina oblong-deltoid, 5–6 in. long, about 3 in. broad, membranous, green and glabrous on both surfaces, cut down to a narrow wing to the main rachis into lanceolate crenate pinnæ ⅓ in. broad, the lowest pair deflexed, as in *P. Phegopteris*. Veins in pinnate groups, with about 2 very ascending veinlets on a side. Sori oblong, in a single series on each side of the midrib of the pinnæ.—Mount Omei, 3500 ft., *Faber* 1036. Nearly allied to the Japanese *P. Krameri* Franch. et Savat., figured lately in Hooker's 'Icones,' tab. 1668.

(13*). **P.** (Phegopteris) **omeiense**, n. sp. Rootstock creeping widely below the surface. Stipe naked, a foot long. Lamina oblong-lanceolate, bipinnatifid, a foot or more long, 4–5 in. broad at the middle, reduced at the base, moderately firm in texture, green and rather hairy on both surfaces; rachis pubescent, not at all paleaceous. Pinnæ sessile, lanceolate, 3–4 in. long, ¾ in. broad, cut down to a narrow wing into entire linear-oblong segments ⅙ in. broad. Veinlets simple, distinct, 8–9-jugate. Sori small, medial.—Mount Omei, *Faber* 1059. Differs from the Himalayan *P. appendiculatum* Wall. by its creeping rootstock and medial sori.

(19*). **P.** (Phegopteris) **braineoides**, n. sp. Lamina very large, oblong, bipinnate, 1½ ft. broad, moderately firm in texture, green and glabrous on both surfaces; rachis quadrangular, not paleaceous. Central pinnæ lanceolate, nearly a foot long, 1¼–1⅓ in. broad, cut down nearly or quite to the rachis into linear-oblong entire segments ⅛ in. broad. Veins simple, about 20 on a side. Sori costal, confluent.—Mount Omei, 2500 ft., *Faber* 1022. Very near the Tropical American *P. decussatum* L.

(26*). **P.** (Phegopteris) **stenopterum**, n. sp. Stipes stramineous, 6–8 in. long, with only a few lanceolate brown paleæ towards the base. Lamina oblong-lanceolate, membranous, bipinnatifid, 1½–2 ft. long, 7–8 in. broad, green and glabrous on both surfaces; rachis stramineous, not paleaceous. Pinnæ lanceolate, sessile, the lower 5–6 in. long, 1¼–1½ in. broad, cut down to a narrow wing into oblong entire or inciso-crenate segments ¼ in. broad. Veins pinnate opposite the lobes of the pinnules, with not more than two simple veinlets on a side. Sori uniserial in the pinnules, subglobose, placed nearer the midrib than the margin.— Patung district, *Henry* 3682. Allied to the common Himalayan *P. distans* Don.

(43*). **P.** (Phegopteris) **alcicorne**, n. sp. Caudex erect. Stipes densely tufted, stramineous, 4–7 in. long, clothed throughout with small ovate unequal-sized brown paleæ. Lamina oblong-lanceolate, 3–4-pinnatifid, 8–9 in. long, about 2 in. broad, firm in texture, green and glabrous; rachis scaly like the stipe. Pinnæ lanceolate, at most 1½ in. long, ⅓ in. broad, much reduced on the lower side; final segments non-contiguous, lanceolate, aristate, 1-nerved, ⅛–⅙ in. long. Sori minute, placed singly near the tip of the segments.—Mount Omei, 3000 ft., *Faber* 1008, 1009. A very distinct and beautiful species, most like *Aspidium fœniculaceum* Hook. in texture and cutting.

*P. amœnum* Wall. Mount Omei, *Faber* 1043, 1094.— *P. erythro-carpum* Mett. Mount Omei, 9000 ft. to summit, *Faber* 1037. A Sikkim species, and gathered lately by Dr. Watt in Manipur. New to China.— *P. calvatum* Baker. Mount Omei, 3500 ft., *Faber* 1049. — *P. Lingua* Sw. Mount Omei, at 6000 ft., on trees, *Faber* 1076. Patung district, *Henry* 3684. — *P. fissum* Baker. Mount Omei, at 3000 ft., on trees, *Faber* 1075. — *P. rostratum* Hook. Patung district, *Henry* 3685.— *P. drymoglossoides* Baker. Mount Omei, *Faber* 1046.— *P. lineare* Thunb. Mount Omei, ascending to the summit, *Faber* 1069, 1070. Patung district, *Henry* 3683.

(304\*). **Polypodium** (PHYMATODES) **asterolepis**, n. sp. Root-stock as thick as a swan's quill, creeping beneath the surface of the soil. Stipe naked, erect, 3–4 in. long. Lamina simple, lanceolate, 12–15 in. long, $1\frac{1}{2}$–$1\frac{3}{4}$ in. broad at the middle, tapering gradually to the base and apex, moderately firm in texture, green and glabrous on both surfaces, with a few scattered brown membranous peltate scales beneath, and numerous transparent dots. Veins anastomosing copiously. Sori large, round, superficial, forming a single row near the margin in the upper half of the frond.—Mount Omei, 4500 ft., on rocks, *Faber* 1063. Differs from all forms of *P. simplex* by its submarginal sori. Another plant included under the same number has nearly medial sori, and may be a form of *simplex*.

*P. ensatum* Thunb. Mount Omei, 3500 ft., *Faber* 1095. The Chinese specimens lately received from various sources quite run together the Himalayan *P. ovatum* Wall. and Japanese *P. ensatum* Thunb.—*P. superficiale* Blume. Mount Omei, *Faber* 1065.

(349\*). **P.** (PHYMATODES) **deltoideum**, n. sp. Rootstock creeping widely below the surface. Stipe naked, erect, 9–12 in. long, winged towards the top. Lamina simple, deltoid, 8–9 in. long and broad, moderately firm in texture, green and glabrous on both surfaces, pinnatifid in the lower half, the two basal lobes much the largest, ovate-acuminate, 2–3 in. long. Main veins produced from the midrib to the edge $\frac{1}{4}$–$\frac{1}{3}$ in. apart; intervening veinlets anastomosing copiously. Sori large, globose, superficial, mainly in regular lax rows on each side of the main veins. Ichang, *Henry* 3279. Near *P. hemitomum* Hance.

*P. hastatum* Thunb. Mount Omei and Min River, *Faber* 1066. — *P. dilatatum* Wall. Mount Omei, 4000 ft., *Faber* 1073. A Himalayan species, gathered previously in Lo-fan-shan by Ford.— *P. himalayense* Hook. Mount Omei, *Faber* 1041. The fuller recent material runs *P. Lehmanni* Mett. into this species. — *P. Fortunei* Kunze. Mount Omei, *Faber* 1072. Patung district, *Henry* 3704. — *P. propinquum* Wall. Mount Omei, *Faber* 1071. A common Himalayan species, now found in China for the first time.

*Gymnogramme involuta* Hook. Mount Omei, 4000 ft., *Faber* 1019. A common East Himalayan species, now found in China for the first time. — *G. macrophylla* Hook. Mount Omei, 3000 ft. and higher, *Faber* 1002. New to China, if *G. Henryi* Baker does not, as I suspect it will be found to, run into it. Not Himalayan. —*G. elliptica* Baker. Mount Omei, 3500 ft., *Faber* 1004. Ichang, *Henry* 3348.

*Vittaria lineata* Sw. (*V. flexuosa* Fée; *V. japonica* Miquel). Mount Omei, 4000 ft., *Faber* 1020.

*Acrostichum flagelliferum* Wall. Mount Omei, 2500 ft., *Faber* 1042.

*Psilotum triquetrum* Sw. Yang-tze-kiang, *Faber* 1102.

*Lycopodium clavatum* L., var. *L. divaricatum* Wall. Mount Omei, 3000 ft., *Faber* 1105. The common East Himalayan form of the species. — *L. annotinum* L. Summit of Mount Omei, *Faber* 1097. — *L. serratum* Thunb. Mount Omei, *Faber* 1100, 1101. Province of Kiu-kiang, *Faber* 1099.

*Selaginella plumosa* Baker. Ichang, *Henry* 3488.—*S. canaliculata* Baker. Mount Omei, *Faber* 1104.— *S. caulescens* Spring. Ichang, *Henry* 3595.— *S. Braunii* Baker. Mount Omei, *Faber* 1103. — *S. Savatieri* Baker. Ichang, *Henry* 3596. A rare Japanese species, now found for the first time in China.

*Azolla pinnata* R. Br. Ichang, *Henry* 3977.

*Equisetum ramosissimum* Desf. Wushan Gorge, *Faber* 583.— *E. diffusum* D. Don. Mount Omei, 3000 ft., *Faber* 1106. A common Himalayan species, now found in China for the first time.

It will be noted that in the mountains of West China there are species of every range of climate from thoroughly tropical types, such as *Acrostichum flagelliferum* and *Gymnogramme involuta*, up to boreal types, such as *Cryptogramme crispa* and *Lycopodium annotinum*. No doubt a great many more Himalayan species and new endemic species will reward the researches of the energetic collectors who are engaged in exploring this rich and, till very lately, almost unknown botanical region.

---

## SOWERBY'S MODELS OF BRITISH FUNGI.

### By Worthington G. Smith, F.L.S.

During the spring of the present year the whole of Sowerby's Models of Fungi, belonging to the Department of Botany, British Museum, South Kensington, have (after thorough cleaning) been painted in oils by me. Previous to this the naming was in an unsatisfactory state, not only on account of the now more or less obsolete nomenclature used by Sowerby, but also on account of several curiously erroneous names, supplied in times long past by persons unacquainted with fungi. The models were in a very mutilated state, and owing to various shiftings and removals certain examples had got free from the stands; in the attempt to replace these, sometimes specimens had got lost or curiously misplaced, and different species were, in some instances, placed in company, and made to do service for a single species. There can be no doubt that the Sowerby collection is somewhat incomplete, and this is not to be wondered at, for many of the models are formed of such dry and brittle materials (as pipe-clay), that if dropped by an attendant (or nervous mycologist) on to the floor, nothing would be left but dust and small fragments such as could not be put together again. In several instances *sections* of agarics, &c., only exist as models, whilst on turning to Sowerby's corresponding plates the general habit is given in one or more figures; and it is not reasonable to suppose that Sowerby, in illustrating fungi by models, would confine himself to a mere section, such, for instance, as that of *Coprinus picaceus*. A section only now exists of this species as a model, and little or nothing can be learned from it. The characters of *C. picaceus* exist in the external aspect, but as no model exists of the general form of this and some others, it is only reasonable to suspect a series of unfortunate and hopeless mycoclysms in times now past. One of the most remarkable changes undergone by the models during the last hundred years has been the change of colour, for Sowerby seems to have used such an uncommonly bad white colour, that all his snow-whites had changed to coal-blacks, or something near it. His sulphur-yellows and some of his pale blues had also become jet-black, and some of his purples had become green. This change of colour was very puzzling and indeed misleading for beginners, and I can well remember how it confused my mind some twenty or thirty years ago. *Agaricus sulphureus*, *A. maximus*, and *A. odorus*, and others were all jet-black. *Cortinarius violaceus* was green. *Fuligo varians*, as it is now called (*Æthalium*), was jet-black; but the changes of colour were too numerous to mention, for none of the colours remained right, and some of the colour-changes were startling. The plates in Sowerby's volumes are better; but even there the whites have, in some instances, changed to dark grey, and other changes, too familiar to artists, may be noted.

An attempt has now been made to set matters a little right, as the models have not only been repainted, but all the names have been looked to, and the plants are arranged in botanical sequence according to modern ideas. It is remarkable that Sowerby, who was able to produce such really excellent models, was so hopelessly lost when he *mounted* his models, for he appears to have merely fixed them on to *sanded blocks of wood or cork*, and surrounded them, as an ornament, with real moss; the absurdity of this mode of mounting will be apparent, when I say that the truffle and other subterranean fungi were so treated and so moss-surrounded. *Coprinus niveus*, admirably modelled in itself, appeared as springing from a squared block of sandstone instead of from horse-dung. *Peziza vesiculosa* was growing on an oblong block of wood. A "natural" mounting has now been adopted, and real dead branches, dead leaves, old tan, horse-dung, beech-nuts, acorns, fir-cones, &c., employed as surroundings of the fungi; this gives them a much more bright and natural appearance, so that the fungus-models have become serious rivals in point of beauty of form, colour, and general surroundings, to the humming-birds themselves in the gallery below.

The work of painting these models was full of surprises; for instance, the Typhulas were found to be made of wire; some of the stems of the Agarics are made of wood or cane; the Clavarias are sometimes twigs, at other times carved in wood; the *Pezizæ* are generally of sheet-iron, zinc or lead. The pilei of nearly all the larger fungi are metal, probably zinc, filled in with pipe-clay; the gills are sometimes of sheet-iron, at other times of card or even paper; some of the veils are of wash-leather, others of gold-beater's skin. Some "models" are undoubtedly the real fungi themselves, dipped in some hardening solution and then painted; this is shown by the exquisite fineness, closeness, and regularity of the gills in some of the specimens, a fineness, thinness, and regularity only seen in nature. In other examples, as *Agaricus velutipes*, the gills are dispensed with altogether, a mere flat surface being made to do service for them. In *Agaricus (Volvaria) volvaceus* Sowerby shows one example, with an ample ring to the stem; this is curious, for according to the sub-generic characters of *Volvaria*, no species should have a ring. The model (if according to nature, as it probably is) shows that *Volvaria*, like its analogue *Amanita*, may produce both ringed and ringless forms. In all, two hundred and fifty-five models are now exhibited.

---

## ON *CALLITRICHE POLYMORPHA* Lönnroth AS A BRITISH PLANT.

### By W. H. Beeby, A.L.S.

The distribution of this species as it is given in Nyman's 'Conspectus' stands thus:—"Suec. Norv. (&c.)."[*] This mode of citation seems to indicate an opinion that the plant had likely been overlooked elsewhere, rather than that it was really confined to the two countries specially named. If this view be correct, its occurrence in Surrey would not be very remarkable, and several (imperfect) gatherings made in this county have been considered by Dr. Hjalmar Nilsson as probably referable to it. Last year, however, I gathered in the Island of Unst, Shetland, a plant which was definitely named *C. polymorpha* Lönnr. by Dr. Nilsson, as recorded in the 'Scott. Naturalist' for last January. Thus the plant may reasonably be searched for throughout the greater part of the kingdom, and a few remarks on its distinctive features may therefore be useful.

*Callitriche polymorpha* was first described by C. J. Lönnroth in a small tract entitled 'Observ. crit. plantas suec. illustrantes,' printed on his inauguration to his degree at Upsala in 1854. The name seems to have escaped the notice of Hegelmaier, as I cannot find that it is mentioned in his 'Monograph,' although this is published ten years later (1864). Lönnroth afterwards published ('Bot. Notiser,' 1867) a paper on the Swedish species of *Callitriche*, illustrated by excellent drawings of the various stages of flower and fruit, &c. This paper I have not seen, but I possess a careful

---

[*] Greenland may be added, *fide* E. Warming in "Berctn. om den bot. expedit. med. 'Fylla' i 1884 ' (Copenhagen, 1887).

tracing of the drawings kindly sent me by Dr. O. Nordstedt.   My
remarks therefore are founded on the description given in the 'Obs.
crit.' and on the drawings alluded to, as well as on a partial comparison of Shetland and Swedish examples with the allied species.
    The most marked character which separates *C. polymorpha* from
the other species of *eu-Callitriche* is the great length of the stigmas.
In all the species these are greatly longer than the ovary in the
early state ; but in none are they more than about one-half longer
(*C. hamulata*) than the ripe fruit, except in *C. polymorpha*.   In this
the stigmas are, according to Lönnroth, 3–4 times as long as the
mature fruit, while the bracts are also persistent.   In the few uninjured fruits which remain on my single sheet of the Shetland
plant the stigmas are 2½–3 times as long as the fruit.   The Shetland plant, generally, bears a resemblance to *C. stagnalis;* but the
fruit is smaller, and "scarcely winged," while the individual carpels
are longer in proportion to their breadth than in that plant.   The
fruits seem to be much the same size as those of *C. hamulata*,
though figured as being smaller by Lönnroth, who considered the
position of *C. polymorpha* to be between *C. vernalis* and *C. hamulata*.
The size and shape of the bracts and other floral parts in the
earlier states appear to afford good characters, but I hesitate to say
anything on this part of the subject from want of experience.
I hope that the above brief notes will be sufficient to enable the
plant to be recognised, if found ; and I may say that it adds greatly
to the certainty of determinations if a few stems bearing ripe fruits
with perfect stigmas be selected from each gathering, and dried
separately, with extra care, on white paper.

---

# NEW MANIPUR FERNS COLLECTED BY DR. WATT.

## By Col. R. H. Beddome, F.L.S.

**Aspidium** (Lastrea) **Wattii**, n. sp. — Rhizome ?   Stipes stramineous, 4–5 in. long, clothed with a few light-coloured deciduous
scales ; fronds 1 ft. or more long by 2–2½ in. broad, tripinnate,
broadest in the centre, gradually narrowed towards the apex and
base ; rachis naked ; texture subcoriaceous ; surfaces glossy ; ultimate segments obovate to lanceolate, sharply acuminate or more
rarely with a rounded apex more or less 2-lobed ; veins 1-forked in
the ultimate segments ; sori apical on the short lower veinlet, often
furnished with a few deciduous hair-like scales.—Manipur, *Dr. Watt*,
No. 6715.
    This species much resembles *fœniculaceum* in its ultimate cutting
and texture, but it is much less compound, with long narrow fronds.
I have not detected an indusium in the few specimens I have seen ;
but I feel certain that it is an *Aspidium*, and that its position is near
*fœniculaceum*.   In my 'Handbook' I followed Mr. Clarke, and removed *Aspidium fœniculaceum* to the genus *Diacalpe*, I have lately
received young fronds of it from Mr. Levinge, collected at Tonglu,

in Sikkim, which show perfect involucres, they are nearly all perfectly reniform, though some few are polystichoid; its position therefore is near *coniifolia* and *aristata* in my first section of *Lastrea*, in which the involucres are sometimes reniform (lastreoid) and sometimes peltate (polystichoid). On a careful examination under a powerful lens of Mr. Clarke's specimens, which were collected at Buckeen, in Sikkim, at 7500 ft. elevation, I find that the involucres, though in appearance much resembling *Diacalpe*, are superior, like a little convex scale, round or oblong, easily removed, leaving the spore-cases on the surface of the frond; in *Diacalpe* the involucres are easily detached as a complete little ball with a short point of attachment underneath, and containing all the spore-cases inside. I consider Mr. Clarke's specimens as abnormal in their involucres, which are certainly not those of *Diacalpe*, nor are they polystichoid; they resemble, I think, those of *Cystopteris* more than any other genus. Mr. Clarke's specimens agree in many other particulars with the *Aspidium fœniculaceum* of Hooker.

**Polypodium** (Phegopteris) **manipurense**, n. sp.— Rhizome ? Stipes 9–12 in. long, furnished with numerous large golden-brown broadly lanceolate acuminated scales; fronds 12–16 in. long, deltoid-ovate, tripinnate, with the tertiary pinnæ pinnatifid; rachises furnished with ferrugineous curled many-jointed hair-like scales, the main one somewhat flexuose, the lower pinnæ the largest 6–8 in. long, ascending; secondary pinnæ 1–2 in. long; tertiary pinnæ about $\frac{1}{2}$ in. long from a broad sessile base pinnatifid nearly half-way down; texture herbaceous, both sides furnished with hair-like scales, similar to those on the rachises; sori small, 1–2 (rarely more) to each ultimate segment or lobe of the tertiary pinnæ, medial, apical or nearly apical on the lower veinlets.—Sirohifurar, 6000–7000 ft. elevation, *Dr. Watt*, No. 6423.

No indusium is to be traced in any of my specimens, so I have placed this fern in *Phegopteris*, and its position is next to *rugulosum ;* should it prove to be a *Lastrea*, which is very probable, it will stand next to *scabrosa*, which it somewhat resembles, only it has not the enlarged pinnules on the lower side of the lower pinnæ, so characteristic of that species.

Polypodium (Gonophlebium) niponicum var. Watth. — Rhizome wide-creeping, brittle, glaucous, naked or nearly so. Stipes 3–6 in. long, hairy; fronds softly hairy, 8–22 in. long by 2–4 in. broad, pinnatifid to within $\frac{1}{4}$ in. of the rachis; segments ciliated, 20–25 pairs, entire or obscurely crenate, oblong from a broad base, blunt at the apex, lowest pair deflexed and slightly reduced; areolæ in a single series; sori in a single row, nearer the midrib than the margin.—Koupra, 4000–6000 ft. elevation, *Dr. Watt*, No. 5852.

Dr. Watt writes that it is a very beautiful fern, with delicate green leaves, the glaucous rhizome creeping on trees, and often suspended in the air. It is too closely allied to *niponicum*, I think, to be considered more than a variety, the only differences being that its rhizomes are more glaucous and glabrous, and the indumentium on the fronds less thick.

# REMARKS ON *PYRUS LATIFOLIA* Syme.

## By T. R. Archer Briggs, F.L.S.

Some remarks on *Pyrus communis* c. *cordata* Desv., were made by me last year in the 'Journal of Botany,' and appended to them will be found a statement (p. 209) respecting some trees of *Pyrus latifolia* Syme, that I had raised from seed of wild Devon bushes, and have growing here at Fursdon, Egg Buckland. Up to that time these cultivated trees, of which I have three, had borne no flowers, but this present year two of them have blossomed freely, one having had over fifty, the other over twenty cymes of flowers. Neither in leaves nor in inflorescence do these seed-produced trees of mine show any departure from the certainly wild, and as I believe, indigenous, *Pyrus latifolia* of Devon and E. Cornwall. The results before us, through the raising of these trees from seed, are, I think, of some importance when taken in connection with some very interesting particulars by the late Dr. Boswell, " On the forms (subspecies or hybrids?) of *Pyrus Aria* Hook.," in the Rep. Bot. Ex. Club, 1872-74, pp. 17-25. In the course of his remarks he says, " There is a very general feeling that the plant, which I believe ought to be called *P. latifolia* " (it was Dr. Boswell's *P. scandica* in E. B., ed. iii.), " is something more than a variety of *P. Aria.*" Still, notwithstanding this, he adds further on, after saying that " Garcke, in his ' Flora of North and Middle Deutchland,' describes *P. latifolia* under the name of *P. Aria-torminalis*," " Certainly in the texture of the leaves and the character of their pubescence when young there is a departure from *P. Aria* in the direction of *torminalis*, and in the broader-leaved specimens the form of the leaf and of the lobes approaches that species, and were *P. latifolia* not so abundant the most probable solution would be that it was a hybrid between *P. Aria* and *P. torminalis*, and there is nothing in its distribution in England to forbid the supposition." The coming of my young trees so perfectly true from seed is opposed to the view of a hybrid origin for *Pyrus latifolia*, as, I may add, is also very markedly its distribution in Devon and E. Cornwall. Dr. Boswell, after remarking on differences between *P. latifolia* and *P. eu-Aria*, adds respecting the former, " The extremes in British specimens lie between specimens sent from Symond's Yatt, Gloucestershire, by Rev. Augustin Ley, in which the leaves are nearly as broad as long, with large and very acute lobes, to the Leigh Wood plant, figured as *P. scandica* in E. B., ed. iii. p. 484, in which the leaves are only about half as broad as long and the lobes short and much blunter." I find the leaves of Devon plants nearly, or quite, as broad as those of specimens from Coldwell Rocks and Symon's Yatt, Gloucestershire, though somewhat longer, more rounded at the base than in the Symond's Yatt specimen (one without inflorescence), more serrate, and less uniformly toothed and lobed. Some description of the fruit, gathered from wild bushes in Devon, will be found in my ' Flora of Plymouth,' to which I may add that it has the mealy character of that of *Aria*, not the softer nature of that of *torminalis;* nor is it acid as in this last.

I regret having no record of the date when I sowed the seeds which produced my trees of *Pyrus latifolia*, but I think it to have been about thirteen or fourteen years ago.   There is at present a good prospect of two of them perfecting fruit this year.   I find young trees of *P. torminalis* progress very slowly in cultivation here, being in this respect unlike those of *P. Aria*.

---

# CATALOGUE OF THE MARINE ALGÆ OF THE WEST INDIAN REGION.

## By George Murray, F.L.S.

(Continued from p. 196.)

### CRYPTONEMIACEÆ.

Gloiosiphonia capillaris Carm.   Bermuda, *Kemp.*
  *Geogr. Distr.*  Atlantic (Europe) and North Sea.
Lygistes vermicularis J. Ag.   Guadeloupe, *Mazé* !
  *Geogr. Distr.*  Atlantic (Spain).
Schizymenia marginata J. Ag.   Guadeloupe, *Mazé* !
  *Geogr. Distr.*  Mediterranean.
Nemastoma multifida J. Ag.   Guadeloupe, *Mazé* !
  *Geogr. Distr.*  Tropical Atlantic.
N. Jardini J. Ag., var. Antillarum Crn.   Guadeloupe, *Mazé.*
Gymnophlœa canariensis Kütz.   Guadeloupe, *Mazé* !
Schimmelmannia Bollei Mont.   Guadeloupe, *Mazé* !
  *Geogr. Distr.*  Canary Islands.
Halymenia dichotoma  J. Ag.   Guadeloupe, *Mazé* !   Bermuda,
  '*Challenger*' !
  *Geogr. Distr.*  Mediterranean and neighbouring Atlantic.
H. decipiens J. Ag.   Florida, *Melvill* !  *Hooper* ! in Farlow, Anderson & Eaton Alg. Exsicc. 80.
  *Geogr. Distr.*  Spain.
H. floresia Ag.   Grenada, *Murray* !   Barbadoes, *Dickie* !  Guadeloupe, *Mazé* !   Florida, *Harvey, Melvill* !
  *Geogr. Distr.*  Atlantic (Europe, Africa, and America), Mediterranean, Red Sea, Australia.
H. ligulata Woodw.   Barbadoes, *Dickie* !   Florida, *Harvey.*
  *Geogr. Distr.*  Atlantic (Europe).
H. ramosissima Suhr ?   Guadeloupe, *Mazé* !
H. pennata Crn.   Guadeloupe, *Mazé* !
Grateloupia dichotoma J. Ag.   Guadeloupe, *Mazé* ! ('West Indian,' *fide* J. G. Agardh).
  *Geogr. Distr.*  Atlantic (Europe) and Mediterranean.
G. filicina Ag.   Grenada, *Murray* !   Guadeloupe, *Mazé* !   St. Thomas, *A. K. Young* !   Florida, *Harvey.*

Var. CONGESTA Crn.  Guadeloupe, *Mazé*!
Var. FILIFORMIS Crn.  Guadeloupe, *Mazé*! *Hohenack*! Meeralgen 380.
Var. ELONGATA Kütz.  Guadeloupe, *Mazé*!
Var. BIPINNATA Crn.  Guadeloupe, *Mazé*!
  *Geogr. Distr.*  Atlantic (as far as the Cape of Good Hope),
    Mediterranean, Indian Ocean.
G. PROLONGATA J. Ag.  Guadeloupe, *Mazé*!
  *Geogr. Distr.*  Pacific (Mexico).
G. GIBBESII Harv.  Guadeloupe, *Mazé*!
G. CUNEIFOLIA J. Ag.  La Guayra (Venezuela), *Binder*.  Guadeloupe,
  *Mazé*!
G. CUTLERIÆ Kütz.  Guadeloupe, *Mazé*!
  *Geogr. Distr.*  Chili.
G. ? AUCKLANDICA Mont.  Guadeloupe, *Mazé*!
  *Geogr. Distr.*  Antarctic Sea.
G. LANCIFERA Mont.  Guadeloupe, *Mazé*!
G. LANCEOLA J. Ag.  Guadeloupe, *Mazé*!
  *Geogr. Distr.*  Atlantic (N. Africa).
G. PURCATA Crn.  Guadeloupe, *Mazé*!
G. SPINULOSA Crn.  Guadeloupe, *Mazé*!
G. SEMIBIPINNATA Crn.  Guadeloupe, *Mazé*!
G. SUBVERTICILLATA Crn.  Guadeloupe, *Mazé*!
CRYPTONEMIA CRENULATA J. Ag.  Florida, *Harvey*! *Melvill*! *Hooper*!
  in Farlow, Anderson & Eaton Alg. Exsicc. Am. Bor. 23.
  Bermuda, *Rein*.
  *Geogr. Distr.*  Atlantic (N. and S. America).
C. LACTUCA J. Ag.  Guadeloupe, *Mazé*.
  *Geogr. Distr.*  Atlantic (Spain).
C. LUXURIANS J. Ag.  Barbadoes, *Dickie*!  Guadeloupe, *Mazé*
  *Geogr. Distr.*  Brazil.


### GIGARTINEÆ.

CHONDRUS CRISPUS Lyngb.  Jamaica?  (Hb. Shuttleworth!)  Ber-
  muda, *Kemp*!
  *Geogr. Distr.*  N. Atlantic.
IRIDÆA LITTORALIS Crn.  Guadeloupe, *Mazé*!
GIGARTINA ACICULARIS Lam.  Guadeloupe, *Mazé*!  Havana, *Melvill*!
  *Geogr. Distr.*  Atlantic (Europe) & Mediterranean.
G. TEEDII Lam.  Bermuda, *Kemp*.
  *Geogr. Distr.*  Atlantic (Europe) and Mediterranean.
AHNFELTIA DURVILLÆI J. Ag.  Guadeloupe, *Mazé*!
  *Geogr. Distr.*  Tropical Pacific.
A. ? PINNULATA Harv.  Florida, *Harvey*.
GYMNOGONGRUS DENSUS J. Ag.  Guadeloupe, *Mazé*!
  *Geogr. Distr.*  Indian Ocean.
G. PYGMÆUS J. Ag.  Guadeloupe, *Mazé*!
  *Geogr. Distr.*  Indian Ocean.
G. FURCELLATUS J. Ag.  Guadeloupe, *Mazé*!  Jamaica, *Wright*!
  Var. PATENS J. Ag.  Guadeloupe, *Mazé*!
  *Geogr. Distr.*  Peru.

G. TENUIS J. Ag. Guadeloupe, *Mazé*! "On the shores of the Mexican Republic," *Liebman*. Agardh gives it as Gulf of Mexico.
Var. ANGUSTA J. Ag. Guadeloupe, *Mazé*!
G. CRENULATUS J. Ag. Guadeloupe, *Mazé*!
*Geogr. Distr.* Atlantic (Spain).
G. CAPENSIS J. Ag. Guadeloupe, *Mazé*!
*Geogr. Distr.* Cape of Good Hope.
G. LINEARIS J. Ag. Guadeloupe, *Mazé*!
*Geogr. Distr.* Pacific (S. America).
G. DILATATUS J. Ag. Guadeloupe, *Mazé*!
*Geogr. Distr.* Cape of Good Hope.
STENOGRAMMA INTERRUPTA Mont. Florida, *Harvey*.
*Geogr. Distr.* Warmer Atlantic (Europe and America), Pacific (California, Corea), N. Zealand, Tasmania.
PHYLLOPHORA BRODIÆI J. Ag. Jamaica, *Chitty*!
*Geogr. Distr.* N. Atlantic.
KALLYMENIA RENIFORMIS J. Ag.? Bermuda, '*Challenger*'!
*Geogr. Distr.* Atlantic (Europe).
K. PAPULOSA Mont. Guadeloupe, *Mazé*!
K. LIMMINGHII Mont. Guadeloupe, *Limminghe, Mazé*!
CALLOPHYLLIS DISCIGERA J. Ag. Guadeloupe, *Mazé*!
*Geogr. Distr.* Cape of Good Hope.
C. LACINIATA J. Ag. Bermuda, *Kemp*.
*Geogr. Distr.* Atlantic and North Sea (? Pacific).
CYSTOCLONIUM DIFFICILE J. Ag. Guadeloupe, *Mazé*!
*Geogr. Distr.* Brazil.

## SPYRIDIEÆ.

SPYRIDIA FILAMENTOSA Harv. Grenada, *Murray*! Barbadoes, *Dickie*! Guadeloupe, *Mazé*! St. Thomas, *A. R. Young*! Jamaica, *Chitty*! Florida, *Harvey*! *Melvill*! Bermuda, *Rein*!
Var. FRIABILIS J. Ag. Guadeloupe, *Mazé*!
Var. VILLOSA Crn. Guadeloupe, *Mazé*!
Var. CUSPIDATA Crn. Guadeloupe, *Mazé*!
Var. REFRACTA Harv. Florida, *Harvey*? *Melvill*!
*Geogr. Distr.* Throughout tropical and subtropical seas.
S. SPINELLA Sond. Guadeloupe, *Mazé*!
*Geogr. Distr.* Australia.
S. ACULEATA J. Ag. Guadeloupe, *Mazé*! Yucatan, *Schott*! Florida, *Harvey*! *Melvill*! Bermuda, *Kemp*!
*Geogr. Distr.* Atlantic (Spain), Mediterranean, Red Sea. (Agardh gives it as West Indian.)
S. INSIGNIS J. Ag. Guadeloupe, *Mazé*!
*Geogr. Distr.* Indian Ocean.
S. CLAVATA Kütz. Guadeloupe, *Mazé*! ( = *S. Montagneana*, Kütz).
*Geogr. Distr.* Senegambia.

## ARESCHOUGIEÆ.

S. COMPLANATA J. Ag. Guadeloupe, *Mazé*! (Agardh gives it as West Indian).
*Geogr. Distr.* Brazil.

RISSOELLA DENTICULATA Mont.   Barbadoes, *Dickie*!
  *Geogr. Distr.*  Peru.

## CHAMPIEÆ.

CHYLOCLADIA ROSEA Harv.   Bermuda, *Kemp.*
  *Geogr. Distr.*  Atlantic (Europe and N. America).
C. RIGENS J. Ag.   Guadeloupe, *Mazé*!
  *Geogr. Distr.*  Warmer Atlantic and Western Pacific.
C. MUELLERI Harv. ?   Guadeloupe, *Mazé*!
  Var. CACTOIDES Crn.   Guadeloupe, *Mazé.*
  *Geogr. Distr.*  Australia.
C. SALICORNIA Crn.   Guadeloupe, *Mazé*!
C. SALICORNOIDES Harv.   Florida, *Harvey, Melvill*!
CHAMPIA PARVULA Harv.   Guadeloupe, *Mazé*!  Florida, *Harvey,
  Melvill*!
  *Geogr. Distr.*  Atlantic (Europe and America), Mediterranean
  (Pacific ?  Australia ?).

## RHODYMENIACEÆ.

HYMENOCLADIA DIVARICATA Harv., var. TROPICA Crn.   Guadeloupe,
  *Mazé*!
  *Geogr. Distr.*  Australia.
CHRYSYMENIA PLANIFRONS Melv. in J. G. Agardh.  Spec. gen. et ord.
  Alg.  Florida, *Melvill*!  Mr. Melvill in his list in ' Journ.
  Bot.' gives this as *C. Agardhii* Harv., var. *planifrons* Melv.,
  but the name is subsequently cited as above by Agardh, to
  whom he sent specimens, and by Mr. Melvill in his own
  herbarium.
C. AGARDHII Harv.   Florida, *Harvey, Melvill*!  Bermuda, '*Challenger*'!
C. HALYMENIOIDES Harv.   Florida, *Harvey*!  *Melvill*!  Bermuda,
  *Kemp*, '*Challenger*'!
C. UVARIA J. Ag.   Grenada, *Murray*!  Barbadoes, *Dickie*!  Guade-
  loupe, *Mazé*!  Florida, *Harvey*!  *Melvill*!  Bermuda, *Kemp,
  Rein, Farlow*!  in Alg. Exsicc. Am. Bor., 150.
  *Geogr. Distr.*  Atlantic (Europe and America), Australia.
C. OBOVATA Sond.   Guadeloupe, *Mazé*!
  *Geogr. Distr.*  Australia.
C. ENTEROMORPHA Harv.   Florida, *Harvey, Melvill*!
C. ? FURCATA Crn.   Guadeloupe, *Mazé*!
C. DICHOTOMO-FLABELLATA Crn.   Guadeloupe, *Mazé*!
C. CHYLOCLADIOIDES Crn.   Guadeloupe, *Mazé*!
C. SUBVERTICILLATA Crn.   Guadeloupe, *Mazé*!
C. TENERA Liebm.   Guadeloupe, *Mazé*!
CORDYLECLADIA IRREGULARIS Harv.   Florida, *Harvey, Hooper*!  in
  Farlow, Anderson and Eaton's Alg. Exsicc. Am. Bor., 18.
RHODYMENIA PALMATA Grev.   Bermuda, *Kemp.*
  *Geogr. Distr.*  N. Atlantic.
R. FLABELLIFOLIA Mont.   Guadeloupe, *Mazé*!
  *Geogr. Distr.*  Pacific (S. America).
R. MAMILLARIS Mont.   Guadeloupe, *Mazé*!  Martinique, *Duperrey*!

R. SUBDENTATA Crn. Guadeloupe, *Mazé*!
Although I take no note of imperfect specimens in this list,
I may observe that there is an imperfect specimen from
Jamaica, *Chitty*! in Herb. Mus. Brit., of a *Rhodymenia*,
which certainly is none of the above.

PLOCAMIUM COCCINEUM J. Ag. ? Jamaica, *Chitty*!
*Geogr. Distr.* N. Atlantic and N. Pacific. (Southern Seas
form probably distinct).

OCHTODES FILIFORMIS J. Ag. Barbadoes, *Dickie*! Guadeloupe, *Mazé*!
Martinique and St. Barthelemy *fide* Agardh.

## SQUAMARIEÆ.

PEYSSONNELIA DUBYI Crn. Grenada, *Murray*! Guadeloupe, *Mazé*!
Florida, *Harvey*, *Melvill*! Bermuda, '*Challenger*'!
*Geogr. Distr.* Atlantic (Europe).

## PORPHYRACEÆ.

BANGIA LUTEA J. Ag. Guadeloupe, *Mazé*!
*Geogr. Distr.* Mediterranean.

B. ATROPURPUREA Ag. Guadeloupe, *Mazé*!
*Geogr. Distr.* Atlantic (Europe & America), Pacific (California).

B. ELEGANS Chauv. Guadeloupe, *Mazé*!

B. DUMONTIOIDES Crn. Guadeloupe, *Mazé*!

B. GRATELOUPICOLA Crn. Guadeloupe, *Mazé*!

PORPHYRA LACINIATA Ag. Bermuda, *Kemp*.
*Geogr Distr.* Temperate Atlantic.

## SPHÆROCOCCOIDEÆ.

CORALLOPSIS SAGRÆANA Mont. Guadeloupe, *Mazé*! Cuba, *R. de la
Sagra*.

GRACILARIA CONFERVOIDES Grev. Grenada, *Murray*! Barbadoes,
*Dickie*! Guadeloupe, *Mazé*! Jamaica, *Chitty*! Florida,
*Harvey*! Bermuda, *Kemp*, '*Challenger*'!
*Geogr. Distr.* Throughout all seas.

G. ARMATA J. Ag. Grenada, *Murray*! Guadeloupe, *Mazé*! Florida,
*Harvey*. Bermuda, *Kemp*, *Rein*.
Var. GRACILIS Crn. Guadeloupe, *Mazé*!
Forma OCEANICA Crn. Guadeloupe, *Mazé*!
*Geogr. Distr.* Mediterranean.

G. FEROX J. Ag. Guadeloupe, *Mazé*! (Martinique, *Duperrey?*)
Bermuda, '*Challenger*'!

G. DAMÆCORNIS J. Ag. Barbadoes, *Dickie*!
Forma MINOR Crn. Guadeloupe, *Mazé*!
*Geogr. Distr.* Atlantic (N. America, but only the warmer
region).

G. USNEOIDES J. Ag. Guadeloupe, *Mazé*! Agardh gives it as West
Indian.

G. BLODGETTII Harv. Florida, *Harvey*.

Guadeloupe, *Mé* !
Guadeloupe, *Ma* !

'rn.   Guadeloupe.*Mazé* !
ʒuadeloupe, *Mazé* ! Gulf of Mexico, *jide Kutzing.*
uadeloupe, *Mazé* !
ı.   Guadeloupe, *Mzé* !
Guadeloupe, *Mazl*   Agardh has published an
pecies under this ame, which will have to give
bove.
n.   Guadeloupe, *Iazé* !
uadeloupe, *Mazé*
Guadeloupe, *Maz* !
ʋopifolius J. Ag. Grenada, *Murray* !   Guade-

Atlantic and Mederrancan.
ɔrmis Crn.   Guadoupe, *Mazé* !
n.   Guadeloupe, *Iazé* !
ʒuadeloupe, *Mazé*
Guadeloupe, *Maz*
.   Guadeloupe, *Mzé* !
deloupe, *Mazé* !
.   Guadeloupe, *Mé* !
Guadeloupe, *Mé* !
.   Guadeloupe, *Mzé* !
Crn.   Guadeloupe*Mazé* !
ı.   Guadeloupe, *Mzé* !
Guadeloupe, *Ma* !
Guadeloupe, *Ma* !
n.   Guadeloupe, *Iazé* !
Guadeloupe, *Mzé* !
Guadeloupe, *Maz* !
Guadeloupe, *Mazl*
ɔs Crn.   Guadeloue, *Mazé* !
tatum J. Ag.   Floda, *Harvey* !
N. Atlantic and ʃediterranean.
. Ag.   Bermuda, *Knp.*
Cape of Good Hopɛ Falkland Islands, Vancouver.
ʏfolia Harv.   Guadeoupe, *Mazé* ! Florida, *Harvey,*

ʋv.   Florida, *Harve* !
ʋam.   Florida, *Mis Reynolds* ! in Farlow, Ander-
ʒaton, Alg. Exsicc. m. Bor. 139.
Warmer Atlantic
ʋeurii J. Ag.   Guadeoupe, *Mazé* ! Florida, *Harvey.*
Throughout warr Atlantic ; Australia and New

(To be contiʒed.)

---

R 2

G. COMPRESSA Ag.  Barbadoes, *Dickie*!  Guadeloupe, *Mazé*!  Vera
   Cruz (*Kutzing in Harvey*).  Florida (*Melvill*!).
   *Geogr. Distr.*  Warmer Atlantic and Mediterranean.
G. SPINESCENS J. Ag.  Guadeloupe, *Mazé*!
   *Geogr. Distr.*  Pacific (New Caledonia), Tasmania.
G. SECUNDATA Harv.  Guadeloupe, *Mazé*.
   *Geogr. Distr.*  Australia.
G. DURA J. Ag.  Guadeloupe, *Mazé*!
   *Geogr. Distr.*  Warmer Atlantic (America and Europe), Medi-
   terranean, Indian Ocean.
G. CAUDATA J. Ag.  Guadeloupe, *Mazé*!  Gulf of Mexico (*Agardh in
   Harvey*).
G. CORNEA J. Ag.  Barbadoes, *Dickie*!  Guadeloupe, *Mazé*.
   *Geogr. Distr.*  Brazil.
G. POITEI Lam.  Grenada, *Murray*!  Guadeloupe, *Mazé*!  Florida,
   *Harvey*, *Melvill*!  Bermuda, '*Challenger*'!
G. WRIGHTII J. Ag. ?  Guadeloupe, *Mazé*!  Agardh possesses speci-
   mens " ex pluribus insulis Indiæ occidentalis."
   *Geogr. Distr.*  Indian Ocean, Red Sea.
G. MULTIPARTITA J. Ag.  Barbadoes, *Dickie*!  Guadeloupe, *Mazé*!
   Jamaica, *Chitty*!  Cuba, *Hb. Montagne*!  Bermuda, *Kemp*.
   *Geogr. Distr.*  Warmer Atlantic (Europe and America), Red
   Sea, N. Zealand.
G. DENTATA J. Ag.  Guadeloupe, *Mazé*!  Agardh gives it as West
   Indian.
   *Geogr. Distr.*  Senegambia.
G. CERVICORNIS J. Ag.  Barbadoes, *Dickie*!  Guadeloupe, *Mazé*!
   Gulf of Mexico (*fide Agardh*).  Bermuda '*Challenger*'!
G. DIVARICATA Harv.  Guadeloupe, *Mazé*!  Florida, *Harvey*!  Ber-
   muda, *Kemp*!
G. CHONDRIOIDES Crn.  Guadeloupe, *Mazé*!
   *Geogr. Distr.*  Brazil.
G. PROLIFICA Crn.  Guadeloupe, *Mazé*!
G. ARCUATA Zan.  Guadeloupe, *Mazé*!
   *Geogr. Distr.*  Red Sea.
G.? TUBERCULOSA Hampe.  Guadeloupe, *Mazé*!
   *Geogr. Distr.*  Peru.
G. OBTUSA Grev.  Guadeloupe, *Mazé*!
   *Geogr. Distr.*  Indian Ocean.
G. DOMINGENSIS Sond.  Barbadoes, *Dickie*!  St. Domingo, *fide
   Dickie.*
G. ACANTHOCOCCOIDES Crn.  Guadeloupe, *Mazé*!
G. APICULATA Crn.  Guadeloupe, *Mazé*!
G. BICUSPIDATA Crn.  Guadeloupe, *Mazé*!
G.? CURTIRAMEA Crn.  Guadeloupe, *Mazé*!
G. CARTILAGINEA Crn.  Guadeloupe, *Mazé*!
G. CIRCINNATA Crn.  Guadeloupe, *Mazé*!
G. CRASSISSIMA Crn.  Guadeloupe, *Mazé*!
G. DICHOTOMO-FLABELLATA Crn.  Guadeloupe, *Mazé*!
G. DENDROIDES Crn.  Guadeloupe, *Mazé*!

G. FLABELLATA Crn.   Guadeloupe, *Mazé*!
G. GREVILLEI Crn.   Guadeloupe, *Mazé*!
   *Geogr. Distr.*
G. LUTEOPALLIDA? Crn.   Guadeloupe, *Mazé*!
G. MEXICANA Crn.   Guadeloupe, *Mazé*!   Gulf of Mexico, *fide Kutzing.*
G. PATENS Crn.   Guadeloupe, *Mazé*!
   Var. GRACILIS Crn.   Guadeloupe, *Mazé*!
G. RAMULOSA Crn.   Guadeloupe, *Mazé*!   Agardh has published an
   Australian species under this name, which will have to give
   way to the above.
G. SECUNDIRAMEA Crn.   Guadeloupe, *Mazé*!
G. SECUNDA Crn.   Guadeloupe, *Mazé*!
G. SQUARROSA Crn.   Guadeloupe, *Mazé*!
SPHÆROCOCCUS CORONOPIFOLIUS J. Ag.   Grenada, *Murray*!   Guade-
   loupe, *Mazé*!
   *Geogr. Distr.*   Atlantic and Mediterranean.
PLOCARIA FLABELLIFORMIS Crn.   Guadeloupe, *Mazé*!
P. CHONDRIOIDES Crn.   Guadeloupe, *Mazé*!
P. DISTICHA Crn.   Guadeloupe, *Mazé*!
P. ACULEATA Crn.   Guadeloupe, *Mazé*!
P. OLIGACANTHA Crn.   Guadeloupe, *Mazé*!
P. VAGA Crn.   Guadeloupe, *Mazé*!
P. COMPLANATA Crn.   Guadeloupe, *Mazé*!
P. LACINULATA Crn.   Guadeloupe, *Mazé*!
P. DACTYLOIDES Crn.   Guadeloupe, *Mazé*!
P. FLAGELLIFORMIS Crn.   Guadeloupe, *Mazé*!
P. DISCIPLINALIS Crn.   Guadeloupe, *Mazé*!
P. DIVARICATA Crn.   Guadeloupe, *Mazé*!
P. SQUARROSA Crn.   Guadeloupe, *Mazé*!
P. TRIDACTYLITES Crn.   Guadeloupe, *Mazé*!
P. POLYMORPHA Crn.   Guadeloupe, *Mazé*!
P. BIPENNATA Crn.   Guadeloupe, *Mazé*!
P. CORTICATA Crn.   Guadeloupe, *Mazé*!
   Var. CHONDROIDES Crn.   Guadeloupe, *Mazé*!
NITOPHYLLUM PUNCTATUM J. Ag.   Florida, *Harvey*!
   *Geogr. Distr.*   N. Atlantic and Mediterranean.
N. PLATYCARPUM J. Ag.   Bermuda, *Kemp.*
   *Geogr. Distr.*   Cape of Good Hope, Falkland Islands, Vancouver.
DELESSERIA TENUIFOLIA Harv.   Guadeloupe, *Mazé*!   Florida, *Harvey,*
   *Melvill*!
D. INVOLVENS Harv.   Florida, *Harvey*!
D. HYPOGLOSSUM Lam.   Florida, *Miss Reynolds*!   in Farlow, Ander-
   son, and Eaton, Alg. Exsicc. Am. Bor. 139.
   *Geogr. Distr.*   Warmer Atlantic.
CALOGLOSSA LEPRIEURII J. Ag.   Guadeloupe, *Mazé*!   Florida, *Harvey.*
   *Geogr. Distr.*   Throughout warm Atlantic; Australia and New
   Zealand.

(To be continued.)

# BIOGRAPHICAL INDEX OF BRITISH AND IRISH BOTANISTS.

## By James Britten, F.L.S., and G. S. Boulger, F.L.S.

(Continued from p. 218).

**Cooper, Daniel** (1817 ?–1842) : d. Leeds, Yorkshire, 24th November, 1842; bur. Quarry Hill Cemetery, Leeds. A.L.S., 1837. Curator Bot. Soc. Lond., 1838. Assistant Zool. Dep. Mus. Brit. Assistant-surgeon in Army. 'Flora Metropolitana,' 1836. Supplement, 1837. Pritz. 68; Jacks. 534; Proc. Linn. Soc. i. 52, 173; Gent. Mag. xix. (1843), 108; R. S. C. ii. 41; Proc. Bot. Soc. Lond. 1839 ; Diet. Nat. Biog. xii. 141. *Cooperia* Herbert = *Sceptranthus* Grah.

**Cooper, Thomas Henry** (fl. 1834). F.L.S. 'Botany of . . . . . . Sussex,' 1834. Pritz. 68 ; Jacks. 260.

**Coote, Dr.** (fl. 1640). Of Shropshire. Sent *Drosera anglica* to Parkinson. Theatr. 1053.

**Copland, William** (fl. 1556–1569) : d. London, 1568 or 1569. Printer. Compiled 'Boke of the Properties of Herbes,' 1552. Herbert in Dibdin's Ames, iii. 133 ; Pult. i. 51 ; Dict. Nat. Biog. xii. 174.

**Corbet, Richard,** alias **Poynter** (fl. 1597). Nurseryman, of Twickenham. Gerard, 'Herbal,' 1269. [Vincent Corbet, a nurseryman of Ewell, was father of Richard Corbet (1582–1635), Bishop of Oxford and Norwich. Dict. Nat. Biog. xii. 203.]

**Corder, Octavius** (fl. 1860). Of Fyfield. Found *Lathyrus tuberosus* in 1859. Contributed to Gibson's 'Flora of Essex.'

**Corder, Thomas** (d. 1874) : d. 15th October, 1874. Of Writtle, Essex. A.L.S., 1833. Local Sec. Bot. Soc. Lond. Found *Bupleurum falcatum* in 1831. E. B. S. 2763. Went to Adelaide, S. Australia, 1843.

**Cornthwaite, Tullie** (d. 1879) : d. Walthamstow? 1st May, 1879. Clerk. M.A. F.L.S., 1864. Of Walthamstow. Formed a herb. and purchased some of Ed. Forster's plants. Herb. now in possession of Hildebrand Ramsden, Esq.

**Corry, Thomas Hughes** (1860 ?–1883) : b. Ireland, 1862 ; drowned in Lough Gill, 4th August, 1883. M.A. F.L.S., 1882. Assistant Curator University Herbarium. Lecturer in the Medical and Science Schools, Cambridge. 'Asclepias Cornuti,' Linn. Trans. ser. 2, vol. ii. 75 ; Journ. Bot. 1883, 313 ; Proc. Linn. Soc. 1883–86, 37.

**Coultas, Harland** (d. 1877) : b. U. S. A.? ; d. London, 2nd February, 1877. Prof. Botany, Penn Medical University, Philadelphia. Lecturer, Charing Cross Hospital. Pritz. 70 ; Jacks. 535 ; Journ. Bot. 1877, 192.

**Coulter, Thomas** (d. 1843) : d. Dublin, 1843. M.D. Explored Central Mexico. Reached Monterey, 1831. Met Douglas in California. Discovered *Pinus Coulteri*. Returned to England, 1833. Curator of Herb. Trin. Coll. Dublin. C. C. Parry,

'Botanical Explorers of Pacific Coast'; Pritz. 70; in 'Overland Monthly,' 1883; Bot. Geol. Survey. Californ. 555; Journ. Roy. Geogr. Soc. 1835, v. 59. *Coulteria*, H. B. K. = *Cæsalpinia* in part.

**Cowell, John** (d. circ. 1730): Of Hoxton, Nurseryman. 'Account of *Aloe Americana*,' 1729. Cott. Gard. viii. 121.

**Cowell, M. H.** (fl. 1839): 'Floral Guide for East Kent,' 1839. Pritz. 71; Jacks. 230, 254.

**Cowley, Abraham** (1618-1667): b. London, 1618; d. Chertsey, 28th July, 1667; bur. in Westminster Abbey. M.A., Camb. 1642. M.D., Oxon, 1657. Poet. F.R.S., 1662. Works, 1668, w. life by Sprat; and portr. Pult. i. 282; Pritz. 71; Jacks. 212; Nich. Illust. iv. 398; Dict. Nat. Biog. xii. 379; Portr. engr. by T. Faithorne in 'Latin Poems,' 1668; another in 'Works.' 1673; and others. Two portrs. at Bodleian; one by Lely in Nat. Portr. Gallery, &c.

**Coyte, William Beeston** (1741-1810): b. 1741; d. Ipswich, 3rd March, 1810; bur. St. Nicholas', Ipswich. M.B., Cambr. 1763. M.D. Clk. of Yarmouth and Halesworth. Afterwards practised as physician at Ipswich. A.L.S. 1788. F.L.S. 1794. 'Hortus Botanicus Gippovicensis,' 1795. 'Index Plantarum,' vol. i. 1807. Pritz. 71; Jacks. 13, 14, 412; Gent. Mag. 1810, i. 389; Rich. Corr, 184; Nich. Illust. vi. 877; Dict. Nat. Biog. xii. 424.

**Coys, William** (fl. 1600): Of Stubbers, N. Ockington, Essex. First flowered *Yucca*, 1604. Lobel, 'Adversaria,' i. 501; ii. 471; Parkinson, Theatrum, 84.

**Crewe, Henry Harpur** (1830-1883): b. 1830; d. Drayton Beauchamp, September 7th, 1883. Clk. M.A. Rector of Drayton Beauchamp, Tring. Entomol. Monthly Mag. xx. 118; Journ. Bot. 1883, 380. *Crocus Crewei* Hook. f.

**Crichton, Sir Alexander** (1763-1856): b. Edinburgh, 2nd December, 1763; d. June, 1856. M.D., Leyden, 1885. L.R.C.P. 1791. F.L.S. 1793. F.R.S. 1800. Knighted 1821. Physician to the Czar. 'On Vegetable remains in Sandstone near Ballisadiere, Co. Sligo,' Trans. Geol. Soc.; Munk. ii. 416; Proc. Linn. Soc. 1857, xxv.; Proc. Roy. Soc. iii. (1856) 269; Quart. Journ. Geol. Soc. xiii. (1857) lxiv.; Dict. Nat. Biog. xiii. 85.

**Croall, Alexander** (1809 ?-1885): d. Stirling, May 19th, 1885. Of Guthrie, Forfar. Local Sec. Bot. Soc. Lond. 1843. 'Plants of Braemar' (dried specimens) 1855; 'Nature-printed Brit. Seaweeds' [with Johnstone], 1859-60. Keeper of Museum and Herbarium, Derby, 1863. Curator, Smith Institute, Stirling, 1873. Herbarium at Stirling. Pritz. 157; Jacks. 242; R. S. C., ii. 95; vii. 460; Trans. Bot. Soc. ed. xvi. 309.

**Crocker, C. W.** (1832 ?-1868): b. Chichester, 1832 ?; d. Torquay, Devon, 19th February, 1868. Foreman at Kew. Afterwards Verger, Chichester Cathedral. 'Germination of *Cyrtandreæ*,' Journ. Linn. Soc. v. 1861; Gard. Chron. 1868, 242.

**Crosfield, George** (1785-1847): b. Warrington, 1785; d. Warrington ?, 15th December, 1847. Secretary, Bot. Soc. Warrington. 'Calendar of Flora,' 1810; Pritz. 72; Jacks. 261; 'Annual Monitor,' 1849; Friends' Books, i. 494; Diet. Nat. Biog. xiii. 213.

**Crowe, James** (d. 1807). Of Lakenham, nr. Norwich. F.L.S., 1788. Studied mosses, fungi and willows. Had a salicetum. Contributed to 'English Botany.' Smith, Lett. i. 17. *Crowea* Sm.

**Crowther, James** (1768-1847): b. Manchester, 24th June, 1768; d. Manchester, 6th January, 1847; bur. St. George's, Hulme. Weaver and porter. Contributed to 'Flora Mancuniensis.' Discovered *Cypripedium* at Malham. Cash, 77; Diet. Nat. Biog. xiii. 245.

**Crozier, George** (1792-1847): b. Eccleston, Lancashire, 1792; d. Peel Street, Hulme, Manchester, April, 1847. Saddler. Pupil of Thomas Townley, at Blackburn. Settled in Manchester, eire. 1831. Contributed to Phyt., vol. i. Left large herbarium. Cash, 119.

**Cruckshanks, Alexander** (fl. 1831): Collected in Chili. R.S.C., ii. 100. *Cruckshanksia* Hook. & Arn.

**Cullen, —.** (fl. 1852): Resident at Court of Travancore. Major-General. Meteorologist. Studied economic botany. Wight Icon. 1761. *Cullenia* Wight.

**Cullen, W. H.** (fl. 1849): M.D. F.B.S.E. Of Sidmouth. 'Flora Sidostiensis,' 1849. Pritz. 73; Jacks. 259.

**Cullum, Sir John** (1733-1785): b. Hawsted, Suffolk, 21st June, 1733; d. 9th October, 1785; bur. Hawsted. Clk. B.A. Cantab., 1756. Bart., 1774. F.R.S., 1775. 'On Cedars in England,' Gent. Mag. 1779, 138. Discovered *Veronica verna*, E.B., 25. Nich. Anecd. vi. 625; viii. 209, 673; Nich. Illust. vii. 408; Diet. Nat. Biog. xiii. 283. Portr. by Angelina Kauffmann, 1778, at Hardwick; engr. by Basire in 'History of Hawsted,' 1813; and in Nich. Anec. viii. 209. *Cullumia* R. Br.

**Cullum, Sir Thomas Gery** (1741-1831): b. Hardwick House, Suffolk, 30th November, 1741: d. Hardwick (?), 8th September, 1831; bur. Hawsted. M.R.C.S., 1800. F.L.S., 1790. F.R.S. Practised at Bury St. Edmunds. Bath King-at-arms, 1771-1800. Bart., 1785. 'Floræ Anglicæ Specimen, 1774. Pritz. 73; Jacks. 232; Dict. Nat.Biog. xlii. 284. *Cullumia* R.Br.

**Culpepper, Nicholas** (1616-1654): b. London, 18th October, 1616; d. 10th January, 1654. Apothecary and herbalist of Spitalfields. 'Herbal,' 1652. 'English Physician,' 1652. 'Physical Directory,' 1649. Pult. i. 180; Pritz. 73; Jacks. 28; Gent. Mag. 1797, i.; Diet. Nat. Biog. xiii. 287. Portr. in 'Directory,' 1649; engr. by Cross, in his 'English Physician,' 1652; another in his 'School of Physic.'

**Cuming, Hugh** (1791-1865): b. West Alvington, S. Devon, 14th February, 1791; d. Gower Street, London, 10th August, 1865. F.L.S., 1832. Collected in South America and Pacific. In the Philippines, &c., 1835-39. Collected 130,000 dried specimens. Pritz. 73; Journ. Bot. 1865, 325; Proc. Linn. Soc. 1865-6, lvii.; R. S. C., ii. 103; Mag. Zool. Bot. 1838, 56. Portr. 'Men of Eminence,' 1864; Gard. Chron. 1865, 824; Dict. Nat. Biog. xiii. 295.

**Cuninghame, James**, or **Cunningham** (d. 1709 ?): Surgeon to Hon. E. I. C., at Amoy, 1698-1703. F.R.S,, 1699. Sent pl. from China, Malacca, Cape, and Ascension to Ray, Petiver, and Plukenet; now in Herb. Sloane, at Mus. Brit. Pult. ii. 59-62; Pritz. 73; Gard. Chron. 1881, ii. 440; Plukenet, 'Amaltheum,' and 'Phytographia'; Petiver, 'Museum'; Sloane MSS., 3322 & 4041; Dict. Nat. Biog. xiii. 312. *Cunninghamia* Schreb. = *Malanea* Aubl. *Cunninghamia* R. Br.

**Cunningham, Allan** (1791-1839); b. Wimbledon, Surrey, 13th July, 1791; d. Sydney, 27th June, 1839; bur. in Scottish Church, Sydney. Employed on 'Hortus Kewensis,' circ. 1808. Kew Collector, 1814-1831. At Rio, with James Bowie, 1814-1816; at Sydney, 1816-1826; in New Zealand, 1826. Returned to England, 1831. Colonial Botanist and Superintendent, Bot. Gard., Sydney, 1836-38. Visited New Zealand, 1838. 'Fl. Ins. Novæ Zelandiæ', Mag. Zool. Bot. 1838, 210, bequeathed to Heward and presented to Kew. Pritz. 73; Jacks. 400. Life by Heward in Journ. Bot. 1842, w. litho. portr. by J. Robinson. Coloured copy at Linn. Soc. Gard. Chron. 1881, ii. 440; Dict. Nat. Biog. xiii. 308. *Allania* Benth. *Cunninghamia* R. Br.

**Cunningham, Richard** (1793-1835): b. Wimbledon, 12th February, 1793; murdered by natives in interior of Australia, circ. April, 1835. Brother of Allan. Employed on 'Hortus Kewensis,' circ. 1808. Colonial Botanist and Superintendent of Bot. Gard. Sydney, 1833-1835. Pritz. 73; Comp. Bot. Mag. ii. (1826) 210, w. litho. portr. fr. one by McNees, belonging to Sir W. J. Hooker; R. S. C., ii., 105. Gard. Chron. 1881, ii. 440; Loud. Gard. Mag. 1836; Mag. Zool. Bot. i. (1837), 210; Dict. Nat. Biog. xiii. 317.

**Curnow, William** (d. 1887); d. Penzance, 24th January, 1887. Of Penzance. Contributed to Phyt. i., and to Keys' Fl. Devon and Cornwall.

**Currey, Frederick** (1819-1881): b. Norwood, Surrey, 19th August, 1819; d. Blackheath, 8th September, 1881; bur. at Weybridge. Fungologist. M.A., Camb., 1844. F.L.S., 1856. Secretary, 1860-80. Treasurer, 1880. F.R.S., 1858. Translated Schacht's 'Das Mikroscop,' 1853, and Hofmeister's 'Higher Cryptogamia,' 1862. Edited Badham's 'Esculent Funguses,' 1863; 'Fungi of Greenwich,' Phyt. 1854. Collection of Fungi at Kew, Pritz. 73; Journ. Bot. 1881, 310; Proc. Linn. Soc. 1881-2, 59; Gard. Chron. 1881, ii. 412; Dict. Nat. Biog. xiii. 341; R. S. C., ii. 108. Photo. portr. at Linn. Soc.

**Curtis, Samuel** (1779-1860): b. Walworth, Surrey, 1779; d. La Chaire, Rozel, Jersey, 6th January, 1860. Florist at Walworth. Married a dau. of William Curtis in 1801. Became proprietor of Bot. Mag. Lived at Glazenwood, Coggeshall, Essex, and at La Chaire. F.L.S., 1810. Pritz. 73; Jacks. 536; Proc. Linn. Soc. 1860, xxii.; Cott. Gard. xxiii. 335; Dict. Nat. Biog. xiii. 349.

**Curtis, William** (1746-1799): b. Alton, Hants, 1746; d. Bromp-

ton, 7th or 27th July, 1799; bur. Battersea. F.L.S., 1788.
Apothecary. Taught by John Lagg, ostler, Crown Inn, Alton.
Demonstrator to Fordyce at St. Thomas's Hospital. Præfectus
Horti, Chelsea, 1772–77. 'Flora Londinensis,' 1777–87. Bot.
Mag. commenced, 1787. 'British Grasses,' 1787. 'Lectures,'
1805. Rees; Pritz. 73; Jacks. 536. Life by Goodenough, Gent.
Mag. vol. 69, pp. 628 & 635. By Thornton, w. portr. in 'Lectures'
(1805); Nich. Illust. vi, 256; Indexes to Bot. Mag. 1828;
Felton, 184; Semple, 104; Annals of Bot. (1805) 189; Friends'
Books, i. 502; Fl. Midd. 393; Cott. Gard. iv. 205; Journ.
Hort. 1876, xxi. 239; Dict. Nat. Biog. xiii. 349. Portr. in
Thornton's 'Botany,' and at Kew. *Curtisia* Aiton.
**Cutler, Catherine** (d. 1866): d. Exmouth 15th April, 1866. Of
Sidmouth. Algologist. Algæ in Herb. Mus. Brit. Pritz. 74;
Journ. Bot. 1866, 238. *Cutleria* Grev.

**Dale, Samuel** (1659–1739): b. Whitechapel (?), 1659; d. Bocking,
Essex, 6th June, 1739; bur. Dissenters' burial-ground, Bocking.
Apothecary and Physician. Practised at Braintree. Friend
and executor of Ray, correspondent of Sloane (Sloane MSS.,
4042). 'Pharmacologia,' 1693. Supplement, 1705. 'History
of Harwich,' 1730. Contributed to Phil. Trans. 1692–1736.
Herbarium bequeathed to Apothecaries' Company, now in Herb.
Mus. Brit. Pult. ii. 122; Rees; Pritz. 75; Jacks. 199, 200;
Semple, 63; Gibson, Fl. of Essex, 446; Journ. Bot. 1883,
193, 225; Dict. Nat. Biog. xiii. 385. Oil portr. at Apothecaries'
Hall. Engrav. by G. Vertue in 'Pharmacologia,' ed. 3, 1737.
Autotyped in Journ. Bot. 1883; copy at Kew. *Dalea* Brown
= *Eupatorium Dalea* L. *Dalea* L.
**Dalton, James** (1765 ?–1843): b. 1765 (?): d. Croft, Yorkshire,
January 2nd, 1843. Clk. M.A. Camb., 1790. F.L.S., 1803.
Contributed to 'E. Botany.' Studied Carices and Mosses. Pritz.
75; Proc. Linn. Soc. i. 172. *Daltonia* Hook. & Tayl.

(To be continued.)

---

## SHORT NOTES.

CERASTIUM PUMILUM IN WILTS.—A *Cerastium* which I found on
the downs in S. Wilts in May last has been certified by Mr. J. G.
Baker to be *C. pumilum* Curtis. I should be glad to correct an
error in my notice of some Wilts plants in Journ. Bot. 1887, p. 55,
where *Hypochœris radicata* should be *H. glabra*.—W. A. CLARKE.

NOTE ON BUCKINGHAMSHIRE RUBI.—So few Rubi are on record
from Buckinghamshire that I send a note of the forms I saw during
a walk on July 12th, in the south of the county: — *R. Lindleianus*,
thickets at Burnham Beeches and by the roadside between Farnham Royal and Stoke Pogis. *R. cordifolius* Angl., the commonest
bramble of the more exposed thickets at Burnham Beeches.

*R. discolor* Angl. (*R. ulmifolius* Schott), hedges between Farnham Royal and Stoke Pogis. *R. thyrsoideus* Angl. (*R. pubescens* W. & N.), lane ̄above the Crown Inn, East Burnham. *R. leucostachys*, Burnham Beeches and between Farnham Royal and Stoke Pogis. *R. Sprengelii*, heathy thickets at Burnham Beeches. *R. Koehleri* var. *R. pallidus* Angl., Burnham Beeches and between Farnham Royal and Stoke Pogis. *R. diversifolius* Lindl., roadside thickets between Farnham Royal and Stoke Pogis. *R. corylifolius*, hedges about Farnham Royal. *R. cæsius*, hedges between Stoke Pogis Church and Slough.—J. G. BAKER.

NOTE ON SALIX FRAGILIS. — I have been much interested in reading Dr. Buchanan White's paper on *Salix fragilis* and *Russelliana*, and hope it will be only the first of a series in which he will tell us of the results of his recent careful investigations into the hybrid British Willows. On this matter we in Britain are at present far behind the continental botanists, and must look to him to bring us up to the position we ought to occupy. Why I wrote now was to tell him that he will find an excellent figure of Smith's *fragilis*, which no doubt is the right thing, in the ' Salictum Woburnense' of Forbes, tab. 27. This gives a far better idea of the plant than any other figure or specimen I have seen. The capsule agrees very well with that of *S. viridis* Fries, but the leaf is further away from that of *alba* than is that of Smith's *Russelliana*, which is well represented in the ' Salictum Woburnense,' at tab. 28. A second form of the continental *fragilis*, with a broader, greener leaf than that of *Russelliana*, is figured by Forbes under the name of *S. montana* at tab. 19. This latter agrees very well with the common Thames-side form of the plant. Of this *S. montana* it is said that the twigs are little inferior to those of *vitellina* for tying and the finer kinds of wicker-work. In an excellent paper in the ' Gardeners' Chronicle' for 1845, p. 69, on the Thames osiers, their local names and relative economic values, it is stated that *fragilis*, *Russelliana*, and *alba* are all three occasionally grown as osiers for the purpose of making eel-wheels and cooper's twigs, but that they are all three much inferior to *rubra* and the various forms of *viminalis* and *triandra* in economic value, whilst *Lambertiana*, *Helix*, and *undulata* are " all rubbish."—J. G. BAKER.

POLYGALA AUSTRIACA Crantz, IN SURREY. — It was my good fortune to meet with this form of *Polygala* on the 2nd of June last, on a roadside bank near Caterham. On the 7th of the present month I obtained additional specimens, with more developed fruit, at the same spot. The specimens have been seen and approved by Mr. J. G. Baker and Mr. W. H. Beeby. According to Hooker (' Student's Flora,' p. 51, 3rd ed.), the Kent form is blue-flowered. In the Caterham form the flowers are of a delicate white.—WILLIAM WHITWELL.

# REPORT OF DEPARTMENT OF BOTANY, BRITISH MUSEUM, FOR 1887.

## By W. Carruthers, F.R.S.

During the past year 60,753 specimens have been mounted, named, and inserted in their places in the Herbarium; of these 16,353 were Phanerogams, and 44,400 were Cryptogams. The Phanerogams have consisted chiefly of specimens collected in Austria by Kerner, in Madras by Gamble, in Malaya by Beccari, in Australia by Baron von Mueller, and the Rev. T. S. Lea, in North America by Orcutt, and in Columbia by Lehmann; and the Cryptogams of Ferns collected in Perak by Scortechini, Mosses from the herbaria of Roemer, Shuttleworth, Spruce, and Schimper, and Algæ from Ceylon, the Red Sea, the Cape of Good Hope, and Guadeloupe. The valuable Herbarium of Fungi bequeathed by the late C. E. Broome has been completely re-mounted, arranged, and made fully accessible to students. The various collections of *Diatomaceæ* have been systematically arranged.

In the progress of incorporating the additions, the following Natural Orders have been more or less completely arranged:— *Anonaceæ*, *Menispermaceæ*, *Berberideæ*, *Nymphæaceæ*, *Cucurbitaceæ*, *Umbelliferæ*, *Araliaceæ*, *Cornaceæ*, *Primulaceæ*, *Nyctagineæ*, *Phytolaccaceæ*, *Podostemaceæ*, *Thymelæaceæ*, *Loranthaceæ*, *Santalaceæ*, *Euphorbiaceæ*, *Urticaceæ*, *Cupuliferæ*, *Cycadaceæ*, *Orchidaceæ*, *Palmæ*, and *Filices*.

The most important addition to the collections during the past year was the Herbarium of the late Dr. Hance, of Whampoa, China, consisting of 22,437 species of plants. The Museum has in this secured an extensive series of plants from various districts in China, as well as the types of all the plants which Dr. Hance had himself discovered and described. Already this collection has been of great service in connection with the 'Flora of China,' now being issued by a Committee of the Royal Society of London.

John W. Miers, Esq., has presented the large collection of Ferns, chiefly rich in South-American forms, that belonged to his late father, John Miers, F.R.S., &c. The Herbarium of flowering plants had already, by bequest on the decease of Mr. Miers, become the property of the Trustees, and by this valuable donation the whole of the plant collections of Mr. Miers has been acquired by the Department.

A valuable selection of the Algæ of Guadeloupe, consisting of 1509 specimens, have been acquired from M. Mazé, representing the species described in Mazé and Schramm's 'Algues de la Guadeloupe.'

The additions to the collections by presentation during the year have consisted of 511 species of Italian plants from H. Groves, Esq., F.L.S.; a collection of Scandinavian Roses from G. Nicholson, Esq.; a set of the plants of the Afghan Boundary Expedition, collected and presented by J. E. T. Aitchison, Esq., M.D.; 747 species of Indian plants from J. S. Gamble, Esq., F.L.S.;

86 species of Indian plants from Dr. King, Calcutta ; 44 species of plants from Perak, collected by the late Rev. F. Scortechini, presented by the Government of Perak ; 100 species of plants from Madras from M. A. Lawson, Esq., F.L.S. ; 109 species of plants from the Cameroons, collected by H. H. Johnston, Esq., presented by the British Association ; 200 species of South African plants from Prof. MacOwan and H. Bolus, Esq., F.L.S. ; 665 species of Australian plants, and eight Orchids and two Palms from New Guinea, from Baron von Mueller ; 54 species of Australian plants and 125 species from Hawaii, from the Rev. T. S. Lea ; 43 species of plants from California, from M. K. Curran ; 173 species of plants from Demerara, from G. S. Jenman, Esq., F.L.S ; 63 species of plants from Brazil, from W. F. Leeson, Esq. ; 750 specimens from Pernambuco, collected and presented by the Rev. T. S. Lea, and Messrs. Ridley and Ramage ; 59 species of Orchids from F. W. Moore, Esq. ; two South African Orchids from H. Bolus, Esq. ; six Orchids from the Organ Mountains, Brazil, from F. M. Pascoe, Esq. ; several species of *Narcissus* and other plants from George Maw, Esq. ; 72 species of *Festuca* from Prof. Hackel ; 122 specimens of *Cyperaceæ* from N. America and New Zealand, from A. Bennett, Esq., F.L.S. ; specimens of *Balanophora indica*, from Prof. Bower and T. G. Millington, Esq. ; a specimen of *Trimorphopetalum dorstenioides* from Madagascar, from J. G. Baker, Esq., F.R.S. ; specimens of cultivated plants from Lord Walsingham, the Hon. and Rev. J. T. Boscawen, Sir Trevor Lawrence, C. B. Clarke, Esq., J. O'Brien, Esq., F. W. Burbidge, Esq., and Prof. Henslow ; six species of Mosses from Prof. Lindberg ; 58 species of Mosses from Travancore from Col. Beddome ; 18 species of Hepatics from W. H. Pearson, Esq. ; 85 species of Algæ from Nova Scotia, and 124 from Vancouver, from Prof. Macoun ; 174 gatherings of Arctic Diatoms and 56 species of Indian Algæ, collected by Dr. Watt, from the Director of the Royal Botanical Gardens, Kew : 64 species of Algæ from the Cape of Good Hope from Capt. Young; 18 species of Algæ from Queensland from W. Alcock Tully, Esq. ; two species of Mediterranean Algæ from Dr. Bornet ; specimens of ·*Valonia* from Bermuda from Mrs. Whelpdale ; a fine specimen of *Cordiceps Taylori* from Victoria, and a drawing of a species of *Agaricus*, from W. G. Smith, Esq. ; three species of Indian Fungi, eight drawings of Fungi, the extensive Herbarium of British Mints, made by the late Rev. Kirby Trimmer, the British Herbaria of Thomas Moore of Chelsea and Mr. Knowlton, and a miscellaneous collection of British plants, from the Director of the Royal Gardens, Kew ; 212 species of British plants from A. Bennett, Esq.; 43 species from W. H. Beeby, Esq. ; 84 species from Wiltshire from the Rev. T. A. Preston ; 196 species of plants from the neighbourhood of London from C. D. Sherborn, Esq. ; a large collection of British plants from the Botanical Record Club, and other British plants from Sir John Lubbock, Bart., the Revs. E. F. Linton and H. P. Reader, Messrs. J. G. Baker, J. Benbow, A. Dymond, D. Fry, H. C. Hart, J. H. A. Jenner, Jas. Saunders, R. T. Towndrow, and R. Weaver, and Miss F. P. Thompson ; 51

species of British Mosses from E. G. Baker, Esq.; 60 species of Algæ from the North Sea from E. M. Holmes, Esq.; 20 species from the North Sea from George Murray, Esq.; 16 species of British Algæ from Capt. Young; 15 species from Alex. Anderson; six specimens of monstrous flowers and stems from Dr. M. T. Masters, F.R.S.; fasciated stem of *Tamus* from W. G. Smith, Esq.; a photograph of a fasciated pineapple from J. J. Quelch, Esq.; and 22 specimens of New Zealand Woods from J. J. Collis, Esq.

The following collections have been acquired by purchase :—360 species of Greek plants collected by Orphanides; 30 species of Swedish Violets; 60 species of *Rubus* from Denmark; 200 species of German plants from Schultz; 100 species from Sicily, collected by Lojacono; 289 species of Galitzian plants, collected by B. Blocki; 100 species of *Junci* from Prof. Caruel; 500 species of German Mosses from Sidow; 100 species of European Mosses; 200 species of freshwater Algæ from France, from Mougeot, Dupray, and Roumeguère; 280 species of freshwater Algæ from France, with notes and drawings, by Desmazières; 50 species of Fungi from Rehm; 642 species from Sumatra, Java, New Guinea, Borneo, and Abyssinia, collected by O. Beccari; 159 species from the Comoro Islands, collected by Humblot; 1459 species of plants from East Tropical Africa, collected by the Rev. W. E. Taylor; 159 species from Lukoma, Lake Victoria, Nyassa, Africa, collected by W. Bellingham; 100 species of North American plants, collected by Curtiss; 696 species, collected by Marcus E. Jones; 683 species from Mexico, collected by Dr. Palmer; 400 species from Southern California, collected by C. R. Orcutt; 348 species of Canadian Mosses, collected by Macoun; 580 species of plants from Columbia, collected by Lehmann.

By exchange, the following collections have been acquired :—489 Austro-Hungarian plants from Kerner; 24 Austrian plants from Fritsche; 198 Portuguese plants from Henriquez; 276 species of Indian plants from J. F. Duthie, Esq.; 870 species of Canadian plants from the Director of the Geological Survey of Canada; and 101 species of plants from Jamaica, from Wm. Fawcett, Esq.

The manuscript records of the distribution of British plants, collected by the late Mr. H. C. Watson, for his ' Cybele Britannica,' have been presented by the Director of the Royal Gardens, Kew; and contributions to the library have been received from Prof. Agardh, J. G. Baker, Esq., C. Bucknall, Esq., Dr. Ernst, H. M. Gepp, Esq., the late Prof. Asa Gray, Dr. King, W. H. Pearson, Esq., the late John Smith, Lieut.-Gen. Strachey, H. J. Veitch, Esq., Dr. Vidal, and the Nederlandische Bot. Verein.

Mr. Henry N. Ridley, an assistant in the Department, obtained an extension of his annual leave of absence to enable him to explore the oceanic island Fernando de Noronha. He secured a large number of specimens illustrating the physical structure and natural history of the island, amounting to 150 specimens of rocks and minerals; 200 species of plants, and 250 species of animals. The whole are being worked out by officers in the Museum, with the view of publication.

NOTICES OF BOOKS.

*Handbook of the Amaryllideæ, including the Alstrœmerieæ and Agaveæ.* By J. G. BAKER, F.R.S. London : Bell & Sons. 8vo, pp. xii. 216. Price 5s.

THE indefatigable Mr. Baker gives us here another of those systematic treatises by which he has done so much to bring into a handy and useable form the scattered observations and descriptions of various writers. To the readers of this Journal this work needs neither introduction nor recommendation, for some of the most useful of Mr. Baker's papers, as well as part of the present volume, have appeared first in these pages.

In the present work, Mr. Baker has " attempted to furnish cultivators and botanists with a compact working handbook, of which the main part consists of characters of genera and species, drawn up from actual specimens." The material on which the descriptions are based includes all the *Amaryllideæ* which have passed through the author's hands during his twenty-three years at Kew, supplemented by an examination of dried specimens of nearly all the species of the order. Their bibliography and pre-Linnean history is left on one side, so that we have here, as Mr. Baker describes it,. " a compact working handbook."

We are sorry that Mr. Baker has followed Bentham and Hooker in adopting *Amaryllideæ* rather than *Amaryllidaceæ* as the name of the order : as a consequence, the first suborder takes the title *Amaryllleæ.* Sixty-one genera and about 670 species are admitted, many of the latter being here first described. Several hybrid Narcissi are here first named, and among the hybrids are placed sundry plants described as species by earlier writers, such as *N. orientalis* L., *N. Macleaii* Lindl., and *N. poculiformis* Salisb. *Leucojum vernum* is not recognised as British. Recent Cape collectors seem still to have done very little for *Gethyllis :* of the nine species described, six have not been collected since the time of Thunberg and Masson. There are interesting and useful notes on the hybrids of *Hippeastrum, Crinum,* and *Nerine.* A new genus, *Stricklandia,* is established for the plant which Mr. Baker had previously referred to *Leperiza* and *Stenomesson,* and which is placed in the ' Genera Plantarum ' under *Phædranassa,* from which "it differs by its monadelphous filaments." The citation of ' Nichols. Dict. Gard.' as the authority for certain names of plants first referred to their genera (in accordance with Bentham and Hooker) by Mr. Nicholson, in his ' Dictionary of Gardening,' will serve to remind folk that this work must not be overlooked in matters relating to synonymy. *Alstrœmeria* is spelt *Alstrœmeria,* we suppose by inadvertence ; although this spelling is adhered to throughout the body of the book ; it is correctly given in the Index.

The volume is, as we have said, a most useful and indeed indispensable handbook to the family of which it treats : and Mr. Baker deserves and will receive the thanks of botanists and cultivators for this last addition to his monographs : to the latter class of workers it will prove perhaps the most useful of any, containing as it does, many genera which are favourites in cultivation. The only improvement we can suggest is typographical. The names of genera, subgenera, and species, are all alike printed in small capitals; and a substitution of thick type for the genera and species would render consultation more easy.

*Nomenclator* '*Floræ Danicæ*' *sive index systematicus et alphabeticus operis, quod* '*Icones Floræ Danicæ*' *inscribitur, cum enumeratione tabularum ordinem temporum habente, adjectis notis criticis.* Auctore JOH. LANGE. Leipsiæ: F. A. Brockhaus. 1887. 4to, pp. viii. 354.

THE publication of the 'Flora Danica' has extended over a period of more than a hundred and twenty years, from the issue of its first fascicle in 1761 to the date of its last supplement in 1883. The extensive changes of synonymy which have taken place during that period will be understood by every one acquainted with botanical literature, and will be fully realised by those among our British botanists who were puzzled by the nomenclature adopted in the last edition of the 'London Catalogue,' and are not unnaturally alarmed at finding that even this extensive revision can hardly be accepted as final or complete. The first part of Prof. Lange's work is devoted to an enumeration in double columns of the plants in their chronological order, the names as published appearing in one column and those now adopted in the other. To this list Prof. Lange has added a number of interesting notes, many of them being corrections of nomenclature and others bearing upon British plants, and deserving the attention of those interested in synonymy. In some cases these notes are of critical value, and we shall take an early opportunity of extracting such for the benefit of our readers, to many of whom Prof. Lange's work will not be readily accessible.

The second index is systematic—the arrangement followed, however, is not that familiar to British botanists—and shows the geographical distribution of each species in Denmark, Sweden, Norway, the Faroes, Iceland, and Greenland, these being the countries represented in the 'Flora Danica.' An alphabetically-arranged list of the species, with references to the Flora, completes the work. It will be seen from this notice that, besides rendering the 'Flora Danica' readily consultable, this 'Nomenclator' contains much information given in a compact form. It is beautifully printed.

NEW BOOKS.—P. PRAHL, 'Kritische Flora der Provinz Schleswig-Holstein' (Kiel, Toeche: 8vo, pp, lxviii. 227).—E. RATHAY, 'Die Geschlechtsverbältnisse der Reben' (Wien, Frick: 8vo, pp. iv. 114, tt. 2).—A. MEUNIER, 'Le Nucléole des Spirogyra' (Lierre, Vaudin: 4to, pp. 79, tt. 2). — M. HOVELACQUE, 'Recherches sur l'appareil vegetatif des Bignoniacées, Rhinanthacées, Orobanchées, et Utriculariées' (Paris, Masson: 8vo, pp. 765).—N. J. C. MULLER, 'Atlas der Holzstructur dargestellt in Microphotographien' (Halle, Knapp: tt. 21, fol.: pp. 110, 8vo.).—P. DIETEL, 'Verzeichnis samtlicher Uredineen' (Leipzig, Serig: 8vo, pp. 48, viii.).— S. A. STEWART & T. H. CORRY, 'A Flora of the North-east of Ireland' (Cambridge, Macmillan: 8vo, xxxiv. 331).

## ARTICLES IN JOURNALS.

*American Naturalist* (May). — M. Rock, 'Guatemala Forests.'— L. Sturtevant, 'History of Garden Vegetables.'

*Annals of Botany* (June). — A. Lister, 'On Plasmodium of *Badhamia utriculatis* and *Brefeldia maxima*' (2 plates).—G. Massee, 'Monograph of *Calostoma* (*C. Berkeleyi*, n. sp. : 1 plate).—Id., 'On the presence of sexual organs in *Æcidium*' (1 plate).—E. H. Acton, 'On formation of sugars in septal glands of *Narcissus.*'—A. Bateson & F. Darwin, 'On a method of studying Geotropism.' — J. R. Vaizey, '*Catherinea lateralis* (*C. anomala* Bryhn), a new British Moss' (1 plate). — F. W. Oliver, 'On the structure, development, and affinities of *Trapella*' (5 plates). — S. H. Vines, 'Systematic position of *Isoetes.*' — J. R. Vaizey, 'Development of root of *Equisetum.*'—M. T. Masters, '*Pinus macrophylla.*'

*Ann. Sciences Nat.* (June). — G. de Saporta, 'Dernières adjonctions à la flore fossile d'Aix en Provence' (10 plates). — P. A. Dangeard, 'Recherches sur les Algues inférieures' (2 plates).— E. Bornet & C. Flahault, 'Revision des Nostocacées hétérocystées.'

*Bot. Centralblatt.* (Nos. 28, 30). — A. Hansgirg, 'Ueber *Bacillus muralis* Tomaschek, nebst Beiträgen zur Kenntniss der Gallertbildungen einiger Spaltalgen.' — A. N. Lundström, 'Ueber die Salixflora der Jenissej-Ufer.' — K. Staback, 'Einige kritische Bemerkungen über *Leptosphæria modesta* Anett.'

*Bot. Gazette* (June). — J. M. Coulter & J. N. Rose, 'Notes on Western Umbelliferæ' (*Eryngium armatum, E. Vaseyi, E. floridanum, Peucedanum Martindalei, P. Donnellii, P. californicum, P. Vaseyi, Selinum Grayi, S. Dawsoni, Coleopleurum maritimum,* spp. nn.). — C. Robertson, 'Zygomorphy and its causes.' — A. F. Foerste, 'Notes on structures adapted to cross-fertilisation' (1 plate).— F. H. Knowlton, 'A new fossil *Chara*' (*C. compressa*, n. sp.).— D. H. Campbell, 'The paraffin-imbedding process.'—Nomenclature of *Disporum.*

*Bot. Zeitung* (June 22).—F. Kienitz-Gerloff, 'Die Gonidien von *Gymnosporangium clavariæforme*' (1 plate). — H. de Vries, 'Ueber eine neue Anwendung der plasmolytischen Methode.' — (June 29). A. Fixler, 'Glycose als Reservestoff der Laubhölzer.' — (July 6). A. Möller, 'Ueber die sogenannten Spermatien der Ascomyceten.' —(July 13, 20). E. Zacharias, 'Ueber Strasburger's Schrift, "Kern und Zelltheilung in Pflanzenreiche."' — (July 27). J. Wortmann, 'Zur Beurtheilung der Krümmungserscheinungen der Pflanzen.'

*Bull. Torrey Bot. Club* (July). — Plants collected by Dr. Rusby in S. America.— J. Macoun, 'Bryological Notes.'— N. L. Britton, Nomenclature of *Disporum.*—E. E. Sterns, 'A suggestion concerning *Smilax herbacea.*'

*Flora* (May 1).—J. Müller, 'Lichenologische Beiträge.'—(May 11, 21).—A. Hansgirg, 'Ueber die Gattungen *Herposteiron* Näg. & *Aphanochæte* Berth., non A. Br.'—(May 28–June 21). T. Wenzig, 'Die Gattung *Spiræa*.'—(May 21, 28). E. Schulz, 'Ueber Reservestoffe in immergrünen Blattern unter besonderer Berücksichtigung des Gerbstoffes.' — (June 11). A. Hansgirg, 'Ueber die aerophytischen Arten der Gattungen *Hormidium, Schizogonium* & *Hormiscia*.'

*Gardeners' Chronicle* (July 7). — *Megaclinium scaberulum* Rolfe, n. sp. — W. G. Smith, '*Peronospora Ficariæ*' (fig. 2).—(July 14). *Thunia candidissima* Rchb. f., *Epidendrum auriculigerum* Rchb. f., *Angræcum tridactylites* Rolfe, spp. nn. — *Pinus Sabiniana* (fig. 4).—(July 21). *Helichrysum devium* J. Y. Johnson, n. sp. — *Ostrowskya magnifica* (fig. 6). — Leaf-cutting of *Sanseviera guineensis* (fig. 7).—(July 28). *Megaclinium oxyodon* Rchb. f., *Aëranthus ophioplectron* Rchb. f., spp. nn.—W. G. Smith, *Puccinia Liliacearum* (fig. 11).

*Journal de Botanique* (July 1). — E. Bureau, 'Sur un Figuier à fruits souterrains' (*F. Ti-Koua*, n. sp. : 1 plate). — N. Patouillard, 'Fragments Mycologiques.' — E. Roze, 'Le Jardin des Plantes en 1636.' — L. Morot, Notice of J. E. Planchon (1823–1888). — (July 15). J. Costantin, 'Observations critiques sur les Champignons Hétérobasidiés.'— —. Masclef, 'Géographie Botanique du Nord de la France.'

*Midland Naturalist* (July).—W. B. Grove & J. E. Bagnall, 'The Fungi of Warwickshire.' — W. Mathews, 'History of the County Botany of Worcester.'

*Notarisia* (July). — P. F. Reinsch, 'Familiæ Polyedriearum Monographia' (*Thamniastrum, Closteridium*, n. g. : 5 plates).— G. B. De Toni, 'Notizie sopra due specie del genere *Treutepohlia*.' —'Algæ novæ.'

*Nuov. Giornale Bot. Ital.* (July).—U. Martelli, 'Sopra una forma singolare di *Agaricus*.' — Id., 'Flora di Massaua.' — L. Macchiati, 'Viti di Arezzo.'—G. Arcangeli, 'Sul Kefir.'—U. Mantelli, 'Webb, 'Fragmenta Florulæ Æthiopico-ægyptiaceæ : continuazione' (*Justicia æthiopica*, sp. n.).— Id., *Phyllosticta Bellunensis* & *P. Venziana*, spp. nn.—Id., 'Sulla *Quercus macedonica*.'—Id., 'Dimorfismo fiorale di alcune specie di *Æsculus*.' — L. Macchiati, 'Diatomacee.' — Id., 'Contribuzione alla Flora del Gesso.' — C. Rossetti, 'Appunti di Epaticologia Toscana.' — C. Boccaccini, 'Sulla resistenza alla stagione e sulla precocità.' –- O. Sommier, 'Una Genziana nuova per l'Europa' (*G. barbata*).

*Œsterr. Bot. Zeitschrift* (July). — L. Simonkai, 'Bemerkungen zur Flora von Ungarn.' — E. Woloszczak, '*Salix bifax* & *S. Mariana*.' A. Hansgirg, 'Kellerbacterien.'—A. v. Degen, '*Botrychium virginianum* im südlichsten Ungarn.' — F. Krasan, 'Reciproke Culturversuche.' — J. Murr, 'Neue Funde in Tirol.' — E. Formánek, 'Zur Flora von Bosnien.'

*Scottish Naturalist* (July). — J. Stirton, 'Lichens' (*Cathisinia*, gen. nov.).

# RECENT TENDENCIES IN AMERICAN BOTANICAL NOMENCLATURE.

## By the Editor.

The vexed questions connected with botanical nomenclature have been, for the most part, settled in Europe by the adoption of the "Laws" formulated by M. DeCandolle. These Laws were discussed by Dr. Asa Gray with his wonted care and ability, and in the main were approved and followed by him, and by his fellow-worker, Dr. Sereno Watson. But there has arisen a new school of able and active botanists, who aim at introducing a system of nomenclature which, as it appears to us, can only result in the introduction of fresh elements of confusion, and must certainly in any case result in an increased synonymy—a contingency far from desirable, except in cases, such as the restitution of Salisbury's genus *Castalia*,\* where it is unavoidable. Even among the very limited number of species comprised in our own Flora, many sweeping changes of name have been made in the last few years, since this question of priority came to the front; the system upon which these alterations have been made has been accepted by British botanists, not without some natural grumbling at the temporary inconvenience caused, because, once established, it can undergo no further change. It is also that laid down in the Laws (art. 48) as follows:—

"For the indication of the name or names of any group to be accurate and complete, it is necessary to quote the author who first published the name or combination of names in question."

This, however, is not the method which commends itself to the new American school, as expounded in several recent papers, and especially in the preface to 'A Preliminary Catalogue of Anthophyta and Pteridophyta reported as growing spontaneously within one hundred miles of New York City,' for a copy of which I am indebted to my almost namesake, Dr. N. L. Britton. The nomenclature of this Catalogue has been entrusted to a sub-committee, consisting of Dr. Britton and Messrs. Stern and Poggenburg; and they have laid down the necessity of following the custom of zoologists in regarding the earliest specific name of any species as absolutely unchangeable, no matter to what genus it may be transferred, excepting, of course, the cases in which that name may have been preoccupied in the genus to which transference is made. It will be best to quote their own statement of the position they have taken up:—

"In the case of nearly every plant it is possible to ascertain positively who first named it in accordance with the Linnæan binomial system. The original author may have failed to refer it to the proper genus, either ignorantly or through a praiseworthy unwillingness to found new genera, except on the strongest grounds.

---

\* Journ. Bot. 1888, p. 8.

But whatever specific or varietal term may have first been applied to the plant belongs to it individually, and the most profound subsequent knowledge of it or of its relations to other plants cannot warrant any essential change in this portion of its name, always having the cases governed by the rule that two species of the same genus must not bear the same name, or by the further rule that the generic and specific names must not be identical. Transfers from genus to genus, and alterations in rank as regards species or variety, must not (except as to gender) in any wise affect the trivial name, which is held to be absolutely fixed by the first publication, so that even the author is not at liberty thereafter to modify it in any way." " The present writers," we read further on, " are convinced that in the not distant future the law they have taken as their guide will be generally accepted as the only one that promises a reasonable fixity of botanical names."

We venture to differ from Dr. Britton and his colleagues, not only with regard to the practice they have adopted, but as to the results which they expect from it. They see that "a considerable and undesirable increase in the number of synonyms and more or less confusion " will result, but they think " these consequences are but temporary, and," comparatively, " of no real moment." This we do not understand: the confusion may, indeed, be but temporary ; but surely the synonyms will remain.

The sub-committee propose to adopt the plan of citing in parenthesis the original author of a name, " in all cases where his plant stands in a genus of rank other than that to which he referred it, and where, consequently, the name as a whole must be credited to some later authority." We open the list at random, and find, under *Pycnanthemum*,

"flexuosum (Walt.).   (P. linifolium, Pursh)."
"Virginicum (L.).   (P. lanceolatum, Pursh)."

Walter described the first plant as an *Origanum (O. flexuosum)*, and Linnæus called the second *Thymus virginicum;* Pursh placed them under their proper genus, and Gray, Chapman, and all authors of note have followed him. It is reserved for Messrs. Britton, Stern, and Poggenburg to "restore " the Linnean names in a genus founded more than thirty years after Linnæus's death.

But this is not all. How are these names—which the sub-committee consider old, but which to us appear new, dating only from the publication of this Catalogue—to be cited in future ? They are anxious to give "due and equal credit to the authors of the original specific name and the present accepted binomial." We have always held this question of "credit" to be purely sentimental; but this in passing. How, then, are the names which we have cited to be quoted from this Catalogue ? " *Pycnanthemum virginicum* (L.) Britton, Stern, & Poggenb." seems somewhat cumbrous ; but we see no alternative—no other way of following the desire of the committee that " due and equal credit " shall be given to the authors of " the present accepted binomial." For a time, indeed, it will be almost or quite necessary to cite also the hitherto accepted synonym, as the sub-committee has done here: and instead of "Pycnanthemum linifolium, Pursh," we shall have to write "Pycnanthemum flexuosum (Walt.), Britton, Stern, & Poggenb. (P. linifolium, Pursh)": ten words instead of three!

A writer, who perversely signs himself "N. or M.," contributes to the 'Botanical Gazette' for June an article, "What shall be done with our *Prosartes?*," in which he points out that Don's genus of that name having been included by Bentham and Hooker in *Disporum*, it is doubtful who should be cited as the authority for the species when placed in that genus. It is not necessary to follow his somewhat lengthy contention, which I only refer to in consequence of Dr. Britton's pronouncement upon it:—"He entirely overlooks the very simple and advantageous method of citing the author of the original name in a parenthesis, thus giving due credit to all concerned." How is this the case? If "credit" is to be assigned, surely Bentham and Hooker may demand their share. Don can scarcely claim much "credit," for, although he founded *Prosartes*, the authors just cited do not allow its right to generic rank. And again, if the "simple and advantageous method" advocated by Dr. Britton be adopted, is *Prosartes Menziesii* Don, for example, always to be printed "*Disporum Menziesii* (Don)," without any other authority? If so, why does not the sub-committee print "*Cimicifuga racemosa* (L.)," instead of "*Cimicifuga racemosa* (L.), Nutt."—the latter being the way in which this and numberless other names are printed in the Catalogue?

Another method of needlessly increasing our synonymy is justified by Prof. E. L. Greene,* the well-known worker on the flora of the Pacific Coast, who animadverts, justly enough, on Salisbury's having given new specific names to the species of *Nymphæa*, placed by him in his genus *Castalia*. By way of putting this right, Prof. Greene promptly adds a fresh combination of names. Salisbury substituted *Castalia pudica* for *Nymphæa odorata* Dryand.†; his change of specific name, we repeat, was regrettable, but, having been made, it must stand. Prof. Greene, however, prefers to write *C. odorata*, and this is cited by the sub-committee of the Torrey Club as "*Castalia odorata* (Dryand.), Greene."

The plant which Gray, Chapman, Sereno Watson, and all recent authors have called *Adlumia cirrhosa* Raf. has now a new name. Prof. Greene has apparently communicated to the sub-committee the fact that Aiton called it *Fumaria fungosa*. We are sure that he would not consider this perfectly familiar piece of synonymy a discovery of note; but so anxious are the sub-committee to give "due credit," that the new name runs "*Adlumia fungosa* (Ait.), Greene in litt. (*A. cirrhosa*, Raf.)." This is pretty well; but it is easy to imagine a still more scrupulous person, in his anxiety to give "due credit" to those who have done science the service of printing this new name, citing it as "*Adlumia fungosa* (Ait.), Greene ex Britton, Stern, & Poggenb. (*A. cirrhosa*, Raf.)."

It will be evident, from the examples given, that it is open to

---

* Bull. Torrey Bot. Club, xv. 85.    † See Journ. Bot. 1888, 9.

any one with a turn for antiquarian research to introduce fresh elements of confusion and complication into botanical nomenclature. It may be urged that we ourselves have on more than one occasion seemed to endorse this course of action ; but such is not the case. We believe that anything like finality in nomenclature can only be obtained by adopting the oldest generic name, and for this we cannot go beyond Linné's ' Genera Plantarum ' (1737). The oldest specific name employed under that genus must follow this, and then that of the author who first published this combination of names : the long commentary on Art. 48 given in the 'Laws'(p. 30) seems to us to explain and conclusively establish this as the only rational position. We are sorry that such earnest and active workers as those who form the "sub-committee on nomenclature" of the Torrey Club should be bent upon a course which, if followed to any extent, will, in our judgment, render the already tangled skein of synonymy even more complicated than it is at present.

The sub-committee have an important supporter in Prof. E. L. Greene, who, in 'Pittonia' (No. 4, Jan.–June, 1888), gives a long and interesting review of the New York Catalogue, in the course of which he emphasises the importance of adopting what he calls "the philosophical and ethical principles of biological nomenclature." With many of his remarks we entirely agree : as, for instance, in urging conformity to the original spelling of a name, and in his consequent objection to the substitution of *Heleocharis* for *Eleocharis*, and *Dicentra* for *Diclytra*. And we concur with him in regretting that the sub-committee should have confined themselves to specific names, leaving those of the genera untouched. Prof. Greene points out that *Hicorius* (Rafinesque, Fl. Ludov. 107 (1817) antedates *Carya* (Nuttall, Genera, ii. 220 (1818): and, had the sub-committee recognised this, they would have avoided the introduction of a new and peculiarly irritating synonymy. As a good example of the confusion resulting from the adoption of the plan proposed by them, the synonymy from the Catalogue may be cited :—

> "Carya, Nutt.
> > alba (L.).   (*C. tomentosa* Nutt.).
> > glabra (Mill.), Torr.   (*C. porcina* Nutt.).
> > microcarpa, Nutt.
> > minima (Marsh.) (*C. amara* Nutt.).
> > ovata (Mill.).   (*C. alba* Nutt.).
> > sulcata (Willd.), Nutt."

Assuming that these names come into general use, there will be special inconvenience in this case, as the trees to which they are applied are so well known. Such names as *Carya alba* are met with in popular or semi-popular as well as in strictly botanical works, and, as is usual in such cases, are cited without the authority being attached. How are we to convey to people's minds the fact that (if the Catalogue be followed) a casual reference to *C. alba* after April, 1888, will mean one tree, while before that date it meant another ? Surely here at least the generally accepted specific names should have been allowed to stand.

But *is Hicorius* the name to be adopted? Bentham and Hooker,[*]
with a carelessness as to details of nomenclature which is too often
apparent in their otherwise admirable work, follow Endlicher[†] in
citing "*Hicorias* et *Scorias* Rafin." as synonyms of *Carya* (*Hicorias*
for *Hicorius* is their own mis-spelling: for *Scorias* they follow
Endlicher). If this be so, and there seems no reason to doubt it,
surely *Scorias* (or rather *Scoria*) is the name which must stand.
Rafinesque says :—

"*Scoria tomentosa, mucronata, alba, pyriformis, globosa ;* ce sont
les *Juglans alba* L., *tomentosa, mucronata* Mich., &c. The Hiccory."[‡]

This is Desvaux's translation of Rafinesque's 'Prospectus,' published in the 'Medical Repository of New York,' v. 350: the original
we have not seen, but Mr. Daydon Jackson has verified the reference;
the names must stand :—

SCORIA Raf., in Med. Repos. of New York, v. 352 (1808).
HICORIUS Raf., Fl. Ludov. 107 (1817).
CARYA Nutt., Gen. ii. 220 (1818).

, We commend this paper of Rafinesque's to the notice of our
American friends, who have apparently overlooked it. They will
see from it that there is no need to apologise for *Echinocystis echinata*
as a "tautological name," for it will be superseded by Rafinesque's
*Micrampelis echinata;* and other interesting points are raised by its
perusal.

Prof. Greene advocates the restoration of Adanson's name *Tissa*
(1763) to replace *Lepigonum* (Fries, 1817). Adanson gave two
names to what are now regarded as species of *Lepigonum*, distinguishing *Tissa* as having five stamens, while *Buda* has ten. The
two occur on the same page,[§] so that neither can claim priority,
and we think *Buda* should be retained, as this has been already
restored by Dumortier,[||] who writes :—

"BUDA.—*B.* et *Tissa*, Adans.
rubra = Arenaria L.
marina = Arenaria L.
media = Arenaria L."[§]

British botanists, please take note !

We have been somewhat led away from the question of the
permanency of the specific name, as to which there is more to be
said. What has become of *Myosotis scorpioides* L.? It is of course
true that this name now represents nearly a dozen species, and that
Linnæus's varietal name *arvensis* became Hill's species, which itself
includes more than Lehmann's restricted *arvensis*, which equals
Link's *M. intermedia;* but should not the sacrosanct specific title be
retained for some one of the forms? Linnæus distinguished and
named two forms—*palustris* and *arvensis :* why does not the sub-committee in some way "credit" him with this? If it is right to print

---

* Gen. Plant. iii. 398.     † Gen. 1126.     ‡ Desvaux, Journ. de Bot. ii. 170.
§ Fam. des Plantes, ii. 507.     || Flor. Belg. 110 (1827),

" Sparganium androcladum (Engelm.), Morong. (*S. simplex* Huds.,
var. *androcladum* Engelm.),"
why not

" Myosotis palustris (L.), Reih. (*M. scorpioides* L. var. *palustris* L.)."

But instead of this, the sub-committee print, " M. palustris, Relh."

The sub-committee cite " Veronica Anagallis, L." and " Alisma
Plantago, L." This is not how Linnæus originally published these
names : he printed in the first, second, and third edition of the
' Species' " *Anagall.* ▽ " and " *Plantago* ▽," and these are correctly
given at length by Mr. Pryor* as " *Veronica Anagallis-aquatica* L."
and " *Alisma Plantago-aquatica* L." The names as now abbreviated
seem to have been first used by Scopoli.†

Sir W. J. Hooker published (Ic. Pl. 547) a plant which he named
*Podocarpus Dieffenbachii.* Shortly afterwards, he discovered that the
supposed *Podocarpus* was a *Veronica*, identical with one which he
had figured (t. 580) as *V. tetragona.* Is the oldest specific name to
be maintained in this case, on account of the " credit" due to the
mistaking a Scrophulariad for a Conifer ?

Yet another terror is foreshadowed for us by Prof. Greene in the
most recent issue of ' Pittonia.' He has discovered that a large
number of binary plant-names, which are familiar to us now,
always presumed to be of Linnæan origin, and always credited
to Linnæus," are employed by older writers ; and he gives a list of
forty-eight such names, to be found in Ray's Catalogue of Cambridge Plants. As far as we can understand, Prof. Greene proposes
that such should be assigned to their " true and original author" ;
and he gives a list of them, beginning :—

" *Allium ursinum*, Fuchs, Historia Plantarum [Stirpium] 739
(1542).
*Alsine media* Camerarius, Hortus Medicus et Philosophicus, 11
(1558) [1588] "

and so on. No one knows better than Prof. Greene that these
names being composed of two words only is a mere accident. Many
cited by Ray are of one word only, others are of three, four, or as
many as was considered necessary for distinguishing the species
from its neighbours. The binomial method, the reduction of
nomenclature to a system, is one of the greatest of the reforms
introduced by Linnæus, and the attempt to deprive him of it is not
likely to be sanctioned by botanists.

It is consolatory to reflect that such tendencies as those to which
we have referred will be counteracted by Mr. Daydon Jackson's great
' Nomenclator,' which is steadily approaching completion, and which,
based as it is on the DeCandollean ' Laws,' will take its place as a
standard of synonymy. None the less is it necessary to enter a
protest against the revolutionary tendencies to which we have
called attention, and to express a hope that the chaotic confusion
which must result from their indulgence may yet be obviated by
considerations of common sense.

---

* Fl. Herts, pp. 306, 383.    † Fl. Carn. ed. 2, i. 13 and 266.

# MOSSES OF MADAGASCAR.

## By C. H. Wright.

In the following list have been brought together the mosses at present recorded from the Island of Madagascar. The 224 species and 8 varieties here enumerated must, from the nature of the island, form but a small proportion of the total number existing there. Of these, 23, collected by Borgen and Borchgrevink (Norwegian Missionaries), were described by Dr. E. Hampe in 'Linnæa,' vol. xxxviii. pp. 207–222. In the 'Reliquiæ Rutenbergianæ,' part iii., Dr. Karl Muller and A. Geheeb enumerated 54 species, of which 38 were new. Bescherelle, in the 'Revue Bryologique,' 1877, p. 15, mentions 10 new species collected in Nossi Bé by Boivin, and describes them in the same Journal for 1880, pp. 17 and 33.

As many of the specimens sent home are marked "Madagascar," without other indication of locality, it is impossible at present to give any satisfactory account of the distribution of the species over the island. Of the two collectors, through whom most of our knowledge is due, the one, Borgen, collected principally (if not entirely) in the Province of Imerina, especially on Mt. Ankaratra; the other, Hildebrandt, in the same province, and also in Nossi Bé, an island off the north-west coast.

The arrangement followed in this list is that of the supplementary part of Jæger's 'Adumbratio Floræ Muscorum.' The number of species in each section may be stated thus:—

|  |  |  |  |
|---|---|---|---|
| Sphagnaceæ | . . | 3 species. |  |
| Acrocarpi | . . . | 119 ,, | 1 variety. |
| Cladocarpi | . . . | 2 ,, |  |
| Pleurocarpi | . . | 100 ,, | 7 varieties. |
| Total . . . | | 224 species. | 8 varieties. |

*Sphagnum madagassum* C. Müll. East Imerina (*Hildebrandt* 2106).

*S. obtusiusculum* Lindb.

*S. Rutenbergii* C. Müll. Ambatondrazaka.

*Anœctangium impressum* Hampe, Linnæa, xxxviii. p. 208 (*Borgen*).

*A. Mariei* Besch. Nossi Bé.

*Weisia apiculata* Kiær. Antananarivo (*Borgen* 11).

*Trematodon pallidus* C. Müll. Nossi Bé.

*T. reticulatus* C. Müll. Rel. Rut. iii. 205. Ambatondrazaka.

*Dicranella limosa* Besch. Amboripossi (*Hildebrandt* 2101). Nossi Bé (*Boivin*).

*D. minuta* Hampe.

*Dicranum dichotomum* Brid. (*D. Boryanum* Schwgr.). Antananarivo (*Meller*).

*D. scopareolum* C. Müll. Alamazantra (*Borgen* 7).

*Leucoloma* (*Pœcilophyllum*) *acutum* Mitt.

*L. arbusculum* C. Müll.

*L. bifidum* Brid. (*Dicranum Commersonianum* C. Müll.

*L. cuniæfolium* Hampe, Rel. Rut. iii. 206. Ambatondrazaka.

*L. dichelymoides* C. Müll. Linnæa, xl. 240. Ambatondrazaka; Mt. Ankaratra (*Borgen* 6).

*L. holomitrioides* C. Müll.

*L. pumilum* C. Müll. Rel. Rut. iii. 206. Maritandrano.

*L. rectum* Lac.

*L. Rutenbergii* C. Müll. Rel. Rut. iii. 206. Ambatondrazaka, Alamazantra.

*L. Sanctæ-Mariæ* Besch.

*L. sinuosum* (Brid.).

*L. squarrosulum* C. Mull. Rel. Rut. iii. 206. Vondruzona.

*L. subchrysobasilare* C. Müll. Mt. Ankaratra (*Borgen* 10).

*L. Thuretii* Besch.

*L. traustum* Hampe, Linnæa, xxxviii. 209. Alamazantraskoven (*Borchgrevink*).

*Campylopus arcuatus* Lac.

*C. capillaceus* (Brid.).

*C. madacassus* Besch. *Bernier.*

*Holomitrium flagellare* C. Müll.

*H. Hildebrandti* C. Müll. East Imerina (*Hildebrandt* 2099).

*Leucobryum Hildebrandti* C. Müll. Andrangoloaka (*Hildebrandt* 2122).

*L.* (*Pegophyllum*) *læve* Mitt. *Meller. Pool.*

*L. madagassum* C. Müll. Rel. Rut. iii. 204. Ambatondrazaka.

*L. selaginoides* C. Müll. Mt. Ankaratra (*Borgen* 3). East Imerina (*Hildebrandt* 2108).

*Ochobryum Rutenbergii* C. Müll. Rel. Rut. iii. 204. Ambatondrazaka.

*Schistomitrium acutiflorum* Mitt. Antananarivo (*Meller*).

*Leucophanes Hildebrandti* C. Müll. Ambatondrazaka.

*Octoblepharum albidum* Hedw. Mohambo (*Gerard* 8). *Borgen* 4.

*Fissidens comorensis* C. Müll. Nossi Bé.

*F. ferrugineus* C. Müll. Antananarivo.

*F. flavolimbatus* Besch. Nossi Bé.

*F. leucocinctus* Hampe, Rel. Rut. iii. 222.

*F. Nossianus* Besch. Nossi Bé.

*F. obsoletidens* C. Müll. Revue Bryologique, 1877, p. 15. Nossi Bé (*Boivin*).

*Conomitrium reflexus* Hampe (Rel. Rut. iii. 221.—*Fissidens*).

*C.* (*Reticularia*) *Mariei* Besch. Nossi Bé.

*Seligeria pallidiseta* C. Müll.

*Garckea Bescherellei* C. Müll. Nossi Bé.

*G. Hildebrandti* C. Müll. ·Nossi Bé (*Hildebrandt* 2061).

*Leptotrichum leptorhynchum* Jæg. Adum. i. 228.

*L. pallidum* Hampe.

*Hyophila Potieri* Besch. Nossi Bé.

*Streptopogon calymperes* C. Müll. Rel. Rut. iii. 207.

*S. Hildebrandti* C. Müll. Andrangoloaka (*Hildebrandt* 2095)

*S. Parkeri* Mitt. Central Madagasear.

*S. Rutenbergii* C. Müll. Rel. Rut. iii. 207. Ambatondrazaka.

*Barbula subrevoluta* Hampe, Linnæa, xxxviii. 218. (*Borgen* 138).

*Syrrhopodon leptodontioides* Besch. Pervillé.

*S. litoralis* C. Müll. Nossi Bé (*Hildebrandt* 2057).

*S. microbolax* C. Müll. Nossi Bé (*Boivin*). Rev. Bry. 1877, 15.

*S. Nossi-Beanus* Besch. Nossi Bé.

*S. Seignaci* Besch.

*Calymperes Borgeni* Kiær. (*Borgen*).

*C. decolorans* C. Müll. Nossi Bé.

*C. Isleanum* Besch. Nossi Bé.

*C. Mariei* Besch. Nossi Bé.

*C. Nossi-Combæ* Besch. Nossi Bé.

*C. Sanctæ-Mariæ* Besch.

*Zygodon madacassus* Hb. Schimp. Pervillé. N.E. Madagascar.

*Schlotheimia excorrugata* C. Müll. Andrangoloaka (*Hildebrandt* 2094).

*S. linealis* C. Müll. Rel. Rut. iii. 208. Ambatondrazaka.

*S. microcarpa* Schimp. Ambatondrazaka.

*S. microphylla* Besch.(*Bernier* 32).

*S. Nossi-Beana* C. Müll. Rev. Bry. 1877, 15. Nossi Bé (*Boivin*).

*S. tenuiseta* C. Müll. Rel. Rut. iii. 208. Ambatondrazaka.

*S. trypanoclada* Schimp. Maritandrano.

*Macromitrium calocalyx* C. Müll. Rel. Rut. iii. 208. Ambatondrazaka.

*M. coarctatum* Schimp. *Pervillé.*

*M. cylindricum* Schimp. *Pervillé.*

*M. longisetum* Schimp. *Pervillé.*

*M. Pervillei* Schimp. *Pervillé.*

*M. rhizomatorum* Schimp. Nossi Bé (*Pervillé*).

*M. subtortum* Schwægr.

*M. urceolatum* C. Müll. Rel. Rut. iii. 208. Ambatondrazaka.

*Ulota fulva* Brid. (*Pervillé*).

*Splachnobryum Boivini* C. Müll. Rev. Bry. 1877, 15. Nossi Bé (*Boivin*).

*S. inundatum* C. Müll. *l. c.* Nossi Bé (*Boivin*).

*Dissodon madagassum* C. Müll.

*Tayloria indica* Mitt. (*Pervillé*).

*Funaria subleptopoda* Hampe, Linnæa, xxxviii. 207. (*Borgen*).

*Bartramia* (*Philonotula*) *byssiformis* C. Müll. Nossi Bé (*Hildebrandt* 2065).

*Philonotis obtusata* C. Müll. (*Borgen* 19).

*P. sparsifolia* Hampe. (*Borgen* 134). E. Imerina (*Hildebrandt* 2035).

*P. tenuicula* Hampe. *Borgen* 20.

*Brachymenium Borgenianum* Hampe. Antananarivo (*Borgen* 14, 136). (*Hildebrandt* 2119, 2139).

*B. madagassum* Hampe. (*Borgen* 137).

*Bryum* (*Apalodictyon*) *naropyxis* C. Müll. Nossi Bé (*Hildebrandt* 2059).

*B.* (*Apalodictyon*) *penicillatum* Hampe. (*Borgen*).

*B.* (*Dicranobryum*) *Baroni* Mitt. (*Baron*).

*B.* (*Dicranobryum*) *philonotula* Hampe. Andragoloaka (*Hildebrandt* 2136).

*B.* (*Dicranobryum*) *roseum* Schreb. (*Baron* 5053).

*B.* (*Orthocarpus*) *radiale* C. Müll. (*Hildebrandt* 2115).—Var. *leptoradiale* C. Müll. (*Hildebrandt* 2118).

*B. alpinulum* Besch. Nossi Bé.

*B. Mariei* Besch. Nossi Bé.

*B. pendulinum* Hampe, Linnæa, xxxviii. 214.

*B. purpureo-nigrum* Duby.

*B. semilimbatum* Kiær. Mt. Ankaratra (*Borgen* 13).

*B. subargenteum* Hampe. (*Borgen* 131).

*Mnium Hildebrandti* C. Müll. E. Imerina (*Hildebrandt* 2109).

*M. madagascariense* Kiær. Mt. Ankaratra (*Borgen* 16).

*Rhizogonium Pervilleanum* Besch. (*Pervillé* 821).

*R. spiniforme* Brid. Vondruzona, Ambatondrazaka, Maritandrano, Ambohimena. *Baron* 230; *Hildebrandt* 2116 a ; *Borgen* 17.

*Pogonatum Hildebrandti* C. Müll. Andrangoloaka (*Hildebrandt* 2133).

*P. madagassum* Hampe, Linnæa, xxxviii. 216.

*Polytrichum* (*Aloidella*) *afroaloides* C. Müll. Rel. Rut. iii. 204. Ambatondrazaka.

*P. juniperellum* C. Müll. Rel. Rut. iii. 205. Ambohimara, Ambatomainty.

*P. longissimum* C. Müll. Andrangoloaka (*Hildebrandt* 2125).

*P.* (*Aloidella*) *obtusatulum* C. Müll. Amboripossi (*Hildebrandt* 2105) ; Ambohimara.

*P. robustum* C. Müll. Andrangoloaka (*Hildebrandt* 2126).

*P.* (*Catharinella*) *Rutenbergii* C. Müll. Rel. Rut. iii. 205. Ambatondrazaka.

*Cryphæa madagassa* C. Müll. Rel. Rut. iii. 210.

*C. Rutenbergii* C. Müll. *l. c.* Ambatondrazaka.

*Lasia Borgeni* C. Müll. Mt. Ankaratra (*Borgen* 24).

*Pterogoniella madagascariensis* (Brid.). *Neckera* C. Müll.

*Pterogonium madagassum* C. Müll. Mt. Ankaratra (*Borgen* 25).

*Leucodon Rutenbergii* C. Müll. Rel. Rut. iii. 210. Ambatondrazaka.

*Rutenbergia madagassa* Geheeb et Hampe, Rel. Rut. iii. 210, t. ii. c. Ambatondrazaka.

*Jægerina solitaria* C. Müll. var. *Nossi-Beana* Besch. Nossi Bé.

*Endotrichum patentissimum* (Hampe) Jæg. *Pilotrichum* Hampe, Linnæa, xxviii. 219. (*Borchgrevink*); Ambatondrazaka; (*Pool*).—Var. *tenue* Müll. et Geheeb, Rel. Rut. iii. 209. Ambatondrazaka.

*Papillaria Ankeriensis* Kiær. Ankerimandinikja (*Borgen*).

*P. floribundula* C. Müll. Mt. Ankaratra (*Borgen* 32).

*P. macrotis* C. Müll. Andrangoloaka (*Hildebrandt* 2103).

*P. perichætialis* (Hampe), Jæg. Alamazantrakoven (*Borchgrevink*).

*P. Rutenbergii* C. Müll. Rel. Rut. iii. 209. Ambatondrazaka.

*Pilotrichella ankaratrensis* Kiær. Mt. Ankaratra (*Borgen* 29).

*P. biformis* (Hampe) Jæg. (*Pilotrichum* Hampe). Ambatondrazaka. Alamazantroskoven (*Borgen*).

*P. Hampeana* Kiær. Mt. Ankaratra (*Borgen* 30).

*P. imbricatula* C. Müll. Rel. Rut. iii. 209. Ambatondrazaka ; S. Betsileo (*Hildebrandt* 2090).

*P. (Orthostichella) obovata* Kiær. Mt. Ankaratra (*Borgen* 28).

*P. subimbricata* Hampe, Linnæa, xxxviii. 216. Alamarantra (*Borchgrevink*).

*Aërobryum subpiligerum* (Hampe) Jæg. Ambatondrazaka (*Borgen*).

*Meteorium imbricatum* Schw. Antananarivo (*Meller*); (*Pool*).

*M. involutifolium* Mitt. Antananarivo(*Meller*); Tanala (*Langley Kitchen*); Central Madagascar (*Pool, Parker*).

*M. involutum* Mitt. (*Pool*).

*M. serrulatum* (Beauv.). (*Pool*).

*M. sylvaticum* Mitt. Antananarivo (*Meller*).

*Trachypus Rutenbergii* C. Müll. Rel. Rut. iii. 209. Ambatondrazaka.

*T. serrulatus* Besch. Mt. Ankaratra (*Borgen* 37); (*Pool*).

*Pilotrichum limbatum* Hampe, Linnæa, xxxviii. 220. Alamazantroskoven (*Borchgrevink*).

*Neckera Boivini* C. Müll. Rev. Bry. 1877, 15. Nossi Bé (*Boivin*).

*N. Borgeniana* Kiær. Mt. Ankaratra (*Borgen* 31, 144); East Imerina (*Hildebrandt* 2085) ; Ambatondrazaka.

*N. madagassa* Besch.

*N. Pervilleana* Besch. Nossi Bé.

*Homalia ankaratrensis* Kiær. Mt. Ankaratra (*Borgen* 38).

*Porotrichum anisopleuron* Kiær. Mt. Ankaratra (*Borgen* 42).

*P. madagassum* Kiær. *Borgen* 146.

*P. subsecundum* Kiær. Mt. Ankaratra (*Borgen* 41).

*P. tamarascinum* (Hampe) Müll. et Geheeb, Rel. Rut. iii. 209. (*Pilotrichum* Hampe, Linnæa, xxxviii. 219). Ambatondrazaka (*Borchgrevink*).

*Daltonia elegantula* Schimp. (*Pervillé*).

*D. madacassa* Schimp. (*Pervillé*).

*Lepidopilum parvulum* Schimp. (*Pervillé*).

*Hookeria lacerans* C. Müll. var. *Nossiana* Besch. Nossi Comba. —Var. *aquilenta* Besch. Nossi Comba.

*Chætomitrium cataractarum* Besch. Nossi Bé.

*Pterigynandrum madagassum* C.
Müll. Rel. Rut. iii. 211. Am-
batomainty.
*Pseudoleskea subatrovirens* Sb.(Jæg.
Adum. ii. 739). (*P. subfila-
mentosa* Kiær.). *Borgen* 47, 145.
*Thuidium Kiæri* C. Müll.; Jæg.
Adum. ii. 740. (*Borgen* 126).
*T. subscissum* C. Müll. Rev. Bry.
1877, 15. Nossi Bé (*Boivin;
Marie*); *Hildebrandt* 2062.
*Rhegmatodon madagassus* Geheeb,
Rel. Rut. iii. 211. Ambaton-
drazaka.
*R. secundus* Kiær. Mt. Ankara-
tra (*Borgen*).
*Leptohymenium fabronioides* Besch.
(*Pterigynandrum* C. Müll.).
Nossi Comba.
*Entodon madagassus* C. Müll. Rel.
Rut. iii. 211. Ambatondra-
zaka (*Borgen* 100).
*E. Rutenbergii* C. Müll. *l. c.* Am-
batondrazaka. Mt. Ankaratra
(*Borgen* 101).
*Lindigia Hildebrandti* C. Müll.
Andrangoloaka (*Hildebrandt
2112*).
*Brachythecium atratheca* Duby.
(*Robillard. Borgen*).
*B. Borgeni* (Hampe) Jæg. Adum.
ii. 389. (*Borgen*).
*B. melanangium* C. Müll. Andran-
goloaka (*Hildebrandt* 2078).
*Eurhynchium spinulænerve* Kiær.
Mt. Ankaratra (*Borgen* 48).
*Rhyncostegium distans* Besch. var.
*breve* Kiær. (*Borgen*).
*R. distans* Besch. var. *cordifolium*
Kiær. Mt. Ankaratra (*Borgen*
49).
*R. Pervilleanum* (Mont.) Jæg.
Adum. ii. 442. *Pervillé*.
*Rhaphidostegium Boivinii*
(Schimp.) Jæg. (*Boivin*).
*R. cuspidatum* Schimp. (*Pervillé*).
*R. Duisabanum* Schimp. (*Borgen*).
—Var. *Nossianum* Besch.
*R. Loucoubense* Besch. Nossi Bé.
*R. microdus* Besch. Nossi Bé
(*Marie*).

*R. ovalifolium* Besch. Nossi Bé.
*R. rubricaule* Besch. Nossi Be.
*R. rufoviride* Besch. Nossi Com-
ba.
*Trichostelium inclinatum* Kiær.
(*Borgen*).
*T. madagassum* C. Müll. Linnæa,
xxxix. 465.
*T. microthamnioides* C. Müll. Rel.
Rut. iii. 212. Vondruzona
(*Borgen*).
*Taxithelium glaucophyllum* Besch.
Nossi Bé; Nossi Comba.
*T. Nossianum* Besch. Nossi Bé.
*T. planulum* Besch. Nossi Bé
(*Marie* 14).
*T. scutellifolium* Besch. Nossi
Comba.
*Microthamnium ankeriensis* Kiær.
(*Borgen* 62).
*M. madagassum* Besch. Mt. An-
karatra (*Borgen*, 59, 60).
*M. mollissimum* C. Müll. Andran-
goloaka (*Hildebrandt* 2081).
*M. nervosum* Kiær. Mt. Anka-
ratra (*Borgen* 61).
*Isopterygium Boivini* Besch. Nossi
Bé (*Marie*).
*I. Combæ* Besch. Nossi Comba.
*I. subleptoblastum* C. Müll. Nossi
Bé (*Marie* 42).
*Ectropothecium Boivini* Besch.
Nossi Bé (*Marie*).
*E. curvulum* Mitt. Antananarivo
(*Meller*).
*E. sphærocarpum* (C. Müll.).
*Amblystegium chlaropelma* C.
Müll. East Imerina (*Hilde-
brandt* 2076).
*Hypnum afrocupressiforme.* Mt.
Ankaratra (*Borgen* 67).
*H. argyroleucum* C. Müll. Mt.
Ankaratra (*Borgen* 46).
*H. (Vesicaria) mundum* C. Müll.
Imerina (*Hildebrandt* 2077).
*H. (Cupressina) angustissimum* C.
Müll. Rel. Rut. iii. 212. Von-
druzona.
*H. (Cupressina) latocespitosum* C.
Müll. S. Betsileo (*Hildebrandt*
2074).

*H.* (*Cupressina*) *nano-crista-castrense* C. Müll. Andrangoloaka (*Hildebrandt* 2127).

*II.* (*Cupressina*) *Pervilleanum* Schimp. Vondruzona, Ambatondrazaka.

*H.* (*Tamariscella*) *Struthiopteris* C. Müll. S. Betsileo (*Hildebrandt* 2075).

*H.* (*Aptychus*) *afrodemissum* C. Müll. Rel. Rut. iii. 212. Vondruzona, Ambohimara, Ambatondrazaka.

*H.* (*Aptychus*) *nanopyxis* C. Müll. *l. c.* Vondruzona.

*H.* (*Sigmatella-Thelidium*) *punctatulum* C. Müll. *l. c.* 213. Ambatondrazaka.

*H.* (*Sigmatella-Thelidium*) *trachypyxis* C. Müll. *l. c.* Ambatondrazaka.

*H. Rutenbergii* C. Müll. Rel. Rut. iii. 213. Ambatondrazaka ; Mt. Ankaratra (*Borgen* 57).

*Rhacopilum Africanum* Mitt. (*Pool*).

*R. prælongum* Schimp. Andrangoloaka (*Hildebrandt* 2124).— Var. *Nossianum*.

*R. tomentosum* C. Müll. (*Borgen* 127).

*Hypopterygium Hildebrandti* C. Müll. S. Betsileo (*Hildebrandt* 2094).

*H. longirostrum* Schimp. (*Pervillé* 805).

*H. Nossi-Beanum* C. Müll. Rev. Bry. 1877, 15. Nossi Bé.

*H. torulosum* Schimp. var. *Nossi-Beanum* Besch. Nossi Bé.

---

# NOTE ON SOWERBY'S MODELS OF BRITISH FUNGI.

## By William Carruthers, F.R.S.

Mr. Smith's interesting notice of these models (p. 231) induces me to add some particulars which he has omitted, as he was not acquainted with them. The collection was made by James Sowerby, the projector and proprietor of ' English Botany,' while he was issuing his ' English Fungi.' His purpose in preparing the models is explained by him in the Introduction to the Supplement to that work. He says, " I intend to finish models of the more particularly poisonous Fungi, and of those which are edible, to prevent, as far as possible, future mistakes, for the use of the public." In order that this object might be accomplished, he opened his collection for public inspection every first and third Tuesday in each month from eleven until three o'clock. Mr. W. G. Smith has more efficiently accomplished this object by the publication of his two large sheets, with the accompanying letterpress, entitled ' Mushrooms and Toadstools ; how to distinguish easily the differences between edible and poisonous Fungi.' It was no more the purpose of Sowerby than of Mr. Smith to illustrate all the large Fungi known to him. Mr. Smith attained his object by giving figures of sixty species, nearly equally divided between the edible and poisonous ; and Sowerby prepared 193 models in accomplishment of his design.

After the death of James Sowerby the models became the property of his son, James De Carle Sowerby, who, in 1831, offered them for sale to the Museum. They were at that time examined by Mr. Brown, and though he recommended their purchase, the trustees did not then acquire them. In 1844 Mr. Sowerby again offered them, and then, on the recommendation of Mr. Brown, they were purchased for £70.

They were received into the Museum in April, 1844, and were in due time mounted by Frederick Nichols, with the care which has characterized all his work in the department, since he was associated with it under Mr. Brown nearly half-a-century ago.

As Mr. Smith says, the greater number are made of pipe-clay, without being burnt, and are consequently very brittle; but though in moving them they suffered some injuries, none of them were destroyed, and all the injuries were dexterously repaired by a *formatore* from Brucciani's, before they were placed in the hands of Mr. Smith. A composition called Cond's Artificial Stone was employed by Mr. Sowerby for some of his models, one or two were made of wax, and in some the stipes were of wood or wire.

The collection as it now exists contains, besides Sowerby's models, sixty published German models, and one of a fine large mushroom, modelled by the late J. C. V. Musgrave, formerly an attendant in the department. The whole have been most carefully painted by Mr. Smith, than whom no one was better fitted to do this work, for in addition to his accurate knowledge of the species, he has a remarkable appreciation of their different colours. He had already prepared a great series of water-colour drawings of the British species, now preserved in the Botanical Department of the British Museum. By his deft manipulation he has now converted these accurate models, which were covered with the dust of nearly a century, and consequently presented a uniform absence of colour, into faithful representations of the living organisms; and having supplied the accidents of their native habitats he has placed in an instructive manner before the student and the public permanent lifelike forms of these evanescent plants.

---

NOTE on the BOTANICAL PLATES of the EXPEDITION of the 'ASTROLABE' and the 'ZÉLÉE.'

By B. Daydon Jackson, Sec. L. S.

The following statement is published in correction of Pritzel's mis-description in his 'Index Iconum,' p. xxvi, and 'Thesaurus,' ed. ii. p. 148, n. 4193. The botanical results of the former voyage of the 'Astrolabe' (1826–29) were worked up by A. Richard, and Pritzel has confounded this production with the later voyage of the 'Astrolabe,' accompanied by the 'Zélée' (1837–40). Furthermore, the collation given in the 'Index Iconum,' on the page cited, is evidently taken from an imperfect copy, of which there appear to be several issued. I have given the contents of each plate, the contents of each number (livraison), and the dates of each number when received at the British Museum, Bloomsbury. The authority " *nob.*" on the plates refers to Hombron, who began the issue of the Botany of the expedition, but, he dying before its completion, the text was issued by Decaisne.

"Dumont d'Urville, J. S. C. — Voyage au pole sud et dans l'océanie sur les corvettes l'Astrolabe et la Zélée, executée par ordre du roi pendant les années 1837–1838–1839–1840, sous le commandement de M. J. Dumont d'Urville, Capitaine de Vaisseau, publié par ordonnance de sa majesté sous la direction superieure de M. Jacquinot, Capitaine de Vaisseau, Commandant de la Zélée.
"Botanique par MM. Hombron et Jacquinot. Tome premier. Plantes Cellulaires, par M. C. Montagne, D. M. Paris : Gide et Cie. 1845. 8vo, pp. xiv, 349."

(Inserted in the British Museum copy is : "Prodromus generum specierumque phycearum novarum in itinere . . . ab illustri Dumont d'Urville peracto, collectarum . . . auctore C. Montagne, D. M. Parisiis, apud Gide, Editorem, 1842. 8vo, pp. 16.")

"Tome Second. Plantes vasculaires, par J. Decaisne, membre de l'Institut. Paris . . . 1853. 8vo, pp. 96."

"Atlas, 1852. fol."

## CRYPTOGAMIE.

1. Hydropuntia Urvillei *Mont.* and Rhodymenia Hombroniana *Mont.*
2. Marginaria Boryana *Mont.*
3. M. Urvilliana *A. Rich.* and M. Boryana (fruit).
4. Scytothalia Jacquinotii *Mont.*
5. Heterosiphonia Berkeleyi *Mont.*, Polysiphonia ceratoclada *Mont.*, and P. punicea *Mont.*
6. Caulerpa læte-virens *Mont.*, C. corynephora *Mont.*, and C. mamillosa *Mont.*
7. Xiphophora Billardierii *Mont.*, Conferva Ægiceras *Mont.*, Rhipidosiphon javensis *Mont.*, Gigartina ancistroclada *Mont.*
8. Delesseria crassinervia *Mont.*, Thamnophora magellanica *Mont.*, Dasyphlœa insignis *Mont.*
9. Hypnea multicornis *Mont.*, Porphyra columbina *Mont.*, Ptilota formosissima *Mont.*
10. Grateloupia ? aucklandica *Mont.*, Sargassum decurrens, *Ag.*, Nothogenia variolosa *Mont.*
11. Rhodymenia ornata *Mont.*
12. Ballia Hombroniana *Mont.*, Halymenia Urvilliana *Mont.*
13. Hypnea rugulosa *Mont.*, Dumontia pusilla *Mont.*, Conferva anisogona *Mont.*, Polysiphonia ? cladostephus *Mont.*
14. Sphacelaria funicularis *Mont.*, Conferva pacifica *Mont.*, Laurencia concinna *Mont.*, Penicillus Arbuscula *Mont.*
15. Sticta orygmæa *Ach.*, S. hirsuta *Mont.*, Parmelia sphinctrina *Mont.*
16. Gottschea Lehmanniana *Nees*, Scapania Urvilliana *Mont.*, Plagiochila pusilla *Mont.*, Scapania clandestina *Mont.*
17. Jungermannia schismoides *Mont.*, Chiloscyphus ? Jacquinotii *Mont.*, Jungermannia punicea *Nees*, Radula physoloba *Mont.*
18. Herpetium australe *Mont.*, H. involutum *Mont.*, Madotheca elegantula *Mont.*
19. Phragmicoma aulacophora *Mont.*, Frullania scandens *Mont.*, F. ptychantha *Mont.*, Herpetium decrescens *Lindenb. & Lehm.*
20. Tortula hyperborea *Mont.*, Orthotrichum magellanicum *Mont.*, Hypnum ? auriculatum *Mont.*

## MONOCOTYLÉDONES CRYPTOGAMES.

1. Asplenium apicedentatum *nob.* [Hombron], A. obtusatum *Forst.*, A. scleroprium *nob.* [Hombron].
2. Lomaria procera *var.* tegmentosa *nob.* [Hombron], Grammitis rigida *nob.* [Hombron], G. australis *R. Br.*, G. humilis *nob.* [Hombron].
3. Asplenium bulbiferum *Forst.*, A. laxum *Br.*, A. viviparum [Hombron].
3. *bis.* A. tremulum *nob.* [Hombron], A. Fabianum [Hombron].
4. Polystichum proliferum [Hombron], P. vestitum *Presl*, Schizæa palmata *nob.* [Hombron].
5. P. venustum *nob.* [Hombron].

## MONOCOTYLÉDONES PHANÉROGAMES.

1. Victoriperrea impavida *nob.* [Hombron].
2. Freycinetia Urvilleana *nob.* [Hombron].
3. Hombronia calathiphora *Gaudich.*
4. Veratrum Dubouzeti *nob.* [Hombron], Lasiorrhiza purpurea *Less.*
5. Philesia buxifolia *Lam.*, Luzula alopecuros *Desv.*
6. Uncinia macrolepis *Decne.*, U. gracilis *Poepp.*
7. Carex Andersoni *Boott*, C. festiva *Dewey.*
8. Festuca scoparia *Hook. f.*, Bromus pictus *Hook. f.*

## DICOTYLÉDONES PHANÉROGAMES.

1. Agalmanthus umbellata *Endl.*
2. Aralia polaris *nob.*
3. Ligusticum antipodum *nob.*
4. Albinea oresigenesa *nob.*
5. Ozothamnus Vauvilliersii *nob.*
6. Calucechinus antarctica *nob.*, Calusparassus Forsteri *nob.* (This plate is erroneously lettered *Monocotylédones*).
7. Calucechinus antarctica *nob.*, Calusparassus betuloides *nob.*
8. Calusparassus Pumilio *nob.*, Calucechinus Montagni *nob.*
9. Veronica decussata *Willd.*, V. finaustrina *nob.*
10. Senecio littoralis *var.* lanata *Gaudich.*, Homanthis echinulatus *Cass.*, Clarionella magellanica *DC.*
11. S. acanthifolius *nob.*, Gnaphalium consanguineum *Gaudich.*, Culcitium magellanicum *nob.*
12. S. verbascifolius *Burm. f.*, S. floccidus *nob.*
13. S. Hookeri *nob.*, S. Danyausi *nob.* (with var.), S. exilis *nob.*, S. Laseguei, *nob.*
14. Colletia discolor *Hook.*, Escallonia serrata *Sm.*
15. Arjoona patagonica *nob.*, Azorella filamentosa *Lam.*, A. trifurcata *Gaertn.*, Bolax glebaria *Comm.*, B. cespitosa *nob.*
16. Valeriana sedoides *Urv.*, V. magellanica *nob.*, Forstera aretiastrifolia *nob.*, F. uliginosa *nob.*, Azorella Ranunculus *Urv.*, Mastigophorus Gaudichaudii *Cass.*
17. Arjoona pusilla *Hook. f.*, Azorella lycopodioides, *Gaudich.*, A. cespitosa *Cav.*, Panargyrum abbreviatum *Hook. & Arn.*, Colobanthus crassifolius *Hook. f.*, C. muscoides *Hook. f.*

18. Drapetes muscosa *Lam.*, Boopis australis *Decne.*, Plantago juncoides *Lam.*
19. Drimys Winteri *Forst.*
20. Berberis ilicifolia *Forst.*, B. empetrifolia *Lam.*
21. B. buxifolia *Lam.*, B. inermis *Pers.*
22. Jacquinotia prostrata *nob.*, Pernettya pumila *Hook.* (with var.), P. mucronata *Gaudich.* (with two vars.)
22. *bis.* J. myrsinites *nob.*, J. volubilis *nob.*, Pernettya ovalifolia *nob.*, P. oblongifolia *nob.*
23. Pernettya rigida *DC.*, P. Gayana *DC.*
24. Acaena multifida *Hook. f.*, A. Sanguisorbæ *Vahl*, A. pumila *Vahl*, A. lucida *Vahl*.
25. A. ovalifolia *Ruiz & Pav.*, A. pinnatifida *Ruiz & Pav.*, A. ascendens *Vahl.*, A. cuneata *Hook. & Arn.*
26. Baccharis patagonica *Hook. & Arn.*, B. magellanica *Pers.*
27. Dracophyllum longifolium *R. Br.*, D. longiflorum var. retortum.
28. Chiliotrichum ovatifolium *nob.*, C. Feliciæ *nob.*
27. [erroneously for 29]. Richea Desgrazii *nob.*, Dracophyllum Lessonianum *A. Rich.*
30. Senecio candidans [*sic*] *DC.*, Panax simplex *Forst.*
31. Gunnera magellanica *Lam.*, Gentiana Campbelli *Griseb.*, Primula magellanica *Lehm.*, Gentiana magellanica *Gaudich.*

The contents of each part was as follows, taken from the tickets in the coloured wrappers; there is no date of publication given :—

Livr. 1. Cryptogames, tt. 1–5 . . . . . . . 11 May, 1845.
„ 2. Monocotylédones Phanérogames, tt. 1–3⎫ „
           „      Cryptogames, tt. 1–2 ⎭
„ 3. Cryptogames, tt. 6–10 . . . . . .
„ 4. Monocotylédones Cryptogames, tt. 3, 4 .⎫
    Dicotylédones Phanérogames, tt. 1, 2 .⎭    ?
„ 5. Cryptogames, tt. 11–14, 16 . . . . 7 Dec., 1845.
„ 6. Monocotylédones Phanérogames, t. 6 .⎫
    Cryptogames, t. 3 *bis.* . . . . . .⎬ 5 May, 1853.
    Dicotylédones Phanérogames, tt. 3, 5, 10 ⎭
„ 7. Cryptogames, tt. 5, 17–20 . . . . . 26 July, 1844.
„ 8. Monocotylédones Phanérogames, t. 4 .⎫ 17 Jan., 1845.
    Dicotylédones         „     tt. 4, 7–9⎭
„ 9.        „        „
              tt. 11, 12, 14, 15, 19   12 June, 1845.
„ 10.
              „
              tt. 13, 16, 20–22 . 14 Feb., 1846.
„ 11.       „         „      tt. 24–28   6 July, 1848.
„ 12.       „     tt. 17, 27, 30, 31 .   ?
„ 13. Monocotylédones Phanérogames,      ⎫
           tt. 5, 6, 7, 8 . . .⎪
    Dicotylédones Phanérogames,      ⎬ 5 May, 1853.
           tt. 18, 22 *bis.* 23  .⎪
       Titre et Table . . . . . . .⎭

# NOTES ON PONDWEEDS.

## By Alfred Fryer.

POTAMOGETON FLUITANS Roth. *Stem* springing from a rootstock which strikes deeply into the mud, and is sometimes thickened into tubers; stout, 2–5 ft. long, usually branching a little below the middle, or simple in the summer shoots, rarely branched from the base. *Leaves of three kinds*, those of the lower part of the stem reduced to phyllodes, those of the middle, *membranous*, elongate, narrowly lanceolate, *submerged;* those of the upper part, *coriaceous, floating*, oblong-lanceolate or elliptical; *all long-stalked*, with *petioles convex above.* *Stipules* large, deeply channelled on the back, *slightly winged*, rather acute, very persistent. *Spikes* densely flowered, 1–1½ in. long, stout, on peduncles which are of equal thickness throughout, or slightly swollen upwards. *Fruit* (immature), 3-keeled, with the central keel rather prominent.

*P. fluitans* starts into growth early in the year; the earliest leaves are reduced to phyllodes, which resemble those of the *lucens*-group, being short, 1½–6 in., *always submerged*, and persistent. Thus differing from the long, floating, quickly-decaying phyllodes of *P. natans.* These phyllodes are produced throughout the whole period of growth, though they are more abundant early in the season, and are sometimes absent in the later shoots. Usually only one or two are present at the base of the stem, but in deep water as many as five or six are occasionally found. Like the true leaves, the phyllodes are stalked, which is apparent in fresh examples by the lower part being *slightly convex*, while the upper part is very slightly *concave;* in dried specimens these minute differences are lost. The phyllodes gradually pass into the ordinary submerged leaves by intermediate *bodkin-pointed leaves*, of which the petiole and midrib are thick and fleshy as in the simple phyllode, but the midrib is expanded into a narrow wing for an inch or two above its junction with the petiole, and then is reduced to a long bodkin-like point. Here again *P. fluitans* approaches *P. lucens*, in which similar bodkin-pointed leaves are constantly produced.

The submerged leaves which next follow strongly characterize the species, as they are unlike those of any other British Pondweed; they are translucent, membranous, *narrowly lanceolate*, 6–12 in. long by ¼–½ in. wide, tapering gradually into the *convex petiole*, which is 2–2½ in. long, slightly folded, with a stout prominent midrib, with two or three faint, translucent, lateral ribs, which are not raised above the surface of the leaf, and which are connected by transverse veins; on each side of the midrib is a narrow band of elongate chain-like aerolations. These elongate secondary leaves are usually submerged, but as the stem lengthens many of them rise to the surface and assume the coriaceous texture of the true floating leaves which they gradually pass into, and together spread like a fan on the surface of the water.

The true floating leaves are thick, coriaceous, with a *thick prominent midrib*, with 5–7 lateral ribs on each side, which are translucent, *immersed in the substance of the leaf*, which is perfectly smooth on both surfaces. These leaves are produced alike on barren and flowering shoots, and are bright green, thick, and remarkably brittle when fresh ; varying in shape from elongate-lanceolate to oblong or elliptical, but *all are gradually narrowed into the petiole*, even when they are slightly auricled or folded towards the base, in which case the lamina is shortly decurrent below the fold. The *convex petiole* is from ¾ to 1½ in., the lamina from 3 to 5 in. long.

The *stipules* are large, 3–4 in. long, *very rarely with a leaf reduced to a phyllode on the back*, closely clasping the stem at first, but afterwards becoming reflexed when a branchlet springs from their axils, horny, hyaline, with two prominent, green, winged ribs on the back and numerous slender green ribs on the sides, which are connected by anastomosing veins ; horn-coloured, or the lowest sometimes tinged with red like the base of the stem. They are curiously persistent even when decayed, not splitting into fibres like those of *P. natans*.

The *flower-spike* is subtended by opposite leaves ; before expansion it is like that of *P. decipiens;* when in flower the sepals are delicate bright green, and the spike closely resembles that of *P. Zizii*. The *immature fruit* is *3-keeled*, with the central keel rather prominent. Mr. W. H. Beeby has sent me some detached fruits, found amongst some Surrey plants of this species recently gathered by him, which are more advanced than any I have seen in the Fens. These when fresh resembled half-grown fruit of *P. natans*, but were very slightly keeled on the back ; when dry they became distinctly keeled. Although they very probably belong to the plants amongst which they were found, the degree of uncertainty which attaches to them renders a more exact description undesirable.

When forsaken by the water *P. fluitans* produces subaerial leaves, and continues to grow until cut down by frost. I have already described this state of the plant in 'Journal of Botany' for 1887, p. 307. This land-form more nearly resembles that of *P. Zizii* than any other species I have observed, but it seems not to produce *phyllodes*, so usually found in land-forms of the *lucens*-group.

The autumnal state of *P. fluitans* is so unlike the early and perfect states as to look like a distinct species. Specimens of this state were first gathered in Surrey in 1886, by Messrs. W. H. Beeby and Arthur Bennett. Dr. Tiselius, to whom they were submitted, thought they might belong to *P. zosteræfolius;* while Mr. Beeby cleverly suggested that they were probably a form of *P. heterophyllus*, with abnormally elongate lower leaves ; a guess only just short of the truth, for *P. fluitans* is undoubtedly closely allied to the section of the genus to which *P. heterophyllus* belongs. The guesses (for they did not claim to be more) of these two skilful botanists will give some idea of the appearance of this autumnal state, so unlike the usual form of the species seen in herbariums. The problem of the specific rank of this puzzling Surrey plant was solved in the autumn of 1887, when I gathered exactly similar specimens from a plant of *P. fluitans* which I had watched from its early summer-state onwards. This latter plant was cut down just as it came into flower in July, when the drain in which it grew was cleared of weeds; I gathered the second growth in August, again cutting it over to the base, and I found by September it had again sprung up, producing shoots exactly like those gathered in Surrey by Mr. Beeby. I submitted examples to him with the suggestion that his doubtful plant was probably *P. fluitans*. After a careful examination of the two sets of specimens my friend concurred with this view, and determined to admit the species in his forthcoming 'Flora of Surrey;' a decision which subsequent observations have amply justified.

In this autumnal state the leaves are all narrowly linear, elongate, and acutely pointed; many of them are so closely folded as to look like phyllodes, but they are really exceedingly delicate grass-like leaves, with the lamina distinctly developed. Further observations made on the growing plant in the present summer show that these grass-like leaves are the normal later growth of *P. fluitans*, and regularly spring from the axils of the lower stipules when the flowering season is past. Similar shoots, but with much shorter leaves, are also produced in the axils of the upper stipules, when branches are severed from the parent plant by accident, or by natural decay of the lower part of the stem. The severed portion of the branch floats on the surface, and in the course of a few days a fascicle of these narrow leaves springs from each joint as the older leaves decay; after a little time rootlets are formed at the base of the new growth, which then falls away from the rotten branch, and ultimately sinks to the bottom to continue the life of the species.

These little fascicles of linear-leaves are analogous to the " winter-buds" that are so freely produced in *P. obtusifolius* and its allied forms. Their development in *P. fluitans* is greatly augmented by the attacks of the larvæ of *Hydrocampa potomegata*, which about the beginning of August often entirely devour the floating leaves. The largest plant in my pond, measuring 4 to 5 yds. across, is at this time reduced to a mere skeleton by these larvæ, and the young fascicles of " winter-buds " are rapidly starting into growth, so that in a few weeks there will be thousands of young plants ready to begin an independent existence. The distribution of the winter-buds of aquatic plants and animals in stagnant waters is greatly assisted by the various species of thread-like Algæ which form the " green-scum" (flannel-weed of fenland speech) that grows so freely in ponds and ditches. These Algæ seem to begin their growth on the very surface of the mud, and as they increase they expand and fill the water with a green cloud-like film, which as it rises lifts up from the bottom vast numbers of mollusca and aquatic animals, and also the seeds and winter-buds of aquatic plants. As the thickly-tenanted mass reaches the surface its growth becomes vesicular, and so buoyant as to be easily driven by the wind. As the " rising of the scum" usually precedes stormy weather, its living burden is soon carried to considerable distances.

In sluggish foul streams, also, species are thus conveyed for miles before their floating cradle is arrested by some obstruction in the water or by the shore itself. Many seeds and delicate " winter-buds " of plants that could not vegetate in the cold depths to which they accidentally sink are thus brought within reach of light and and warmth. To such an extent is this transportation of the floor-life of stagnant waters carried, that I have seen ditch-bottoms that were completely hidden with Algæ early in spring, bright and clean by Midsummer. It is especially instructive to see beds of bivalve mollusca, such as *Cyclas*, rise and float away. Distribution, " cross-fertilization," and change of soil, are all secured by this simple means for the most inert forms of lower life.

But to return to the subject of this note :—Through the kindness of my friend, Mr. Arthur Bennett, I possess a very similar plant to the autumn state of *P. fluitans*, but with the addition of a few floating coriaceous leaves. This specimen was gathered in Sweden by Dr. Tiselius in 1881, and doubtfully referred by him to *P. fluitans*, a name in which Mr. Bennett did not concur. Afterwards Dr. Tiselius suggested that this plant might be a hybrid between *P. natans* and *P. heterophyllus*, and that it was probably the same form as Messrs. Linton's Irish " *P. sparganifolius.*"

These suggestions of the learned Swedish botanist, who is one of the greatest living authorities on the genus Potamogeton, bring us to the consideration of the specific rank and alliances of *P. fluitans*. Is it a hybrid ? Or does it differ sufficiently from other forms of the genus to entitle it to specific segregation, and where is its place in the genus ?

In favour of the first question : the flower-spikes are usually barren, perhaps always so in our British plant, of which no certainly fertile fruit has yet been seen ; the plant is also of extraordinarily vigorous growth, and can propagate itself readily by other means than seed ; further than this, in facies it is considered by some botanists to be intermediate between *P. natans* and forms of *P. lucens ;* and also it has never been found in this country except in waters where these two species grow.

On the other hand it may be urged that barrenness is no proof of hybridity in the genus Potamogeton ; nor does the fact that a form of Potamogeton has never been known to produce seed make it probable that it will not, under favourable conditions, ripen seed abundantly. The Rev. Thomas Morong tells me that *P. varians* has never ripened its fruit in America, although carefully observed for forty years. In England it ripens its seed freely. *P. Zizii* too, on the same authority, is always barren in some of the North-American lakes, in others it is fertile. Here in the Fens it is always abundantly fruitful, except in one isolated locality, where I have never seen it produce even a flower-spike. Again, for how many years was fruit of *P. lanceolatus* (one of the best and most distinct species of Potamogeton known) sought for without success ? We have, therefore, some reason to suppose that our English form of *P. fluitans* may ultimately be obtained in a fruiting condition, just as the Continental form of the species (closely agreeing with ours !), though usually barren, is sometimes obtained in good fruit.

Perhaps the strongest argument against hybridity in the plant in question is to be found in its uniformity of character over widely separated localities. At present in England it is absolutely *without a variety*, the plant of Sussex cannot be distinguished from the plant of Huntingdonshire! Of course it would be quite possible for a vigorous hybrid form of this genus to be become distributed over very wide areas indeed by offsets from the original plant, but the peculiar distribution of *P. fluitans* in the Fens, *strongly points to its being distributed by seed rather than by division ;* hence I confidently expect to obtain ripe seed from our plant some day.

One other suggested reason for its hybrid origin remains to be examined :—its being " intermediate " in character between *P. natans* and *P. lucens*. From the above given description, it will be seen that I have been quite unable to detect any real resemblance to *P. natans* whatever, while on the other hand I have pointed out numerous close resemblances to *P. lucens*. For the rest, if a Potamogeton being intermediate in facies between two other allied forms is a proof of hybridity, I am afraid nearly all our British species must be called hybrids !

On present knowledge, therefore, I think it will be safest to assume that our *P. fluitans* is a good species, and that it is identical with the Continental plant so named ; of this latter proposition my friend, Mr. Arthur Bennett, has given sufficient proofs.

Perhaps a glance at the characters which distinguish *P. fluitans* from its nearest (or *supposed* nearest) congeners, will enable the student to form some opinion of its true position in the genus. From *P. natans*, it is separated by its *short phyllodes always submerged*, by its *membranous submerged* secondary leaves, and by the *absence of a joint* on the upper part of the petiole. From *P. polygonifolius*, by the translucent nerves of the full-grown upper coriaceous leaves, by its much *stouter peduncles* and *larger flower-spikes*, and by its much *longer-stalked, narrowly-lanceolate submerged leaves*. From forms of *P. Zizii*, with the floating leaves well developed, it is sufficiently distinguished by its *elongate-lanceolate long-stalked* secondary leaves. From any known form of typical *P. lucens* the *floating coriaceous upper leaves* at once form a wide ground of separation. *P. fluitans* is, however, most closely allied to *P. lonchites ;* and with that plant, and possibly with the Irish *P. sparganifolius*, it would, perhaps, be most naturally placed in that section of the group of which *P. heterophyllus* and *P. lucens* are the extremes.

The first certain record of this plant was from Ramsey, Huntingdonshire, June 29th, 1884 ; afterwards it was found in Cambridgeshire, in two localities in 1886, and in a third in 1887 ; all these stations being near Chatteris. In Mr. Beeby's herbarium I found a specimen collected by him in West Sussex, in June, 1880; and the examples from Surrey, collected by Mr. Beeby and Mr. Bennett in 1886, have been already referred to. But since I began this note, Mr. Beeby has again collected very characteristic specimens in quite a new Surrey locality, and also in the original Sussex station. No doubt *P. fluitans* will be found in many of our southern counties, where it has likely been overlooked from its superficial resemblance to certain states of *P. natans* and *P. poly-gonifolius*. In Ireland too, which has afforded the closely-allied *P. lonchites*, it ought to be carefully looked for.

## BIOGRAPHICAL INDEX OF BRITISH AND IRISH BOTANISTS.

By JAMES BRITTEN, F.L.S., AND G. S. BOULGER, F.L.S.

(Continued from p. 248).

**Dalton, John** (1766–1844): b. Eaglesfield, nr. Cockermouth, 6th Sept. 1766 ; d. Manchester, 28th July, 1844 ; bur. Ardwick Cemetery, Manchester. LL.D., Edin., 1834. D.C.L. Oxon, 1832. F.R.S., 1822. Chemist. Pupil of John Gough. Herbarium at Manchester Public Library. Memoirs by R. Angus Smith, w. portr. engr. after Allen. Mem. Manchester Lit. Phil. Soc. 1856; and, w. portr., by Dr. W. C. Henry, 1854. ' Annual Monitor,' 1845. ' Friends' Books, i. 506 ; Dict. Nat. Biog. xiii. 429 : R. S. C., ii. 122 Portr. by Allen, 1814, at Manchester Phil. Soc. Marble statue by Chantrey at Manchester Roy. Instit., and bronze copy at Royal Infirmary.

**Dalzell, Nicholas Alexander** (1817–1878): b. Edinburgh, 21st April, 1817; d. Edinburgh, Jan. 1878. M.A. Edin., 1837. In India, 1841–1870. Conservator of Forests, Bombay. ' Bombay Flora ' (w. A. Gibson), 1861. Contributed to Linn. Trans.; Journ. Bot. 1850–1857 ; Proc. Bot. Soc. Ed., &c. R. S. C. ii. 135 ; vii. 479 ; Pritz. 75 ; Jacks. 387, 390, 392 ; Dict. Nat. Biog. xiii. 448. Plants at Kew.

**Dampier, William** (1652–1715): b. East Coker, Yeovil, 1652 ; d. Coleman Street, London, March, 1715. Circumnavigator. Collected in Brazil, Australia, Timor, New Guinea, &c. ' New Voyage round the World' (1697). Pritz. 75 ; Rees ; Ray's Hist. Plant. iii. (Suppl.) 225 ; Lasègue, 360; Dict. Nat. Biog. xiv. 2. Portr. by T. Murray in Nat. Portr. Gallery. Plants in Brit. Mus. (Herb. Sloane, 93, 94). *Dampiera* R. Br.

**Dancer, Thomas** (1755 ?–1811): b. 1755 ? ; d. Kingston, Jamaica, 1st Aug. 1811. M.D. ·Succeeded Dr. Clark as Curator Bot. Gard. Bath, Jamaica, 1787. ' Catalogue of ...... Bot. Gard. Jamaica,' 1792. ' Cinnamon trees in Jamaica,' Trans. Soc. Arts, vol. 8. Forsyth MSS. in Cott. Gard. viii. 159 ; Pritz. 75 ; Jacks. 449 ; Gent. Mag. 1811, pt. ii. 390 ; Dict. Nat. Biog. xiv. 14. Plants at Kew.

**Dandridge, Thomas** (fl. 1723–1730): Of Stoke Newington. Fungologist, ornithologist and lepidopterist. Friend of W. Sherard. Corresponded with Petiver. Nich. Illust. i. 357 ; iii. 782 ; Rich. Corr. 204.

**Daniel, Henry** (fl. 1379): Dominican friar. ' Aaron Danielis, de re herbaria, de arboribus, fructicibus ' ... MS. in Bodleian. Pult. i. 23 ; Tanner, 218 ; Haller, i. 232 ; Dict. Nat. Biog. xiv. 24.

**Daniel, Samuel** (ᵰ. 1695). Surgeon. Sent plants from Greece to Petiver (Mus. Pet. 211, 624).

**Daniell, William Freeman** (1818-1865): b. Liverpool, 1818; d. Southampton, June 26th, 1865; bur. Kensal Green. M.D. F.L.S., 1855. Resided seventeen years on West Coast of Africa, and visited Jamaica, Bahamas, and, in 1860, China. Pl. in Herb. Mus. Brit. Pritz. 75; Jacks. 368; Pharm. Journ. vols. ix.–xviii. and 2nd ser. i.–iv.; Ann. & Mag., 1862; Proc. Linn. Soc. 1865-6, p. lix.; Journ. Bot. 1863, 294; R. S. C. ii. 146; Dict. Nat. Biog. xiv. 35. *Daniellia* Benn.

**Danvers, Henry, Earl of Danby** (1573-1644): b. Dauntsey, Wilts, 28th June, 1573; d. Cornbury Park, Oxford, 20th January, 1644; bur. in Dauntsey Church. Founded Oxford Physic Garden, 1632. Pult. i. 165; Diet. Nat. Biog. xiv. 37. Portr. at Dauntsey Rectory; and one engr. in Lodge. Mezzo. by V. Green, after A. van Dyck.

**Dare, George** (ᵰ. 1690?): Apothecary of London. Discovered *Hymenophyllum* at Tunbridge Wells, prior to 1698. Petiver, Museum, p. 73; Plukenet, Almagestum, 10. *Darea* Juss.

**Darwin, Charles Robert** (1809-1882): b. The Mount, Shrewsbury, 12th Feb. 1809; d. Down, Kent, 19th April, 1882; bur. Westminster Abbey. M.A., Camb., 1831. LL.D., 1877, &c. F.L.S., 1854. F.R.S., 1839. Naturalist, H.M.S. 'Beagle,' 1831-1836. 'Journal,' 1839. 'Origin of Species,' 1859. 'Fertilisation of Orchids,' 1862. 'Climbing Plants,' Journ. Linn. Soc. ix. 1875. 'Variation,' 1868. 'Insectivorous Plants,' 1875. 'Cross and Self Fertilisation,' 1876. 'Life and Letters,' with 3 portr., bibliog., and list of portr., 1887. Oil portr. by J. Collier, at Linn. Soc., etched by L. Flameng. Statue by Boehm, Mus. Nat. Hist., Kensington. Pritz. 76; Jacks. 537; Journ. Bot. 1882; Proc. Linn. Soc. 1881-2, 60; Gard. Chron. w. portr. 1882, i. 535; Diet. Nat. Biog. xiv. 72; Portr. in Ipswich Museum series. Copies at Kew and Linn. Soc.

**Darwin, Erasmus** (1731-1802): b. Elston Hall, Notts., 12th Dec. 1731; d. Derby, 18th April, 1802. M.B. Camb., 1755. M.D. Edin. F.R.S. F.L.S., 1792. Poet. Practised at Lichfield till 1781; then at Derby. 'Botanic Garden,' 1781. 'Phytologia,' 1801. Translated Linnæus's 'Systema Vegetabilium' as a 'System of Vegetables' (1783), and his Genera and Mantissæ as 'The Families of Plants,' 1787. Pritz. 76; Jacks. 537. Memoirs by Miss Seward, 1804. Life by Charles Darwin, 1879. Nich. Anecd. ix. 75; Diet. Nat. Biog. xiv. 84; Felton, 164; Cott. Gard. iv. 43. Portr. in Thornton's Botany; one by Wright, 1770; photo'd. in Life by Charles Darwin, 1879; one by S. J. Arnold, engr. by B. Pym, 1801. Wedgwood medallion; one at Kew. *Darwinia* Rudge. *Darwinia* Dennstadt = *Litsæa* Juss.

**Darwin, Robert Waring** (1724-1816): b. Elston Hall, nr. Newark; d. Elston, 1816. 'Principia Botanica,' 1787, ed. 2, 1807; ed. 3, 1810. Pritz. 76; Jacks. 537.

**Daubeny, Charles Giles Bridle** (1795-1867): b. Stratton, Gloucestershire, 11th Feb. 1795; d. Oxford, 13th Dec. 1867;

bur. Magdalen College. M.A., Oxon, 1817. M.D., 1821. F.L.S., 1830. F.R.C.P., 1822. F.R.S. Prof. Chemistry, Oxford, 1822; of Botany, 1834. ' Action of Light on Plants,' Phil Trans. 1836. ' Roman Husbandry,' 1857. ' Trees of the Ancients,' 1865. ' Miscellanies,' 1867. Pritz. 77 ; Jacks. 537 ; R. S. C. ii. 155; vii. 488 ; Journ. Bot. 1868, 22 ; Gard. Chron. 1867, 1294; Trans. Bot. Soc. Ed. ix. 267; Proc. Roy. Soc. xvii, lxxiv; Munk, iii. 254; Journ. Hort. xiii (1867) 462; Dict. Nat. Biog. xiv. 94. Oil portr. at Oxford Bot. Garden. *Daubenya* Lindl.

**Davall, Edmund** (1763–1798): b. in England, 1763; d. Orbe, Switzerland, 26th Sept. 1798. Lived in Switzerland from 1788. Correspondent of Smith. Left incomplete ' Illustrations of Swiss Plants.' Pl. in Smith's Herbarium, Annals of Bot. i. (1805), 576. Rees; Pritz. 77 ; Smith Lett. ii. 1–4, 70, &c. Dict. Nat. Biog. xiv. 99. *Davallia* Sm.

**Davies, Hugh** (1739 ?–1821): b. Anglesey, 1739 ?; d. Beaumaris, 16th Feb. 1821. Clerk. M.A., Cantab. Rector of Aber. F.L.S., 1790. Friend of Hudson. ' Welsh Botanology,' 1813. Contributed to ' Flora Anglica,' ' Flora Britannica,' and ' Eng. Bot.' Herbarium in Mus. Brit. Pritz. 77 ; Jacks. 247 ; Dict. Nat. Biog. xiv. 138 ; R. S. C. ii. 166. *Daviesia* Sm.

**Davies, John** (circ. 1570–1644): b. Llanrhaiadar, Denbigh, circ. 1570; d. Mallwyd, Merioneth, 15th May, 1644 ; bur. in Mallwyd Church. Clerk. D.D., Oxon, 1616. Welsh lexicographer. ' Antiquæ linguæ Britannicæ et . . . Latinæ Dictionarium,' including ' Botanologium,' 1632. Dict. Nat. Biog. xiv. 144.

**Davyes, Robert** (1684–1728): b. 1684; d. 22nd May, 1728. Of Guissancy, Flintshire. Antiquary. Catalogue of British names of plants in Johnson's Gerard. Dict. Nat. Biog. xiv. 154. Monument in Mold Church.

**Dawes, John Samuel** (1802–1878): b. Birmingham, 1802 ; d. Edgbaston, 20th Dec. 1878. Ironmaster. F.G.S. 1842. ' Sternbergia,' Quart. Journ. Geol. Soc. 1845; ' Halonia.' *ib.* 1848 ; ' Calamites,' *ib.* 1849–51. Proc. Geol. Soc. 1878–9.

**Deakin, Richard** (d. 1873): d. Tunbridge Wells, 18th February, 1873. M.D. Practised at Sheffield. ' Florigraphia Britannica,' 1837–1848 (re-issue, 1857). ' Flora of Colosseum,' 1855. ' Flowering Plants of Tunbridge Wells,' 1871. Pritz. 77 ; Jacks. 537 ; R. S. C. ii. 185; vii. 500 ; Journ. Bot. 1873, 128.

**Dede, James** (fl. 1809). ' The English Botanist's Pocket Companion,' 1809. Jacks. 36.

**Deering. George Charles** (1695 ?–1749): b. in Saxony, 1695 ?; d. Nottingham, 12th April, 1749 ; bur. St. Peter's Churchyard, Nottingham. M.D., Leyden and Rheims. At Leyden, 1708– 1711. Pupil of Boerhaave and Bernard de Jussieu. Member of Bot. Soc. Lond. 1721–1726. Practised at Nottingham from 1736. ' Catalogue . . . . Nottingham,' 1738. Assisted Dillenius in ' Historia Muscorum.' Herbarium bought by Hon. Rothwell Willughby. Pult. ii. 257 ; Pritz. 78 ; Jacks. 257 ; Nich. Illust. i. 211, 220 ; iii. 571 ; Dict. Nat. Biog. xiv. 279. *Deeringia* Br. = *Celosia*.

**Delany, Mary,** *née* **Granville,** formerly **Pendarves** (1700-1788):
b. Coulton, Wilts, 14th May, 1700; d. St. James' Place, Westminster, 15th April, 1788; bur. St. James', Westminster; m. 1,
Alexander Pendarves; m. 2, Rev. Patrick Delany, Dean of Down.
Executed a Herbal in 9 vols. folio, of 100 plates each, in coloured
papers, having "the exactness of Botany," 1778. Pritz. 78;
Jacks. 44; Autobiog. 1861-2; W. Gilpin, 'Picturesque Beauty,'
ii. 190; Nich. Anecd. iv. 715; Gent. Mag. lviii. 371, 462; Diet.
Nat. Biog. xiv. 308. Portr. by Opie at Hampton Court.

**Denham, Dixon** (1786-1828): b. London, 1st January, 1786;
d. Free Town, Sierra Leone, 8th May, 1828. Lieut.-col.
African traveller. In Africa, with Clapperton and Oudney from
1822-1825. 'Narrative of Travels . . . . . in . . . . . and Central
Africa, 1826, with an 'Appendix of Natural History.' Diet.
Nat. Biog. xiv. 341. *Denhamia* Schott = *Calcasia.*

**Dennes, George Edgar** (fl. 1838-8444). F.L.S., 1838. Sec. Bot.
Soc. Lond. 1839. Editor, 'London Cat. British Plants' (ed. 1).
Phytol. i. 1014, 1098 (1844); Proc. Bot. Soc. Lond. *Vicia
Dennesiana* Wats.

**Denson, John** (fl. 1822-1838). A.L.S. Curator, Bot. Gard.
Bury St. Edmunds. Did the botanical part of Loudon's 'Ar-
boretum.' 'Arboretum,' p. viii. Cat. Bot. Gard. Bury St.
Edmunds, 1822; Jacks. 409; R. S. C. ii. 240.

**Dent, Peter** (d. 1689): b. Cambridge?; d. Cambridge, 1689;
bur. St. Sepulchre's, Cambridge, 5th Oct. Physician and
Apothecary. M.B., Lambeth, 1678; Cambridge, 1680. Friend
of Ray. Assisted in the 'Historia.' Pult. i. 200; Pryor, Flora
Herts. l.; Dict. Nat. Biog. xiv. 378.

**Dewhurst, John** (fl. 1810-1835). Fustian-maker. Of Manchester.
President of Lancashire Botanists, 1810-1835. Cash, 16.

**Dick, Robert** (1811-1866): b. Tullybody, Clackmannanshire,
Jan. 1811; d. Thurso, 24th Dec. 1866; bur. same place.
Rediscovered *Hierochloe borealis.* Cash, 170; Life by S. Smiles,
with portr. Herbarium at Free Library, Thurso. Dict. Nat.
Biog. xv. 16.

**Dickenson, Samuel** (fl. 1777-1799). Clerk. LL.B. Rector of
Blymhill, Staffordshire. Correspondent of Withering, and
contributed notes on *Agrostis,* &c., to 3rd ed. of his 'Arrangement.' With. Arr. 3rd ed., preface. Contributed list of plants to
Shaw's 'Hist. of Staffordshire.' Letters in Bot. Dept. Brit. Mus.

**Dickie, George** (1812-1882): b. Aberdeen, 23rd Nov. 1812;
d. Aberdeen, 15th July, 1882. M.A., Aberdeen, 1830. M.R.C.S.,
Lond., 1834. M.D., Aberd., 1842. A.L.S., 1863. F.R.S., 1881.
Algologist. Lecturer on Bot., King's Coll., Aberd., 1839. Prof.
Nat. Hist. Belfast, 1849. Prof. Bot. Aberdeen, 1860-1877.
'Flora Aberdonensis,' 1839. 'Botanist's Guide to . . . . . .
Aberdeen . . . . . ', 1860. 'Flora of Ulster,' 1864. Pritz. 82;
Jacks. 539; Dict. Nat. Biog. xv. 32; R. S. C. ii. 283; vii. 531;
Top. Bot. ed. i. 522; Trans. Bot. Soc. Ed. 1839; Proc. Linn. Soc.
1882-3, 40; Stewart & Corry, Fl. N.E. Ireland, xx. Algæ in Herb.
Mus. Brit. *Dickieia* Berkl. & Ralfs. *Cystopteris Dickieana* Hook.

**Dickinson, John** (fl. 1699). Sent plants from Bermuda to Petiver. Journ. Bot. 1881, 258.

**Dickinson, Joseph** (d. 1865): d. Liverpool, 21st or 26th July, 1865. M.A. and M.D., Dublin, 1843. F.L.S. Lecturer on Physic and Botany, Liverpool School of Medicine, 1839. ' Flora of Liverpool,' 1851; Supplement, 1855. Pritz. 82; Jacks. 255; R. S. C. ii. 285; Dict. Nat. Biog. xv. 36.

**Dickinson, William** (1799?-1882): d. Thorncroft, near Workington, Cumberland, 1882. Land-surveyor. Compiled ' Cumberland Glossary.' Baker, ' Flora of Lake District,' p. 12; R. S. C. ii. 285.

**Dickson, Alexander** (1836-1887): b. Edinburgh, 21st Feb. 1836; d. Hartree, Peebles, 30th Dec. 1887. M.D., Edinb., 1860. M.D., Dublin. LL.D., Glasgow. Pupil of J. H. Balfour. Deputy Prof. Bot., Aberdeen, 1862. Prof. Trin. Coll. Dublin, 1866; Glasgow, 1868; Edinburgh, and Regius Keeper, Bot. Gard., 1879. Pritz. 82; Jacks. 91; R. S. C. ii. 285; vii. 532; Journ. Bot. 1888, 64; ' Nature,' Jan. 5, 1888; Ann. Bot., 1888, 396 (bibliog.); Dict. Nat. Biog. xv. 41.

(To be continued.)

---

# SHORT NOTES.

ULOTA PHYLLANTHA IN FRUIT FROM KILLARNEY.—In the Schimper collection, presented to the Kew Herbarium by the Baroness Burdett Coutts, in June, 1880, is a specimen of the above moss, labelled in Schimper's own handwriting, " Muckross, Killarney, Hibern. legi. Junio, 1865." On one of the largest tufts I have found five capsules; three quite old; one in good condition, with operculum still on; and one quite young, with a small calyptra. The capsules show the same characters as those recently found by me in the collections of Thomas Howell, from Oregon, U.S.A., and described by M. Cardot (Revue Bryol. xv. no. 3). The spore-sac is quite short and rounded below, showing conspicuously through the walls of the capsule, which are ribbed when dry; the operculum is apiculate and slightly oblique, and the calyptra is sparsely hairy. The peristome has not been examined, as it seemed a pity to destroy the only good specimen for that purpose. It seems strange that Schimper should have overlooked them, but as the capsules are short-pedicelled—only a few of them, and these were partly hidden by an hepatic (a form of *Frullania germana* Tayl.) —it is not to be wondered at, especially when we consider that the Orthotrichæ nearly always grow mixed, and fruiting specimens of other species are frequently found with *Ulota phyllantha*. Also it must be remembered that this species has been known for years from the coasts of Northern Europe and America, but always sterile, so that bryologists had grown accustomed to classing it with those mosses that never fruit, and ceased to expect it.—ELIZABETH G. BRITTON.

HANTS PLANTS. — A few days' visit of the Toynbee Natural History Society to the New Forest and the adjoining coast in August was productive of the following finds :—*Isnardia palustris, Spiranthes æstivalis* (in a new locality), *Ranunculus tripartitus* (with capillary leaves), *Carex filiformis, Drosera anglica, Eriophorum gracile.* On the shore near Christchurch, *Scirpus parvulus, Lotus hispidus,* and *Elymus arenarius.* The last, first discovered in Hants by the Rev. W. Moyle Rogers near Bournemouth (see Journ. Bot. 1886, 284, where "Wilts" is misprinted for "Hants"), has thus its county range considerably extended. Owing to the submersion of part of the sand here, *Diotis maritima* and *Polygonum maritimum* are probably extinct in this locality.—BOLTON KING.

---

## NOTICES OF BOOKS.

*A Flora of the North-east of Ireland, including the Phanerogamia, the Cryptogamia, Vascularia, and the Muscineæ.* By SAMUEL ALEXANDER STEWART, and the late THOMAS HUGHES CORRY. Published by the Belfast Naturalists' Field Club. Cambridge: Macmillan & Bowes. 8vo, pp. xxxiii. 331.

WE are very glad to receive this valuable addition to our knowledge of the botany of our sister island. Many of us have of late years found our knowledge of Ireland and Irish affairs considerably enlarged and extended ; and it is gratifying to find that the more peaceable pursuits of science are not neglected, in spite of the more exciting allurements of national struggle and political debate. For many years the Belfast Naturalists' Club has pursued its labours in the investigation of the botany of the counties of Antrim, Down, and Derry ; chief among the labourers being Mr. S. A. Stewart, who, in the present volume, brings together the results of many years' diligent work; and Mr. T. H. Corry, whose premature death inflicted a loss upon general as well as local botany. We have, then, now presented to us in an attractively-printed book, which is not too large for the pocket—a rare qualification in works of the kind—what may be accepted as a full and trustworthy account of the native vegetation of the counties referred to.

From the interesting introduction (which contains biographical sketches of Templeton, Threlkeld, Corry, David Moore, and others), we learn that the Flora, as here recorded, contains 1169 species— flowering plants and higher cryptogams, 803 ; mosses, 293 ; hepatics, 73. This excludes 271 species which have been erroneously recorded or require confirmation. Of the two plants peculiar to the district, one, *Carex Buxbaumii*, is said to be on the verge of extinction. When last noticed by Mr. Stewart in 1886, on the small island in Lough Neagh, which is its only locality in the United Kingdom, only " one little patch of about two feet square was seen, on which there were a number of stems, some immature, some trampled down, or eaten by cattle, and a few perfect.    The dense, almost impenetrable, thicket of shrubby wood, which a few years since afforded excellent cover for many aquatic birds, and also sheltered *C. Buxbaumii*, has recently been cut down.    The little island is now a bare exposed pasturage where, in the struggle for existence, only the more hardy and aggressive are likely to maintain their hold."    The other peculiar species, *Calamagrostis Hookeri*, first found on the same or a neighbouring island, has disappeared thence ; but although rare, still occurs on the Antrim and Derry shores of Lough Neagh.    Many plants, indeed, are extinct or on the way to extinction in the north-east of Ireland ; such are *Cladium germanicum* (" not found recently "), *Hypericum hirsutum*, *Adoxa Moschatellina*, *Arabis hirsuta*, *Subularia aquatica*, *Andromeda Polifolia*, both species of *Rhamnus*, *Epipactis palustris*, and others.

One or two known introductions have thoroughly established themselves.    *Hottonia*, which was planted by Templeton at Cranmore, has been " more recently introduced to the bog-meadows, and has spread amazingly in the drains there.    Still later it has been brought to Holywood and to Cushendun.    The origin of this plant in Down was most probably through human agency, at no remote date."    The cowslip is wild only in one locality, Rostrevor, and even there is said to be " possibly introduced ;" in its several other localities it was " probably introduced by design or accident."    The plants excluded from the ' Flora ' are put by themselves in an appendix—a practice which, while showing a laudable tendency not to swell the numerical estimate of the work, is, we think, of doubtful convenience.    So carefully are these suspicious characters isolated, that even a separate index is provided for them, which is almost certain to be overlooked.    These unfortunate plants are of two grades : the first consisting of such as are " not indigenous and not naturalised ;" the second of " plants erroneously recorded."    No fewer than 141 are in the first class and 129 in the second.    Mr. Stewart has certainly not erred on the side of leniency, and it seems to us that some of the first group might have claimed admission to the body of the book as naturalised.    The second grade is unusually large, and much care is manifested in the notes stating why certain plants are excluded. Templeton's memory is satisfactorily cleared from a large number of errors for which he has been held responsible ; but many plants which found their way into the Irish flora on the authority of Mr. D. Orr are shown to require confirmation, and indeed it is not obscurely hinted that the records of this collector are not to be accepted as trustworthy.

Mr. Stewart has secured the help of specialists in the more critical groups, whose help is duly acknowledged ; and the local naturalists have actively co-operated.    Mr. G. A. Holt, of Manchester, has critically examined the Mosses and Hepaticæ; the enumeration of which is very extensive.    The Algæ and Fungi are not included in the volume.    The local names for many species are given: "Michaelmas Daisy " for *Matricaria inodora* is probably a slip.

Mr. Stewart does not seem to be aware that Robert Brown, during his residences in Ireland in 1797 and 1800, paid much attention to the botany of the districts in which his regiment was stationed. I was fortunate enough to pick up on a book-stall, four years ago, two folio volumes containing most minute and careful descriptions of 406 species of plants, originally drawn up by Brown and subsequently transcribed by him. Among these are many species collected by him to which the localities are appended; and these are in many instances additional to those given in the 'Flora.' *Cakile maritima*, for example, was gathered by Brown on the beach at Ballantoy, in September, 1797 : *Blysmus rufus* in the same locality, "on salt and marshy ground occasionally covered at high-water." Mr. Stewart quotes Templeton's localities (the only ones) for *Carex stricta*, but adds : " The localities may be right, but there are no specimens by which the correctness of the determination may be tested." Brown's authority may be cited in confirmation ; his description of the plant is from specimens " cultivated in Mr. Templeton's garden, but found wild in the neighbourhood." *

We have also in the Botanical Department of the British Museum, where I have deposited the volumes above referred to, Brown's journal for the year 1800, during which he was staying at Londonderry ; this contains many botanical notes, and much interesting personal matter, of which I hope some day to make an abstract. In the Herbarium are Irish specimens, collected by Brown, which relate to the 'Flora'; *Drosera anglica*, for example, he collected in August, 1795, " in bogs between Newton-Limavady and Coleraine"—a more northerly station for Derry than those given by Mr. Stewart. We have also Brown's own specimens and others sent by him to Banks, of *Sagina maritima*—a plant which, although credited to George Don, who published the species under that name in his 'Herbarium Britannicum' (fasc. vii., No. 155), was first described by Brown in his MSS. in 1797 from specimens " growing in bare spots by the sea-side at Ballycastle, where it occurs in great abundance, flowering from July till October or November. Observed also by the water-edge at Larn on August 7th, 1797 ; in this last place it grew mixed with other herbage, and the stems were always solitary." His own specimens from the Antrim coast are dated 1795–1797, and others from him are the types of ' English Botany,' t. 2195. This was not published until 1810; I

---

* While writing of Robert Brown, I cannot help regretting the inadequacy of the sketch of his life published by Dr. Robert Hunt in the ' Dictionary of National Biography.' It is little more than a clumsy paraphrase, with some added errors, of Mr. J. J. Bennett's charmingly-written obituary notice in the Linnean Society's ' Proceedings' for 1859—a notice which is not mentioned in the list of authorities given by Mr. Hunt. No reference is made to Humboldt's well-known designation of Brown as " facile botanicorum princeps "; a paper read by Brown in 1791 (really in 1792 : see Journ. Bot. 1871, pp. 321—332, where the paper is printed *in extenso*) is said to have been " used by Dr. Withering, who was *at this time* engaged in preparing the second edition of his ' Arrangement '"—a work which appeared in 1787 ; and other mistakes might be pointed out. It is to be regretted that no English botanist should have been entrusted with the life of the greatest botanist whom England has produced.

do not know the date of Don's seventh fascicle, but his first was not issued until 1804, seven years after Brown had drawn up his MS. description. Both in his herbarium and MSS. Brown calls the plant " *Sagina maritima* Nost.," thus showing that he considered it new.

One or two points suggest criticism. A map would greatly add to the usefulness of the work. There are five indexes—at least three too many, for the "topographical index," a new and useful feature in works of this kind, may well stand by itself. No special authority seems to have been followed for the nomenclature ; a large number of specific names which require capitals are printed with small initials ; the objectionable practice animadverted upon elsewhere in this number of citing two authorities for a species (" *Glyceria aquatica* (Linn.) Smith ") is sometimes followed ; and the arrangement might be improved in many places, where the name of a genus stands at the foot of one page, and that of the species begins the next, or, worse still, where the latter name ends a page and all the localities occur overleaf (e. g. *Pyrola media*, p. 91). There are also some mis-spellings. On the whole, however, this latest of our Floras is highly creditable to all who have been concerned in its production.

JAMES BRITTEN.

*General Index to the First Twenty Volumes of the Journal (Botany), and the Botanical portions of the Proceedings, Nov.* 1838 *to June,* 1886, *of the Linnean Society.* Edited by B. DAYDON JACKSON, Sec. L. S. London : Burlington House, Piccadilly. 8vo, pp. vii. 427.

THIS is a model index, and one of the most useful works which has been published by the Linnean Society. A good index is a blessing to mankind, and a good index-maker is entitled to, and receives, the gratitude of his race. " A great book," we are told, " is a great evil ;" but it ceases to be so if it is properly indexed. On the other hand, the most valuable work is to a great extent deprived of its worth if it does not possess a good index. A bad index is worse than none at all. To use a formula familiar to many, it "leaves undone the things which it ought to have done, and it does those things which it ought not to have done." The inadequate index is only too well known to all of us ; but perhaps the too complete one is even more bewildering, though happily less common. It may be seen at its worst in some of Mr. F. G. Heath's books—in his edition of Gilpin's ' Forest Scenery ' for example, from which the following specimen may be extracted :—

| " Setting sun and autumn leaves, The | . | 331 |
| „ and internal forest scenes, The | | 331 |
| „ and shadow, The . | . | . 331 |
| „ and summer leaves, The. | | . 331 |
| „ Brightened gloom of the | . | 331 |
| „ Glory of the parting rays of the | | 331 |
| „ Glowing colours of the . | . | . 331" |

Then there are persons who index matters under the definite and indefinite adjectives: in fact, there is no end to the misplaced ingenuity which index-makers have displayed, apparently with the object of showing how useless an index can be made.

The Linnean Society's Journals have exemplified both the inadequate and superfluous kinds of index; and it was high time that some one should have produced a satisfactory substitute for these. No one more fitted for the task could have been found than Mr. Daydon Jackson, who is a born indexer—we sometimes think that indexers, like poets, are born, not made—and who has, we are sure, entered upon his task with a zest which must have relieved it of much of its monotony. Every item is arranged in one alphabetical series—a benefit for which only those who have been condemned to consult indexes of many alphabets will be sufficiently grateful. Even those who do not possess the whole series of the Society's Journals and Proceedings, will be glad to have at hand a trustworthy record and indication of what has appeared in them. Mr. Jackson has done much good work since he has filled the post of Secretary to the Linnean Society, but he has undertaken nothing more useful than this. If some day the Society should see its way to issue a list of its Fellows and Associates during the first hundred years of its existence, which would, we think, be of interest to many, we trust the task will be placed in Mr. Jackson's hands.

### ARTICLES IN JOURNALS.

*Ann. Sciences Naturelles* (Aug.). — E. Bornet & C. Flahault, 'Revision des Nostocacées hétérocystées contenues dans les principaux herbiers de France.' — L. Courchet, 'Recherches sur les Chromoleucites (6 plates). — P. van Tieghem, 'Sur le réseau de soutien de l'écorce de la racine.'

*Bot. Centralblatt* (Nos. 28, 29). — A. Hansgirg, 'Ueber *Bacillus muralis*.' — (Nos. 29, 30). —. Eichelbaum, 'Mykologische Beobachtungen.'— K. Starbück, *Leptosphæria modesta.* — (Nos. 31–34). R. Keller, 'Wilde Rosen des Kantons Zürich.' — (Nos. 31, 32). A. N. Lundström, 'Ueber farblose Oelplastidin und die biologische Bedeutung der Oeltropfen gewisser Potamogeton-Arten.' — E. Ljungström, 'Eine Primula-Excursion nach Möen.'—S. Berggren, 'Ueber Apogamie des Prothalliums von *Notochlæna*.' — (No. 33). H. Molisch, 'Zur kenntniss der Thyllen.'—(No. 34). F. Areschoug, 'Ueber *Trapa natans* var. *conocarpa* F. Aresch.'

*Botanical Gazette* (July).—F. C. Newcombe, 'Spore-dissemination of *Equisetum*' (1 plate). — C. V. Riley, 'Personal Reminiscences of Asa Gray.' — M. S. Bebb, 'Notes on American Willows' (*S. phylicoides:* 1 plate). — J. D. Smith, 'Undescribed Plants from Guatemala' (*Gonzalea thyrsoidea, Mikania pyramidata, Zexmenia guatemalensis, Enclia pleistocephala, Lamourouxia integerrima, Pitcairnia Turckheimii, Zanthoxylum costaricense*, spp. nn.). — W. Trelease, 'Subterranean shoots of *Oxalis violacea*' (1 plate). — A. Gattinger, *Diervilla rivularis*, n. sp.

*Bull. Soc. Bot. France* (xxxiv. Comptes rendus 8 : Aug. 1).—
H. Coste, ' Herborisations sur le Causse Central.' — Memoir of R.
Caspary (29 Jan. 1818-18 Sept. 1887).—R. de Nanteuil, ' Quelques
plantes rares ou nouvelles pour la flore de Paris.' — H. Douliot,
' Sur la periderme des Rosacées.'—P. Brunand, ' Champignons des
environs de Saintes.' — G. Rouy, ' Plantes de Gibraltar et d'Alge-
ciras ' (*Senecio gibraltaricus, Mercurialis Reverchoni*, spp. nn.). —
P. van Tieghem, ' Sur l'exoderme de la racine des Restiacées.'—
D. Bois, ' *Trapa verbanensis*.' — L. Dufour, ' Quelques expériences
relatives à des germinations de Fève. — L. Morot, ' Variation de
forme du *Pleurotus ostreatus*.'— G. Bonnier, ' Sur des cultures com-
parées des mêmes espèces à diverses altitudes.'— —. Hue, ' Quelques
Lichens intéressants pour la Flore Française.'

*Bull. Torrey Bot. Club* (Aug.). — E. E. Sterns, ' Fruit of *Caly-
canthus*.' — B. D. Halsted, ' Abnormal Ash-leaves.' — V. Havard,
' Distribution of *Buchloë dactyloides*.'

*Gardeners' Chronicle* (Aug. 18). — *Aloe penduliflora* Baker, *Masdevallia platyrachis* Rolfe, *Iris cypriana* Baker & Foster, *I. Barnumi*
Baker & Foster, spp. nn. — W. G. Smith, *Peronospora elliptica*
(fig. 21).—Proliferous Strawberry (fig. 23).—(Aug. 25). *Saccolabium
cerinum* Rchb. f., *Bollea hemixantha* Rchb. f., spp. nn.

*Journal de Botanique* (Aug. 1). — A. G. Garcin, ' Sur le genre
*Euglena* et sur sa place dans la classification.' — A. Masclef,
· Géographie botanique du Nord de la France.' — P. Vuillemin,
' L'*Ascospora Beijerinckii* et la maladie des Cerisiers.' — (Aug. 16).
E. Bondier, ' Sur le vrai genre *Pilacre*.' — P. Maury, *Eranthemum
plumbaginoides*, n. sp. — N. Patouillard, *Prototremella*, n. g. — A.
Masclef, ' Flore des collines d'Artois.'

*Journ. Linn. Soc.* (xxiv. 163 : Aug. 8). — S. le M. Moore,
' Influence of Light upon Protoplasmic Movement ' (3 plates).—
H. N. Ridley, ' Self-fertilization and Cleistogamy in Orchids '
(1 plate). — H. J. Veitch, ' Fertilization of *Cattleya labiata* var.
*Mossiæ* ' (14 cuts).

*Oesterr. Bot. Zeitschrift* (Aug.). — H. Braun, Memoir of Josef
Pančić (May 6, 1814-March 8, 1888). — K. Fritzsch, *Verbascum
Styriacum*, sp. n.—A. Hansgirg, ' Kellerbacterien.'—J. Bornmüller,
*Verbascum Pancicii*, hybr. nov.

---

THE following recent appointments have not been recorded in
these pages :—Dr. SYDNEY VINES succeeds Professor Balfour, who
has left Oxford for the Chair of Botany at Edinburgh. Mr. F. W.
OLIVER succeeds his father, Professor Daniel Oliver, at University
College, London. Mr. H. N. RIDLEY goes to Singapore as Director
of the Gardens and Forests of the Straits Settlements. Mr. H. O.
FORBES has been appointed Director of the Museum at Christchurch, New Zealand. Dr. TRIMEN has been elected a Fellow of
the Royal Society.

# BOTANICAL NOMENCLATURE.

[WE have received the following letter from M. Alphonse DeCandolle, which will be read with interest.—ED. JOURN. BOT.]

J'AI lu avec intérêt l'article que vous m'avez communiqué du 'Journal of Botany,' Sept., 1888, p. 257. Vous critiquez, avec raison ce me semble, des usages qui s'introduisent en Amérique pour la citation des auteurs, mais je ne connais pas assez exactement les publications dont vous parlez pour les examiner en détail. Il me parait qu'elles sont réfutées déjà, par des bonnes raisons, soit dans mon 'Commentary of the Laws of Nomenclature,' pag. 57 et suivantes des 'Laws of Botanical Nomenclature,' London, 1868, que vous possedez sans doute, et j'ajouterai que la question a été traitée plus à fond dans mes 'Nouvelles Remarques sur la Nomenclature Botanique,' Genève, 1883, pag. 25 à 27, et p. 53, 54. Je vous envoie cet opuscule.

Comme il n'a pas été traduit en anglais, vous pourriez peut-être avancer la question en reproduisant dans votre journal en anglais les pages indiquées tout à l'heure, suivant 25 à 27.

J'ai combattu dans ma 'Phytographie,' Paris, 1880, p. 259, 272, et ailleurs, l'emploi de formules ou de signes ou de procédés typographiques qu'on ne comprend pas à première vue, sans explication. C'est le cas des parenthèses. Quand on voit un nom encadré, comme (Pursh), personne ne comprend ce que cela signifie à moins d'aller chercher dans la préface ou dans un autre ouvrage le sens attribué à cette forme. La clarté, condition essentielle d'un livre scientifique, exige qu'on puisse lire à haute voix et comprendre à livre ouvert. Les idées du savant sont déjà un peu difficiles à saisir : il faut au moins que les mots et les formes typographiques soient clairs.

Sous le rapport de la clarté et de l'exactitude aucun botaniste n'a dépassé Asa Gray. Ses ouvrages sont des modèles, et les Américains devraient en être fiers. Pourquoi cherchent-ils à faire autrement que lui ?

Je suis surpris que vous n'adoptiez pas la règle ordinaire de conserver un nom spécifique (à moins de cas exceptionnels) lorsque l'espèce est transportée dans un autre genre. Le principe admis est la fixité des noms (art. 3 des Lois adoptées à Paris) et l'article 62 le confirme en indiquant les exceptions à la règle. Ainsi, dans le cas du *Castalia*, Salisbury n'aurait pas dû changer le nom spécifique *odorata*, qui n'a rien de faux ou impossible dans le genre *Castalia*. Après lui tout botaniste qui admet le genre *Castalia* doit dire *Castalia odorata*. Je n'ai jamais hésité sur ce point. La fixité des noms d'espèces est une des choses les plus importantes. La seule manière d'empêcher la création inutile de noms spécifiques nouveaux et, en général, de noms nouveaux est de ne pas les adopter. La Prodromus a rétabli une quantité de noms d'espèces détruites, on ne sait pourquoi, lorsqu'on avait porté l'espèce dans un autre genre.

ALPH. DE CANDOLLE.

[The following is a translation of the portion of the 'Nouvelles Remarques' to which M. DeCandolle refers above.]

### ARTICLE 48.—CITATION OF NAMES OF AUTHORS.

The essential principle, which ought to govern every citation of the name of an author, is this : *Never to make an author say what he has not said;* we might even add, what he has not said *clearly*. This is a much-extended application of the principle, Never do to others what you would not have them do to you. *

Many naturalists do not observe this rule, sometimes from carelessness, and sometimes, which is more strange, from an erroneous notion of justice. Thus, a family is often attributed to an author, although he has only made a tribe of the group in question ; a sub-genus is ascribed to him when he has constituted it a genus, or *vice versâ*. These are causes of error and obscurity against which I have already pronounced. †

The purpose of the citation of the author's name is misunderstood by some. This is merely an abridged form of a bibliographical indication, intended to establish, without lengthy research, the date which fixes the priority of a name. Dr. Asa Gray has manifested to me his wish that this had been expressed in the article, and I now comply with his desire by the addition of these few words.

Sometimes one is puzzled as to the citation of the author of a combination (*groupe*). I will indicate a practical method, applicable in all cases :

Write the name of the combination (*groupe*), with the citation of the work in which it was first published: thus—

*Bidens* Linné Genera, no. 932.
*Bellevalia romana* Reichenbach, Fl. germ. excurs., p. 105.

Take away what follows the name of the author, and you have:

*Bidens* Linné.
*Bellevalia romana* Reichenbach.

The advantage of this proceeding is, that it never attributes to an author what he has not published, for, if he had not published it, you could not cite title nor page.

Inversely, we can verify the accuracy of the citation of an author by searching in what publication he has given the name or names attributed to him. If it is not found there, the citation is erroneous. Many modern citations will not stand this test.

I have often combatted ‡ the innovations which consist in mixing with the names of the species and its author the history of names which have preceded it. I have not had the good fortune to convert some naturalists whose merit I fully recognise, but a large

---

* A. DeCandolle, Bull soc. roy. de bot. de Belgique, xv. 1876.
† *Ibid.*
‡ Commentaire, pp. 45—55 ; Phytographie, pp. 360, 464.

number of' others * have warmly supported me. I wish to show now, without going again to the bottom of the matter, that the proceedings of which I speak tend to destroy two of the principal advantages of the Linnean nomenclature of species—clearness and brevity. This nomenclature is styled binominal, but, as the name of the author is nearly always added, it is rather trinominal. But the methods invented by some zoologists, and imitated in botany first by M. Bubani † and afterwards by others, render nomenclature quadrinomial, and sometimes even longer, seeing that there are many ways of mixing the history of a species with its description.

The committee of zoologists of the British Association, in 1842, recommended that the name of the original author of the species should be given in a parenthesis, no matter to what genus the species might afterwards be transferred. *Muscicapa crinita* Linné would become *Tyrannus crinitus* Linné (sp.), or, if preferred, *Tyrannus crinitus* (Linné).

M. Bubani has followed a more explicit form, " *Thlaspi rivale* (Cupani) Presl," showing that Cupani first distinguished the plant, and that Presl referred it to the genus *Thlaspi*.

Further on, he gives " *Helianthemum croceum (Clusii, Cupani, Micheli)* Persoon," in which he shows himself to be more logical than any other adherent of the new systems. For, if we wish to intercalate in the title of a species the authors who deserve recognition, we must cite the one who first described it (perhaps very badly), he who first placed it in its right genus, he or those who have published the best description, he who has given a good figure, and, in some cases, the collector who has risked his life to bring the plant from a distant land.

Other botanists content themselves with a history more developed than that proposed by the British Association, but more clear than that of *Thlaspi* cited from Bubani : e. g., " *Evax exigua (Sibthorp) sub Filago*," or " *Matthiola tristis Linné (Cheiranthus).*"

The zoologists sometimes ‡ say " *Crania craniolaris* Nilsson ex Linné," which may be interpreted in two ways : either that Linné spoke of a genus *Crania* and of Nilsson, which is impossible, or that something concerning the species has been found in his works.

Let us now read aloud these designations of species, and count how many words they necessitate :

## Linnean method.

Muscicapa crinita Linné      ...      ...      ...      ...      3 words.

---

* The Commission of the Bulletin of the Botanical Society of France, 1860, p. 430; Caruel, *ibid.*, 1864, p. 11; Malinvaud, *ibid*, 1881, p. 10; Caruel, Journal of Botany, 1877, p. 282; Trimen, Journal of Botany, 1878 [1877], p. 189; 1878, p. 170; D Jackson, *ibid.*, 1881, p. 76; not to mention a host of writers of Floras, Monographs, or ' Genera,' who have followed the old method of citation.

† Bubani, Dodecanthea, Florentiæ, 1850.

‡ This proceeding is recommended by M. Crépin (La nomenclature au Congrès de Paris) in a fair and extended discussion, in which he arrives at conclusions opposed to our own.

New methods.

| | |
|---|---|
| Tyrannus crinitus, in parenthesis Linné ... ... | 5 words. |
| Tyrannus crinitus Linné, in parenthesis species ... | 6 „ |
| Thlaspi rivale, in parenthesis Cupani, Presl ... | 6 „ |
| Evax exigua, in parenthesis Sibthorp sub Filago ... | 7 „ |
| Matthiola tristis Linné, in parenthesis Cheiranthus Brown ... ... ... ... ... ... | 7 „ |
| Helianthemum croceum—open parenthesis, Clusii, Cupani, Micheli—close parenthesis, Persoon ... | 10 „ |

This is nothing less than a return to *phrases*, from which the eminently practical genius of Linné delivered natural history.

---

THE learned editor of this Journal has discussed (pp. 257–262) at considerable length, under the title "Recent Tendencies in American Botanical Nomenclature," certain mooted questions regarding plant-names, with especial reference to two papers in which I have been concerned; and, as he has, with his customary courtesy, invited me to reply, I cheerfully take advantage of the opportunity. As I am much occupied with other matters during my present stay in England, I must content myself with answering his criticisms only in part, leaving the remainder, and perhaps a fuller statement of the position which he has assailed to be taken up in the future, and by others better qualified than myself.

We are all agreed, in a general way, that priority of publication is the only test to be applied in determining the name an organism should bear; we are also pretty much of one opinion as to what constitutes publication, but we are not yet all of one mind as to what is to be regarded as constituting a name for this purpose. Is it the specific name only, or is it the binomial? The fact that some of us have agreed to thus regard the former, and to cite the original founder of the species as authority for it in whatever genus or rank it may be placed, has been the cause of a discussion, of which the present paper forms a part. It seems strange that this simple question should not have been decided long ago, and stranger yet when we consider that with nearly all biologists, except some students of the Flowering Plants and the Fern cohort, it has been agreed to accept the system to which Mr. Britten now takes such vigorous exception. No valid reason has yet been assigned why these anthologists and pteridologists have refused to join in this practice their not less able nor worthy colleagues. We are told, it is true, that desperate confusion will be evolved by the increased number of synonyms, and that the other way is so much more convenient, but no systematist should contend that these are really sufficient excuse for a bad practice, and we are almost forced to the conclusion that a mistaken conservatism has more to do with it than anything else.

Extending our inquiry a little beyond England, we find that even many students of the higher plants have adopted and are now following the system which is so strongly condemned. Mr. Britten remarks at the outset of his argument, "The vexed questions connected with botanical nomenclature have been for the most part settled in Europe by the adoption of the 'Laws' formulated by M. DeCandolle," but, if these recommendations actually mean what he supposes they do, we can hardly regard this statement as justified while anthologists at Berlin, Paris, and other continental centres of botanical activity, several of the monographers of the new DeCandolle series, and of that greatest botanical work, Martius' 'Flora Brasiliensis,' Boissier, a countryman and neighbour of DeCandolle, Lindberg in Finland, Braithwaite and Spruce in England, and indeed all bryologists, mycologists, and algologists the world over use the earliest specific name as permanent! Nor can it, I think, be argued that it is being settled in the way he would have it, for the recommendations of the Paris Congress date from 1867, while there is at the present time an increasing tendency towards the recognition and maintenance of the earliest specific name. Indeed the practice has been followed in the most recent British Local Flora.

But do these laws prohibit the invariable use of the earliest specific name, or the parenthetical citation of the original author? Article 48 is quoted to show that they do, and as it is the only one which bears directly on the question, I will quote it also :—" For the indication of the name or names of any group to be accurate and complete, it is necessary to quote the author who first published the name or combination of names in question." Will our learned critic perhaps tell us in what way we have transgressed this law? It seems to me that, in citing both the original author of the specific name and that of the accepted binomial, we obey that law to the letter, for we quote both the authors of the name and of the combination of names.

It is true that in his commentary on this article M. DeCandolle personally does not favour the practice, but the only objection he has to make to the form *Matthiola tristis* (L.), Brown, is that "the parenthesis has first to be explained." Certainly we have to explain all symbols to a person first learning, but once explained, it is understood that Linnæus made the name and Brown the binomial. He further states that "the reader, having learnt that Linnæus made only the specific name, wishes to know under what generic name." An excellent result, if it really has that effect, which we somewhat doubt.

But the whole matter had been ably discussed long before the time of the Paris Congress. The Stricklandian Code adopted by the British Association for the Advancement of Science in 1842 was in 1845 discussed by the Association of American Geologists and Naturalists, and adopted with but few changes; and in the published report of their Committee (reprint, p. 7) we find :—

§ 9. "It is recommended that the original authority of a species should always follow the name in brackets, and if the name be subsequently altered, the authority for the same be added without brackets. It has been common for systematists to change a generic name, and then to add their own name to all the species. To prevent this injustice, which is no less than a kind of piracy, the above rule is proposed. As an example, the *Tyrannus crinitus* of Swainson is the *Muscicapa crinita* of Linnæus; to distinguish here the author of the former name and give due justice to Linnæus it may be written *Tyrannus crinitus* (Linn.), Swain."

Now our friends the zoologists, who, as has been recently justly remarked, have always been a little ahead of us in matters of this kind, and the cryptogamists have long been working on the basis of these rules and recommendations (not without certain modifications, however), and their nomenclature has assumed a gratifying degree of stability. Yet Mr. Britten awards us the honour of having introduced the system! This is rather hard on Elias Fries, Boissier, and Richard Spruce, not to mention a host of others who long ago practised it.

We may now advantageously consider what degree of stability there may be for the binomial. Mr. Britten thinks that "when once established it can undergo no farther change." Has that been his experience with names apparently settled? As long as anyone who can show with good reason that a genus should be divided, or two or more united, may give any names he pleases to the species, and have them maintained by botanists generally, it is certain that nothing like stability has been reached. The wide differences of opinion regarding generic limits must always render that process inevitable. It is true that the Laws of the Botanical Congress do not approve of changing specific names under these circumstances, but they do not forbid it. The plan of fixing the earliest specific name gives us, on the other hand, an excellent basis for stability.

Surely the editor understands the custom of omitting the author's name after a binomial in the place where it is first created. That is sufficiently shown in all reputable periodicals by printing the name of the writer at the beginning or end of the article. Yet he asks why I write *Disporum Menziesii* (Don), in a recent paper where I transferred the species of *Prosartes* to that genus, and restored their original names.* Were I referring to that plant again, I should write *Disporum Menziesii* (Don), Britton. If I should write as he desires, it might be *D. Menziesii*, Britton, making it appear as if I had actually founded and described the species, whereas Don established it forty years or more ago. I have simply examined it, and placed it in a genus where I conceive it should rest. Or, worse still, should I follow his wishes, it might be *D. californicum*, or any other name not already taken up in the genus, and Don's very good species might have lost in name all trace of his original appellation. Does he argue that this would be conducive to stability? Or does he consider it right? He appears to think that Bentham and Hooker have something to do with it under *Disporum*. Nothing could be more erroneous. How are we to know what name they might have given it under *Disporum?* Mr. Bentham, as we know from his published statements, was a firm believer in the oldest binomial, and he might have called the plant anything he considered appropriate. In the 'Genera Plantarum' it is simply indicated that the authors did not regard the genera as separable.

---

* Bull. Torrey Club, xv., 187.

It is inconsequent to argue at the present stage of botanical literature on the terrible "confusion" arising from increased synonymy. Mr. Britten cheerfully admits the necessity of taking up the old generic names, *Castalia, Hookera, Buda*, and such, and realizes that their acceptance in the place of more familiar ones will involve great changes in names. Certainly it will; but what possible excuse can there be for not taking up the old specific ones at the same time? According to him, Prof. Greene "has needlessly increased our synonymy" by restoring Dryander's name to our delicious aquatic, *Castalia odorata*. Does he make out that it is any more confusing to replace *Nymphæa odorata* in his brain by *Castalia odorata* than by *Castalia pudica*? The irritating example of *Carya* given shows very well how Nuttall coined names for our Hickories. Without realizing it, Mr. Britten has, in this example, hit on a grave mistake made by the Torrey Club Committee on Nomenclature, who erred in writing *Carya alba* for a plant when there was already such a binomial belonging to a different species. Fortunately, one or the other of Rafinesque's genera (are they not meant for the same?) has abundant priority, and everything in *Carya* must go to the limbo of synonymy; we hope that the new binomials will be less annoying.

A great fuss is made about the apparent cumbrousness of the citation of the Committee who edited the nomenclature of the New York Catalogue, and still the editor must be aware of the common custom of abbreviating authors' names. We do not write out Humboldt, Bonpland, and Kunth, or Casimir DeCandolle, every time we cite these authors, but instead H. B. K., and C. D. C., and so, while I am assured that the Committee are painfully aware of their slight importance compared with such masters of the science as these, we still find it convenient to write *Pycnanthemum virginicum* (L.), B. S. P.

As is distinctly stated in the preface to their Preliminary Catalogue, the Committee decided not to consider the question of generic names at the time the work was done, and the reasons for this apparent neglect are there given, and need not be repeated here. In accepting for the most part those taken up by Bentham and Hooker, their work was appreciably shortened, as was quite necessary under the circumstances.

And, finally, as regards this Catalogue, on which so much notice has now been bestowed, the circumstances are simply these:—A Committee was appointed by the Torrey Botanical Club to prepare a check-list of plants growing naturally within 100 miles of New York City; the nomenclature was delegated to a Sub-committee; the members did not consider the ordinary method of citing authors as just, rational, nor stable, and they adopted what they believed to be a better plan. Certainly it is not incumbent on others to take the same view. As to such we can only regret their refusal to accept a system which our common sense of justice and right assures us to be the better.

N. L. BRITTON.

Kew Gardens, Sept. 5, 1888.

Without wishing to prolong this discussion, I may briefly comment on one or two of the points raised above. The whole matter, however, has been so fully dealt with in this Journal that it was perhaps hardly necessary to reopen the matter. As I have done so, I may briefly cite one or two passages which support the views which I ventured to express in this Journal for September.

The view advanced in the last paragraph of M. DeCandolle's letter has been already formulated by him in these pages; * and Dr. Trimen's note thereupon so exactly expresses the objection I still venture to feel with regard to it that I transcribe it here. Dr. Trimen says † :—" Probably all botanists are agreed that it is very desirable to retain, when possible, old specific names, but some of the best authors do not certainly consider themselves bound by any generally accepted *rule* in the matter. Still less will they be inclined to allow that a writer is at liberty, as M. DeCandolle thinks, to reject the specific appellations made by an author whose genera are accepted, in favour of older ones in other genera. It will appear to such that to do this is needlessly to create another synonym." This is what I have said (p. 259) with reference to *Castalia pudica;* and it is still my opinion that *"Castalia pudica* Salisb.," not *"Castalia odorata* (Dryand.) Greene," must stand as the name of that plant. Prof. Caruel ‡ and Mr. Hiern § support this view, and their arguments seem to me unanswerable.

I find little in Mr. Britton's ably-written paper which has not been answered by anticipation. One or two little points seem to me a little captions—Mr. Britton could hardly have thought I meant to award to the young American school what he styles "the honour of having introduced the system" to which I have taken exception. With regard to *Prosartes*, I do distinctly think that "Bentham and Hooker have *something to do with it* under *Disporum*," inasmuch as they were the first to place it there; but I was then showing the absurdity of the notion of "credit" being connected with name-giving, and I have elsewhere pointed out the inaccuracy of appending "Benth. & Hook. f." to species transferred in this manner. When Mr. Britton says that by writing " D. Menziesii Britton" he would make it appear that he had " actually founded and described the species," he seems to me to beg the whole question.

Mr. Britton claims what he calls "the most recent British local flora" in support of his view. By this he means Messrs. Stewart and Corry's ' Flora of the North-east of Ireland,' noticed last month. I pointed out to him that this only supported his position occasionally, and suggested the omission of the reference; but Mr. Britton insisted on the retention of the passage. I am bound to admit that the objectionable practice is more widely adopted in that book than I had thought; but such citations as *Sisymbrium Alliaria* Linn., *Erodium cicutarium* L'Herit., *E. moschatum* L'Herit., *E. maritimum* L'Herit., *Melilotus officinalis* Willd., *Lychnis Githago* Linn., *Stellaria media* Linn., *Vicia tetrasperma* Moench, and a host of others, seem to show that Mr. Stewart was halting between two opinions. It is

---

* Journ. Bot. 1887, 242.    † *Ibid.*    ‡ *Ibid.* 282.    § Journ. Bot. 1878, 73.

also apparent that the new quadrinomial nomenclature is not free from complications: here are a few names from Mr. Stewart's book, with their equivalents in what Mr. Britton wants us to cite as B.S.P.:—

| B.S.P. | STEWART. |
|---|---|
| Glaucium luteum, Scop. | Glaucium flavum Crantz (G. luteum Scop.). |
| Nasturtium palustre, (L.), DC. | Nasturtium palustre (Willd.) De Candolle. |
| sylvestre, (L.), R. Br. | sylvestre R. Brown. |
| Barbarea præcox, (Smith), R. Br. | Barbarea præcox R. Brown. |
| Arabis hirsuta, (L.), Scop. | Arabis hirsuta (Linn.) R. Brown. |
| Sisymbrium Alliaria, Scop. | Sisymbrium alliaria Linn. |
| Thaliana, (L.), Gay. | thalianum (Linn.) Gaud. |
| Senebiera Coronopus, (L.), Poir. | Senebiera coronopus (Gært.) Poiret. |

Mr. Stewart also gives "*Lepidium Smithii* (Linn.) Hooker," as to which I would only remark that this species was first described by Hooker in 1835,* nearly eighty years after the death of Linnæus.

JAMES BRITTEN.

---

## NOTES ON PONDWEEDS.

### By ALFRED FRYER.

POTAMOGETON FLABELLATUS Bab. — Rootstock stout, tuberous, with far-creeping stolons. Stem stout, round or slightly two-edged, much branched from the base when rising from a tuber or old rootstock, but usually simple below when springing from a newly-produced stolon, ultimately much branched above, with the branches spreading on the surface of the water like a fan. Leaves all similar, submerged, alternate, linear, flat, with the lower part united to the stipule, and so forming a sheathing petiole, which is elongated into an obtuse scarious ligule, free at its apex. Lamina of the *lower leaves broadly linear*, abruptly cuspidate or acute, or rarely rounded and slightly concave or hooded at the tip, 4-6 in. long by ⅛ in. wide, *3-5-ribbed, with transverse veins*, decayed on the fertile stems at the time of flowering; sheathing petiole ¼–⅓ as long as the lamina. *Upper leaves 3-ribbed*, narrowly linear, slightly channelled, tapering to an acute or acuminate point. Peduncles filiform, equal, lateral, or rarely terminal, exceeding the subtending foliage, 2-4 or many times as long as the interrupted few-flowered spike. Fruiting spike interrupted below or throughout its whole length, or with the fruits more thickly clustered above. *Fruit large, nearly straight on the inner margin*, which is terminated by a short, almost lateral beak; outer margin *gibbous above*, with a *prominent central keel, and obscure lateral ridges*. *Nut with a prominent*

---

* British Flora, ed. 3, 300.

*keel.* Whole plant dull olive-green, drying darker, of very vigorous growth, 3–10 ft. long.

The broad, lower leaves are submerged on the fruiting stems, and on these are usually (or always?) much decayed by the time of flowering. They are occasionally produced late in the summer on barren shoots, and are then to be found on the same rootstock as the fruit-bearing stems. Sometimes in rapid streams, which are too swift to allow the flower-spikes to emerge and become fertilized, they are persistent, and sometimes they are quite absent for one or more seasons. These broad, 3–5-ribbed leaves are usually *stem-leaves,* but in barren states of the plant they occur on the upper branches, and then form the outer or *sheathing leaf of each branchlet,* the whole of the leaves of which in such cases are often of a somewhat thickened, stouter growth; rarely this state of the plant produces flower-spikes, when it is, I believe, the *P. Vaillantii* of some authors.

In stagnant water all the branches ascend to the surface, and spread out like a fan; but in tidal rivers or running streams the branchlets have a more elongate, parallel growth, and the tips of the leaves rise slightly above the surface of the water, looking like short blades of grass. Like all other Potamogetons, this species bears the flower-spikes above water until fertilization has taken place, when the peduncle sinks back, and the spike remains barely submerged on the surface of the foliage until the fruit is full-grown; but I believe the final ripening of the fruit takes place at the bottom of the water after the fruiting branchlets have rotted off. As far as I have been able to observe, pondweeds vary considerably in the depth at which they mature their seeds. *P. natans* sometimes ripens its fruit fully exposed to the air; *P. heterophyllus* bends back its spikes, and the seeds are matured at the depth of two or three inches; and *P. prælongus* sinks to the bottom even before the fruits are full-grown. I wish to call the attention of students to these variations of habit, as they may possibly afford valuable diagnostic characters when properly worked out.

Some confusion has arisen in the minds of botanists through a state of *P. flabellatus,* named *P. scoparius,* having been wrongly placed under *P. pectinatus* as a variety of that species; this form certainly has a superficial resemblance to the latter plant in its slender growth, and finely setaceous leaves; but these leaves have the structure of those of *P. flabellatus,* and the fruit is absolutely identical in character with that of the latter species. "*P. scoparius*" is a form that inhabits brackish and stagnant waters: through the kindness of Mr. J. Owles, of Great Yarmouth, I have been able to cultivate a salt-water form of this variety, which has very fine bristle-like leaves, the internodes between which are so shortened as to give the branchlets a fasciculated appearance. To this form I believe the varietal name of *pseudo-marinus* has been given. Under cultivation in a stagnant pond I find these several forms of *P. flabellatus,* while retaining their individual characteristics to some extent, do certainly approximate to the fine-leaved form named "*scoparius,*" and that they do not at all approach *P.*

*pectinatus* growing in the same pond. Hence, as far as my observations have gone, I am inclined to consider *"scoparius,"* " *Vaillantii*," and "*pseudo-marinus*" as mere states of *P. flabellatus;* I retain that name for the species because its author, Prof. Babington, was the first to clearly separate it from the Linnean aggregate, *P. pectinatus.*

*P. flabellatus* differs from *P. pectinatus* by its more robust and less submerged growth, and by the more fan-like expansion of its branchlets; by its broad, flattened, 3–5-veined lower leaves, and above all, by its prominently keeled fruit. In *P. pectinatus* the fruit has no central keel, and the lateral ridges are usually very conspicuous. On transverse section the difference in the fruits of the two species becomes very apparent, the dorsal margin of the nutlet forming an elliptical arch in *P. flabellatus,* while in *P. pectinatus* it forms a rounded arch.

*P. flabellatus* inhabits both fresh and brackish water, and, judging from its frequency in herbariums, is probably the commoner form throughout Great Britain.

My thanks are especially due to Prof. Babington for his careful examination of the many specimens of this group which I have submitted to him, and for allowing me to study the type-specimens on which he founded this species. By this means I have been able to identify the fruit of *P. flabellatus* with that of *P. scoparius.* To Messrs. Charles Bailey and Arthur Bennett my thanks are also due for the loan of their valuable series of the plants of this group collected in all parts of the world.

By these aids, and by continued observations made year after year in the field, and also by cultivation of various forms of the *pectinatus* group, I have come to the conclusion that *P. flabellatus* is a *good species*, bearing, perhaps, the same relation to *P. pectinatus* as *P. heterophyllus* bears to *P. Zizii* or the latter to *P. lucens.*

## JOHN GOLDIE.

ON the death of this veteran botanist in June, 1886, an interesting biographical sketch appeared in the ' Botanical Gazette' for October of that year, which we then marked for abstract in these pages, but which has hitherto been held over from want of space. At the same time we had occasion to refer to the description of plants published by him in the ' Edinburgh Philosophical Journal' for 1822 (pp. 319–333), which is prefaced by an account of his first visit to North America. This narrative is not referred to in the ' Botanical Gazette;' and, as it presents many points of interest, we reproduce it here.

John Goldie was born near Maybole in Ayrshire, on March 21st, 1793, and was brought up a gardener. He married early, in 1815, and became connected with the Glasgow Botanic Garden, where, in company with David Douglas, he studied botany under Sir William Hooker, whose life-long friendship he afterwards enjoyed. In 1817 he paid his first visit to America, of which he published, as already stated, the following account :—

" Having had for many years a great desire to visit North America, chiefly with a view to examine and collect some of its vegetable productions, I contrived, in 1817, to obtain as much money as would just pay my passage there, leaving, when this was done, but a very small surplus. In the month of June I sailed from Leith, and landing at Halifax, remained for some days botanizing in the neighbourhood of that place, where I met with several plants which were interesting to me, especially a yellow-flowered variety of *Sarracenia purpurea*, which I have never since seen elsewhere. From hence I went to Quebec, carrying with me all the roots and specimens that I had obtained, which, together with the produce of two weeks' researches in the neighbourhood of Quebec, I put on board a vessel which was bound for Greenock, but never heard of them afterwards. Hence I proceeded to Montreal, where, meeting with Mr. Pursh, author of the North American Flora, he advised me to turn my course towards the north-west country in the following spring, and promised to procure me permission to accompany the traders leaving Montreal. I travelled on foot to Albany, and then proceeded by water to New York. I remained but a short time in this last place, for I explored the eastern part of New Jersey,— a country which, though barren and thinly inhabited, yet presents many rarities to the botanist, and gave me more gratification than any part of America that I have seen. At a place called Quaker's Bridge I gathered some most interesting plants, and having accumulated as large a load as my back would carry, I took my journey to Philadelphia, where I staid but a very short time ; for knowing that a ship was about to sail from New York to Scotland, I hastened to return thither ; and having again entrusted my treasures to the deep, I had again, as the first time, the disappointment of never obtaining any intelligence whatever of them.

" My finances being now extremely low, and winter having commenced, I hardly knew what to do ; but after some delay, went up to the Mohawk river, where I found employment during the season as a schoolmaster. I quitted this place in April, 1818, and proceeded to Montreal, expecting to be ready to depart on my journey towards the north-west country. I was disappointed in finding that Mr. Pursh had left Montreal for Quebec, and that even if present, his interest would scarce have been sufficiently strong to have obtained for me the assistance and protection which I desired. My only alternative was now the spade, at which I worked all summer, excepting only two days in each week which I devoted to botanizing, and went also a little way up the Otowa, or Grand River, the only excursion of any length which I accomplished. In the autumn I shipped my collection of plants, and in two months had the mortification to learn that the vessel was totally wrecked in the St. Lawrence. Thus did I lose the fruit of two years' labour.

" During the next winter I did little, except employing myself, with such small skill as I was able, in designing some flower-pieces, for which I got a trifle. Early in the following spring I commenced labour again, and by the beginning of June had amassed about fifty dollars, which, with as much more that I borrowed from a friend, formed my stock of money for the next summer's tour. I started in the beginning of June from Montreal, and passing through Kingston, went to New York, then visited the Falls of Niagara and Fort Erie, and crossed over to the United States. Keeping along the eastern side of Lake Erie for ninety miles, I afterwards took a direct course to Pittsburgh on the Ohio, which, owing to the advanced state of the season, was the most distant point to which I could attain. On my return I kept along the side of the Alleghany river to Point Ollean, in the State of New York, then visited the salt-works of Onondago and Sackett's Harbour on Lake Ontario, whence, proceeding to Kingston, I packed up my whole collection, with which I returned to Montreal, and embarking in a vessel which was bound for Greenock, got safely home; the plants which I carried with myself being the whole that I saved out of the produce of nearly three years spent in botanical researches."

The descriptions of twenty-three species follow. Of these, fifteen were considered new, including *Viola Selkirkii*, named in MS. by Pursh, and *Aspidium Goldianum* of Hooker's MSS.; *Ranunculus rhomboideus*, *Stellaria longipes*, *Drosera linearis*, *Corydalis canadensis* (now referred to *Dicentra*), *Xylosteum oblongifolium* (*Lonicera*), and *Primula pusilla*, still retain specific rank.

The account of the 'Botanical Gazette' gives the following summary of the remainder of Goldie's life :—

" On returning to Scotland, after this second American tour, he was, in the year 1824, recommended by Mr. McNab, of the Edinburgh Botanical Gardens, to collect and take charge of a vessel-load of plants to be taken to St. Petersburg for the starting of a botanical garden there, in which mission he acquitted himself to the satisfaction of his employers. On his return from this expedition he settled down with his family in the nursery business, but returned to Russia again in 1830, and made a collecting excursion through the country, amongst some of the fruits of which was the introduction to the English horticultural world of such plants as the *Picea pictita*, *Pavenia tenuifolia plena*, &c. From this time till the year 1844 he followed the business of nurseryman and florist at the old home near by to the birthplace of the poet Burns, a few miles from the town of Ayr.

" In 1844, having formed a favourable opinion of Canada West as a place of emigration, in which he might have a chance to better the circumstances of himself and family, he took ship with his entire household for Montreal, and from there journeyed westward, and chose as a resting-place a spot near some of his old-world neighbours, about a mile from Ayr, in the county of Waterloo, when he died, surrounded by children, grandchildren, and great-grandchildren, last June, in his ninety-fourth year."

# RANZANIA: A NEW GENUS OF BERBERIDACEÆ.

## BY TOKUTARO ITO, F.L.S.

SOME years ago, while occupied in studying certain Japanese species of Berberidaceæ, a family more magnificently represented in Japan than in any other parts of the world, I came across some imperfect specimens of a plant which appeared to me as yet undescribed. Finding that this plant showed a closer relation in many respects to *Podophyllum* than to any of the other genera of Berberidaceæ, I appended to one of the specimens the new specific name of *Podophyllum japonicum*, and sent it to M. Maximowicz, of St. Petersburg, for further consideration. Consequently this name appeared in the sixth part of Maximowicz' ' Diagnoses plantarum novarum Asiaticarum,' as one of the new plants lately discovered in Japan.* Afterwards, when I made a systematic study of Japanese Berberidaceæ, I entertained much doubt with regard to its generic position, and stated as follows :— †

"Hæc tamen est manifeste nova species, cui ceteræ *Podophylli* species de toto dissimiles sunt ternatis foliis. Mihi antem a speciminum paucitate positio generis adhuc dubia videtur."

Since I made this statement, I have been able to examine more specimens, some of which have either solitary or a few flowers, and thus the "generis positio" of this plant became much more manifest. It appears that the plant, at the time of flowering, has the leaves still immature, but soon after the whole plant, especially the leaves, makes enormous development, becoming several times larger than at the time of flowering ; and finally the ovoid fruits, each containing numerous minute seeds, become ripe. The plant has often solitary, few, or sometimes umbellate flowers, each flower having a fine trimerous perianth. The mode of dehiscence of the anthers, the sessile stigma, ternate leaves, and other characters, seem to show the necessity for its distinction from *Podophyllum*, and for the establishment of a new genus.

I propose to call the new genus after the name of the late Ono Ranzan, the most eminent of Japanese naturalists, rightly denominated by Siebold as the "Linné du Japon." Our genus approaches, and perhaps comes between, *Podophyllum* and *Diphylleia*.

## RANZANIA *T. Ito*, nov. gen.

**R. japonica** mihi.—*Podophyllum japonicum* T. Ito in Maximowicz, Mél. Biol. xii. 1886, 417 ; Journ. Linn. Soc. Bot. xxii. 1887, 434.

Hab. in Japoniæ principali insula : in monte Togakushi, prov. Shinano.

Japonice : Togakushi-sō, *i. e.*, Planta e Togakushi (mihi).

---

* Mélanges Biologiques tirès du Bulletin de l'Acad. Impériale des Sci. d. St. Pétersbourg, vol. xii. 1886, p. 417.

† T. Ito, Berberidearum Japoniæ Conspectus, in Journ. Linn. Soc. London, Bot. vol. xxii. 1887, p. 434.

I may take this opportunity to state that the occurrence in Japan of the much-doubted *Podophyllum peltatum* L. is beyond doubt, and that the real habitat of this species is also on Mount Togakushi, in the province of Shinano, thus manifesting another habitation of this beautiful plant besides those which are on the other side of the Pacific.

PODOPHYLLUM L.— *P. peltatum* L. ex Maximowicz, *loc. cit.* 418 ; T. Ito, *loc. cit.*, 434.—Honzō Dsufu, vol. xxiii. fol. 7 recto.

Hab. in Japoniæ principali insula, m. Togakushi, prov. Shinano. Japonice : Momidsiba Sankayō (mihi).

The full description of *Ranzania japonica* mihi, with diagnosis of the genus, will, I hope, shortly be published.

---

# CATALOGUE OF THE MARINE ALGÆ OF THE WEST INDIAN REGION.

## By GEORGE MURRAY, F.L.S.

(Continued from p. 243.)

### HELMINTHOCLADIACEÆ.

HELMINTHOCLADIA CASSEI Crn. Guadeloupe, *Mazé* !

H. SCHRAMMI Crn. Guadeloupe, *Mazé* !

HELMINTHORA DIVARICATA Ag. Florida, *Harvey, Melvill* ! Bermuda, *Kemp.*

  *Geogr. Distr.* Atlantic (Europe and America), Mediterranean, and Australia.

H. GUADELUPENSIS Crn. Guadeloupe, *Mazé.*

H. ANTILLARUM Crn. Guadeloupe, *Mazé* !

H. DENDROIDEA Crn. Guadeloupe, *Mazé* l

NEMALION LIAGOROIDES Crn. Guadeloupe, *Mazé* !

SCINAIA FURCELLATA Bivon. Guadeloupe, *Mazé* ! Florida, *Harvey.* Bermuda, '*Challenger.*'

  *Geogr. Distr.* Atlantic (Europe and America), Mediterranean, and Australia.

LIAGORA CHEYNEANA Harv. Florida, *Hooper* ! in Farlow, Anderson and Eaton, Alg. Exsicc. Am. Bor., No. 71.

  *Geogr. Distr.* Australia.

L. LEPROSA J. Ag. Guadeloupe, *Mazé* ! Vera Cruz, *Liebman* (*fide Harvey*). Florida, *Harvey, Melvill* !

L. PULVERULENTA Ag. Grenada, *Murray* ! Barbadoes, *Dickie* ! Guadeloupe, *Mazé* ! Vera Cruz, *Liebman* (*fide Harvey*). Bermuda, *Kemp.*

  f. TENUIOR Crn. Guadeloupe, *Mazé* !

L. TURNERI Zan. Guadeloupe, *Mazé* !

  *Geogr. Distr.* Red Sea, Indian Ocean, and Australia.

L. PINNATA Harv. Guadeloupe, *Mazé* ! Florida, *Harvey.*

  Var. ARBUSCULA Crn. Guadeloupe, *Mazé* !

L. valida Harv.    Florida, *Harvey*! *Melvill*, *Hooper*! in Farlow, Anderson and Eaton, No. 70.    Bermuda, *Kemp*, *Rein*, '*Challenger*'!

L. rugosa Zan.    Guadeloupe, *Mazé*!
  *Geogr. Distr.*  Red Sea.

L. viscida Ag.    Grenada, *Murray*!    Jamaica, *Chitty*!
  Var. gracilis Crn.    Guadeloupe, *Mazé*!
  Var. laxa Kütz.    Guadeloupe, *Mazé*!
  Var. coarctata Kütz.    Guadeloupe, *Mazé*!
  *Geogr. Distr.*  Warmer Atlantic, Mediterranean, Australia, and Tasmania.

L. fragilis Zan.    Guadeloupe, *Mazé*!
  *Geogr. Distr.*  Red Sea.

L. ceranoides Lam.    Guadeloupe, *Mazé*!  Vera Cruz, *Liebman* (*fide Harvey*).
  *Geogr. Distr.*  Warm Atlantic and Mediterranean.

L. distenta J. Ag.    Guadeloupe, *Mazé*!
  Var. complanata J. Ag.    Guadeloupe, *Mazé*!
  *Geogr. Distr.*  Warm Atlantic and Mediterranean.

L. albicans Lam.    Guadeloupe, *Mazé*!
L. brachyclada Decne.    Guadeloupe, *Mazé*!
L. decussata Mont.    Guadeloupe, *Mazé*!
  *Geogr. Distr.*  Cape Verde.

L. farionicolor Melv.    Florida, *Melvill*!
L. Cayohuesonica Melv.    Florida, *Melvill*!
L. patens Crn.    Guadeloupe, *Mazé*!
L. prolifera Crn.    Guadeloupe, *Mazé*!
L. ? dendroidea Crn.    Guadeloupe, *Mazé*.
L. bipinnata Crn.    Guadeloupe, *Mazé*!

Galaxaura obtusata Lam.    Barbadoes, *Dickie*!  Guadeloupe, *Mazé*!  Jamaica, *Chitty*!
  *Geogr. Distr.*  Warm Atlantic (Brazil).

G. umbellata Lam.    Guadeloupe, *Mazé*!
  *Geogr. Distr.*  Warm Atlantic, Australia.

G. cylindrica Lam.    Grenada, *Murray*!  Barbadoes, *Dickie*!  Guadeloupe, *Mazé*!  Jamaica, *Sloane*!  *Chitty*!
  *Geogr. Distr.*  Red Sea.

G. fragilis Lam.    Guadeloupe, *Mazé*!
  *Geogr. Distr.*  Tropical seas.

G. fastigiata Harv.    Bermuda, *Rein*.
  *Geogr. Distr.*  There is no *G. fastigiata* Harv. known to me, and the plant so quoted by Rein may be identical with *G. fastigiata* Decaisne, from the Moluccas.

G. rugosa Lam.    Grenada, *Murray*!  Barbadoes, *Dickie*!  Guadeloupe, *Mazé*!  Santa Cruz (*fide Dickie*).  Jamaica, *Sloane*!  Danish West Indian Islands [*Hb. Mus. Brit.*!].  Bermuda, '*Challenger*'!
  *Geogr. Distr.*  Warm Atlantic [Pacific? and Indian Ocean?].

G. annulata Lam.    Barbadoes, *Dickie*!
  *Geogr. Distr.*  Pacific and Indian Oceans.

G. LAPIDESCENS Lam. Grenada, *Murray*! Barbadoes, *Dickie*! Guadeloupe, *Mazé*! Jamaica, *Chitty*! Mexico (*fide Dickie*). Tortugas, *Bailey*! Florida, *Hooper*! in Farlow, Anderson & Eaton, No. 16. Bermuda, '*Challenger*'!
*Geogr. Distr.* Throughout all tropical seas.
G. INDURATA Lam. Danish West Indian Islands (*Hb. Mus. Brit.*!). Bahamas, *Ellis & Solander*.
G. FRUTICULOSA Lam. Bahamas, *Ellis & Solander*.
G. LICHENOIDES Lam. Bahamas, *Ellis & Solander*.
G. VALIDA Crn. Guadeloupe, *Mazé*!

## CHÆTANGIEÆ.

ZANARDINIA MARGINATA J. Ag. Barbadoes, *Dickie*! Guadeloupe, *Mazé*! Antigua (*fide Kützing*).
*Geogr. Distr.* Warm Atlantic, Indian Ocean, Australia.
ACROTYLUS CLAVATUS Harv. Guadeloupe, *Mazé*! Florida, *Harvey*!

## GELIDIEÆ.

WURDEMANNIA SETACEA Harv. Guadeloupe, *Duchassaing* (fide J. Ag.). Florida, *Harvey*! *Melvill*! *Hooper*! in Farlow, Anderson & Eaton, No. 14. Bermuda, *Rein*.
GELIDIUM VARIABILE J. Ag. Guadeloupe, *Mazé*!
*Geogr. Distr.* Indian Ocean.
G. RIGIDUM Mont. Grenada, *Murray*! Barbadoes, *Dickie*! Guadeloupe, *Mazé*! Martinique (*Hb. Brongniart*!). Cuba, *R. de la Sagra*. Bermuda. *Rein.*
Var. RADICANS J. Ag. Guadeloupe, *Mazé*! Cuba, *R. de la Sagra*. Bermuda, *Farlow*! (in Farlow, Anderson & Eaton, No. 142). *Reliquiæ Brebissonianæ Ser. 2, Nos. 129 & 290*!
*Geogr. Distr.* Throughout warm seas.
G. CORNEUM Lam. Grenada, *Murray*! Danish West Indian Islands, *Hohenack*! No. 559. Cuba, *R. de la Sagra*. Bermuda, *Kemp*.
Var. CRINALIS J. Ag. Guadeloupe, *Mazé*!
Var. SETACEA J. Ag. Guadeloupe, *Mazé*!
Var. CÆSPITOSA J. Ag. Guadeloupe, *Mazé*!
Var. PRISTOIDES J. Ag. Guadeloupe, *Mazé*!
*Geogr. Distr.* Throughout all oceans.
G. SERRULATUM J. Ag. La Guayra, *Liebman* (fide Harvey).
G. CARTILAGINEUM J. Ag. Florida [*Hb. Mus. Brit.*!].
*Geogr. Distr.* Indian and Pacific Oceans.
G. FASTIGIATUM Kütz. Guadeloupe, *Mazé*! = DICURELLA FLABELLATA J. Ag.
*Geogr. Distr.* Cape of Good Hope.
G. NANUM Grev. Guadeloupe, *Mazé*!
G. CÆRULESCENS Kütz. Guadeloupe, *Mazé*!
G. DELICATULUM Crn. Guadeloupe, *Mazé*!
G. LIGULATONERVOSUM Crn. Guadeloupe, *Mazé*!
G. SPINESCENS Crn. Guadeloupe, *Mazé*!
G. RAMULOSUM Mart. Martinique, *Hb. Montagne*! (? *Duperrey*).

## HYPNEACEÆ.

HYPNEA MUSCIFORMIS Lam.   Grenada, *Murray*!   Barbadoes, *Dickie*!
Guadeloupe, *Mazé*!   Martinique, *Hb. le Jolis*!   Jamaica,
*Chitty*!   Cuba, *R. de la Sagra*.   Tortugas, *Hb. Dickie*!   Santa
Cruz, *Miss Dix*!   Florida, *Harvey, Melvill*!   Danish West
Indian Islands, *Hohenack.*!   No. 591.   Bermuda, *Kemp, Rein.*

Var. SPINULOSA Mont. et Maill.   Guadeloupe, *Mazé*!   Cuba,
*R. de la Sagra*.

f. TENUIS Hohen.   St. Thomas, *Hohenack*!

*Geogr. Distr.*   Warm Atlantic, Indian and Southern Oceans.

H. NIGRESCENS Grev.   Guadeloupe, *Mazé*!
*Geogr. Distr.*   Indian Ocean.

H. ARMATA J. Ag.   Guadeloupe, *Mazé*!
*Geogr. Distr.*   Cape of Good Hope.

H. SPICIFERA J. Ag.?   Guadeloupe, *Mazé*!
*Geogr. Distr.*   Cape of Good Hope.

H. HAMULOSA Mont.   Guadeloupe, *Mazé*!   Martinique, *Hb. Montagne*!
(? *Duperrey*).
*Geogr. Distr.*   Red Sea, Indian Ocean, Cape of Good Hope.

H. RISSOANA J. Ag.   Barbadoes, *Dickie*!   Guadeloupe, *Mazé*!
*Geogr. Distr.*   Mediterranean.

H. DIVARICATA Grev.   Guadeloupe, *Mazé*!   Gulf of Mexico, *Liebman*
(*fide J. Ag.*).
*Geogr. Distr.*   Mascarene Islands?

H. CORNUTA J. Ag.   Guadeloupe, *Mazé*!   Santa Cruz, *Miss Dix*
(*fide Harvey*).   Florida, *Harvey, Melvill*!   Bermuda, ' *Challenger* '!
*Geogr. Distr.*   China Seas?

H. VALENTIÆ Mont.   Grenada, *Murray*!   Guadeloupe, *Mazé*!
*Geogr. Distr.*   Red Sea, Indian Ocean?

H. CERVICORNIS J. Ag.   Guadeloupe, *Mazé*!   Martinique, *Hb. le
Jolis*!   Gulf of Mexico, Mexican Coast, *Liebman* (*fide J. Ag.*).
*Geogr. Distr.*   Mauritius.

H. SPINELLA J. Ag.   Grenada, *Murray*!   Guadeloupe, *Mazé*!   Cuba,
*R. de la Sagra*.

f. MAJOR Crn.   Guadeloupe, *Mazé*!

H. SETACEA, Kütz.   Guadeloupe, *Mazé*!

H. SECUNDIRAMEA Mont.   Martinique, *Duperrey*.

H. ARBORESCENS Crn.   Guadeloupe, *Mazé*!

H. CORYMBOSA Crn.   Guadeloupe, *Mazé*!

H. ACANTHOCLADA Crn.   Guadeloupe, *Mazé*!

H. GRACILARIOIDES Crn.   Guadeloupe, *Mazé*!

MYCHODEA SCHRAMMI Crn.   Guadeloupe, *Mazé*!

M. GUADELUPENSIS Crn.   Guadeloupe, *Mazé*!

M. POLYACANTHA Crn.   Guadeloupe, *Mazé*!

M. PENNATA Crn.   Guadeloupe, *Mazé*!

## SOLIERIEÆ.

GELINARIA DENTATA Crn.   Guadeloupe, *Mazé*!

CATENELLA OPUNTIA J. Ag., var. PINNATA J. Ag. (= C. PINNATA Harv.).
Florida, *Harvey*! Bermuda, *Farlow*! (in Farlow, Anderson and Eaton, No. 149). *Reliquiæ Brebissonianæ*, Ser. 2, No. 207!
*Geogr. Distr.* Atlantic, Mediterranean, Pacific, Australia.
C. IMPUDICA J. Ag. Guadeloupe, *Mazé*!
*Geogr. Distr.* Indian Ocean.
RHABDONIA TENERA J. Ag. Barbadoes, *Dickie*! Guadeloupe, *Mazé*!
Florida, *Harvey, Melvill*!
*Geogr. Distr.* Atlantic (N. America).
R. RAMOSISSIMA Harv. Guadeloupe, *Mazé*! Florida, *Harvey, Melvill*!
Var. LATIFRONS Crn. Guadeloupe, *Mazé*!
R. DURA Zan., var. GRACILIS Crn. Guadeloupe, *Mazé*!
*Geogr. Distr.* Red Sea.
EUCHEUMA SPINOSUM J. Ag. Guadeloupe, *Mazé*!
*Geogr. Distr.* Indian Ocean.
E. NUDUM J. Ag. Guadeloupe, *Mazé*!
E. ACANTHOCLADA Harv. Barbadoes, *Hb. J. E. Gray*! Florida,
*Harvey, Melvill*!
E. ISIFORME Ag. Cuba, *Tuomey* (*fide Harvey*). Florida, *Harvey*!
*Melvill*! *Hooper*! (in Farlow, Anderson and Eaton, No. 12).
Bermuda, *Kemp, Rein, ' Challenger'*!
E. GELIDIUM J. Ag. Guadeloupe, *Mazé*!
E. SPECIOSUM J. Ag. Grenada, *Murray*!
*Geogr. Distr.* Australia.

### WRANGELIEÆ.

WRANGELIA PLEBEIA J. Ag. Guadeloupe, *Mazé*! Vera Cruz, *Liebman*.
W. PENICILLATA Ag. Guadeloupe, *Mazé*! Florida, *Harvey*! *Melvill*!
Bermuda, *Kemp, Rein*.
*Geogr. Distr.* Atlantic (Europe and America), Mediterranean,
Australia?

(To be continued.)

---

# BIOGRAPHICAL INDEX OF BRITISH AND IRISH BOTANISTS.

By JAMES BRITTEN, F.L.S., AND G. S. BOULGER, F.L.S.

(Continued from p. 282).

**Dickson, James** (1738–1822): b. Traquhair, Peebles, 1738; d. Broad Green, Croydon, 14th August, 1822. Cryptogamist and nurseryman. F.L.S., 1788. Orig. Memb. R. H. S. (1804). Published fascicles of British cryptogams (1785–1801) and flowering plants (1793–1802). Pritz. 82; R. S. C. ii. 285; Trans. Hort. Soc. v., appendix 1; Smith Lett. ii. 234; Felton, 186; Gent. Mag. 92, 376; Dict. Nat. Biog. xv. 44. Portr. in oil in Lindley Library, and one in possession of family; one at Kew. *Dicksonia* L'Heritier.

**Dickson, Joseph** (d. 1874): d. Berwick-on-Tweed, 4th March, 1874. M.D., Edin., 1842. Lived long in Jersey. R. S. C. ii. 286.

**Dickson, Robert** (1804–1875): b. Dumfries, 1804; d. 13th Oct. 1875. M.D., Edin., 1826. F.L.S., 1831. Practised in London. Bot. Lect. St. George's Hospital. 'Lecture on Dry Rot,' 1838. Dict. Nat. Biog. xv. 44.

**Dickson, R. W.** (pseud. **Alexander M'Donald**) (fl. 1806). M.D.? Of Hendon, Middlesex. 'Complete Dict. of Practical Gardening,' 1807. Johnson, 282.

**Dillenius, John James** (1687–1747): b. Darmstadt, 1687; d. Oxford, 2nd April, 1747; bur. St. Peter's-in-the-east, Oxford. M.D., Giessen. M.D , Oxon, 1735. Came to England, 1721. Edited Ray's 'Synopsis,' 1724. 'Hortus Elthamensis,' 1732. 'Historia Muscorum,' 1741. First Sherardian Prof. Bot. Oxford. Pult. ii. 153; Rees; Pritz. 84; Jacks. 539; Dict. Nat. Biog. xv. 79; Linn. Letters. ii. 82; Rich. Corr. 209; Druce, Fl. Oxford, 381. Portr. at Oxford Bot. Gard. *Dillenia* L.

**Dillwyn, Lewis Weston** (1778–1855): b. Ipswich, 1778; d. Skethy Hall, Swansea, 31st Aug. 1855. Of Walthamstow and (from 1803) Swansea. China manufacturer. F.L.S., 1800. F.R.S., 1804. M.P. for Glamorgan, 1832–1841. 'British Confervæ,' 1802–1807. 'Botanist's Guide' (with Dawson Turner), 1805. 'Hortus Collinsonianus,' 1743. Pritz. 84; Jacks. 540; R. S. C. ii. 295; Dict. Nat. Biog. xv. 90; Linn. Trans. vi. (1802); Proc. Linn. Soc. 1856, xxxvi.; Friends' Books, i. 532. Portr. at Kew. *Dillwynia* Sm.

**Dixon, David** (fl. c. 1800). Of Wakefield. Vet. Surgeon. Lees, 'Fl. West Yorks.' p. 99.

**Dodsworth, Matthew.** Clerk. Of Yorkshire. Corresponded with Plukenet (Alm. 180, 201, 251). Mentioned in R. Syn. Pult. ii. 121.

**Don, David** (1800–1841): b. Doo Hillock, Forfar, 21st Dec. 1800; d. Soho Square, 8th Dec. 1841; bur. Kensal Green. A.L.S., 1823. F.L.S. Came to London, 1819. Employed in Chelsea Garden. Librarian to Lambert, and, from 1822, to Linn. Soc. Prof. Bot. King's College, 1836–1841. 'Prodromus Floræ Nepalensis,' 1825. Conducted 'Sweet's British Flower Garden' from about 1830. Pritz. 89; Jacks. 540; R. S. C. ii. 312; Phyt. i. 133, with bibliog.; Ann. & Mag. viii. 1842, 397, with bibliog., and 478; 'Florist's Journal,' no. xxiv.; Dict. Nat. Biog. xv. 204. *Donia* R. Br. = *Grindelia*.

**Don, George** (1798–1856): b. Doo Hillock, Forfar, 17th May, 1798; d. Bedford Place, Kensington, 25th Feb., 1856. A.L.S., 1822. F.L.S., 1831. Brother of preceding. Foreman, Chelsea, 1816–1821. Collected in 'Iphigenia' for Roy. Hort. Soc. in Brazil, W. Indies, and Sierra Leone, 1822. Trans. Roy. Hort. Soc. vols. v., vi., &c. Edited Sweet's 'Hortus Britannicus,' ed. 3, and prepared 1st Supp. to Loudon's 'Encyclopædia.' 'General System of Gardening and Botany,' 1831–1837. Pl. bought from Roy. Hort. Soc. by Mus. Brit. Pritz. 89; Jacks.

540; R. S. C. ii. 314; Proc. Linn. Soc. 1856, xxxix.; Cott. Gard. xvi. 152; Dict. Nat. Biog. xv. 206.

**Don, George** (d. 1814): b. Kincardineshire; d. Forfar, Jan. 1814. A.L.S., 1803. Father of preceding. Curator, Edinburgh Bot. Gard. Afterwards nurseryman, of Doo Hillock, Forfar. Found various rare Highland plants. Mem. Wernerian Soc. vol. iii. 1821. R. S. C. ii. 314. *Donia* G. & D. Don = *Clianthus*. *Jungermannia Donniana* Hook.

**Donald, James** (1815-1872): b. Forfar, 1815; d. Hampton Court, 13th Dec. 1872. At Chiswick, 1839-1842. Pupil of Lindley. Superintendent, Hampton Court, from 1856. Left an herbarium. Wrote on Begonias. Gard. Chron. 1873, 46. R. S. C. ii. 314. *Donaldia* Klotzsch = *Begonia*.

**Donn, James** (1758-1813): b. 1758; d. Cambridge, 14th June, 1813. A.L.S., 1795. F.L.S., 1812. Under Aiton at Kew. Curator, Cambridge Garden, 1796. 'Hortus Cantabrigiensis, 1796. Pritz. 89; Jacks. 409; Dict. Nat. Biog. xv. 222.

**Donovan, Edward** (1798-1837): b. 1798; d. Kennington, London, 1st Feb. 1837. F.L.S. Pritz. 90; Jacks. 540; Dict. Nat. Biog. xv. 236.

**Doody, Samuel** (1656-1706): b. Staffordshire, 1656; d. London, 1706. Apothecary. F.R.S., 1695. Keeper of Chelsea Garden from 1692. Assisted Ray in 'Historia,' vol. ii., and in 'Synopsis.' Left MS. on Mosses, Sloane MS. 2315. Corresponded with Petiver and Plukenet. Herbarium in Sloane's at Mus. Brit. "The Dillenius of his time," Pult. "Inter pharmacopœos Londinenses sui temporis coryphæus," Juss. Rich. Corr. 11; Sloane MSS. 2972, 4043; Phil. Trans. 1697, 390; Fl. Midd. 376; Dict. Nat. Biog. xv. 236. *Doodia* Roxb. = *Uraria*. *Doodia* R. Br.

**Douglas, David** (1798-1834): b. Scone, Perth, 1798; killed, Sandwich Isles, 12th July, 1834; bur. on spot. A.L.S., 1824. F.L.S.,1828. Apprenticed at Scone. With W. J. Hooker in Highlands. Sent to America by Hort. Soc., 1823. At Rio, 1824; in British Columbia, 1825-1827; in California, 1830-1832; on Fraser River, 1832-1833. Pl. at Kew, Mus. Brit., and Cambridge. Collected 800 species in California. Introduced 217 new species into England. Parry, 'Early Botanical Explorers of . . . . . Pacific Coast,' 'Overland Monthly,' 1883; Loud. Gard. Mag. xi. (1835), 271; Cott. Gard. vi. 263; Gard. Chron. 1885, vol. xxiv. 173, with engr. portr.; Pritz. 570; Dict. Nat. Biog. xv. 291; R. S. C. ii. 327. Original portraits at Kew and Linn. Soc. *Douglasia* Lindl. *Pseudotsuga Douglassii* Carr.

**Douglas, Francis** (fl. 1830-1841). M.D. Pres. Borwick Nat. Club, 1841-1842. Proc. Berw. Club, i. 132; R. S. C. ii. 327.

**Douglas, James** (1675-1742): b. Scotland, 1675; d. Red Lion Square, London, April, 1742; bur. St. Andrew's, Holborn. M.D., Rheims. F.R.S., 1706. F.R.C.P., 1721. 'Lilium sarniense,' 1725. 'Arbor Yemensis,' 1727. "Vir eruditus et solers," Haller. Phil. Trans. 1725, 380; Pult. ii. 234; Pritz. 90; Munk, ii. 77; Dict. Nat. Biog. xv. 329. *Douglassia* Houston = *Volkameria* L.

**Dovaston, John Freeman Milwood** (1782–1854): b. Westfelton, Shrewsbury, 30th Dec. 1782; d. same place, 8th Aug. 1854. M.A., Oxon, 1807. R. S. C. ii. 329; Dict. Nat. Biog. xv. 376.

**Drummond, James L.** (fl. 1814–1851). Of Belfast. M.D. Pres. Belfast Nat. Hist. Soc. 'Directions for Preserving Sea-plants,' Mag. Zool. Bot. ii. 144. Caused establishment of Belfast Garden, 1830. Loudon, Encycl. Gardening, 283; Pritz. 91; Jacks. 540; R. S. C. ii. 347.

**Drummond, James** (1784?–1863): d. Perth, West Australia, 27th March, 1863. A.L.S., 1810. Curator, Bot. Gard. Cork, 1809. Discovered *Spiranthes gemmipara* at Cork in 1810. Went to W. Australia, 1829; issued six sets of specimens, beginning 1839. Gard. Chron. 1841, 341; Munster Farmer's Mag. vi. & vii. (1818–1820); Lasègue, 282; R. S. C. ii. 346; Proc. Linn. Soc. 1864, xli.; Dict. Nat. Biog. xvi. 33. Plants at Brit. Mus., Kew, &c. *Drummondita* Harv. (dedicated to the two Drummonds, with the termination *ita*—"an I for James, and a T for Thomas").

**Drummond, Thomas** (d. 1835): d. Havana, Cuba, March, 1835. A.L.S., 1830. Brother of preceding. Succeeded G. Don in nursery at Forfar. Curator, Belfast Garden, 1828–1829. Curator, Belfast Bot. Gard., 1828–1829. Assistant-naturalist to 2nd Land Arctic Expedition under Franklin. Collected in N. America, Canada, and Texas, for Glasgow Garden. Issued fascicles of 'Musci Scotici,' and two series of American Mosses, the first in 1828, the second published after his death by Wilson and Hooker in 1841. Bot. Misc. i. 178; Comp. Bot. Mag. i. 39; Journ. Bot. 1834, 50, 183; 1843, 663; Bot. Mag. t. 3441; Comp. Bot. Mag. i. 16; Lasègue, 196, 204; Stewart & Corry, Fl. N.E. Ireland; R. S. C. ii. 347; Dict. Nat. Biog. xvi. 41. Plants at Brit. Mus., Kew, &c. Portr. at Kew. *Drummondia* DC. = *Mitella. Drummondia* Hook. *Drummondita* Harv. *Phlox Drummondii* Hook.

**Dryander, Jonas** (1748–1810): b. Sweden, 1748; d. Soho Square, London, 19th Oct. 1810. F.L.S., 1788. Pupil of Linnæus. Librarian to Banks, 1782; to Linn. Soc., 1788. 'Catalogus Bibliothecæ Josephi Banks.' Rees; Pritz. 91; Jacks. 541; R. S. C. ii. 347; Smith Lett. i. 165; Nich. Anecd. ix. 43; Dict. Nat. Biog. xvi. 64. Portr. Kew. *Dryandra* Thunb. = *Aleurites. Dryandra* Br. = *Josephia*.

**Dubois, Charles** (1656–1740): d. Mitcham, Surrey, 20th Oct. 1740; bur. Mitcham. Had a botanic garden at Mitcham. Treasurer, East India Company. London merchant. Sent plants to Petiver and Plukenet. Herbarium at Oxford of 70 vols. Plants in Hb. Sloane, 32, 59, 89, &c. Loudon, Arboretum, 62; Gent. Mag. 82 (1812), pt. i. 205; Journ. Bot. 1854, 249; Dict. Nat. Biog. xvi. 77.

**Duck, J. N.** 'Nat. Hist. of Portishead,' Bristol, 1852. Contains botanical lists. Jacks. 258.

**Duguid, Alexander.** MS. 'Flora Orcadensis' [with Dr. Gillies], 1832. Top. Bot. ed. 2, 545.

**Dunbar, George** (1774–1851): b. Coddingham, Berwick, 1774

d. Rose Park, Edinburgh, 6th Dec. 1851; bur. Greyfriars, Edinburgh. Began life as a gardener. M.A., Edinb., 1807. Prof. of Greek, Edinb. Univ., 1805. Cott. Gard. vii. 187; Dict. Nat. Biog. xvi. 153. *Dunbaria* Wight & Arn.

**Duncan, Andrew** (1744-1828) : b. Pinkerton, St. Andrew's, 17th Oct. 1744; d. Edinburgh, June, 1828; bur. Edinburgh. M.A., St. Andrew's, 1762. M.D., 1769. Felton, 190; Pritz. 94; R. S. C. ii. 401; Dict. Nat. Biog. xvi. 61. Portr. by Raiburn, engr. Mitchell. *Duncania* Rchb.

**Duncan, James.** Clerk. MS. Flora of Jedburgh, circ. 1830. Top. Bot. ed. ii. 543.

**Duncan, James** (1802-1876): b. Aberdeen, Oct. 1802; d. Calne, Wilts, 11th Aug. 1876; bur. Calne. Curator Bot. Garden, Mauritius, 1849. Catalogue of Garden, 1863. Pritz. 94; Jacks. 448.

**Duncan, John** (1794-1881): b. Stonehaven, Kincardine, 1794; d. Alford, Aberdeen, 1881. Weaver. Herbarium of over 1100 pl. presented to Aberdeen Univ. Journ. Bot. 1881, 64, 287; Nature, 20th Jan. 1881. Life, by W. Jolly, with portr., 1883.

**Duncan, John Shute** (fl. 1831). M.A., Oxon. Keeper of Ashmolean Museum. 'Botanical Theology.' Pritz. 94; Jacks. 19.

**Dundas, Maria,** afterwards **Calcott** [*see* CALCOTT].

**Dunstall, John** (fl. 1644-1675). Engraver. 'A booke of flowers .... exactly drawne.' Pritz. 95; Dict. Nat. Biog. xvi. 221.

**Duppa, Richard** (1770-1831): b. Culmington, Shropshire, 1770; d. Chesney Longueville, Radnor, 23rd Feb. 1831. LL.B., Cambridge, 1814. 'Elements of Science of Botany.' 'Classes and Orders of Linnean System' (an abridgment of preceding), 1816. Pritz. 95; Jacks. 541; Dict. Nat. Biog. xvi. 243.

**Dyer, Thomas Webb** (fl. 1800). Of Bristol. M.D. F.L.S., 1799. Discovered *Arabis stricta*. E. B. 614; Dr. John Evans, 'Picture of Bristol,' ed. 4 (1828).

(To be continued.)

SHORT NOTES.

ALCHEMILLA VULGARIS L. IN KENT. — This plant was found by me in June of last year on the borders of Broadhoath Wood, near Seal, Kent. It is not recorded, for Kent in 'Topographical Botany.'—H. W. MONINGTON.

POLYGONUM MARITIMUM STILL IN S. HANTS. — Mr. Bolton King, as well as the readers of the Journal generally, will be glad to learn that this plant is not yet extinct on the coast near Christchurch. After a comparatively short search, I had the good fortune to see it there to-day,—one vigorous branching plant, which may well become the parent of many more. Others, I should hope, may still exist here and there along so great a stretch of suitable ground.—W. MOYLE ROGERS.

EAST KENT PLANTS.—A week in this vice-county during last June yielded the following records, apparently new to it:—*Papaver*

*Lecoqii* Lamotte. Bank below the downs, about a mile north of Shorncliffe Station. *Viola Reichenbachiana* Bor. Copses, &c., between Westenhanger and Folkestone; moist copse at the north end of Ham Ponds. *Arenaria serpyllifolia* L., b. *glutinosa* Koch. Plentiful at intervals from Sandwich to Deal; also near Walmer, on shingle. *Festuca ambigua* Le Gall. In profusion upon several parts of the sandhills from Sandwich to Deal; also sparingly on old walls in both towns; doubtless overlooked from its withering early. *Melampyrum pratense* L., b. *latifolium* (Syme). Abundant and well-marked in woods near Wye. In the ditches about Sandwich *Carex stricta* Good, is exceedingly variable. The flowers of *Polygala austriaca* Crantz, are usually described as " blue;" " bluishlilac " would be nearer the colour of the Wye plant, which is recognizable at once by its peculiar tint.—EDWARD S. MARSHALL.

ELYMUS ARENARIUS L. IN DORSET. — On September 10th, my son and I had the good fortune to find this grass on the Dorset coast, between Poole and Canford Chine. We saw three patches near together, and in one of them several of the plants had flowered. In the ' Flora of Dorset,' Mr. Mansel-Pleydell states that Pulteney had a Dorset specimen of it, but was not certain whether it had been gathered at Weymouth or in Purbeck; and he adds, " Not confirmed since Pulteney's time." The present locality is distinct from the two referred to, though at no great distance from Purbeck.—W. MOYLE ROGERS.

RUMEX MARITIMUS AND R. PALUSTRIS IN EAST SUSSEX. — The only existing specimen of *Rumex maritimus* from East Sussex, as far as I know, is the plant in Mr. Borrer's herbarium at Kew, labelled " West Dean, near Seaford, 1855 "; and although I, as well as Messrs. Unwin and Jenner, of Lewes, and Mr. Ellman, of Berwich, have many times examined the valley of the Cuckmere in search of it, it has never been rediscovered until last year, when Mr. Ellman met with a single plant in the marsh between Charlston and the Cuckmere. This year, however, it has appeared in considerable quantity both at Charlston Pond and a pond at West Dean, where Borrer probably first found it. I also found one plant of *R. palustris* by Charlston Pond. This occurs occasionally in the Lewes Levels, but Mr. Jenner writes me it only seems to appear at intervals; and the same thing applies perhaps to *R. maritimus*, as I have many times been to Charlston Pond before, and could not have failed to see so conspicuous a plant as the golden heads of *R. maritimus*, had it been there. It is the first time *R. palustris* has been found in this district.—F. C. S. ROPER.

HIERACIUM TRIDENTATUM IN WORCESTERSHIRE. — I met with this plant, on July 30th, growing with *H. vulgatum* and *H. umbellatum* (the latter a rare plant here) in a small wood at Powick, near Worcester. It has not hitherto been recorded from this county. The station is in the Malvern and Teme division adopted by Mr. Edwin Lees in his ' Botany of Worcestershire.' — RICHARD F. TOWNDROW.

HELIANTHEMUM POLIFOLIUM Pers. IN N. SOMERSET.—As this very rare plant has been regarded hitherto as exclusively confined in N. Somerset to the southern slopes of Brean Down, it may, perhaps, be worth recording that I have found it plentifully during the present month (Sept.) at another locality in the same vice-county, namely, Purn Hill, Bleadon, an elevation of the carboniferous limestone situated inland at a distance of between two and three miles from Brean Down, and to the S.E. of that promontory. Considering the extremely-narrow limits previously assigned to this species in N. Somerset, its abundant occurrence at Bleadon marks a somewhat important extension of its range.—DAVID FRY.

---

*NOTICES OF BOOKS.*

*The Origin of Floral Structures through Insect and other Agencies.* By the REV. GEORGE HENSLOW. Kegan Paul, Trench & Co. 8vo, pp. xx. 350. Price 5s.

THIS last volume of the International Scientific Series is of much interest to English botanists for several reasons. It will help them to see, for one thing, what a misnomer is the popular term "Darwinism," when used as absolutely convertible with "Evolution." It is not sufficiently recognised by those who use the word in this crude fashion, that there is an increasing number of men of science who are thorough evolutionists, and who yet reject partially or wholly many of the most important Darwinian hypotheses. Professor Henslow, for example, is a well-known upholder of the principle of evolution; but in the present work he vehemently combats two of the theories which are most closely associated with the great name of Darwin. Asa Gray wrote in 1874, " The aphorism, 'Nature abhors close fertilisation,' and the demonstration of the principle, belong to our age and to Mr. Darwin. To have originated this, and also the principle of Natural Selection, . . . and to have applied these principles to the system of nature, in such a manner as to make, within a dozen years, a deeper impression upon natural history than has been made since Linnæus, is ample title for one man's fame." The two principles here selected as Darwin's chief titles to fame are in the work before us severely criticised. In Chap. xxxi. Prof. Henslow sums up the arguments which he had brought forward on various previous occasions in support of the belief that self-fertilisation is *per se* in no wise injurious. The facts that he adduces are weighty and well arranged, the observations by Messrs. Forbes, Ridley, and Veitch, on the self-fertilisation of Orchids, being of special interest and value. In Chap. xxxii. Natural Selection itself is boldly attacked. Prof. Henslow points out that in the case of plants, the " struggle for life is mainly during the early period of growth, before any varietal or specific characters of the flowers have put in an appearance at all;" and he asks, if that is so, how can a plant be selected " because it has some floral structure more appropriate than others." He regards, therefore, the survival of the fittest, as first issuing from " *Constitutional* Selection." It must be borne in mind that Darwin himself towards the close of his life seemed distinctly to move in the direction of these criticisms, and, with the absolute justice and candour which always distinguished him, readily admitted that he had in his earlier works stated too strongly the case for " cross-fertilisation" and " Natural Selection." Followers of the great naturalist will not need to be reminded that a rejection of some of his theories is quite compatible with intense admiration for his life and labours in the cause of science.

But the present work is by no means merely negative and destructive of previously accepted opinions. It gives a novel theory, which is, to say the least, most intelligible, suggestive, and capable of explaining very many of the phenomena of floral structure. This new theory ought to be specially welcome to evolutionists, since it relieves them from a difficulty which has always been felt as to the reason of variation. To say that the seedlings of a particular plant vary from their parent, because they have a tendency to variation, is a most unsatisfactory and evasive explanation, indeed simply amounts to saying " they vary because they vary." Those who have read (who has not ?) the delightful ' Life and Letters of Charles Darwin,' will remember his characteristic reply, when acknowledging the justice of one of Huxley's criticisms on the Origin of Species :—" If, as I must think, external conditions produce little *direct* effect, what the devil determines each particular variation ? What makes a tuft of feathers come on a cock's head, or moss on a moss-rose ? There is ' Much virtue in *If.*'" Prof. Henslow cuts, instead of untying, this Gordian knot. External conditions, he tells us, *do* produce great effect, and are, in truth, the direct cause of variation. He looks lovingly towards that "new Lamarckism" which was frowned upon by the President of the Biological Section at the recent meeting of the British Association : he goes back to the " Monde ambiant" of Geoffroy St. Hilaire, the surrounding circumstances and conditions of life, in a word, the " environment." His formula for variation is that it is the result of the " responsive power of protoplasm to external stimuli ;" and he ·finds the stimulating agency, as far as flowers are concerned, to be chiefly that of insects. For example, with regard to secretive tissues he tells us, " The simple origin of nectaries is that insects, having been attracted to the juicy tissues of flowers, by perpetually withdrawing fluids, have thereby kept up a flow of the secretion which has become hereditary, while the irritated spot has developed into a glandular secreting organ." Irregularity or " zygomorphism " is ascribed to a change wrought by generations of insects on originally regular whorls. Every one has noticed how in many *Orchideæ*, *Labiatæ*, and papilionaceous *Leguminosæ*, the flower seems as if specially constructed so as to afford a secure landing-place for insects. Such structures are explained as the resultant of various forces brought to bear upon the flower by " the weight of the insect in front, the local irritations behind, due to the thrust of the insect's head and probing for nectar, coupled with the absence of all strain upon the sides." The "guides" and "pathfinders" which show the way to the treasury of nectar in many flowers, the different enations and trichomes which in the form of hairy whorls, woolly tangles, and processes producing viscid secretions, bar the path of unwelcome guests—all these and many other puzzling structures are described as the immediate results of the irritations set up by the insects themselves.

But enough has been said to show the nature of the ingenious theory which runs all through the book, which lends itself to innumerable applications, and which seems to form a very plausible solution of many botanical difficulties. The strangest part of this startling hypothesis is that it makes the floral world of to-day with its brilliant hues and infinitely varied beauty of structure, to result from wounds and bruises, malformation and deformity, in the parent flowers of the olden time. The fanciful analogist might find pleasure in reflecting that the strength and freedom of some great nations are the result to some extent of the sword-cuts, rifle-wounds, and bayonet stabs bravely endured by the creators of those states, and that therefore it is entirely in accordance with the fitness of things that the present state of the vegetable kingdom should have been moulded by the stabs and punctures, blows and scars, patiently received by the ancient heroes of the floral world. The paradox of yesterday is the truism of to-day, and so in nature what is a startling abnormality in one age, becomes the recognised norm of the next. Prof. Henslow himself speaks of one variation as a " pathological phenomenon which has become fixed and hereditary," and in truth his theory would show that the Pathology of the progenitors is the Physiology of their posterity, and that what is included in our present Morphology belonged in ages gone by to the domain of Teratology. Prof. Henslow's theory is in fact a distinct adoption of the suggestion made in the last words of Dr. Masters' classic work on Teratology, so long ago as 1869 :—" That monstrosities so-called may become the starting points of new forms is proved by the circumstance that, in many cases, the peculiarities are inherited so that a new 'race' is produced and perpetuated ; and if a new race, why not a new species ? The difference is one of degree only." The obligations (always duly acknowledged) of the present work to Dr. Masters are very numerous ; and it is much to be hoped that, as one result of the attention which the *Insect-selection* theory is sure to receive, we may have a new edition of the "Teratology," which will give us in English the valuable additions now to be found only in the German edition of 1886. This would be a great boon to many students who read English more readily than German, and who are unable to obtain the scarce first edition published by the Ray Society.

There is but little space left to chronicle the many other merits of Prof. Henslow's book. Quite apart from the interesting theory which gives life and colour to the old work, it has much value as a reliable compendium of the best and latest information available on floral structures and fertilisation, and it also supplies some inter-

esting chapters on the subject of degeneracy, in which it is gratifying to find that some of the too little known researches of Mr. C.
F. White on pollens are utilised.   There are, of course, some
slight defects, such as misprints and inaccuracies in scientific
nomenclature, and an occasional haziness of expression, probably
originating in the author's fulness of the subject, and forgetfulness
that his readers cannot know what is in his mind, only what falls
from his pen : but these blemishes will doubtless be removed in a
second impression.   Like the other volumes of the valuable series
to which it belongs, the ' Origin of Floral Structure ' is well
printed, well illustrated, handy and cheap ; and is deserving of a
place in the library of even the humblest British botanist, who will
find a fresh interest given to his observations by searching for facts
which may throw light on this important subject, and tend to corroborate or disprove the ingenious theory which Prof. Henslow
modestly describes as " at most only a ' working hypothesis ' for
future investigations."                                 PERCY MYLES.

*Index Generum Phanerogamarum usque ad finem anni* 1887 *promulga-
torum in Benthami et Hookeri* ' *Genera Plantarum* ' *fundatus
cum numero specierum synonymis et area geographica conscripsit*
TH. DURAND *subcustos* Herb. Hort. Bot. Publ. Bruxellensis.
London : Dulau.  8vo, pp. xxii., 723.  £1 1s.

THIS is a noteworthy addition to the limited number of books
which may be regarded as indispensable to the systematic botanist :
indeed, it is perhaps more really entitled to that epithet than any
volume which has appeared since the ' Genera,' upon which it is
based.   Here, in small space, we have a condensed history of all
the genera of flowering plants, or rather, we have their names,
with an indication showing where their history may be found.   It
is thus of value to the humbler student, whose shelves are few and
already well-nigh filled, and who, with this at his elbow, can
readily obtain references to works which he will be able to consult
in our large botanical libraries ; and it forms an index and a supplement to the ' Genera Plantarum.'   Many helpers have combined
to render M. Durand's work a success ; but the labour imposed
on him must nevertheless have been very great, and the
thoroughness with which the plan has been carried out reflects
the highest credit upon him.

Every one must have felt the need of some reference list of the
additions which have been made to our knowledge of plants since the
publication of Bentham and Hooker's great work, the first part of
which was issued in 1862.   Those who prudently interleaved their
copy of that book have in some measure supplied this want
for themselves, but botanical literature is so extensive that private
individuals can hardly take stock of it all.   The additions in some
orders are inconsiderable, in others there are none ; but there are
those in which the number of generic types has been very largely
increased, as in *Anonaceæ*, where the 40 genera of Bentham and
Hooker are augmented by 21 ; in *Sapotaceæ* (the genera of which

have been rearranged for this Index by Prof. Radlkofer) where the numbers stand as 24 against 38 ; and in *Asclepiadeæ*, where they are 146 against 204. M. Durand has of course simply taken these genera on the faith of their authors, and it is more than probable that many of them would not be accepted by Bentham and Hooker as coming up to their standard of distinctness; but critical revision is no part of M. Durand's work, and he has wisely confined himself to the lines he has laid down.

The plan of the book will be best shown by an extract, which we take from the first page :—

## " DICOTYLEDONES.

### POLYPETALÆ.

ORDO I. **RANUNCULACEÆ.** ([1])

TRIBUS I. **Clematideæ.**

1. 1. **Clematis** L. G. i. 3 et 953.—Sp. descript. ultra 200, a cl. Kunze ad 66 reducti. Orbis fere tot. reg. temp. et trop.

Sect. 1. *Viticella* DC., Viticella Mönch.

Sect. 2. *Cheiropsis* DC., Atragene L. Cheiropsis et Viorna Spach.

Sect. 3. *Flammula* DC., Meclatis Spach.

2. 2. **Naravelia** DC. G. I. 4.—Sp. 2 v. 3. Asia trop."

" G." stands, of course, for Bentham and Hooker's 'Genera'; the other abbreviations need no explanation. There are two series of numbers : one running throughout the book, a useful return to the plan adopted by Endlicher ; the other peculiar to each order. Every care has been taken to insert 'additions in the place they hold in relation to their allies, and this, again, is useful to those whose libraries are limited : it is, indeed, to this care in small details that much of the excellence of this Index is due. The place and year of publication of new genera is, of course, indicated ; thus :

" Physotrichia Hiern in Journ. Bot. 161 (1873), Sp. 2. Angola, Zambesia."

" Phellolophium Bak. in Journ. Linn. Soc. xxi. 349 (1884). Sp. 1. Madagascar."

As might be expected, a certain number of names are here proposed for the first time, owing for the most part to the double occupation of certain generic titles. These occur in their proper places, but a list of them is thoughtfully placed by M. Durand at the beginning of the book. It may be useful to cite them :—

*Eeldia* * = *Candollea* Labill. Ill. Pl. N. Hort. ii. 33, non ejusd. in Ann. Mus. Paris, vi. 453.

*Pirea* = *Dityosperma* Rgl. non Wendl. et Dr.

_____

" [1] Sp. descript. ultra 1350, verisimiliter ad 680 reduct., per totum fere orbem terrarum dispersæ."

* " Dicat. S. van Eelde, uxori meæ et adjutrici indefessæ laborum."

*Valetonia* = *Martia* T. Valeton, non Benth.
*Strailia* = *Cercophora* Miers, non Fresenius.
*Nostolachma* = *Lachnostoma* Korth., non H. B. K.
*Buseria* = *Leiochilus* H. f., non Kn. et Weste.
*Lagoa* = *Zygostema* E. Fourn., non Benth.
*Thuspeinantha* = *Tapeinanthus* Boiss., non Herb.
*Lophotocarpus* = *Lophiocarpus* M. Micheli, non Turcz.
*Boeckeleria* = *Decalepis* Böck., non W. et A.

The other new genera are—six proposed by Radlkofer in *Sapindaceæ*—*Athyana*, *Aphanococcus*, *Melanodiscus*, *Tristiropsis*, *Tinopsis*, and *Conchopetalum*—and one (*Schwackæa*) by Cogniaux in *Melastomaceæ*. Prof. Radlkofer has appended a short diagnostic phrase to each of his novelties, which will entitle them to be quoted by future writers ; but *Schwackæa* is a *nomen nudum*. The *Sapotaceæ* are revised by Radlkofer, and Beccari's help is acknowledged for the Palms. The authors who have monographed for the ' Suites au Prodromus,' are followed for these orders ; and the admirable work now publishing in parts, ' Die natürlichen Pflanzenfamilien,' has been consulted for the Palms, Grasses, and other groups.

The Appendix of " Genera dubiæ sedis vel non satis nota," occupies six pages, much less than might perhaps have been expected. One or two names may be removed from it, in addition to *Aphloia*, which must have slipped in by accident, as it is duly placed in the body of the book. *Micræa* of Miers has already been shown in this Journal* (on Mr. Miers's own authority) to be identical with *Ruellia dulcis* Cav. "*Leptophragma* R. Br. et Benn., Pl. Jav. Rar. 185," = *Turræa Brownii* C. DC., Mon. Phan. i. 442. We have in the Department of Botany, (1) Brown's MS. description of the plant, which was first referred by him to *Turræa*, and afterwards separated as *Leptophragma*, the fruit being " essentially different " from that of *Turræa ;* (2) a specimen collected by Brown, and named by him " *Turræa*," with a reference to the above MS. description ; (3) a beautiful drawing by Bauer, labelled *Leptophragma denudatum*, to which Brown refers in the above MS. A specimen collected by Banks in the Bay of Inlets, to which Dryander has attached the above name, may be a different species. There are at the Natural History Museum specimens of some of Aublet's and Loureiro's doubtful genera, which, although mostly of a fragmentary nature, may help to throw light upon some of these. We are glad to see that the Gymnosperms occupy their proper position, after the Monocotyledons.

The following table of the numbers of orders, genera, and species—the latter such as were considered " bene distinctæ, linneanæ dictæ "—will be noted with interest :—

|  | | Order. | Genera. | Species. |
|---|---|---|---|---|
| Dicotyledones . | Polypetalæ    . | 90 | 3050 | 28,300 |
| | Gamopetalæ    . | 46 | 2885 | 37,800 |
| | Monochlamydeæ | 36 | 849 | 12,100 |
| | | 172 | 6784 | 78,200 |

| | | | | | | |
|---|---|---|---|---|---|---|
| Monocotyledones | . | . | . | 35 | 1587 | 19,600 |
| Gymnospermæ | . | . | . | 3 | 46 | 2420 |
| | | | | 210 | 8417 | 100,220 |

M. Durand's painstaking thoughtfulness extends even to the index of this 'Index,' in which he includes many synonyms which are cited neither by Bentham and Hooker, nor in the body of this work. There are thus three classes of names, rightly placed in one index : first, those retained, to which the name of the order and the number in the body of the book is added, thus :

"*Abatia* R. & P. (Samyd.) 255."

next, the synonyms cited in the book, each followed by the genus to which it is referred, and the number of that genus :—

" Aa Rchb. f. (Altensteinia H. B. K.) 6981."

and lastly, the additional synonyms above referred to, to which no number is attached :

" Aakesia Tuss. (Cupania L.)."

The only point which may be criticised is the printing (in the index) the names retained in italics, and those rejected in roman. This is, of course, exactly opposed to the plan of the ' Genera Plantarum,' and the two books will be used so largely by the same set of people, that uniformity in this respect would have been desirable. M. Durand has, however, produced a work, the usefulness of which cannot well be exaggerated, and his prompt and successful accomplishment of a tedious undertaking entitles him to the thanks of all botanists.

JAMES BRITTEN.

## ARTICLES IN JOURNALS.

*American Naturalist* (July). — C. E. Bessey, ' A miniature Tumble-weed ' (*Townsendia sericea*).

*Annals of Botany* (August). — T. Johnson, ' *Arceuthobium Oxycedri*' (1 plate). — A. B. Rendle, ' Development of Alemone-Grains in Lupin ' (1 plate).—G. Murray & L. A. Boodle, ' Structure of *Spongycladia* ' (*S. dichotoma, S. neocaledonica*, spp. nn.).--C. Reid, ' Geological History of Recent Flora of Britain.' — M. M. Hartog, ' Recent Researches on *Saprolegnieæ*.'—H. Marshall Ward, ' Structure and Life-history of *Puccinia graminis* ' (2 plates). — L. H. Vines, ' Systematic position of Isoetes.'—A. B. Rendle, ' Occurrence of Starch in the Onion.'—S. Schönland, ' Modification of Pagan's Growing Slide.'

. *Botanical Gazette* (Sept.).—S. Schönland, ' The Botanical Laboratory at Oxford.'—C. Robertson, ' Zygomorphy and its Causes.'

*Bot. Centralblatt* (Nos. 35, 36).—R. Keller, ' Wilde Rosen des Kantons Zürich.' — (No. 36). C. J. Johanson, ' Einige Beobachtungen über Torfmoore im südlichen Schweden.'—(No. 38). T. Wenzig, ' Nova ex Pomaceis.' — K. F. Dusén, ' Ueber einige Sphagnum-Proben aus der Tiefe südschwedischer Torfmoore.'— O. F. Andersson, ' Uber *Palmella uvaformis* Kg. und die Dauersporen von *Draparnaldia glomerata* Ag.'

*Bot. Zeitung* (Aug. 3). — J. Wortmann, ' Zur Beurtheilung der Krümmungserscheinungen der Pflanzen.' — (Aug. 10, 17, 24, 31). H. Vöchting, ' Ueber˙ die Lichstellung der Laubblätter ' (1 plate). — (Sept. 7, 14). G. Karsten, ' Ueber die Entwickelung der Schwimmblätter bei einigen Wasserpflanzen.' — (Sept. 21). A. de Bary, ' Species der Saprolegnieen ' (2 plates).

*Bull. Soc. Bot. France* (xxxv. Comptes rendus, 3 : Sept. 1). —Fliehe, ' Sur les formes du genre *Ostrya*.' — P. Maury, ' Sur les Cypéracées du Mexique.'—A. Franchet, ' Sur le *Cheilanthes hispanica*.' —P. A. Dangeard, ' Sur l'anatomie des *Salsoleæ*.'—E. Wasserzug, ' Recherches sur un Hyphomycète.' — M. Gomont, ' Sur les enveloppes cellulaires des Nostocacées filamenteuses ' (2 plates).— L. du Sablon, ' Sur les antherozoides du *Cheilanthes hirta*.' — P. Duchartre, ' Sur l'enracinement de l'albumen d'un *Cycas*.' — F. Constantin, ' Sur quelques parasites des Champignons supérieurs.' —E. Roze, ' *Galanthus nivalis* aux environs de Paris.' — P. A. Dangeard, ' Nouvelles observations sur les *Pinguicula*.' — P. Duchartre, ' Sur un cas de l'abolition du Géotropisme.'—P. van Tieghem, ' Sur le Réseau sus-endodermique de la racine chez les Legumineuses et les Ericacées.' — E. Roze, ' L' *Ustilago caricis* aux environs de Paris.'—P. van Tieghem & H. Douliot, ' Sur les Plantes qui forment leurs radicelles sans poche.' — G. Chastaingt, ' Deux Rosiers ' (*R. suzilliacensis* Chast. and *R. superba* Chast.). — D. Clos, ' Dodart et les deux Marchant.'—J. Costantin, ' Recherches sur un *Diplocladium*.'—Id. & Rolland, ' Sur le developpement d'un *Stysanus* et d'un *Hormodendron*.' — H. Jumelle, ' Sur les graines à deux téguments.' — Devaux, ' De l'action de la lumière sur les racines croissant dans l'eau.'—P. Duchartre, ' Fleurs prolifères de Bègonias tubéreux.'—A. Pomel, ' *Evacidium Heldreichii*' (= *Evax Heldreichii*). —A. Lothelier, ' Sur les piquants de quelques plantes.' — G. Rouy, ' *Teucrium majorana* and *T. majoricum*.'

*Bull. Torrey Bot. Club* (Sept.).—G. E. Davenport, ' Fern Notes ' (*Cheilanthes mexicana*, sp. n.). — E. E. Sterns, ' The Nomenclature Question and how to settle it.'

*Gardeners' Chronicle* (Sept. 1).—*Oncidium Hrubyanum* Rchb. f., sp. n. — ' Droppers in *Tulipa sylvestris* ' (fig. 30). — (Sept. 8). *Pentstemon rotundifolius* (fig. 31).—M. T. Masters, ' *Pinus pyrenaica* (*vera*) ' (fig. 32). — (Sept. 15). *Phalænopsis Buyssoniana*, sp. n., Rchb. f. —(Sept. 22). *Masdevallia punctata* Rolfe, sp. n. — (Sept. 29). *Passiflora Miersii* (fig. 46).

*Journal de Botanique* (Sept. 1, 16). — E. Bonnet & P. Maury, ' D'Ain-Lefra à Djenien-bon-resq, Voyage botanique dans le Sud-Oranais.'—(Sept. 1). P. Maury, *Prasophyllum Laufferianum*, sp. n. (fig.).—(Sept. 16). A. Franchet, ' Les Saussurea du Yun-nan.'— L. Quelet, ' Sur les genres *Ombrophila* et *Guepinia*.'

*Oesterr. Bot. Zeitschrift* (Sept.). — F. Krasan, ' Weitere Bemer-kungen über Parallelformen.' — B. Blocki, ' *Hieracium gypsicola*, sp. n.'—J. Murr, ' Zur Diluvialflora des nördlichen Tirols.' — L. Simonkai, *Genista nervata* and *Erysimum Banaticum*.—E. Formánek, ' Zur Flora von Bosnien ' (*Carlina semiamplexicaulis*, sp. n.).

# *CAREX* NOTES FROM THE BRITISH MUSEUM.

## By L. H. Bailey.

To the student of American Carices the National Herbarium at South Kensington possesses peculiar interest, because it contains, among other treasures, the herbarium of Edward Rudge, and the types of the species described by Robert Brown from the collections made in the high north by the expeditions of Captain Parry and others. Although the species described by Rudge and Brown are comparatively few, they are important from the earliness of their publication and the considerable obscurity in which they have remained.

Rudge described and figured five new species of American Carices in 1804 in the Linnean Transactions : *C. intumescens*, *C. ovata*, *C. tenuis*, *C. gigantea*, and *C. flexilis*. *C. intumescens* and *C. tenuis* have been properly understood, the former being maintained, and the latter falling as a synonym to Michaux's *C. debilis*, which has a year's priority. *C. ovata* was reduced by Dr. Boott to a variety of *C. atrata* Linn., but its character does not appear to have been fully apprehended. Certain very marked specimens from the Rocky Mountains were referred here, although there has been no evidence that these western plants occur in Newfoundland, whence Rudge obtained his specimens, nor could they be satisfactorily referred to the species characterized by Rudge. I have long felt that the so-called *C. atrata* of the White Mountains of New Hampshire is at least a distinct variety from the species as it occurs in Europe and the Rocky Mountains, and that Rudge must have had the same from Newfoundland. Rudge's specimens are the same as those from the White Mountains, although more slender than ordinarily found there, and the perigynia are so young that the spikes had not assumed their cylindrical character. In my opinion, Rudge's name should be applied to the eastern plant which he characterized, and the western plant should be separated from it :—

C. ATRATA Linn., var. OVATA Boott, Ill. 114, in part. *C. ovata* Rudge, Linn. Trans. vii. 96, t. 9.—Distinguished from the species by its habitually more slender habit, its long-peduncled and more or less drooping spikes, which are reddish brown in colour and smaller, and scales usually shorter and blunter.—White Mountains of New Hampshire and Newfoundland.

Var. DISCOLOR. *C. atrata* var. *ovata* Boott in part; and Bailey, Proc. Am. Acad. Arts and Sci. xiv. (n. s.) 77. — Spikes long-cylindrical, drooping ; perigynium very broad and thin, commonly broader and often longer than the black scale, white or whitish ; staminate flowers very numerous, and covered by long and acute scales. — Mountains of Colorado and Utah and southward. This plant approaches *C. Mertensii* Presc., and has been confounded with it. It is not improbable that it is specifically distinct from *C. atrata*. Certain European forms of *C. atrata* resemble this plant, but none of them possess its peculiar perigynia, and particularly not its numerous and long staminate scales.

*C. atrata* var. *nigra* Boott does not appear to occur in America. The plant which has been referred here by various botanists is specifically distinct, and may be briefly characterized from its most striking features as follows:—

**Carex nova,** n. sp. — Spikes three or four, globular or nearly so, closely sessile and aggregated, the staminate portion inconspicuous; perigynium very broad and conspicuously spreading, usually light-coloured; scales broad, often terminating in a cusp.— Mountains of Wyoming and Colorado and southward; evidently rare.

Rudge's *C. gigantea* does not appear to be the same as the plant usually referred to that species, but a fuller examination of materials in various herbaria must be made before definite notes can be had concerning it. His specimen of *C. folliculata* Linn. is *C. Michauxiana* Boeckl. (*C. rostrata* Michx.)!

Of Robert Brown's northern Carices, all but *C. attenuata* are in the herbarium of the Natural History Museum. This species was seen in Herb. Hooker by Dr. Boott, and referred by him to *C. rupestris* All.

C. AFFINIS R. Br. Frankl. Narr. App. 763, was referred by Dr. Boott to *C. obtusata* Lilj. Three sheets were placed together in the herbarium, the plants all collected on the same expedition, of which two sheets are not labelled, but are evidently *C. obtusata*, although too young for positive determination. The third sheet is Brown's type of *C. affinis*, and it is *Kobresia scirpina* Linn.

C. CONCOLOR R. Br. Suppl. App. Parry's Voy. 218, which I doubtfully referred to *C. lenticularis* Michx. (Proc. Am. Acad. Arts and Sci. xiv. (n. s.) 86) is *C. vulgaris* var. *alpina* Boott (*C. rigida* Gooden.).

C. PODOCARPA R. Br. Frankl. Narr. App. ed. ii. 36, has been entirely misunderstood. It is apparently only an aberrant form of *C. atrofusca* Schkuhr (*C. ustulata* Wahl.). From most specimens of *C. atrofusca* it differs more or less in the entire orifice of its perigynium, which is not at all puncticulate, somewhat narrower spikes, and broader leaves. But *C. atrofusca* is a variable species, and I see no constant characters to separate Brown's plant. The plant which has passed for *C. podocarpa* R. Br. (Boott, Ill. 197) is *C. macrochæta* Meyer.*

C. PEDATA Linn. Sp. Pl. ed. ii. 1384, is an enigma. The name alone is indication that Linnæus meant to refer to *C. ornithopoda* Willd., the only other European species to which the name will apply being *C. digitata*, which he himself described. Moreover, Micheli's figure to which he refers (Mich. Gen. t. 32, f. 14) is unmistakably *C. ornithopoda*. Unfortunately the plant does not exist in Linnæus's herbarium, although, if it did exist, it might afford little aid, for his specimens are often anything but the ones he described. In the present case, although the name and reference point unmistakably to a known species, his description applies

---

* Two plants, in fact, have passed under the name of *C. podocarpa*, one of which was separated as *C. invisa* Bailey, in Proc. Am. Acad. Arts and Sci. xiv. (n. s.) 82.

rather, as Fries has stated, to *C. globularis* or its allies. *C. pedata* is undoubtedly a fictitious plant, and the name, as used by Linnæus, cannot be pressed into service. It is interesting in this connection, however, to find in the herbarium prepared to represent the Hortus Cliffortianus, which is deposited at South Kensington, a specimen of *C. ornithopoda* which is labelled *C. pedata*. How much importance can be attached to this specimen is uncertain, as it is not known whether this herbarium passed through Linnæus's hands.

The curious herbarium of Thomas Walter, author of 'Flora Caroliniana' (1788) is also at South Kensington, but its collection of *Carex* is confined to a single unnamed scrap of *Carex riparia* Curt.

---

## ON A THIRD COLLECTION OF FERNS MADE IN WEST BORNEO BY THE BISHOP OF SINGAPORE AND SARAWAK.

### By J. G. BAKER, F.R.S.

THE present collection is a continuation of two previous ones, reported upon in the 'Journal of the Linnean Society,' vols. xxii., p. 222, and xxiv., p. 256. West Borneo is evidently very rich in ferns, and no doubt a large number of new species still remain to be discovered. The numbers are Dr. Hose's collecting numbers, and those within brackets indicate the position of the new species.

**217.** *Dicksonia sorbifolia* Smith (*D. papuana* F. M.).—Miri, Sarawak, *Chas. Hose.* A very rare species, known only in New Guinea, besides C. Smith's and Roxburgh's original localities.

**218.** *Trichomanes rigidum* Sw.—Matang, Sarawak.

**219 (17\*).** **Davallia** (LEUCOSTEGIA) **Hosei,** n. sp. — Rootstock creeping, $\frac{1}{8}$ in. diam., clothed with minute adpressed linear brown paleæ. Stipes naked, brownish, 12–16 in. long. Lamina oblong-lanceolate, bipinnate, firm in texture, green and glabrous on both surfaces, $1\frac{1}{2}$–2 ft. long, 5–6 in. broad. Pinnæ distant, lanceolate, ascending, subsessile, $\frac{3}{4}$–1 in. broad, cut down to the rachis into unequal-sided oblong-lanceolate deeply crenate pinnules $\frac{1}{6}$ in. broad, produced on the upper and cut away on the lower side at the base. Veins indistinct, one to each final lobe. Sori small, one at the base of each final lobe. Indusium as broad as long, rigid, glabrous, persistent, free at the sides.—Lambur, Sarawak, *Chas. Hose.* Allied to *D. Kingii, nephrodioides,* and *ciliata.*

**220 (13\*).** **D.** (LEUCOSTEGIA) **oligophlebia,** n. sp. — Rootstock slender, creeping, clothed with minute adpressed linear brown paleæ. Stipe slender, naked, under an inch long, minutely scaly towards the brown base. Lamina lanceolate, rigid, pinnate, green and glabrous on both sides, $1\frac{1}{2}$–2 in. long, $\frac{1}{2}$ in. broad. Pinnæ sessile, rhomboid, obtuse, 3-lobed, with only one vein to each lobe. Sori usually only one to a pinna, placed in the small anterior lobe. Indusium suborbicular, rigid, glabrous, free at the sides. — Laupi,

Sarawak, *Chas. Hose.* Not nearly allied to anything already described.

221 (28*). **Lindsaya** (Isoloma) **indurata**, n. sp. — Rootstock creeping, 1-12th in. diam., clothed at first with very minute spreading linear brownish-black paleæ. Stipe wiry, naked, castaneous, 8-9 in. long. Lamina lanceolate, simply pinnate, about a foot long, ¾-1 in. broad, firm in texture, green and glabrous on both surfaces. Pinnæ oblong, obtuse, sessile, subequilateral, entire, ⅓ in. broad, the edge rather reflexed. Midrib distinct; the other veins obscure and immersed. Sori continuing all round the sides and tip of the pinnæ. Indusium with two narrow valves.—Niah, Sarawak, *Chas. Hose.* Allied to *L. divergens* Wall.

222 (16*). **Adiantum Hosel**, n. sp. — Rootstock not seen. Stipe slender, castaneous, a foot long, with a few small reflexed linear brown paleæ towards the base. Lamina rhomboid, bipinnate, moderately firm in texture, green and glabrous on both surfaces; rachis naked, castaneous. Pinnæ lanceolate, 4-9 in. long, ¾-1 in. broad. Segments close, numerous, rhomboid-dimidiate, obtuse, shortly petioled, the central ones ¾-1 in. long, ¼-⅓ in. broad, minutely crenate when sterile. Sori 1-2, placed on the upper margin of the segments, at most ⅙-¼ in. long. Indusium broad, firm, glabrous.—Pendulous on limestone cliffs, Sarawak. Near *A. affine* Willd.

224 (29*). **Pteris** (Eupteris) **Walkeri**, n. sp. — Rootstock not seen. Stipes naked, castaneous, 6-8 in. long. Lamina rhomboid, irregularly tripinnate, about a foot long, moderately firm in texture, green and glabrous on both surfaces; rachises castaneous. Pinnæ usually caudate; upper and terminal pinna simply pinnate; several lower forked at the base; final segments linear, obtuse, at most ¾ in. long, 1-12th in. broad. Veins copiously pinnate in the final segments; veinlets indistinct. Sori occupying nearly the whole margin of the final segments. Indusium narrow, glabrous.—Found by Dr. Hose on Banggi Island, and previously by Mr. H. Walker at Silam, in North Borneo. Allied to *P. quadriaurita* Retz.

223 (22*). **P.** (Eupteris) **furcans**, n. sp. — Rootstock not seen. Stipe naked, slender, castaneous, 5-6 in. long. Lamina deltoid, 6-8 in. long, moderately firm in texture, green and glabrous on both surfaces; rachises castaneous. Pinnæ lanceolate, simply pinnate, ¾-1 in. broad; lower forked at the base. Final segments linear, entire, obtuse, adnate to the rachis, 1-12th in. broad. Veins copiously pinnate in the final segments. Sori falling short of the tip of the segments. Indusium narrow, glabrous.—Baram, Sarawak. Near *P. quadriaurita.*

225. *P. quadriaurita* var. *digitata* Baker. — Sarawak. These fine specimens show that this is substantially identical with the Himalayan *P. Grevilleana* Wall.

226. *Asplenium Phyllitidis* D. Don.—Niah, Sarawak, *Chas. Hose.*

227. *A. bantamense* Baker.—Langir, Sarawak, *Chas. Hose.*

229. *Nephrodium borneense* Hook.—Sarawak.

228. *N. vastum* Baker.—Lambir, Sarawak.

230. *N. dissectum* Desv. — Entrance of the Sarawak limestone-caves.

231 (182\*). **N.** (EUNEPHRODIUM) **simulans**, n. sp. — Rootstock erect. Stipes tufted, naked, nearly a foot long. Lamina oblong-lanceolate, bipinnatifid, 1–1½ ft. long, 2–3 in. broad, moderately firm in texture, green and finely pubescent on both surfaces. Pinnæ sessile, linear-oblong, obtuse, shallowly pinnatifid, ½ in. broad; lower sharply deflexed. Veins pinnate opposite each final lobe; veinlets 4–5-jugate. Sori medial on all the veinlets. Indusium obscure, minute, fugacious. — Limestone hills, Sarawak. Closely resembles the West Indian *Polypodium reptans* Sw.

232 (201\*). **N.** (SAGENIA) **pteropodum**, n. sp.—Rhizome wide-creeping, hypogæous, ⅙ in. diam. Stipe with a narrow wing beginning about 2 in. from the base, about ¼ in. broad 1–1½ ft. above it. Lamina oblong-lanceolate, acuminate, entire, about a foot long, 3–4 in. broad, narrowed at the base very gradually. Main veins distinct nearly to the edge, about ½ in. apart; areolæ with copious free included veinlets. Sori minute, irregularly scattered, sometimes confluent. Indusium minute, obscure.— Nearly allied to *N. singaporianum* Baker.

233 (221\*). **N.** (SAGENIA) **melanorachis**, n. sp. — Stipe black, 1½ ft. long, beset in the lower half, more densely towards the base, with large spreading linear brown paleæ. Lamina oblong-deltoid, bipinnatifid, 1½–2 ft. long, about a foot broad, moderately firm in texture, green and finely pubescent on both surfaces. Pinnæ in about 3 pairs below the pinnatifid apex, the lowest much the largest, deltoid, produced on the lower side, deeply pinnatifid, with lanceolate lobes. Main veins distinct from the midrib to the margin; areolæ with copious free included veinlets. Sori biserial, approximate to the main veins. Indusium small, reniform, membranous, moderately persistent. — Paku, Sarawak. Near *N. cicutarium* Baker.

242 (5\*). **Polypodium** (GONIOPHLEBIUM) **holophyllum**, n. sp. — Rootstock erect. Stipes crowded, naked, erect, ½ ft. long. Lamina oblong-lanceolate, entire, acuminate, truncate at the base, 6–8 in. long, 1½–2 in. broad, moderately firm in texture, slightly hairy on the midrib beneath. Main veins distinct from the midrib to the margin, parallel, about ¼ in. apart; cross-veinlets 10–12, arcuate. Sori 2 to each cross-veinlet, minute, superficial, globose. —Niah, Sarawak, *Chas. Hose*. Habit of *Meniscium simplex*.

234. *P.* (*Dictyopteris*) *Barberi* Hook.—Lambir, Sarawak.

236. *P. stenophyllum* Blume.—Lambir, Sarawak, *Chas. Hose*.

235 ex parte. *P. decorum* Brack.—Langir, Sarawak, *Chas. Hose*.

235 ex parte. *P. repandulum* Mett.—With the last.

236. *P. campyloneuroides* Baker.—Sarawak.

243. *Vittaria scolopendrina* Thwaites. — Baram, Sarawak, *Chas. Hose*. A very robust, broad-leaved form.

239 (64\*). **Gymnogramme** (SYNGRAMME) **valleculata**, n. sp.— Rootstock woody, short-creeping, 1-12th in. diam., clothed with short stiff subulate brown paleæ. Stipe glossy, naked, castaneous, 6–10 in. long. Lamina oblong, entire, thick, rigid, glabrous,

4–5 in. long, 1½–2 in. broad at the middle, deltoid or rather rounded at the base, deltoid or obtuse at the apex, with a castaneous midrib and revolute margin. Veins immersed, obscure, anastomosing only near the margin. Sori immersed, erecto-patent, falling a little short of the margin.—Lambir, Sarawak, *Chas. Hose.* Near *G. alismæfolia* Hook.

240. *G. alismæfolia* Hook.—Baram, Sarawak, *Chas. Hose.*
241. *G. Feei* Hook.—Sarawak, *Chas. Hose.*
238 (71*). **G.** (SELLIGUEA) **acuminata**, n. sp.—Rootstock wide-creeping, ⅙ in. diam., clothed with minute linear black clathrate paleæ. Stipes remote, slender, naked, 4–5 in. long. Lamina simple, oblong-lanceolate, very acuminate, and tapering very gradually to the base, 6–8 in. long, 1½–2 in. broad at the middle, membranous, green, glabrous. Main veins fine, parallel, erecto-patent, ¼ in. apart, produced nearly to the edge; intermediate areolæ with free included veins. Sori linear, much interrupted, confined to the upper third or half of the frond.—Next *G membranacea* Hook.

244 (107*). **Acrostichum** (GYMNOPTERIS) **exsculptum**, n. sp. —Rootstock creeping. Stipe of sterile frond slender, naked, ½ ft. long. Sterile lamina lanceolate, simply pinnate, 9–12 in. long, 2–2½ in. broad, rigid, green, glabrous. Pinnæ multijugate, sessile, lanceolate, acute, crenate, truncate at the base, 1–1¼ in. long, ⅓ in. broad; lower not reduced. Veining of *Goniopteris;* main veins distinct to the edge, 1-12th to 1-8th in. apart; veinlets trijugate. Fertile frond with a longer stipe; lamina 1 ft. long, 1¼ in. broad; Pinnæ sessile, remote, ⅙ in. broad. — Niah, Sarawak, *Chas. Hose.* Allied to *A. virens* Wall.

---

# BOTANICAL NOMENCLATURE IN NORTH AMERICA.

## BY EDWARD L. GREENE.

LIGHT and help on the subject of nomenclature we long since learned to expect from every paragraph thereon which might emanate from the editorship of this Journal. Certain recent animadversions* are not in this regard a disappointment. The article upon which it seems needful to offer one more word of comment is, upon the whole, a very instructive one, and we have welcomed it, notwithstanding that it bears rather heavily upon some of us in America.

Without asking for space in which to discuss a number of interesting propositions set forth by Mr. Britten in the body of his article, I must be permitted to try to correct a wrong impression which will have been made by his opening paragraph, feeling confident that he, no less than others, will welcome the correction.

It is quite erroneous to say, as the Editor does say, implicitly, if not in just so many words, that, while an older generation of

---

* Journ. Bot. 1888, p. 257.

American botanists have been and are governed by established laws in nomenclature, a new school has arisen whose aim is to introduce a new system, one which is thought objectionable as bringing in " fresh elements of confusion."

Not to pause for a moment in explanation or defence of a system which, so far from being new, our esteemed critic himself knows to have been long recognised and adhered to as the correct one, in almost every one of the great branches of systematic biology outside of the one department of phanerogamic botany ; in which latter branch, even, it has had respectable advocates ; I am only called upon to show that no contrast quite so striking really exists between the practices of ourselves of the " new school," if so we are to be called, and those of our elders.

We are censured in this ; that we suffer ourselves to be governed by the principle of priority in relation to specific, as well as generic, names. Since we had to be subjected to an ordeal so rather trying as that of a comparison of our own wisdom and discretion with those of our fathers,—for by such comparison the younger inevitably, and perhaps always more or less justly, suffers,—it might have been well to mention the one thing wherein we should seem commendable above those who have gone before us, *i. e.*, our resolute defence of, and abiding by, the law of priority in generic names. The earlier race of American botanists herein exhibit a laxity of view, with which our own strictness forms a contrast ; and it is not from any representatives of an old school in America that such genera as *Hookera* and *Castalia*, which the Editor of this Journal has so clearly shown to be of obligation under the law of priority, will meet with approval and adoption. The remnant of that party here resists these reinstatements with whatever it has retained of its former influence and authority.

Against the practice of restoring old specific names in those genera when new ones had been made to replace them, it must be admitted that Dr. Gray sometimes argued, " with his wonted care and ability," in divers journalistic paragraphs ; and our friends in England, not having looked into his books to see how very often, through successive pages of plant-naming and describing, he adopts the very practice which he disapproves in others, imagine that here they have made a point against us. We would, therefore, invite attention to Dr. Gray's nomenclature of any of the genera of the Synoptical Flora, in which there are Linnæan species now placed in other genera. Take the *Ericaceæ* for an example. There is *Rhododendron*, at present made to include the species of the Linnæan *Azalea*. There are named and described five species of the *Azalea* subdivision. Every one of them had received its first specific name under *Azalea*. To four out of the five, new specific names had been given upon their introduction into *Rhododendron ;* but, in each of these four instances, our author has rejected the " first name which the species received under its proper genus," adopting that more recent combination which embraces the old specific name under *Azalea*. One of my colleagues in America has lately adverted, incidentally, to the case of *Moneses*,* in which Dr. Gray, as long ago as 1847, set aside his English namesake's *M. grandiflora* (S. F. Gray, Nat. Am. ii. 403), and rehabilitated the little plant in its old Linnæan (yea, pre-Linnæan) specific name, making the new combination *Moneses uniflora* A. Gray. And these which I cite are but fair samples of Dr. Gray's occasional practice when the species are old, and have received several specific names. Whatever may have induced him now and then, in critical essays, to write in disparagement of this usage, one who studies him in his books and monographs must see that he not only had a very strong predilection for the oldest specific names, but was willing to transgress rules which he professed to respect and be governed by, in order to keep such names in use.

Dr. Watson, who is also cited as if exemplifying more approved methods in nomenclature, has made himself, in some of his pages, a luminous example of our "new school" usage. For a good illustration, we have but to advert to his readjustments in the specific nomenclature of *Onagraceæ*, in the first volume of the Botany of California. Spach, in proposing the genera *Godetia* and *Boisduvalia*, had dropped a number of very old specific names which the plants had been known by under *Œnothera;* and Dr. Watson, with what we, his American colleagues, consider a commendable zeal for thorough priority, restored those old neglected names, every one ; and so we read, in the place referred to, *Godetia purpurea* Watson instead of the much older combination *G. Willdenoviana* Spach, *G. tenella* Watson instead of *G. Cavanillesii* Spach, *Boisduvalia densiflora* Watson in place of *B. Douglasii* Spach, and so on.

I shall be far from asserting that our elders have followed this rule. On some of their pages they conform to this, on others to some other, and the having of so many rules is equivalent to having none at all. That this is the true condition of botanical nomenclature in America, with all authors, up to a somewhat recent date, one has but to look into our most pretentious treatises to see. I have been constrained lately to remark this unhappy fact.† For any two or three botanists to have settled down to any one particular usage, or to have subjected themselves to any code whatever, would have been to form, in America, a "new school." A number of us younger workers have, in so far as I know, without any mutual understanding or agreement, one after another, placed ourselves under obedience to the simple law of priority in nomenclature ; and, be our action commendable or be it deprecable, it does, we confess it, place us in contrast with the earlier generation, whose misfortune it may have been to have had us in training.

---

* Bull. Torr. Club, 1888, p. 230.    † 'Pittonia,' i. p. 185.

# SOUTH DERBYSHIRE PLANTS.

## BY THE REV. W. R. LINTON, M.A.

THE district in which the following observations have been made lies west of Ashbourne, having Shirley as its centre, with a radius of four or five miles. Gravel, sand, red marl, and clay in places compose the soil, limestone occurring at one spot only in the south-west.

Following the order of the 8th edition of the 'London Catalogue,' the first plant which demands notice is a large form of *Ranunculus penicillatus* Hiern, which occurs plentifully in Brailsford Brook. In several spots it produces no floating leaves; where produced, they are remarkably thick and fleshy in texture, with cuneate segments.

*Viola Reichenbachiana* Bor. occurs in two places in the district, having only been recorded hitherto in localities to the north of it. *Stellaria neglecta* Weihe occurs at Brailsford and Shirley, but is not nearly so frequent as its ally, *Stellaria umbrosa* Opiz. *Medicago maculata* Sibth. is a casual on cultivated land, but scarcely to be reckoned to the county Flora. *Prunus Cerasus* L. grows at intervals along one lane in Brailsford, where it is well-established, if not wild. In the same lane occurs one bush of the yellow-fruited form of *Prunus insititia* L., the form with black fruit being plentiful everywhere in the district.

Of Rubi, several have been recorded and commented on in the recently-issued 'Report of the Botanical Exchange Club.' I may add that *R. carpinifolius* W. & N. is quite one of the brambles of the district; *R. fissus* Lindl. is plentiful in Shirley Wood, and occurs in Bradley Wood; and *R. amphichloros* Focke I find now in Brailsford, in addition to its habitat in Shirley.

Of Rosæ, *R. mollis* var. *cærulea* Woods grows in a few places, Bradley and Hognaston, to wit; *Rosa micrantha* (Sm.) is frequent; *R. canina* var. *surculosa* (Woods), of excellent character, in Bradley; var. *Malmundariensis* Desegl. in Yeldersley; varr. *verticillacantha* (Mérat), *collina* (Jacq.), and *Koscinciana* (Besser) in several localities.

The Epilobia furnish material for very interesting study, especially on account of the frequency of the occurrence of hybrids, mainly between *E. parviflorum* Schreb. and *E. obscurum* Schreb., and between the latter and *E. montanum* L. In the case of some of the hybrid forms, both Mr. Towndrow and M. W. Barbey, to whom they have been submitted, have suggested *E. tetragonum* L. as one of the parents. But at the close of another season's investigation of the district I can only regret what appears in the Exchange Club Report, that *E. tetragonum* seems absent from this part of the county. Of *E. obscurum*, both the broad-leaved and the rare narrow-leaved forms occur.

*Ribes Uva-Crispa* L. occurs in a rough lane at Atlow, and I have also noticed it near the head of Dovedale. *Taraxacum officinale* Web., var. *erythrospermum* Andrz., occurs on steep grassy slopes in Shirley. *Sonchus arvensis* L., var. *glabra* Lond. Cat., may be found occasionally with the type. About a dozen plants occurred in a field in

Shirley this summer. *Primula vulgaris* Huds., var. *caulescens* Bab., as well as *P. veris* × *vulgaris*, are found sparingly in the spring. *Cuscuta Trifolii* Bab. has been seen on cultivated land in Bradley. *Mentha viridis* L. and *M. rubra* Sm. occur by Shirley Brook, and *M. Piperita* Huds. at Yeaveley, the first and the last being not indigenous. *Betula glutinosa* Fr. is to be seen not unfrequently in hedges.

Salices of the *Smithiana* group are found in several places, some agreeing more closely with *S. Smithiana* Willd. proper, others coinciding with *S. rugosa* Leefe. In them the foliage varies very little, the main differences being exhibited by the stigmas, which are linear and long, or oval and shorter, entire bifid or bipartite in different plants examined. *S. repens* L. though not observed in the district, has been seen by me further south in the county, by the road-side near Etwall. The usual form of *Populus tremula* L. which occurs about here is the var. *glabra* Syme; and *Iris Pseudacorus* L. only occurs as *acoriformis* Bor. *Juncus diffusus* Hoppe has not been noticed, I believe, in many places in the county, and has only been recently added to the county Flora; I have seen it in Shirley growing plentifully, and still more abundantly some few miles to the north, near Corley. In both cases it appears to be a species, as described in Bab. Man., rather than a hybrid, as described in Hooker's 'Students' Flora.' *J. supinus* Mœnch, var. *Kochii* Syme, I have seen in Shirley Wood. *Potamogeton serratus* Huds. is the common form which *P. crispus* takes in brooks and streams in the district.

*Carex curta* Good. is to be found, but not plentifully, in a marsh in Yeldersley. One patch of *C. Goodenowii* J. Gay, var. *juncella* Fr., occurs in Shirley Wood. *C. præcox* Jacq. almost covered the ground in several fields, in 1887, in Shirley; this year I have seen no trace of it. *C. fulva* Good., I believe the type, is found in one meadow in Shirley. *C. paludosa* Good. very frequently takes the form *Kochiana* Gaud.

Among grasses, I have seen the form *Arrhenatherum nodosum* Rchb., and *Bromus arvensis* L. as a casual by the road-side. *Agropyron repens* Beauv. frequently occurs as var. *barbata* Duval-Jouve. I can see no *Characeæ* in the district.

The more or less boreal character of the district is shown by the absence of *Clematis Vitalba* L., *Epilobium tetragonum* L., *Bryonia dioica* L., and *Lysimachia Nummularia* L.; and by the presence of several members of the subcristate section of *Rosa canina*, *Circæa alpina* L., which is more frequent than *C. lutetiana* L., *Campanula latifolia* L. and *Salix pentandra* L., which last I have noticed in three places. Appended is a list of such of the plants mentioned above as are, so far as I can find, additions to the County Flora :—

*Stellaria media* var. *neglecta* Weihe.
*Rubus carpinifolius* W. & N.
*Rosa mollis* var. *cærulea* Woods.
*R. canina* var. *surculosa* (Woods).
      ,,       *Malmundariensis* Desegl.
      ,,       *verticillacantha* (Mérat).

*R. canina* var. *collina* (Jacq.).
     ,,           *Koscinciana* (Besser).
*Ribes Grossularia* var. *Uva-Crispa* L.
*Taraxacum officinale* var. *erythrospermum* (Andrz.).
*Sonchus arvensis* var. *glabrä* Lond. Cat.
*Cuscuta Trifolii* Bab.
*Betula glutinosa* Fr.
*Salix repens* L.
*Iris Pseudacorus* var. *acoriformis* (Bor.).
*Juncus supinus* var. *Kochii* Syme.
*Carex paludosa* var. *Kochiana* Gaud.
*Bromus arvensis* L.
*Agropyron repens* var. *barbata* Duval-Jouve.

---

# CATALOGUE OF THE MARINE ALGÆ OF THE WEST INDIAN REGION.

## By George Murray, F.L.S.

(Continued from p. 307.)

(Plate 284).

### Chondrieæ.

Laurencia implicata J. Ag. Guadeloupe, *Mazé* ! Florida, *Harvey, Melvill*! New Grenada, *Hb. Dickie*! (but possibly Pacific coast).

L. cervicornis Harv. Guadeloupe, *Mazé*. Florida, *Harvey, Melvill*!

L. arbuscula Sond. (= L. heteroclada Harv. ?). Guadeloupe, *Mazé*.
     *Geogr. Distr.* Australia.

L. corymbosa J. Ag. Guadeloupe, *Mazé* !
     *Geogr. Distr.* Cape of Good Hope.

L. scoparia J. Ag. La Guayra, *Hb. Binder*. Grenada, *Murray*!
     Barbadoes, *Dickie* ! Guadeloupe, *Mazé* ! Bermuda, *Kemp*.
   f. minor Crn. Guadeloupe, *Mazé*!

L. intricata Lam. Grenada, *Murray* ! Guadeloupe, *Mazé* ! Cuba,
     *R. de la Sagra.*

L. divaricata J. Ag. Guadeloupe, *Mazé* !
     *Geogr. Distr.* Red Sea, Indian Ocean, Cape of Good Hope?

L. dendroidea J. Ag. Guadeloupe, *Mazé* !
   Var. tenuifolia Crn. Guadeloupe, *Mazé* !
   Var. corymbifera Crn. Guadeloupe, *Mazé* !
   f. denudata Crn. Guadeloupe, *Mazé* !
     *Geogr. Distr.* Warm Atlantic (Brazil), Australia, Corea.

L. paniculata J. Ag. Guadeloupe, *Mazé* !
     *Geogr. Distr.* Warm Atlantic (Europe), Indian Ocean ?

L. thuyoides Kütz. (= L. paniculata J. Ag. ?). Guadeloupe, *Mazé* !

L. papillosa Grev. Grenada, *Murray* ! Barbadoes, *Dickie* ! Gua-
     deloupe, *Mazé* ! Danish West Indian Islands, *Hohenack.* !
     No. 571. Jamaica, *Chitty* ! Cuba, *R. de la Sagra.* Mexico
     (*fide Dickie*). Florida, *Harvey* ! *Melvill* ! Bermuda, *Kemp, Rein.*

Var. SUBSECUNDA Kütz.    *Hooper*! (in Farlow, Anderson & Eaton,
No. 63).
*Geogr. Distr.*    Warm Atlantic, Mediterranean, Red Sea, Indian
Ocean, Pacific.
L. OBTUSA Lam.    Barbadoes, *Dickie*!    Guadeloupe, *Mazé*!    Danish
West Indian Islands, *Hohenack.*! No. 570.    Jamaica, *Chitty*!
Florida, *Harvey, Melvill*!    Bermuda, *Kemp, Rein, 'Challenger'*!
Var. GELATINOSA J. Ag.    Guadeloupe, *Mazé*!
Var. RACEMOSA Kütz.    Guadeloupe, *Mazé*!
Var. MICROCLADIA Kütz.    Danish West Indian Islands, *Hohenack.*!
No. 385.
*Geogr. Distr.*    Throughout all seas.
L. VIRGATA J. Ag.    Guadeloupe, *Mazé*!
f. DENUDATA Crn.    Guadeloupe, *Mazé*!
*Geogr. Distr.*    Cape of Good Hope, Australia, New Zealand?
L. HYBRIDA Lenormand.    Guadeloupe, *Mazé*!
*Geogr. Distr.*    Atlantic (Europe).
L. CÆSPITOSA J. Ag.?    Guadeloupe, *Mazé*!
*Geogr. Distr.*    Atlantic (Europe).
L. PINNATIFIDA Lam.    Guadeloupe, *Mazé*!    Jamaica, *Chitty*!
*Geogr. Distr.*    Atlantic, Mediterranean.
L. TUBERCULOSA J. Ag.    Guadeloupe, *Mazé*!    Vera Cruz, *Liebman*.
Var. GEMMIFERA J. Ag. (= L. GEMMIFERA Harv.).    Guadeloupe,
*Mazé*!    Florida, *Harvey*! *Melvill*!    Bermuda, *Rein, 'Challenger.'*
L. FLEXUOSA J. Ag.    Guadeloupe, *Mazé*!
*Geogr. Distr.*    Cape of Good Hope.
L. CONCINNA Mont.    Grenada, *Murray*!    Guadeloupe, *Mazé*!
*Geogr. Distr.*    Australia (Port Natal?).
L. CANARIENSIS Mont.    Guadeloupe, *Mazé*!
f. MAJOR Crn.    Guadeloupe, *Mazé*!
*Geogr. Distr.*    Warm Atlantic.
L. VAGA Kütz.    Guadeloupe, *Mazé*!
*Geogr. Distr.*    New Caledonia.
L. CHONDRYOPSIDES Crn.    Guadeloupe, *Mazé*!
L. CRASSIFRONS Crn.    Guadeloupe, *Mazé*!
f. DENDROIDES Crn.    Guadeloupe, *Mazé*!
ASPARAGOPSIS DELILEI Mont.    Barbadoes, *Dickie*!    Guadeloupe,
*Mazé*!    Bermuda, *Rein*.
*Geogr. Distr.*    Atlantic, Mediterranean, Indian Ocean, Aus-
tralia? Philippines.


RHODOMELEÆ.

CHONDRIOPSIS LITTORALIS J. Ag.    Guadeloupe, *Mazé*!    Vera Cruz,
*Liebman*.    Florida, *Harvey*! *Melvill*!    Bermuda, *Rein*.
C. ATROPURPUREA J. Ag.    Florida, *Harvey*.    (I have seen Harvey's
specimen from S. Carolina.)
*Geogr. Distr.*    Brazil.
C. CAPENSIS Harv.    Guadeloupe, *Mazé*!
*Geogr. Distr.*    Cape of Good Hope.

Tab 284

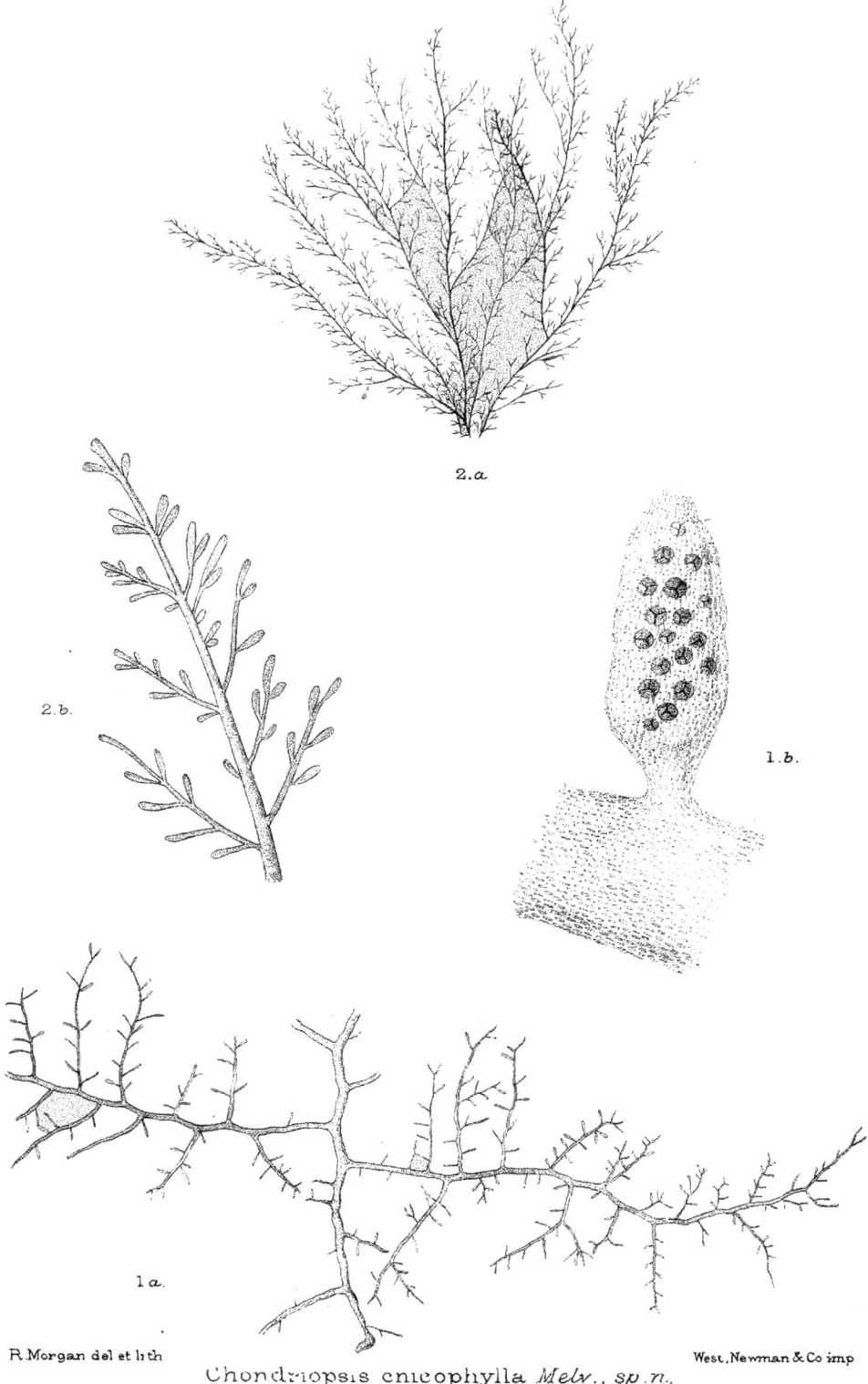

2.a

2.b.

1.b.

1a.

R.Morgan del et lith

West, Newman & Co imp

Chondriopsis cnicophylla *Melv.*, *sp.n.*
C.        leptacremon    „      „  „

C. TENUISSIMA J. Ag. Guadeloupe, *Mazé*!
  f. CRASSA Crn. Guadeloupe, *Mazé*!
    *Geogr. Distr.* Atlantic (Europe and America), Mediterranean, Australia.
C. HARVEYANA J. Ag.? Guadeloupe, *Mazé*!
    *Geogr. Distr.* Tasmania.
C. DASYPHYLLA J. Ag. Grenada, *Murray*! Guadeloupe, *Mazé*!
    Florida, *Harvey*! *Melvill*!
  Var. SEDIFOLIA J. Ag. Florida, *Harvey*, *Melvill*!
  f. GRACILIS J. Ag. Guadeloupe, *Mazé*!
    *Geogr. Distr.* Atlantic (Europe and America), Australia.
C. TENERA Crn. Guadeloupe, *Mazé*!

**Chondriopsis cnicophylla** Melv., sp. nov.—Fronde robustâ, divaricato-ramosâ, ramis decompositis, alternatis, ramulis squarrosa-patentibus, apicibus plus minus clavatis, interdum attenuatis, basi constrictâ.

Root a small disc. Main stem robust; zigzagged frond with divaricate spreading branches, alternate, decompound; the ultimate ramuli squarely patent; apices of the ramuli more or less clubbed, but occasionally attenuate, constricted at the base. Tetraspores immersed in the ramuli; conceptacles not present; colour, when fresh, dark purplish. (Plate 284, figs. 1 *a*, 1 *b*).

This species differs from all the varieties of *C. dasyphylla* Woodw. or *C. Baileyana* Harv. I have seen, in its divaricate growth, and squarely-set patent ramuli, giving a thorny character to the frond. One specimen alone was found on the sandy beach at Cedar Keys, West Florida, on the Gulf of Mexico, after a somewhat severe gale, 8th March, 1872.—J. C. M.

**Chondriopsis leptacremon** Mclv., sp. nov. — Fronde tenui, subtili, simpliciter ramosâ, ramulis alternatis, elongatis, apicibus subattenuatis, basi constrictâ.

Plant somewhat cæspitose at base; frond extremely slender and graceful, simply branched; ramuli elongate; apices of the ramuli slightly clavate or subattenuate, constricted at the base. (Plate 284, figs. 2 *a*, 2 *b*).

This is the smallest and slenderest species of the genus, bearing the same proportion to *C. tenuissima* Good. in this respect that *Ceramium byssoideum* Harv. does to *C. gracillimum* Ag. The fructification is unknown. Colour pale brownish pink, becoming dark brown when dry, and staining the paper the same colour. Of this species but two specimens were gathered, both growing at low-water on a coral slab below Fort Taylor, at the S.W. extremity of the Island of Key West, S. Florida: one of the specimens was submitted to Prof. J. G. Agardh, who pronounced it almost certainly an undescribed new form.—J. C. M.

ACANTHOPHORA MUSCOIDES Bory. Barbadoes, *Dickie*! Guadeloupe, *Mazé*! Santa Cruz, *Miss Dix*! Cuba, *R. de la Sagra*. Florida, *Harvey*! *Palmer*! *Hooper*! (in Farlow, Anderson & Eaton, No. 8).
    *Geogr. Distr.* Warm Atlantic (Africa and America).

A. DELILEI Lam. Florida, *Harvey, Melvill*!
 *Geogr. Distr.* Mediterranean, Red Sea.
A. THIERII Lam. Grenada, *Murray*! Barbadoes, *Dickie*! Guadeloupe, *Mazé*! Martinique, *Duperrey.* Santa Cruz, *Miss Dix (fide Harvey).* St. Thomas, *Oersted, Duchassainy, Hohenack.*! No. 389. Cuba, *R. de la Sagra.* Mexico, *Liebman.* Florida, *Harvey, Melvill, Hooper*! (in Farlow, Anderson and Eaton, No. 9). Bermuda, *Kemp, Rein.*
 f. GRACILIS Crn. Guadeloupe, *Mazé*!
 *Geogr. Distr.* Brazil, Europe (Biarritz).
A. INTERMEDIA Crn. Guadeloupe, *Mazé*!
MARTENSIA PAVONIA J. Ag. Guadeloupe, *Mazé*! Santa Cruz, *Duchassaing, Hb. Rosenberg.*
ALSIDIUM BLODGETTII Harv. Florida, *Harvey.* Bermuda, *Rein.*
DIGENEA SIMPLEX Ag. Barbadoes, *Dickie*! Guadeloupe, *Mazé*! Florida, *Harvey*! *Melvill*! Bermuda, *Kemp, Rein.*
 *Geogr. Distr.* Warm Atlantic, Mediterranean, Red Sea, Indian Ocean, Japan.
BRYOTHAMNION SEAFORTHII Kütz. Grenada, *Murray*! Barbadoes, *Dickie*! Guadeloupe, *Mazé*! Santa Cruz and Jamaica (*fide J. Ag.*). Cuba, *R. de la Sagra.* Mexico (*fide J. Ag.*). "West Indies," *Hb. Waller*!
 *Geogr. Distr.* Brazil.
B. TRIANGULARE Kütz. Barbadoes, *Dickie*! Guadeloupe, *Mazé*! Jamaica, *Chitty*! Cuba, *R. de la Sagra.* Danish West Indian Islands, *Hohenack.*! Florida, *Harvey*! *Melvill*!
 *Geogr. Distr.* Brazil.
B. HYPNOIDES Kütz. Grenada, *Murray*! Guadeloupe, *Mazé*! Cuba, *Hb. Montagne* (? R. de la Sagra).
BOSTRYCHIA RIVULARIS Harv. Guadeloupe, *Mazé*! Florida, *Harvey.* (I have seen Harvey's specimen from S. Carolina.)
 *Geogr. Distr.* Australia.
B. PERICLADOS J. Ag. Guadeloupe, *Mazé*! Santa Cruz, *Hb. Hofman Bang.* Florida, *Harvey.*
 *Geogr. Distr.* Friendly Islands.
B. MORIZIANA J. Ag. Guadeloupe, *Mazé*! "Antilles," *Moritz.*
 *Geogr. Distr.* Guiana.
B. SCORPIOIDES Mont. Bermuda, *Kemp.*
 *Geogr. Distr.* N. Atlantic (British Islands to Spain), Mediterranean.
B. TENELLA J. Ag. Guadeloupe, *Mazé*! Martinique, *Hb. Le Jolis*! Cuba, *R. de la Sagra.* Florida, *Harvey*! *Melvill*! Bermuda, *Farlow*! (in Farlow, Anderson and Eaton, No. 137).
 *Geogr. Distr.* Cape of Good Hope, Indian Ocean, Pacific.
B. MONTAGNEI Harv. Cuba, *Tuomey* (*fide Harvey*). Florida, *Harvey*! Bermuda, *Rein*! *Farlow*! in (Farlow, Anderson and Eaton, No. 136); *Reliquiæ Brebissonianæ*! (Ser. 2, Nos. 203 & 298).
B. SERTULARIA Mont. Guadeloupe, *Mazé*! Bermuda, *Farlow*! (in Farlow, Anderson and Eaton, No. 138).
 *Geogr. Distr.* Guiana.

B. PILIFERA Kütz.  Guadeloupe, *Mazé* !
  *Geogr. Distr.*  Senegambia.
B. ELEGANS Crn.  Guadeloupe, *Mazé* !
B. POLYSIPHONIOIDES Crn.  Guadeloupe, *Mazé* !
B. MUSCOIDES Crn.  Guadeloupe, *Mazé* !
B. DASYÆFORMIS Crn.  Guadeloupe, *Mazé* !
B. GUADELUPENSIS Crn.  Guadeloupe, *Mazé* !
B. MAZEI Crn.  Guadeloupe, *Mazé* !
B. CAPILLACEA Crn.  Guadeloupe, *Mazé* !
POLYSIPHONIA SECUNDA Zan.  Grenada, *Murray*! Guadeloupe, *Mazé*!
  Cuba, *R. de la Sagra*.  Florida, *Harvey*, *Melvill* !  Bermuda,
  *Farlow* !  (in Farlow, Anderson and Eaton, No. 134) ; *Herb.*
  *Greville* !
  *Geogr. Distr.*  Atlantic, Mediterranean.
P. PECTEN-VENERIS Harv.  Florida, *Harvey*, *Melvill* !
P. RUFOLANOSA Harv.  Guadeloupe, *Mazé* !
  *Geogr. Distr.*  Australia.
P. MONOCARPA Mont.  Guadeloupe, *Mazé*.
  *Geogr. Distr.*  Cape of Good Hope.
·P. OBSCURA J. Ag.  Guadeloupe, *Mazé* !  Santa Cruz, *Oersted*.
  *Geogr. Distr.*  Atlantic and Mediterranean.
P. EXILIS Harv.  Florida, *Harvey*.  Bermuda, '*Challenger*' !
P. FRACTA Harv.  Grenada, *Murray* !  Florida, *Harvey* !
P. ECHINATA Harv.  Florida, *Harvey*.
P. DICTYURUS J. Ag.  Guadeloupe, *Mazé* !
  *Geogr. Distr.*  Pacific (Mexico).
P. THYRSIGERA J. Ag.  La Guayra (*fide Kützing*).  Grenada, *Murray* !
  Guadeloupe, *Mazé* !  [Harvey says "La Guayra, Mexico,"
  and quotes the locality from Kützing, who, however, says
  nothing of "Mexico."  No doubt it is Venezuela.]
P. VERTICILLATA Harv.  Guadeloupe, *Mazé* !
  *Geogr. Distr.*  Pacific (California).
P. PULVINATA J. Ag.  Guadeloupe, *Mazé* !
  *Geogr. Distr.*  Atlantic, Mediterranean.
P. INCOMPTA Harv.  Guadeloupe, *Mazé* !
  *Geogr. Distr.*  Cape of Good Hope.
P. HAVANENSIS Mont.  Guadeloupe, *Mazé* !  Santa Cruz, *Oersted*.
  Cuba, *R. de la Sagra* !  Florida, *Harvey* ! *Melvill* !
  Var. BINNEYI J. Ag.  Guadeloupe, *Mazé* !  Florida, *Harvey* ! *Melvill* !
  *Geogr. Distr.*  Atlantic (France).
P. SUBTILISSIMA Mont.  Grenada, *Murray* !  Guadeloupe, *Mazé* ! (?).
  St. Thomas, '*Challenger*' !  Bermuda, '*Challenger*.'
  *Geogr. Distr.*  Atlantic (Cayenne and U. S. America).
P. UTRICULARIS Zan.  Grenada, *Murray* !  Guadeloupe, *Mazé* !
  *Geogr. Distr.*  Suez.
P. MOLLIS Hook. et Harv.  Grenada, *Murray* !  Guadeloupe, *Mazé* !
  *Geogr. Distr.*  Australia.
P. CAMPTOCLADA Mont.  Guadeloupe, *Mazé* !
  *Geogr. Distr.*  Pacific (Peru).
P. FUNEBRIS De Not.  Guadeloupe, *Mazé* !
  *Geogr. Distr.*  Mediterranean.

P. GORGONIÆ Harv.    Florida, *Harvey, Melvill* !

P. FERULACEA Suhr.    Barbadoes, *Dickie* !    Guadeloupe, *Mazé* !    Vera Cruz, *Liebman*.    Florida, *Harvey, Melvill* !
*Geogr. Distr.*    Australia.

P. HAPALACANTHA Harv.    Florida, *Harvey, Melvill*.

P. VIOLACEA Grev.    Guadeloupe, *Mazé* !
*Geogr. Distr.*    Atlantic (Europe and America).

P. FIBRILLOSA Grev.    Bermuda, *Kemp, Rein*.
*Geogr. Distr.*    Atlantic (Europe and America).

P. ELONGATA Grev.    Bermuda, *Kemp*.
*Geogr. Distr.*    Atlantic (Europe, America, and N. Africa), Mediterranean.

P. HIRTA J. Ag.    Florida, *Harvey, Melvill* !
*Geogr. Distr.*    Mediterranean.

P. COLLABENS Kütz.    Guadeloupe, *Mazé* !
*Geogr. Distr.*    Mediterranean and neighbouring Atlantic.

P. FURCELLATA Harv.    Guadeloupe, *Mazé* !
*Geogr. Distr.*    Atlantic and Mediterranean.

P. VARIEGATA Zan.    Guadeloupe, *Mazé* !
*Geogr. Distr.*    Atlantic (Europe and America), Mediterranean.

P. NIGRESCENS Grev.    Bermuda, *Rein*.
*Geogr. Distr.*    Atlantic (Europe and N. America), New Zealand.

P. CALLITHAMNIODES Crn.    Guadeloupe, *Mazé* !

P. BOSTRYCHIOIDES Crn.    Guadeloupe, *Mazé* !

P. MUCOSA Crn.    Guadeloupe, *Mazé* !

P. CAPUCINA Crn.    Guadeloupe, *Mazé* !

AMANSIA MULTIFIDA Zan.    Guadeloupe, *Mazé* !    Martinique, *Duperrey* !    Jamaica, *Chitty* !    Florida, *Harvey* !    *Melvill* !
*Geogr. Distr.*    Brazil, New Zealand.

A. DUPERREYI J. Ag.    Guadeloupe, *Mazé* !    Martinique, *Duperrey*.

VIDALIA OBTUSILOBA J. Ag.    Guadeloupe, *Mazé* !

POLYZONIA ? DIVARICATA Crn.    Guadeloupe, *Mazé* !

DASYA PELLUCIDA Harv. ?    Guadeloupe, *Mazé* !
*Geogr. Distr.*    Cape of Good Hope, Australia.

D. WURDEMANNI Bailey.    Guadeloupe, *Mazé* !    Florida, *Harvey* !    *Melvill* !
*Geogr. Distr.*    Atlantic, Mediterranean.

D. GIBBESII Harv.    Nassau (Bahamas), *Palmer* !    Florida, *Harvey* !    *Melvill* !    *Hooper* !    in (Farlow, Anderson and Eaton, No. 3).

D. HUSSONIANA Mont.    Guadeloupe, *Mazé* !
*Geogr. Distr.*    Red Sea.

D. ELEGANS Ag.    Florida, *Harvey*.    Bermuda, *Rein*.
Var. SCOTIOCHROA Melv.    Florida, *Melvill* !
*Geogr. Distr.*    Atlantic (Europe and N. America), Mediterranean.

D. MOLLIS Harv.    Florida, *Harvey, Melvill* !
*Geogr. Distr.*    Australia.

D. CORYMBIFERA J. Ag.    Grenada, *Murray* !    Guadeloupe, *Mazé* !
*Geogr. Distr.*    Atlantic (Europe), Mediterranean.

D. ARBUSCULA Ag.    Guadeloupe, *Mazé* !
*Geogr. Distr.*    Atlantic (Europe and Africa), Mediterranean.

D. MUCRONATA Harv.  Grenada, *Murray* !  Florida, *Harvey* ! *Melvill* !
*Hooper* ! (in Farlow, Anderson and Eaton, No. 2).  Bermuda,
*Kemp.*

D. RAMOSISSIMA Harv.  Grenada, *Murray* !  Florida, *Harvey* ! *Melvill* !  *Hooper* ! (in Farlow, Anderson and Eaton, No. 1).

D. HARVEYI Ashm.  Florida, *Melvill* !

D. TRICHOCLADOS Mart.  Hayti, *Hb. Maille.*  Florida, *Harvey.*
Var. OERSTEDI J. Ag.  Guadeloupe, *Mazé* !  Santa Cruz and St.
Thomas, *Oersted.*

D. LALLEMANDI Mont. ?  Guadeloupe, *Mazé* !
*Geogr. Distr.*  Red Sea, Australia.

D. TUMANOWICZI Gatty.  "West Indies," *Tumanowicz.*  Guadeloupe,
*Mazé* !  Florida, *Harvey, Melvill* !  Bermuda, '*Challenger*' !

D. DICHOTOMO-FLABELLATA Crn.  Guadeloupe, *Mazé* !

DASYA MAZEI nob. = EUPOGODON MAZEI Crn.  Guadeloupe, *Mazé* !

D. GRANDIS nob. = EUPOGODON GRANDE Crn.  Guadeloupe, *Mazé* !

DICTYURUS OCCIDENTALIS J. Ag.  Barbadoes, *Dickie* !  Guadeloupe,
*Mazé* !  Jamaica, *Chitty* !  Vera Cruz, *Liebman.*

SARCOMENIA MINIATA J. Ag.  Guadeloupe, *Mazé* !
*Geogr. Distr.*  Atlantic (Cadiz).

### CORALLINEÆ.

HAPALIDIUM CONFERVICOLA J. Ag.  Guadeloupe, *Mazé.*
*Geogr. Distr.*  Atlantic, Mediterranean.

MELOBESIA MEMBRANACEA Lam.  Guadeloupe, *Mazé* !
*Geogr. Distr.*  Atlantic, Mediterranean.

M. FARINOSA Lam.  Guadeloupe, *Mazé* !
*Geogr. Distr.*  Atlantic, Mediterranean, Australia.

M. VERRUCATA Lam.  Grenada, *Murray* !  Barbadoes, *Dickie* !  Guadeloupe, *Mazé* !
*Geogr. Distr.*  Atlantic, Mediterranean.

M. PUSTULATA Lam.  Grenada, *Murray* !  Barbadoes, *Dickie* !  Martinique, *Hb. Le Jolis* !  Bermuda, '*Challenger*' !
*Geogr. Distr.*  Atlantic, Mediterranean, Pacific.

M. AMPLEXIFRONS Harv.  Guadeloupe, *Mazé* !
*Geogr. Distr.*  Cape of Good Hope.

LITHOTHAMNION POLYMORPHUM J. Ag.  Grenada, *Murray* !  Guadeloupe, *Mazé* !
*Geogr. Distr.*  General.

MASTOPHORA LAMOUROUXII Harv.  Guadeloupe, *Mazé* !
*Geogr. Distr.*  Cape of Good Hope, Java (not Australia).

AMPHIROA FRAGILISSIMA Lam.  Grenada, *Murray* !  Guadeloupe,
*Mazé* !  Jamaica, *Sloane* !  Florida, *Harvey, Melvill* !  *Hooper* ! (in
Farlow, Anderson & Eaton, No. 15).  Bermuda, '*Challenger*' !
*Geogr. Distr.*  Mediterranean.

A. BREVIARTICULATA Aresch.  "In mari Indiæ occidentalis," *Hb. Sonder.*

A. TRIBULUS Lam.  Guadeloupe, *Mazé* !  Santa Cruz, *Oersted.*

A. DILATATA Lam.  Guadeloupe, *Mazé* !
*Geogr. Distr.*  Cape of Good Hope (not Australia).

A. ANCEPS Decne.  Guadeloupe, *Mazé* !
*Geogr. Distr.*  Norfolk Island.

A. CHAROIDES Lam.   Guadeloupe, *Mazé* !
  *Geogr. Distr.*   Australia.
A. FOLIACEA Lam.   Guadeloupe, *Mazé*!
  *Geogr. Distr.*   Marianne Islands.
A. BRASILIANA Decne., var. MAJOR Crn.   Guadeloupe, *Mazé* !
  Var. UNGULATA Crn.   Guadeloupe, *Mazé* !
  *Geogr. Distr.*   Brazil.
A. CRASSA Lam.   Guadeloupe, *Mazé* !
A. DUBIA Kütz.   Guadeloupe, *Mazé* !
  *Geogr. Distr.*   Cape of Good Hope.
A. NODULOSA Kütz.   La Guayra (*J. Smith in Hb. Sonder*).   Guade-
  loupe, *Mazé*!
A. IRREGULARIS Kütz.   Guadeloupe, *Mazé* !
  *Geogr. Distr.*   Mediterranean.
A. VERRUCOSA Lam.   Guadeloupe, *Mazé* !
  *Geogr. Distr.*   Australia.
A. DEBILIS Kütz.   Cuba, *Hb. Binder.*   Florida, *Harvey, Melvill*!
  Bermuda, *Rein.*
JANIA FASTIGIATA Harv.   Guadeloupe, *Mazé* !
  *Geogr. Distr.*   Cape of Good Hope.
J. RUBENS Lam.   Grenada, *Murray* !  Guadeloupe, *Mazé* !  Jamaica,
  *Sloane* !  *Chitty* !  Florida, *Harvey, Melvill* !
  *Geogr. Distr.*   General.
J. PYGMÆA Lam.   Guadeloupe, *Mazé*!
  *Geogr. Distr.*   Cape of Good Hope.
J. PUMILA Lam.   Guadeloupe, *Mazé* !
  *Geogr. Distr.*   Red Sea and Indian Ocean.
J. CUBENSIS Mont.   Grenada, *Murray* !  Guadeloupe, *Mazé* !  Cuba,
  *R. de la Sagra.*   Vera Cruz, *Hb. Le Jolis* !  Florida, *Harvey,
  Melvill* !  Bermuda, *Rein.*
J. LONGIFURCA Zan.   Guadeloupe, *Mazé* !
  *Geogr. Distr.*   Mediterranean.
J. TENELLA Kütz.   Guadeloupe, *Mazé* !  "Ad oras Mexicanas,"
  *Kützing.*
  *Geogr. Distr.*   Mediterranean.
J. CAPILLACEA Harv.   Cuba, *Tuomey* (*fide Harvey*).
J. COMOSA Crn.   Guadeloupe, *Mazé* !
CORALLINA OFFICINALIS L.   Bermuda, *Kemp.*
  *Geogr. Distr.*   Arctic Sea and N. Atlantic, Mediterranean,
  Black Sea.
C. SUBULATA Ellis et Sol.   Barbadoes, *Dickie* !  Jamaica, *Sloane* !
  "West Indies," *Ellis & Solander.*
  *Geogr. Distr.*   Brazil, Mediterranean, and Australia.
C. CUVIERI Lam.   Guadeloupe, *Mazé* !
  *Geogr. Distr.*   Australia.
C. PLUMIFERA Kütz.   Guadeloupe, *Mazé* !
  *Geogr. Distr.*   Australia.
C. TRICHOCARPA Kütz.   Guadeloupe, *Mazé* !
  *Geogr. Distr.*   Australia.
C. CERATOIDES Kütz.   Guadeloupe, *Mazé* !   Mexico, *Leibold.*

(To be continued.)

# THE DESMIDS OF MAINE.

## By Wm. West, F.L.S.

PROF. AUBERT, of Maine State College, in answer to my request, has kindly sent me two small gatherings of Desmids which he made this summer in the neighbourhood of Orono. I have come across some Algæ belonging to other groups during the examination of them, which I shall not now enumerate.

When I had just about finished my examination of the gatherings, I received a paper entitled ' The Fresh-water Algæ of Maine,' by F. L. Harvey (Bulletin of the Torrey Botanical Club, June, 1888), in which the writer enumerates for the first time the Fresh-water Algæ of that State. Among these are forty-seven species and two varieties of Desmids. Of these I have so far noticed in Prof. Aubert's gatherings thirty-five of the species and one of the varieties, *in addition to which* I have also noticed those enumerated in the following list, comprising seventy-three species and five varieties and forms, none of which are contained in Mr. Harvey's list, so that the total number of the Desmids known for Maine consists of 120 species and seven varieties and forms.

The gatherings cannot be said to be thoroughly exhausted until more of the material is worked up. I have to acknowledge my indebtedness to my son, G. S. West, for valuable assistance rendered during the preparation of this paper.

One form of *Staurastrum pygmæum* Breb. was noticed several times, the end view of which had the angles much more pointed than usual. Some of the *Cosmarium Meneghinii* Breb. noted were not more than 11 μ broad, though typical as to shape. Examples of *Staurastrum saxonicum* Buln. were seen with the spines much stouter than they are shown in Wolle's figure. Zygospores of *Cosmarium amœnum* Breb. were seen.

*Desmidium aptogonium* Breb.
*Sphærozosma pulchrum* Bail., var. *planum* Wolle.
*S. filiforme* Raben.
*S. excavatum* Ralfs.
*Penium crassa* De Bary.
*P. spirostriolatum* Bark.
*Closterium obtusum* Breb.
*C. juncidum* Ralfs.
*C. gracile* Breb.
*C. didymotocum* Corda.
*C. angustatum* Kutz.
*C. acerosum* Ehrenb.
*C. turgidum* Ehrenb.
*C. Cornu* Ehrenb., form. *major* Wolle. This has not been recorded as yet for the United States; it agrees exactly with

Wille's fig. 81, tab. xiv. ' Nordenskiold's Expedition, 1875.'
*C. acutum* Breb.
*C. Dianæ* Ehrenb.
*Docidium crenulatum* Rabenh.
*D. truncatum* Breb.
*D. Archeri* Wolle.
*Calocylindrus Ralfsii* Kirch.
*C. connatus* Breb.
*Cosmarium De Baryi* Arch. This is not like Wolle's figure, but exactly like that of Cooke in his ' British Desmids.'
*C. quadratum* Ralfs.
*C. globosum* Buln.
*C. bioculatum* Breb.
*C. tinctum* Ralfs.
*C. sejunctum* Wolle.

*C. pseudonitidulum* Nord.

*C. læve* Rab.

*C. sexangulare* Lund.

*C. polygonum* Naeg. This was noticed up to 27 μ in breadth.

*C. undulatum* Corda, var. *crenulatum* Wolle.

*C. pseudopyramidatum* Lund.

*C. pseudotaxichondrum* Nord.

*C. triplicatum* Wolle.

*C. punctulatum* Breb.

*C. conspersum* Ralfs.

*C. Portianum* Arch.

*C. orbiculatum* Ralfs.

*C. suborbiculare* Wood.

*C. Hammeri* Reinsch.

*C. cruciatum* Breb., forma *minor*, 14 μ broad.

*Xanthidium asteptum* Nord.

*X. fasciculatum* Ralfs. This is typical as the figure given by Cooke in his ' British Desmids' with six pairs of spines to each semicell, but the spines not at all approaching the figure of Wolle as regards length, which, moreover, has but four pairs of spines.

*Euastrum oblongum* Ralfs.

*E. ansatum* Ralfs.

*E. didelta* Ralfs.

*E. crassicolle* Lund.

*E. Nordstedtianum* Wolle, var. *minor* Wolle.

*E. binale* Ralfs, forma *minor*, 9 μ

broad, 11 μ long, with the apex quite rounded, and the notch not so gaping as in the type.

*Micrasterias papillifera* Breb.

*M. truncata* Ralfs.

*Staurastrum muticum* Breb., var. *minor* Wolle.

*S. orbiculare* Ralfs. This was also seen with zygospores.

*S. dejectum* Breb. — Var. *mucronatum* Ralfs.

*S. brevispina* Breb. — Var. *inerme* Wille.

*S. Dickiei* Ralfs.

*S. brachiatum* Ralfs.

*S. margaritaceum* Ehrenb.

*S. tricorne* Breb.

*S. alternans* Breb.

*S. arachne* Ralfs.

*S. fasciculoides* Wolle.

*S. pusillum* Wolle.

*S. gracile* Ralfs.

*S. ranum* Wolle.

*S. anatinum* Cooke et Wills.

*S. aculeatum* Ehrenb.

*S. teliferum* Ralfs.

*S. echinatum* Breb. This was seen up to 86 μ broad.

*S. hirsutum* Breb.

*S. Ravenellii* Wood.

*S. eustephanum* Ralfs.

*S. pseudofurcigerum* Reinsch.

*S. furcatum* Breb.

*S. enorme* Ralfs.

---

# ON THE TWO VALERIANS.

## By W. H. Beeby, A.L.S.

One of the points which have for some time attracted attention in connection with Surrey Botany, is the relationship existing between the two Valerians, *Valeriana Mikanii* and *V. sambucifolia.* For a good number of years I have been accustomed to distinguish the rare *Mikanii* of our chalk-hills from the common *sambucifolia* without much difficulty ; but the opinions held by various botanists in this country have been so diverse, that I was induced to enter upon a series of experiments and observations in the hope of getting at some definite conclusion. Before proceeding to give the result of these experiments, I may briefly repeat some of the views which have been held concerning these plants.

Watson (Cyb. Brit. ii. p. 26) speaks of the " two alleged species " as very doubtfully distinct, adding that he is not prepared to separate the specimens contained in his own herbarium, " most or all of which probably belong to V. sambucifolia." Subsequently, in various places, he seems to recognise the two plants as being distinguishable in some way, and (Compend. Suppl. p. 52) gives the aggregate V. officinalis as occurring in Sub-prov. " 1–37," and states that V. sambucifolia has " probably the same area." On the other hand, Mikanii is given for only Sub-prov. " 4, 7, 8, 12, and [28]." In Top. Bot. (ed. i. and ii.) it is mentioned that the " Mikanii of Eng. Bot., ed. iii., is very sparingly on record hitherto." This quite accords with my own observations as to the relative rarity of the two plants, so far as a few of the southern counties are concerned, but not with Sir J. D. Hooker's experience (Stud. Flo., ed. iii. p. 197) that V. sambucifolia is " very local." As Watson has remarked, V. sambucifolia is the V. officinalis of Smith, who has, however, a var. β. " V. sylvestris major montana, Bauh. Pin.," which answers to Mikanii. Syme (Eng. Bot., ed. iii. vol. iv. p. 236) can " scarcely separate V. sambucifolia even as a mere variety," and adds—" the ripe fruit is said to be different in the 'Flora of Essex'; it varies slightly in both varieties, but I can see no constant difference." Professor Babington (Man. Brit. Bot., ed. viii. p. 177) separates the plants as varieties, adding under the type ('V. officinalis L.'), " with suckers, not stoles "; and under sambucifolia, " stoles long." Reichenbach, whose 'Icones' are here referred to, treats V. officinalis L. and V. exaltata Mikan as the same species, the description being that usually restricted to the latter plant. Prof. Babington's use of the character " suckers " (" Surculi progeniei approximati, nec stolones"; Reich.) suggests the possibility that V. exaltata may be a native of this country, perhaps of the eastern counties. The two plants treated of in this paper both produce long stolons, and accord therein with Koch's description of his ' V. officinalis L.' and V. sambucifolia Mikan. Whether Koch is correct in his appreciation of the name V. officinalis, or whether Reichenbach in applying the same name to V. exaltata, I am unable to say; but there is this in favour of the latter view,—that Fries (Mant. iii. p. 1) speaks of " officinalis " as being everywhere conspicuous for its want of stolons, about Upsala, &c., and he admits sambucifolia as a distinct species on this very ground. This, however, has only to do with the name; the subject of this paper is the difference between the two stoloniferous plants. Turning to some local Floras, we find that Gibson (Flo. Essex) separates two species as " alike in general appearance, but differing in the shape of the fruit and number of leaflets." Mr. Archer Briggs (Flo. Plymouth) has failed to distinguish the two. Mr. Townsend (Flo. Hants) treats the plants as distinct species, and doubtless the Hants officinalis is identical with the Surrey Mikanii of this paper, as the only localities given are a few in " woods on a chalky soil." In one of the most recently published Floras, that of West Yorkshire, the subject is more fully discussed. Dr. Arnold Lees there considers *Mikanii* to be a dry soil state, and nothing more ; and adds—" where a *dry* stretch of soil slopes gradually down to a ditch or stream, numerous intermediates may be observed in every stage of transition, from luxuriant, succulent, elder-like leaved plants by the water, to others a foot high at most, with linear, deeply cut, coarsely-serrate leaflets, in the driest situation." Such "intermediates" show very little approach to the *Mikanii* now meant, which is not characterised by its coarsely-serrate leaflets, or by its dwarf habit. It is, in fact, when growing on a chalk-bank, as tall a plant as *sambucifolia* when growing in a wet ditch ; or, if its height be regarded in proportion to the thickness of its stem, it must be considered the taller plant. Dry-place states of *sambucifolia*, which would seem to answer to the description of Dr. Lees' intermediates, are not uncommon ; such a state may be seen (or might have been a couple of years ago) on the wall of Kew Gardens moat, by the Thames tow-path.

The above selection affords, I believe, a fair view of existing opinions, and their diversity seems a sufficient reason for giving a somewhat detailed account of recent experiments.

In the spring of 1887 I brought into the Reigate garden healthy roots of the two plants—*Mikanii* from a hedgerow on the chalk, by Farthing Downs, near Coulsdon, and *sambucifolia* from a ditch near Reigate. They were planted in the same bed, at a distance of about two feet from each other, in the ordinary sandy soil of the district, and the conditions to which they were exposed were in all respects identical. The plants did not recover sufficiently to be available for observations that year ; but in the spring of the present year they both showed that they were in a vigorous state and well-established. By May 6th they had made a good growth, and the first fact of importance noted was that while the root-leaves of *Mikanii* spread flat on the ground, those of *sambucifolia* were erect or suberect ; and this difference was also strongly marked in the lower stem-leaves. When the stems were about half-grown the uppermost pair of leaves in *Mikanii* had a tendency to spread at once almost horizontally, even before their individual leaflets had become unfolded ; while in *sambucifolia* they remained for some time not only erect, but with their tips converging over the terminal bud. By the 19th May *Mikanii* had no less than eighteen stems showing good trusses of buds, most of these arising from the newly-formed stolons ; in *sambucifolia* none were yet visible, except a couple of stems from the old roots. Later on, the stolons of *sambucifolia* also produced flower-stems, and that plant seemed now to grow more rapidly, so as almost to catch up *Mikanii*, which latter (June 11th) opened its first flower exactly one week earlier than the other. But I think that there is probably more difference in the time of flowering than is, in this instance, represented by the opening of the first blossom, for *Mikanii* was seeding pretty well when *sambucifolia* was still in good flower. Koch gives the flowering-time as follows:—" *V. officinalis*," May, June ; *V. sambucifolia*, June to August.

On June 11th the tallest stem of *Mikanii* measured 4ft. 2 in., that of *sambucifolia* 4 ft. 5 in.; ultimately two or three stems of *Mikanii* exceeded the tallest of *sambucifolia* by a few inches. The stem of *Mikanii* is much (probably one-third) more slender than that of the other. The colour of the foliage, stems, &c., was very different throughout the growth of the two plants. In *Mikanii* the leaves were of a dark, opaque, somewhat bluish green; in *sambucifolia* of a beautiful, bright, clear green, much like that of *Cardamine. amara.* The more divaricate habit of *Mikanii*, already mentioned when speaking of the lower leaves, was characteristic throughout its growth; it is seen in the more widely-spreading main branches of the inflorescence, and reaches a climax in the fruiting cymes. In the ultimate cymes the branchlets are more than divaricate—they may almost be called scorpioid; in *sambucifolia* the ultimate branchlets remain simply erect. There is an appreciable difference in the shape of the ripe fruit; that of *sambucifolia* is actually larger, considerably broader at the base in proportion to the apex, and the seed does not fill the cavity, so that, the empty margins of the fruit being appressed, it has a kind of false wing on each side in the lower part. The fruit of *Mikanii* is smaller, more oblong in form, less tapering upwards, so that there is less difference in its breadth at base and apex; this is partly owing to absence of the wing-like expansion of the pericarp, which is only sufficiently wide to enclose the seed which entirely fills the cavity. (No fruits should be compared which do not contain ripe seed; sometimes many are sterile.) Strictly speaking, *V. sambucifolia* is normally a *soboliferous* plant, and I have never seen it with aërial stolons; once in a dried-up swamp, where the ground was very hard and caked, the soboles crept along the surface, but half underground, as though they would have been quite underground, if possible. *V. Mikanii* seems more variable; it is often soboliferous, but some-times, on the chalk, produces stolons which are not only aërial, but somewhat arching, so that they only touch the soil again at the rooting apex. In cultivation both plants were at first soboliferous in habit, and *sambucifolia* remained so; but in *Mikanii*, after some growth had been made at the end of the soboles, the *intermediate* part rose above ground and became strongly arched, and often con-torted in fantastic shapes. This, and the arched stolons sometimes seen on the chalk, seem to point to some different inclination in the two plants, but I am unable to frame any character therefrom. The leaflets of *sambucifolia* are variable, and there is a sub-entire leaved form; in *Mikanii* the autumnal states and the young plants on barren stolons, sometimes produce leaves which are a good deal more cut than is normal; I think there is less variation in the number of the leaflets than in their cutting. In *sambucifolia* I find the well-developed leaves to have commonly 4–6 pairs of leaflets, and I cannot discover, in my large collection of forms of this plant from both wet and dry places, a single example in which there are more than six pairs; in *Mikanii* they are commonly 8–10, though sometimes fewer in ill-developed leaves; the root-leaves of both have fewer leaflets than those of the lower stem-leaves; the leaves on barren stolons fewest of all.

There is one other point of view from which the two plants may be regarded, *viz.*, their different officinal qualities. That a difference exists, is mentioned by several authors of the earlier part of the present century; but it will suffice to quote Smith (Eng. Flo. i. p. 43), who says—"var. *β*. more aromatic and preferred. for medical use." This statement has received an unexpected confirmation. In the spring I removed from both plants some stolons and early leaves for preservation, when the cats, being thus made aware of the presence of their favourite plant, at once attacked *Mikanii*, grubbing up the roots, and so on. Both plants were then protected with stakes, but without avail as regards *Mikanii*. Finally, both were surrounded with wire-netting, and the cats effectually excluded. All this time no attempt had been made on *sambucifolia*, and shortly afterwards I had the netting and all other protection removed from it. It being still untouched, several stolons were dug up, and, together with some leaves, bruised, and left lying on the ground round the plant. Nothing, however, would induce the cats, although deprived of *Mikanii*, to pay any attention to *sambucifolia*, which has remained, without exception, unheeded by them from the time it appeared above ground until now. Concerning this character, I have made no systematic observations on the wild plants; but on more than one occasion the odour of *Mikanii* has proclaimed itself when the plant has been dug up, and this has not occurred in the case of *sambucifolia*.

It will be inferred from the above account that I do not hesitate to follow those authors who class these plants as distinct species. They answer admirably to Koch's description of *V. officinalis* and *V. sambucifolia*; and doubtless these are the plants intended by those names in Pohl's ' Tent. Flo. Bohemiæ,' but the descriptions there are most meagre. In the last-named work, the Eng. Bot. plate 698 is referred to *officinalis*, and the right-hand leaf in that plate may possibly belong to that plant. I have examined a good number of continental examples of " *V. officinalis*" and *V. sambucifolia*, and I have no doubt of their identity with the two plants now described; and this is especially the case with the latter. I have this year brought into cultivation several more forms or states of these two species, and should any further light be thus thrown on the subject, I shall hope to report thereon on some future occasion. It only remains to add that the garden observations have been, in the main, abundantly confirmed from the wild plants; in numerous examples in the case of the common *sambucifolia*.

# BIOGRAPHICAL INDEX OF BRITISH AND IRISH BOTANISTS.

## By James Britten, F.L.S., and G. S. Boulger, F.L.S.

(Continued from p. 311).

**Eagle, Francis King** (1788?–1856): b. Lakenheath, Suffolk?, 1788?; d. Bury St. Edmunds, 8th June, 1856. LL.B., Camb., 1809. F.L.S., 1807. Bencher, Middle Temple. Muscologist. Contributed to Eng. Bot. Proc. Linn. Soc. 1857, xxvii.

**Eales, Dr.** (fl. 1690). Of Welwyn, Herts. Correspondent of Ray. R. Syn. ed. iii. 25, &c.; Pryor, Fl. Herts, xxxix.

**Edgeworth, Maria** (1767–1849): b. Black Bourton, Oxon, 1st Jan. 1767; d. Edgeworthstown, Ireland, 24th May, 1849. Novelist. 'Dialogues on Botany.' Pritz. 98; Jacks. 86; Dict. Nat. Biog. xvi. 380.

**Edgeworth, Michael Pakenham** (1812–1881): b. 24th May, 1812; d. Eigg, Inverness, 30th July, 1881. F.L.S., 1842. Half-brother of preceding. Bengal Civil Service, 1831. Contributed to Linn. Trans. and Journ. Linn. Soc. from 1843. Caryophyllaceæ in 'Flora of India,' 1874. 'Pollen,' 1877. Herbarium at Kew. Pritz. 98; Jacks. 542: R. S. C. ii. 444; vii. 594; Journ. Bot. 1881, 288; Proc. Linn. Soc. 1881–2, 63; Dict. Nat. Biog. xvi. 382. *Edgeworthia* Falc. = *Reptonia*. *Edgeworthia* Meisn.

**Edmond, James Williamson** (d. 1875): d. Inveresk, 22nd March, 1875. M.B., Edin., 1870. Assisted Carrington with 'British Hepaticæ.' Trans. Bot. Soc. Edin. xii. 409.

**Edmonston, Thomas** (1825–1846): b. Buness, Shetland, 20th Sept. 1825; d. Sua, Atacamas, Ecuador, 24th Jan. 1846, and bur. there. 'List of Shetland Plants,' Ann. & Mag. vii. (1841), 287. Contributed to 'Phyt.' i. 1844. 'Flora of Shetland,' 1845. 'Flora Shetlandica, 1837' (MS.) in Bot. Dept., Mus. Brit. Discovered *Arenaria norvegica*. Naturalist to H.M.S. 'Herald,' 1845–6. Pritz. 98; Jacks. 259; R. S. C. ii. 496; Seeman, 'Voyage of Herald,' i. 67. Dict. Nat. Biog. xvi. 398. *Edmonstonia* Seem. = *Tetrathylacium*.

**Edwards, Edward** (1812–1886). Assistant in library, Mus. Brit. till 1846. Contributed to Phyt. o. s. & n. s. 'Lives of Founders of Brit. Mus.' Pryor, Fl. Herts, li.; R. S. C. vii. 448.

**Edwards, John** (fl. 1775). 'British Herbal,' 1775 (plates and text). Pritz. 98; Jacks. 232.

**Edwards, Teak Sydenham** (1769?–1819): b. Abergavenny, 1769?; d. Chelsea, 8th Feb. 1819; bur. Chelsea Church. F.L.S., 1804. Botanical artist. Illustrated 'New Botanic Garden,' 1812, 'Bot. Mag.,' 1799–1814, and 'Bot. Register,' 1815–1819. Indexes Bot. Mag. p. x. (1828). Faulkner, Chelsea, ii. 10. Pritz. 98; Jacks. 542. *Edwardsia* Salisb.

**Edwards, William Frederick** (1776–1842): b. Jamaica, 6th April, 1776; d. Paris, 24th Aug. 1842; bur. Paris. Pritz. 28.

**Ehret, George Dionysius** (1708?-1770): b. Erfurt, Saxony, 1708?; d. Chelsea, 9th Sept. 1770. Employed in Oxford Garden, 1750. F.R.S., 1757. Botanical artist. Correspondent of Linnæus. 'Plantæ Selectæ,' 1750-1773. Mezzo. by J. Haid, after A. Heckell; copy at Kew. Drawings and MS. autobiog. in Bot. Dept., Mus. Brit. Pult. ii. 284; Proc. Linn. Soc. 1883-6, 42; Pritz. 97; Jacks. 110. *Ehretia* Trew., Linn.

**Ellis, Daniel** (1772?-1841): b. Gloucestershire, 1772?; d. Edinburgh, 17th Jan. 1841. Articles 'Vegetable Anatomy' and 'Veg. Physiology' in Encyc. Brit. Suppl., ed. 6. Gard. Chron. 1841, 87; Loud. Gard. Mag. 1841, 188.

**Ellis, John** (c. 1710-1776): b. in Ireland, c. 1710; d. London, 15th Oct. 1776. London Merchant. F.R.S., 1754. Copley Medal, 1768. Memb. Roy. Soc. Upsal. 'Natural History of Corallines,' 1755. Agent for West Florida, 1764; for Dominica, 1770. Imported many American seeds. Correspondent of Linnæus. "The main support of Natural History in England," Linnæus. Rees; Pritz. 100; Jacks. 543; Linn. Letters, i. 79; Nich. Anecd. ix. 331; Loud. Arboret. 70. *Ellisia* P. Browne = *Duranta* L. *Ellisia* L.

**Ellis, Robert** (fl. 1700). Collected in S. Carolina, 1700, and sent plants to Petiver. Mus. Pet. 79. Plants in Hb. Sloane, 159.

**Ellis, William** (1795-1872): b. London, 1795; d. Hoddesdon, Middlesex, 9th June, 1872. Clerk. Missionary. Introduced *Ouvirandra, Grammatophyllum Ellisii*, &c. Gard. Chron. 1872, 806.

**Embleton, Robert Castles** (1806-1877): b. Berwick-on-Tweed, 14th Dec. 1806; d. Beadnell, Northumberland, 6th Jan. 1877. Surgeon. Orig. Member Berwickshire F. Club. Fellow-student of H. C. Watson. Formed a herbarium and entomological collection. Top. Bot. ii. 543; Proc. Berwickshire Field Club, viii. 373.

**Elsey, Joseph Ravenscroft** (1833?-1857): b. 1833?; d. Springfield, St. Kitt's, 31st Oct. 1857. Surgeon and naturalist to North Australian Expedition. Went to West Indies to collect for Grisebach. Gard. Chron. 1858, 112.

**Empson, James** (d. 1765). Keeper of Nat. Hist. Dept., British Museum. Edited Petiver's works, ed. 1767. Jacks. 32.

**English, James Lake** (1820-1888): b. Epping, 21st Aug. 1820; d. Epping, 12th Jan. 1888; bur. Epping. Umbrella-mender, taxidermist, entomologist. Assisted the Doubledays. 'Manual for Preservation of Fungi,' 1882. Issued fascicles of Epping Forest Mosses, 1883-5. R. S. C. vii. 616.

**Evans, Joseph** (1803?-1874): b. Tyldesley?, 1803?; d. Boothstown, Manchester, 23rd June, 1874; bur. St. Mark's, Worsley. "Medical botanist." Friend of Caley, Dewhurst, &c. Cultivated 300 spp. of medicinal plants. Gard. Chron. 1874, ii. 614.

**Evans, William Wilson** (1820-1885): b. Dysart, Fifeshire, 1st Jan. 1820; d. Edinburgh, 5th May, 1885. Assoc. Bot. Soc. Ed. 1841. Curator, 1843. Trans. Bot. Soc. Ed. xvi. 1886, 311.

**Evelyn, John** (1620-1706): b. Wotton, Surrey, 31st Oct. 1620;

d. 27th Feb. 1706; bur. Wotton. F.R.S., 1662. 'Sylva,' 1664, with portr. by Nanteuil. Memoirs by Bray, with engr. portr. by Nanteuil and by Kneller, 1818 ; Pritz. 103 ; Jacks. xxxiv. 206 ; Cott. Gard. v. 57 ; Journ. Hort. 1875, xxix. 249, with portr. ; Felton, 98. Portr. at Kew ; and, after Kneller, at Roy. Soc. *Evelyna* P. & Endl.

**Ewer, Samuel** (fl. 1808). F.L.S., 1789. 'Manuale sive Compendium botanices,' 1808. Pritz. 103 ; Jacks. 35.

**Eyre, J.** (fl. 1851). General, R.A. Collected and made drawings in Hong Kong. Sent plants to Kew. Fl. Hongkong, pref. 11 ; Journ. Bot. 1851, 331. *Eyrea* Champ. = *Turpinia* Vent.

**Fairbairn, John** (d. 1814): d. Chelsea, Dec. 1814. Curator of Chelsea Garden, 1784–1814. F.L.S., 1788. Eng. Bot. 1719, 2081 ; Semple, 119.

**Fairchild, Thomas** (1667 ?–1729): b. 1667 ? ; d. Hoxton, 10th Oct. 1729 ; bur. Poor's ground, Hackney Road. Gardener of the City Gardens, Hoxton, 1690–1722. 'The City Gardener,' 1722. Experiments on sap, Phil. Trans. xxxiii. Introduced *Pavia rubra, Cornus florida*, &c. Correspondent of Linnæus. Pult. ii. 238 ; Cott. Gard. vi. 143 ; Nich. Illustr. i. 371 ; Sir Henry Ellis, Hist. of Shoreditch, 288 ; Felton, 60.

**Falconer, Hugh** (1808–1865): b. Forres, Morayshire, 29th Feb. 1808 ; d. London, 31st Jan. 1865. M.A., Aberd. M.D., Edin., 1829. F.R.S. In India, 1830–1855. Superintendent, Saharunpur Gard., 1832; Calcutta, 1848. In Cashmere, with Burnes, 1838. Introduced tea and cinchona. 'Aucklandia,' Linn. Trans. 1840. 'Cryptolepis,' *ib.* 1841. Pritz. 104 ; Jacks. 413 ; R. S. C. ii. 551 ; vii. 636 ; Journ. Bot. 1865, 101 ; Mag. Zool. Bot. 1839, iii. 195 ; Journ. Hort. viii. (1865), 234 ; Proc. Linn. Soc. 1864–5, xc.; Proc. Geol. Soc. 1865, xlv. *Falconeria* Royle.

**Falconer, John** (d. 1547): d. Ferrara, 1547. Communicated pl. to Amatus Lusitanus. Pult. i. 71 ; Turner, 'Herbal'; Haller, i. 266 ; Druce, Fl. Oxford, 372 ; Pritz. 104.

**Falconer, Randle Wilbraham** (1816 ?–1881): b. Somerset, 1816 ? ; d. Bath, 6th May, 1881. M.D. 'Catalogue of Tenby Plants,' 1848. President, Roy. Med. Soc. 1837. Twice Mayor of Bath. Trans. Bot. Soc. Ed. xiv. 303 ; Pritz. 104 ; Jacks. 260 ; Journ. Bot. 1881, 192.

**Falconer, William** (1741 ? or 1743–1824): b. Chester, 1741 ? or 1743 ; d. Circus, Bath, 30th Aug. 1824. Father of preceding. Physician to Bath Hospital, 1789. 'History of Sugar,' Mem. Manchester Phil. Soc. 1796. 'Miscellaneous Tracts,' 1793, including list of pl. known to Greeks. Pritz. 104 ; Jacks. 21, 213 ; R. S. C. ii. 552 ; Cott. Gard. viii. 299 ; Felton, 183 ; Dict. of Living Authors. New Monthly Mag. xii. Portr. by Daniel, of Bath, 1791 ; engr. by Fitler in 'Influence of the Passions.'

**Farquhar, Jane** (*née* **Colden**). Daughter of Cadwallader Colden. One of the first ladies to study the Linnean System. MS. 'Flora Nov-Eboracensis,' in Bot. Dep. Mus. Brit., with drawings. Pritz. 65 ; Linn. Letters, i. ; 'Memorials of Bartram,' p. 400.

**Fennell, James H.** (fl. 1840). 'Drawing-Room Botany' (1840). Jacks. 39.

**Ferguson, William** (1820–1887): b. July, 1820; d. Colombo, Ceylon, 31st July, 1887. Surveyor. In Ceylon from 1839. 'Timber-trees of Ceylon,' 1863. 'Notes on Ceylon Ferns,' 1880. 'Ceylon Grasses,' Journ. Ceylon Branch, R. A. S., 1880. Pritz. 106; Jacks. 396; Journ. Bot. 1887, 320; Ann. Bot. 1888, 403, with bibliog. *Fergusonia* Hook. fil.

**Field, Barron** (1786–1846): b. London, 23rd Oct. 1786; d. Torquay, 11th April, 1846. F.L.S., 1825. Judge Supreme Court N. S. Wales, 1816–1824. Proc. Linn. Soc. i. 298; Jacks. 400; *Fieldia* Gaud. = *Vanda*. *Fieldia* A. Cunn.

**Fielding, Henry Barron** (d. 1851): d. 21st Nov. 1851. Of Bolton Lodge, Lancashire. F.L.S., 1838. 'Sertum plantarum' [with G. Gardner], 1844. Herbarium of 70,000 specimens bequeathed to Univ. of Oxford. Pritz. 107; Jacks. 117; Proc. Linn. Soc. ii. 188; Phytol. iv. 655; Cott. Gard. vii. 188; Lasègue, 330.

**Fifield** (fl. 1702). Surgeon. Sent plants from Campeachy to Petiver. Mus. Petiv. 94.

**Finlayson, George** (fl. 1821). Of Thurso. Surgeon and naturalist to Siam Expedition, 1821. Friend of Wallich. Wall. Pl. Asiat. ii. 49; Lasègue, 132, 141. Plants at Kew. *Finlaysonia* Wall.

**Firminger, Thomas A. E.** (1812?–1884): d. Edmonton, Middlesex, 18th Jan. 1884. Clerk. M.A. 'Manual of Gardening for Bengal,' 1863; ed. 2, 1869; ed. 3, 1874; Gard. Chron. xxi. (1884) 124.

**Fishwick, John** (fl. 1696). Sent plants from Andalusia and the Mediterranean to Plukenet. Almagest. 18, 54, 85, 221.

**Fitt, George** (fl. 1844–1847). Of Great Yarmouth. Contributed to Phyt. i. and ii.; Journ. Bot. 1847, 287; R. S. C. ii. 628.

**Fitton, Elizabeth** (fl. 1817): b. Dublin. Sister of Dr. W. H. Fitton. 'Conversations on Botany,' 1817. Pritz. ed. 1, 86; Jacks. 43.

**Fitton, Sarah Mary** (fl. 1817–1866): b. Dublin. 'Conversations on Botany' (1817), (with foregoing—her sister). 'The Four Seasons' (1865). Pritz. ed. 1, 86; Jacks. 43, 45; Proc. Geol. Soc. 1862, xxxiv.

(To be continued.)

---

# SHORT NOTES.

ARUM ITALICUM (Mill.). — On May 15th last, I traced this *Arum* along the base of the chalk-hills known as the Sugar Loaf and Cæsar's Camp, near Folkestone, for a considerable distance; but it grew much more sparingly than the common *A. maculatum* (L.), which are very abundant. Occasionally some specimens were found that now appear *hybrids*, partaking of certain of the individual characters of both species. Is it not within the bounds of possibility that the two may be extremes of *one* variable plant? I have not been able to collect satisfactory specimens this season, but, as I frequently visit Folkestone, I shall hope to do so shortly. Some few years ago I chronicled in the 'Journal of Botany' the discovery of *A. italicum* in this neighbourhood, these first specimens being found mostly nearer Cheriton. The neighbourhood of Paddlesworth, on the summit of the chalk-downs overlooking Folkestone, requires minuter research for this plant than it has yet received, the ordinary Arum being very frequent in the neighbourhood.—J. Cosmo Melvill.

East Kent Plants. — About Whitstable I gathered, this September, *Polypogon monspeliensis, Agrostis nigra, Rubus rusticanus,* and, between Whitstable and Canterbury, *Vicia gracilis, Epilobium obscurum, E. lanceolatum,* and a hybrid (*lanceolatum × obscurum* teste Haussknecht), which were apparently additions to East Kent Botany. *Bupleurum tenuissimum, Peucedanum officinale, Glyceria distans* var. *glauca, Hordeum marinum, Lactuca saligna, Vinca minor, Chenopodium olidum, Trifolium scabrum,* and *Populus canescens* were also noted in the same localities.—G. Claridge Druce.

---

## NOTICES OF BOOKS.

*The British Moss-Flora.* By R. Braithwaite, M.D., F.L.S. Part XI. Fam. X. Grimmiaceæ. I. (The Author, 303, Clapham Road.) 8*s.*

With this Part begins the second volume of the completest and most beautifully illustrated work on the British Mosses that has yet been published, and it fully sustains the high reputation of the author. An eminent botanist has wisely said, "Good plates are amongst the best means of promoting the progress of Botany. When they represent the form of the plant according to Nature, and especially when they develop the characters of the genus and species, even to their minutest parts, they fulfil all that can be desired." The work before us not only fulfils all these conditions, but gives also full and careful descriptions of each plant delineated. Thus it affords to the student of Bryology help such as is rarely given in even the best of monographs.

As some guide to a proper appreciation of the contents of this part, it may be mentioned that it contains fifty-six pages of analytical or descriptive letterpress, and eight plates. On these are delineations of forty-five species and six varieties of the genus *Grimmia* and its allies, giving 484 figures, which represent faithfully and in exquisite style not only the natural habit of the plant, but also the minuter details. It is a complete monograph of the British species and varieties of that difficult genus, *Grimmia,* and its allies.

The part commences with a scientific dissertation on the *Grimmiaceæ* as a whole, with observations on the comparative distinctions of the two great genera, *Grimmia* and *Orthotrichum.*

This is followed by a full description of the genus *Grimmia*, and a carefully drawn up clavis or analytical key to the species of greatest
help to the student, whether tyro or advanced.  The genus *Grimmia*
is divided into four natural sections : 1, *Schistidium;* 2, *Eu-Grimmia;*
3, *Dryptodon;* and 4, *Trichostomum;* the latter section containing
those mosses formerly placed by the older botanists in the separate
genus *Racomitrium;* in combining these mosses with *Grimmia*, the
author has wisely followed those eminent botanists, Lindberg,
C. Mueller, and Mitten.   In addition to the genus *Grimmia*, we
have descriptions and illustrations of the genera *Coscinodon, Glypho-
mitrium,* and *Anæctangium.*

Several species are figured and described, some for the first
time, that have been added to our flora since the publication of
Wilson's 'Bryologia,' such as *Grimmia anodon, G. crinita, G.
incurva, G. Hartmanii, G. subsquarrosa, G. elatior, G. montana,
G. ovata,* and *G. elongata.*   The two varieties of *Racomitrium
heterostichum—gracilescens* and *alopecurum*—are raised to specific
rank under the new names *G. obtusa* and *G. affinis;* and *R. micro-
carpon* is admitted doubtfully as a native, as *G. ramulosa.  Coscinodon
cribrosus* is also fully described, and well illustrated.

A notice of this work would scarcely be complete without some
reference to the very complete synonymy which is given with each
plant, in which each plant is traced through the various published
works on Bryology, from the older authorities to the most recent.
In conclusion, we may say with confidence that British Bryology is
greatly benefited by Dr. Braithwaite's valuable addition to our
botanical literature.                                        J. E. B.

*British Mosses: their Homes, Aspects, Structure, and Uses.*  By F. E.
    Tripp.  2 vols., super-royal 8vo, pp. 42–301 ; 37 coloured
    plates.  New edition.  Bell & Sons.  Price 42s.

This work has been written by a lady whose mind is deeply
imbued with religious instincts, poetical ideas, and genuine
enthusiasm for mosses and moss-studies.  Her aim has been to
present to her readers " descriptions of British mosses which would
aid the amateur wishing to ascertain the names of the species he
meets with.   The language has been simplified as much as possible,
and reference to minute detail and intricacy of structure has been
avoided where practicable."

The author appears to make a mistake too frequently made by
the writers of popular books, that of imagining that vague and
inexact descriptions are more helpful to the beginner than are the
more exact descriptions of our text-books.  A description to be of
service to the amateur should be, at any rate, intelligible to the
advanced student ; unfortunately this is not the case in the work
before us.

The work is divided into two parts, " The Introduction " and
the " Explanation of the Plates."   In the Introduction, Section I.
treats of " The Homes of Mosses," and opens with a detailed
account of the third day of creation.  Afterwards the author leads

us through lovely bits of Cornish, Devon, and Hampshire scenery, giving by the way pretty and fanciful descriptions of the various wild flowers, ferns, and the like that haunt those scenes, but very little of the mosses. Still, this chapter is a very pleasant one. Section II. treats of the "Characteristics of Mosses." In this the mosses are arranged under three heads : "star-mosses," of which *Tortula muralis* is given as typical; "feather-mosses," *Hylocomium splendens* serving as an example ; and "candelabra,"—of this *Sphagnum* is the type. A short quotation will show the author's mode of simplifying :—

"'In His Hands are all the corners of the earth'; a corner humble enough is a bank in a moorland district. This is sometimes covered with a star-moss, *Polytrichum piliferum*, whose leaves, dark, rigid, thickly beset the stem ; from the middle rises the fruit-stalk, orange-coloured, deepening near the top to brown, and wearing a conical cap covered with silky yellow hair, the golden hair and orange stalks and green leaves shining beneath the blue sky. This moss is an aloe in miniature, and on many of the stems, in the centre of the circlet of leaves, is a crimson cup, as perfectly like a cactus-blossom. A sage-green moss, *Hedwigia ciliata*, is common on rocks, has at the end of every branchlet its fruit, like a coral bead. The fruit of another is at once described by its specific name 'Apple Moss,' *Bartramia pomiformis*. One family of star-mosses is appropriately designated the 'Swan-necked,' *Mnium*, so lovely is the curve of its fruit-stalk; another might be called the 'Crane's-billed,' *Atrichum*. But strangest of all, to a lowly moss, *Mnium undulatum* has been given the form of that which is at once the stateliest and the loveliest of all green things, the date-palm tree.''

Section III. treats of "The Structure of Mosses," which may be read with advantage by young students ; Section IV., "Mode of collecting and examining Mosses," in which the author wisely advises her readers to avoid as much as possible the use of lenses ; Section V., "The Use of Mosses," gives a full account of this portion of the subject, and ends with a short dissertation on the uses of mosses in decoration, or as types for decoration.

Following this is a "Synopsis of the Genera of British Mosses, based on the System of Dr. W. P. Schimper." In this there are a few inaccuracies of translation, but it will give the student a very good idea of the classification adopted, *viz.*, that of Schimper's 'Synopsis Muscorum Europæorum,' second edition, 1876.

This is followed by the "Explanation of the Plates," which occupies the rest of the work ; the descriptions are often too vague to help the student very much, but possibly by means of these and the plates themselves he may get some help. Many of the latter are excellent, and give very nicely the true characteristics of the mosses depicted. The whole work is tastefully got up, the paper good, and the printing of a good readable type ; and the author at any rate deserves our thanks for the great labour she has bestowed on a work intended for the help of the amateur in the study of Bryology.                                    J. E. B.

ARTICLES IN JOURNALS.

*Botanisches Centralblatt.* (Nos. 40–43). — J. Bornmüller, 'Zur Kenntniss der Flora des bulgarischen Küstenlandes.' —(No. 40). R. Keller, 'Doppelspreitige Blätter von *Valeriana sambucifolia* Mik.' — (No. 42). V. F. Brotherus, 'Musci novi exotici' (*Arthrocormus africanus, Splachnobryum Baileyi, Breutelia Wainioi, Papillaria Baileyi, Isopterygium robustum,* spp. nn.).

*Bot. Zeitung* (Sept. 28, Oct. 5, 12). — A. de Bary, 'Species der Saprolegnieen.'—(Oct. 9). T. W. Engelmann, 'Die Purpurbacterien und ihre Beziehungen zum Licht.'

*Bull. Torrey Bot. Club* (Oct.).—'Plants collected by H. H. Rusby in S. America' (*Acrostichum Eatonianum* E. G. Britton, sp. n.).— D. H. Campbell, 'Systematic Position of *Rhizocarpeæ.*' — W. M. Beauchamp, 'Onondaga Indian names of Plants.' — T. Meehan, 'Irregular Tendencies in Tubifloral Compositæ.'

*Gardeners' Chronicle* (Oct. 6). — *Pterocarya fraxinifolia* (fig. 52). —(Oct. 13). *Pseudophœnix Sargenti* (fig. 56).

*Journal de Botanique* (Oct. 1).—J. Vallot, '*Juniperus phœnicea* à forme spiculaire.' — A. Franchet, 'Les *Saussurea* du Yun-nan' (*S. ciliaris, S. edulis, S. spatulifolia, S. romuleifolia, S. yunnanensis,* spp. nn.). — E. Boudier & N. Patouillard, *Clavaria echinospora & C. cardinalis,* spp. nn.). — A. Masclef, 'Flore des collines d'Artois.' — (Oct. 16). E. G. Camus, '× *Orchis Timbaliana* (*O. Morio* × *O. maculata*) (1 plate).— P. A. Dangeard, 'La sexualité chez quelques Algues supérieures.' — A. Franchet, 'Les *Saussurea* du Yun-nan' (*S. villosa, S. longifolia, S. grosseserrata, S. Delavayi, S. likiangensis, S. radiata, S. lampsanifolia, S. peduncularis, S. vestita, S. chetchozonsis,* spp. nn.).

*Magyar Növénytani Lapok* (131, 132).—J. Csató, 'Kirándulás a Bulla völgyén Keresztül a Lêgoj Kúpjához.'

*Notarisia* (Oct.). — G. B. De Toni, 'Sopra un nuovo genere di Trentepohliacee' (*Hansgirgia*). — A. Hansgirg, 'Synopsis generum subgenerumque Myxophycearum (Cyanophycearum) hucusque cognitorum cum descriptione generis novi, *Dactylococcopsis.*'—G. Lagerheim, 'Sopra alcune alghe d'acqua dolce nuove o rimarchevoli.'

*Oesterr. Bot. Zeitschrift.* (Oct.). — K. Vandas, 'Flora von Süd-Hercegovina.' — F. Krasan, 'Weitere Bemerkungen über Parallelformen.' — B. Blocki, '*Rumex Skofitzii* (*R. conferto* × *crispus*).'— L. Simonkai, 'Zur Flora von Ungarn.'—E. Formánek, 'Zur Flora von Bosnien.'

*Scottish Naturalist* (Oct.). — W. Wilson, Botany of Alford.— J. W. H. Traill, 'Fungi of East of Scotland, 1888.'—J. F. Grant & A. Bennett, 'Flora of Caithness.'

---

Mr. ALFRED BARTON RENDLE, B.A., B. Sc., has been appointed Assistant in the Department of Botany, Natural History Museum.

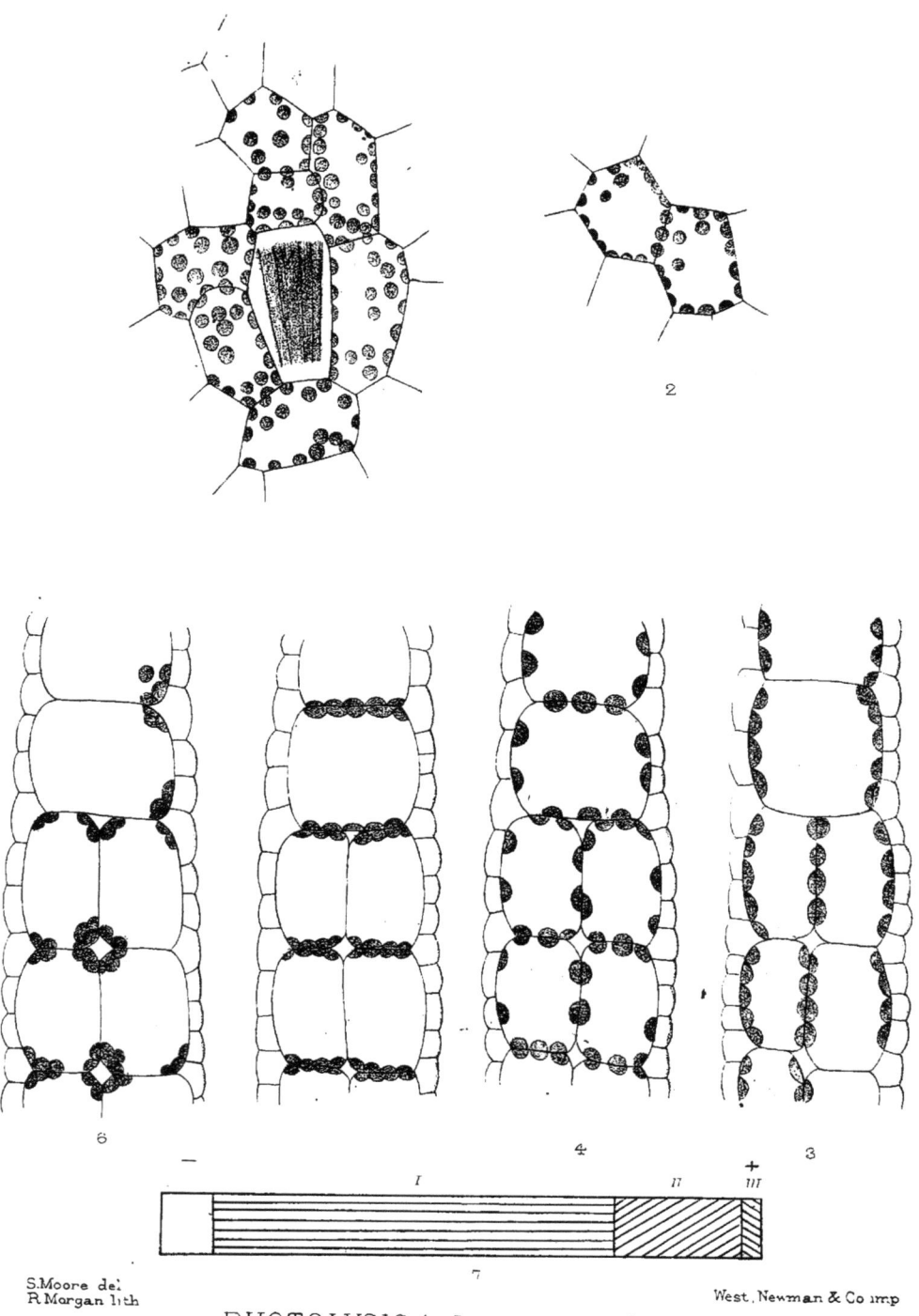

S.Moore del.
R.Morgan lith

West, Newman & Co imp

PHOTOLYSIS in Lemna trisulca

# PHOTOLYSIS IN *LEMNA TRISULCA*.

## By Spencer Le M. Moore, F.L.S.

### (Plate 285.)

In illustration of the variations in position undergone by the chlorophyll of *Lemna trisulca* in consequence of the alternation of day and night (photolysis), certain figures have been published by Stahl,* which I am unable to acquiesce in as representing the facts according to my impression of them. Although perhaps a point of minor interest, it may be well to call attention to this, inasmuch as the error—as it seems to be, crops up in the recently-published text-books on Vegetable Physiology. Thus Sachs † reproduces Stahl's figures, with implicit acceptance of the Jena Professor's conclusions, and Pfeffer ‡ and Vines § do the same.

Borodin ‖ was the first to study photolysis in our type. He found that in diffused light the grains range themselves upon the two walls of each cell lying parallel to the frond's surface, whereas by sunlight they are first driven on to the side walls, afterwards collecting in little heaps in the angles of the cells. It will be remembered that the first of these positions is known as " Epistrophe " and the second as " Apostrophe " (Frank), while for the massing Schimper has coined the word " Systrophe." Stahl ¶ took up the subject in more detail than did his predecessor, since he brought the effects of darkness within the scope of his research. He maintains that a difference in the distribution of their chlorophyll is shown by the cells in the thin marginal portion of the frond, as compared with those of the thicker parts. It may be remarked that this marginal portion is mainly made up of a single layer of large chlorophyll-containing cells, but such cells lie first in two, further in in several layers, in the thicker part, which is traversed to a certain distance by the delicate vascular bundle, and also contains wide intercellular spaces ; over the whole frond extends the wavy-bordered epidermis from which chlorophyll is absent. Now in rather poor diffused light the grains in both portions of the frond take up their position upon superficial walls, to shift therefrom in stronger light on to the perpendicular walls. It is, however, not in accordance with my experience that apostrophe should be described as the result of the action of *intense* light. In intense light, *i. e.*, direct sunlight, the grains are at first apostrophised, it is true, but they are soon afterwards systrophised. Moreover, as was shown in my first memoir ** on this subject, apostrophe is induced by grades of illumination considerably lower than sunlight. Indeed, on a bright day, it is a somewhat difficult matter to select, in a well-lighted laboratory, a place where illumination is of poor enough quality to fully epistrophise the chlorophyll. Finally, apostrophe can be brought about in low light. It may be incidentally remarked that we have

---

* Bot. Zeitung, 1880.  + Vorlesungen, No. xxxv.  ‡ Pflanzenphysiologie, ii. p. 397.  § Physiology of Plants, p. 300.  ‖ Bull. Acad. d. Sc. d. St. Petersbourg, 1869.  ¶ *Op. cit.* pp. 332—335.  ** Journ. Linn. Soc. Bot. vol. 24.

no evidence as to positive apostrophe (*i. e.*, apostrophe caused by strong light) being a natural event; and when it is remembered that *L. trisulca* usually, if not always, grows immersed to some depth in the water of shaded ponds, we may well doubt whether the indicated movement is of ordinary occurrence. Perhaps some microscopist living near the habitat of this duckweed will enlighten us upon this point.

It is, however, when nocturnal effects come under consideration that current opinion most readily admits of challenge. Stahl's figure shows the chlorophyll of the thin part of the frond ranged upon the side walls during the night, while in the cells of the thick part the inner wall also is studded with chlorophyll, the superficial wall being bare. According to Schimper,* however, while all the grains upon the wall abutting upon the epidermis are apostrophised during the night, a few of those ranged during the day upon the inner wall still remain in epistrophe.

If Stahl be correct, an ugly hole is made in a hypothesis for which I am responsible. In my first memoir,‡ it was shown that sun-loving aerophytes differ photolytically from aquatics, inasmuch as while the chlorophyll of the former is apostrophised during the night, in the latter epistrophe obtains, negative apostrophe not setting in until after some days' or weeks' continuous darkness, and an endeavour was made to explain this discrepancy on the view that, aquatics being adapted to a lower intensity of light than aerophytes, greater effect—in the form of mechanical strain—is produced in their protoplasm, and so, longer time is required by them in order that the motile effects of light may soak out. It was explained that *L. trisulca* is a partial exception to this rule, since most of the chlorophyll grains are apostrophised during the night, although a few may still remain in epistrophe, even after exposure to darkness for several days. On re-examining the matter, however, I find that the exception is not so striking as it appeared to be two years ago. I have thus the misfortune of being again at variance with the respected German authority above-mentioned.

Wishing to get to the bottom of this question, I took the trouble of making a number of calculations to determine, firstly: the number of chlorophyll grains in the cells both of the thick and the thin portions of the fronds; secondly, the proportion of apostrophised to still epistrophised grains in the cells of plants set overnight in darkness; and, thirdly, the further apostrophising effect of a few days' seclusion from light. The number of grains was determined upon specimens growing in light of intensity corresponding to that at a point well within the epistrophic interval; and since the result of a preliminary estimate taught that approximately the same number of grains were ranged upon both walls parallel to the surface, only one wall was brought into focus, and by doubling th number found thereon, the total was arrived at. The following table, dealing with the marginal cells, requires no

---

* Jahrb. f. Wiss. Bot. 1885, p. 231.    ‡ *Op. cit.*

further explanation ; in estimating the average number of grains per cell fractions, both here and elsewhere, have been omitted.

| No. of plant. | No. of cells examined. | Total No. of grains. | Maximum in a cell. | Minimum in a cell. | Average per cell. |
|---|---|---|---|---|---|
| I. | 12 | 516 | 56 | 34 | 43 |
| II. | 12 | 444 | 52 | 26 | 37 |
| III. | 12 | 386 | 46 | 20 | 32 |
| IV. | 12 | 366 | 52 | 18 | 30 |
| V. | 12 | 426 | 44 | 20 | 35 |
| VI. | 12 | 354 | 36 | 22 | 29 |

The last three plants upon this table were examined for me by my brother. His total for the 36 cells is 1146, giving an average of 31 grains per cell; mine is just 200 more, which means an average of 37; the grand average obtained from both is thus 34, and this may be regarded as approximately correct.

Cells of the thick part similarly examined by myself alone give the following numbers :—

| No. of plant. | No. of cells examined. | Total No. of grains. | Maximum in a cell. | Minimum in a cell. | Average per cell. |
|---|---|---|---|---|---|
| I. | 12 | 423 | 40 | 30 | 35 |
| II. | 12 | 442 | 62 | 34 | 45 |
| III. | 12 | 596 | 62 | 38 | 49 |

That is, 36 cells yield a total of 1561 grains, or an average of 43 per cell.

The two following tables refer to plants set overnight in darkness; as inspection will immediately show, I am unable to bear out Schimper's before-mentioned statement that the superficial wall abutting on the epidermis is deprived of chlorophyll as the result of one night's darkness. In these cases the grains upon the lower wall were counted, after focussing through the upper wall; and no difficulty was met with in the course of this. The upper three columns deal with upper, the lower three with under walls. The first table refers to marginal cells.

| No. of plant. | No. of cells examined. | Total No. of grains still in epistrophe. | Maximum in epistrophe. | Minimum in epistrophe. |
|---|---|---|---|---|
| I. | 12 | 50 | 7 | 0 |
| II. | 12 | 69 | 9 | 2 |
| III. | 12 | 103 | 12 | 6 |
| I. | 12 | 46 | 6 | 0 |
| II. | 12 | 87 | 11 | 6 |
| III. | 12 | 91 | 9 | 6 |

This gives a total of 222 for the 36 upper, and 224 for the lower surface-walls, or an average of 6 per wall, which is equivalent to 12 grains per cell. *Consequently the effect of night is to transfer to the side-walls only 22 of the 34 grains of a cell, leaving 12 of them still in epistrophe,* and I feel convinced that this is a correct statement so far as refers to healthy cells, a representation of one night's darkness upon which will be found at fig. 1 of Tab. 185. If the cells be old, or lie in the neighbourhood of parts which have been injured, apostrophe will, of course, be much more pronounced, as will appear on reference to Fig. 2, which is an attempt to show the result of one night's darkness upon two such cells; but even here one or more of the grains are almost always still in epistrophe in the morning. It is significant that, of the 36 cells dealt with in the last table, only one verified the dictum of Stahl by having all its grains in apostrophe.

In the next table, cells from the thick part are similarly dealt with :—

| No. of plant. | No. of cells examined. | Total No. of grains in epistrophe. | Maximum in epistrophe. | Minimum in epistrophe. |
|---|---|---|---|---|
| I. | 12 | 138 | 18 | 9 |
| II. | 12 | 121 | 15 | 6 |
| III. | 12 | 119 | 16 | 7 |
| I. | 12 | 142 | 18 | 9 |
| II. | 12 | 122 | 17 | 7 |
| III. | 12 | 105 | 13 | 6 |

From this we gather that 36 upper (outer) walls had between them 378, and a similar number of lower (inner) walls 369 grains still in epistrophe, in spite of their exposure to the night's darkness. This gives an average of 10 grains per wall or 20 per cell; and since the average number of grains in these cells was found to be 43, *it appears that during the night rather more than 50 per cent. move on to the side walls, the remainder being fairly equally distributed upon both upper and lower walls.* For this condition the term " merostrophe." may perhaps be approved.

It is maintained therefore that photolytic effects are approximately the same in all the cells of *L. trisulca.* In poorish diffused light the grains collect upon the two walls lying parallel to the plane of the frond, to pass in stronger illumination on to the lateral walls, and mass under the influence of still intenser light in the cells' corners. Slow movement towards the side walls is caused by darkness; but in accordance with the rule as to aquatics, negative apostrophe is not complete until after a considerable lapse of time. These four statements are diagrammatically illustrated in figs. 3--6.

To show the slow rate at which negative apostrophe proceeds, we counted the number of grains still in epistrophe after four days in darkness. My brother found upon the upper wall of 12 marginal cells of each of three plants 48, 24 and 23 grains, and on the corresponding lower walls 56, 45 and 48—an average of 6 per cell; my own figures were 61, 42 and 51 upon the upper, and 56, 41 and 36 upon the under walls, or just 8 per cell. Striking the balance between these computations, we find 7 out of 34 grains —or more than 20 per cent.—not yet apostrophised, even by the close of the fourth day. I counted the still epistrophised grains in 12 cells of the thick part of three plants under similar circumstances, and got an average of 9 grains per cell, while my brother's average for cells set in darkness six days comes out fractionally less.

Before closing, it may be well briefly to refer to one more point in connection with the present subject. In my first memoir on Light and Protoplasm, it was shown that in some cases chlorophyll is apostrophised in low as well as in strong light; so that by graphically representing all grades of illumination upon the " photrum," it is possible to mark off a space therein containing àll degrees of light capable of bringing positively or negatively apostrophised grains into epistrophe. This space was called the " epistrophic interval "; but with a view of shortening this terminology, it is now proposed to apply to it the term "epistrophion." The epistrophion of *L. trisulca*, as determined upon a brilliant day of last September, is shown at Fig. 7, as the space barred with horizontal lines over which the mark I. is placed. To the right of this another space will be observed, shaded with diagonal lines and with the mark II. over it. This contains all grades of illumination sufficient to apostrophise the chlorophyll, but not of intensity equal to the task of systrophising the grains; for it the term "apostrophic interval" or "apostrophion" may be reserved. At the extreme right end of the photrum there is a small space over which the mark III. is set. This represents all the intensities of light capable of systrophising the grains during the four hours devoted to the experiment, it thus appearing that even systrophe can be induced in diffused light of the highest quality. Except, however, quite at verge of the photrum, systrophe is not very strongly pronounced after the four hours, and may concern only a small percentage of the cells. This space may be designated as the " systrophic interval " or "systrophion."

Lastly, a few words upon an observation of Borodin's,[*] that the apostrophised grains of cells bounding those which contain the bundles of raphides are systrophised in corners remote from the raphides. There can be no doubt that such is indeed the rule, although I have observed occasional exceptions. Now if the wall abutting on a raphidal cell be longer than the others, and make a wide angle with its meeting walls, it might happen that, in the course of their ante-systrophic movement round the cell (slow rotation), the grains might acquire sufficient momentum to carry them round the two corners; but this explanation is obviously insufficient, since it sometimes happens that the wall in contact with the raphidal cell is smaller than the rest—often markedly so—and yet its angles may be denuded of systrophised chlorophyll. For want of a better explanation, I am inclined to regard as the cause of this curious fact the reflection of heat from the faces of the crystals: this heat would clearly make its existence evident upon the walls bounding the cells containing the crystals, and we know that photolysis is in some way related to heat, having in all probability minimum, optimum and maximum points, though this requires further investigation.

---

[*] *Op. cit.*

EXPLANATION OF PLATE 185.—Fig. 1, superficial view of the marginal portion of a frond (epidermis omitted) kept in darkness overnight, showing the chlorophyll in merostrophe × 4')0. Fig. 2, two cells from the injured part of a frond after a night's darkness; most of the grains are apostrophised × 400. Figs. 3—6, diagrams illustrating photolysis: 3, medium diffused-light position of chlorophyll (epistrophe); 4, night position (merostrophe); 5, permanent effect of strong diffused light, and temporary of sunlight (apostrophe); 6, effect of continued sunlight and diffused light of the highest quality (systrophe). Fig. 7, the photrum of *L. trisulca* on a bright September day, showing the epistrophion, apostrophion and systrophion.

---

# CATALOGUE OF THE MARINE ALGÆ OF THE WEST INDIAN REGION.

## BY GEORGE MURRAY, F.L.S.

(Continued from p. 338.)

## II. PHÆOPHYCEÆ.

### FUCACEÆ.

SARGASSUM INCISIFOLIUM J. Ag.  Guadeloupe, *Mazé*!
  *Geogr. Distr.*  Cape of Good Hope.
S. POLYPHYLLUM J. Ag.  Guadeloupe, *Mazé*!  St. Thomas, *Hohenack*!
  *Geogr. Distr.*  Pacific.
S. ANGUSTIFOLIUM Ag.  Grenada, *Murray*!  Guadeloupe, *Mazé*!
  *Geogr. Distr.*  Indian Ocean.
S. FILIPENDULA Ag.  Gulf of Mexico (*fide J. G. Agardh*).  Florida, *Melvill*!
S. DENTIFOLIUM Ag.  Florida (*fide Farlow* in U. S. Com. Fish. Report, 1876, p. 706).
  *Geogr. Distr.* · Red Sea.
S. PTEROPLEURON Grunow.  Bahamas, ' *Novara*.'  Florida, *Palmer*! (in Farlow, Anderson and Eaton, No. 104).
S. HYSTRIX J. Ag.  Gulf of Mexico (*fide J. G. Agardh*).
  *Geogr. Distr.*  Atlantic (N. America).
S. PLATYCARPUM Mont.  Grenada, *Murray*!  Barbadoes.  *Dickie?*
  Guadeloupe, *Mazé*!  Martinique, *Duperrey*.  " Ad alias insulas Indiæ occidentalis (Mus. Petropolit. !)," J. G. Ag.
  *Geogr. Distr.*  Warm Atlantic.
S. LIEBMANNI J. Ag.  Guadeloupe, *Mazé*!
  *Geogr. Distr.*  Pacific (Mexico).
S. DESFONTAINESII Ag.  Guadeloupe, *Mazé*!  " Ex insulis Indiæ occidentalis," *Hb. Binder*!  J. G. Ag.
S. LENDIGERUM Ag.  Guadeloupe, *Mazé*!  Cuba, *R. de la Sagra*.
  Bermuda, *Rein*.
  Var. FISSIFOLIUM Harv.  Bermuda, *Rein*.
  *Geogr. Distr.*  Atlantic (Senegambia, Ascension, Teneriffe).
S. CYMOSUM Ag.  Barbadoes, *Dickie*!  Guadeloupe, *Mazé*!  Cuba, *R. de la Sagra*.
  Var. DICHOTOMUM Mont.  Guadeloupe, *Mazé*!
  *Geogr. Distr.*  Warm Atlantic (Brazil and Africa).

S. LEPTOCARPUM Kütz. Guadeloupe, *Mazé*! "Ad Antillas," *Kütz.*
S. LINIFOLIUM Ag. Bermuda, *Rein.*
  *Geogr. Distr.* Mediterranean (Canary Islands, rare).
S. VULGARE Ag. Grenada, *Murray*! Barbadoes, *Dickie*! Guade-
  loupe, *Mazé*! Jamaica, *Sloane*! *Chitty*! Cuba, *R. de la
  Sagra.* Florida, *Harvey, Melvill*! Bermuda, *Kemp, Rein.*
  f. FOLIOSISSIMA = S. FOLIOSISSIMUM Lam. Barbadoes, *Dickie*!
  *Geogr. Distr.* Warm Atlantic.
S. POLYCERATIUM Mont. Guadeloupe, *Mazé*! Cuba, *R. de la Sagra.*
S. FURCATUM Kütz. Guadeloupe, *Mazé*! St. Thomas (*fide Kützing*).
S. AFFINE J. Ag. Guadeloupe, *Mazé*! Danish West Indian Islands,
  *Hohenack*! "Inter Cubam et Jamaicam," *Liebman.* Ber-
  muda, *Rein.*
  Var. ANGUSTIFOLIA. West Indies, *Hohenack*!
S. BACCIFERUM Ag. Grenada, *Murray*! Barbadoes, *Dickie*! Gua-
  deloupe, *Mazé*! Jamaica, *Sloane*! *Chitty*! Vera Cruz,
  *Hohenack*! Florida, *Harvey, Melvill*! Bermuda, *Kemp, Rein.*
  *Geogr. Distr.* Warm Oceans.
S. ESPERI Ag. Cuba, *R. de la Sagra.*
  *Geogr. Distr.* Brazil.
S. TRACHYPHYLLUM Kütz. Guadeloupe, *Mazé*! Danish West Indian
  Islands, *Hohenack*! (No. 434). "Ad Antillas," *Kütz.*
S. CHEIRIFOLIUM Kütz. Guadeloupe, *Mazé*!
  *Geogr. Distr.* Senegambia.
S. DIVERSIFOLIUM Kütz. Guadeloupe, *Mazé*!
  *Geogr. Distr.* Mediterranean and Atlantic (Canary Islands).
S. INTEGRIFOLIUM Kütz. Guadeloupe, *Mazé*!
  *Geogr. Distr.* Brazil.
S. DICHOCARPUM Kütz. Guadeloupe, *Mazé*!
  *Geogr. Distr.* Senegambia.
S. BREVIPES Kütz. Barbadoes, *Dickie* Vera Cruz (*fide Kützing*).
S. TURNERI Kütz. Guadeloupe, *Mazé*!
  *Geogr. Distr.* Mediterranean, Red Sea.
S. SPINULOSUM Crn. Guadeloupe, *Mazé*! "Ex India occidentali,"
  *Kütz.*
S. MONTAGNEI Bailey. Guadeloupe, *Mazé*!
  *Geogr. Distr.* Atlantic (N. America).
TURBINARIA VULGARIS J. Ag. Guadeloupe, *Mazé*! Cuba, *R. de la
  Sagra.* Jamaica, *Sloane*!
  Var. DECURRENS J. Ag. Guadeloupe, *Mazé*!
  Var. TRIALATA J. Ag. Guadeloupe, *Mazé*!
  *Geogr. Distr.* Warm Oceans.
CYSTOSEIRA MYRICA J. Ag. Nassau (Bahama Islands), *Palmer*!
  (No. 8, Algæ Bahamenses).
  *Geogr. Distr.* Red Sea and Persian Gulf.
FUCUS CERANOIDES L. Bermuda, *Kemp.*
  *Geogr. Distr.* North Sea, Atlantic, Europe, and N. America.
F. DISTICHUS L. Bermuda, *Kemp.*
  *Geogr. Distr.* North Sea and N. Atlantic (Faroe, Iceland,
  Greenland, Newfoundland, &c.).

## DICTYOTACEÆ.

DICTYOTA DICHOTOMA J. Ag. (D. ATTENUATA Kütz.). Grenada, *Murray*!
  Guadeloupe, *Mazé*! Jamaica, *Chitty*! Florida, *Harvey*,
  *Melvill*! Bermuda, *Kemp, Rein*, '*Challenger*.'
  Var. IMPLEXA J. Ag. Guadeloupe, *Mazé*!
  Var. CURVULA Crn. Guadeloupe, *Mazé*!
  Var. LATIFRONS Crn. Guadeloupe, *Mazé*!
  Var. INTRICATA Grev. Bermuda, *Kemp*.
  *Geogr. Distr.* Warm and temperate oceans.
D. RADICANS Harv. Guadeloupe, *Mazé*!
  *Geogr. Distr.* Australia.
D. PATENS J. Ag. St. Thomas (*fide J. Agardh*).
  *Geogr. Distr.* Friendly Islands.
D. CILIATA J. Ag. Barbadoes, *Dickie*! Guadeloupe, *Mazé*! Martinique,
  *Hb. LeJolis*! Florida, *Harvey*! *Melvill*! Bermuda, *Kemp, Rein*.
  *Geogr. Distr.* Indian Ocean? Australia?
D. LIGULATA Kütz. Bermuda, '*Challenger*'!
  *Geogr. Distr.* Atlantic (France) and Mediterranean.
D. NÆVOSA Suhr. Guadeloupe, *Mazé*!
  *Geogr. Distr.* Cape of Good Hope.
D. LITURATA J. Ag. Guadeloupe, *Mazé*!
  *Geogr. Distr.* Cape of Good Hope.
D. BARTAYRESIANA Lam. Barbadoes, *Dickie*! Guadeloupe, *Mazé*!
  St. Thomas, *Young*! Danish West Indian Islands, *Hohen-
  ack*! (Meeralgen, Nos. 427, 428). Vera Cruz, *Liebman*.
  Florida, *Melvill*. Bermuda, *Kemp*, '*Challenger*.'
  Var. DIVARICATA J. Ag. Guadeloupe, *Mazé*!
D. DENTATA Lam. Barbadoes, *Dickie*! Guadeloupe, *Mazé*! Cuba,
  *R. de la Sagra*.
  *Geogr. Distr.* Brazil.
D. BRONGNIARTII J. Ag. Guadeloupe, *Mazé*! Martinique, *Duperrey*.
  *Geogr. Distr.* Brazil.
D. CRENULATA J. Ag. Guadeloupe, *Mazé*! Bermuda, *Kemp*.
  *Geogr. Distr.* Pacific (Mexico), Indian Ocean (Tuticorin).
D. SANDVICENSIS Kütz. Guadeloupe, *Mazé*!
  *Geogr. Distr.* Pacific.
D. PARDALIS Kütz. Barbadoes, *Dickie*! "E mari Antillarum,"
  *Kütz*.
D. PINNATIFIDA Kütz. Guadeloupe, *Mazé*! Antigua, *Hb. Sonder*.
  *Geogr. Distr.* Warm Pacific.
D. LINEARIS Ag. Guadeloupe, *Mazé*! Cuba, *R. de la Sagra*.
  *Geogr. Distr.* Mediterranean and neighbouring Atlantic. [Red
  Sea?]
D. ACUTILOBA J. Ag. Barbadoes, *Hb. J. E. Gray*! Guadeloupe,
  *Mazé*! St. Thomas, *Hohenack*! Florida, *Melvill*!
  *Geogr. Distr.* Pacific and China Sea.
D. FURCELLATA Kütz. Guadeloupe, *Mazé*!
  *Geogr. Distr.* Australia.
D. FASCIOLA Lam. Guadeloupe, *Mazé*! Danish West Indian Islands,
  *Hohenack*! (Meeralgen, No. 429). Florida, *Harvey, Melvill*!

*Hooper*! (in Farlow, Anderson and Eaton, No. 161). Bermuda, *Rein*.

Var. ABYSSINICA J. Ag. Guadeloupe, *Mazé*!
*Geogr. Distr.* Mediterranean, Red Sea.

D. CERVICORNIS Kütz. Guadeloupe, *Mazé*! Florida, *Harvey*. [Recorded by Harvey as *D. fasciola*.]

D. PANICULATA J. Ag. ? Guadeloupe, *Mazé*!
*Geogr. Distr.* Australia.

D. ANTIGUÆ Kütz. Barbadoes, *Dickie*! Guadeloupe, *Mazé*! Antigua, *Hb. Sonder*.

D. INDICA Sond. Barbadoes, *Dickie*! Guadeloupe, *Mazé*! Cuba, *J. Smith* (*Hb. Sonder*).
    f. TORTA Crn. Guadeloupe, *Mazé*.

D. ÆQUALIS Kütz. Guadeloupe, *Mazé*!
*Geogr. Distr.* Mediterranean.

D. CUSPIDATA Kütz. Guadeloupe, *Mazé*! Vera Cruz, *Liebman*.

D. BIPINNATA Crn. Guadeloupe, *Mazé*!

D. VARIABILE Crn. (= D. POLYCARPA Kutz. SPATOGLOSSUM VARIABILE Fig. et De Not.). Guadeloupe, *Mazé*!
*Geogr. Distr.* Cape of Good Hope.

D. PROLIFERA Lam. Guadeloupe, *Mazé*!
*Geogr. Distr.* Australia.

D. SUHRII nob. (= D. PROLIFERA Suhr, non Lam.). Guadeloupe, *Mazé*!
*Geogr. Distr.* Cape of Good Hope.

DILOPHUS GUINEENSIS J. Ag. Guadeloupe, *Mazé*! St. Croix, *Oersted*.
*Geogr. Distr.* The "St. Thomas" quoted by Kützing and Agardh is the island in the Gulf of Guinea.

D. ALTERNANS J. Ag. Florida, *Farlow*.

TAONIA SCHROEDERI J. Ag. Guadeloupe, *Mazé*! Vera Cruz, *Liebman*, *Hohenack*! Jamaica, *Chitty*! Bermuda, *Kemp? Farlow*! (in Farlow, Anderson and Eaton, No. 159).
*Geogr. Distr.* Brazil.

T. ATOMARIA J. Ag. ? Dominica, *Hb. Mus. Brit.*! [no collector's name. Ex Herb. Harv. ?].
*Geogr. Distr.* N. Atlantic, Mediterranean.

PADINA PAVONIA J. Ag. Grenada, *Murray*! Barbadoes, *Dickie*! Guadeloupe, *Mazé*! St. Thomas, *Young*! Jamaica, *Sloane*! *Chitty*! *Wright*! Cuba, *R. de la Sagra*. Florida, *Harvey*! *Melvill*! Bermuda, *Kemp, Rein,* '*Challenger*'!
*Geogr. Distr.* General in temperate and warm oceans.

P. COMMERSONI Bory. Guadeloupe, *Mazé*! Cuba, *R. de la Sagra*. St. Croix and Florida (*fide J. Agardh*).
*Geogr. Distr.* Indian Ocean, Australia, Pacific.

ZONARIA VARIEGATA Lam. Guadeloupe, *Mazé*! Martinique, *Duperrey*. Danish West Indian Islands, *Hohenack*! (Meeralgen, No. 516. Cuba, *R. de la Sagra*. Bermuda, '*Challenger*'!
*Geogr. Distr.* Warm Atlantic.

Z. LOBATA Ag. (= Z. FULIGINOSA Mart.). Barbadoes, *Dickie*! Guadeloupe, *Mazé*! [with "forme padinoide"]. Florida, *Harvey*! *Melvill*! *Hooper*! (in Farlow, Anderson and Eaton, No. 92). Bermuda, *Kemp, Rein,* '*Challenger*'!
*Geogr. Distr.* Atlantic (Cape of Good Hope, Brazil, Canaries).

Z. CRUSTACEA Crn.   Guadeloupe, *Mazé*!

Z. GYMNOSPORA Kütz.?   Guadeloupe, *Mazé*!   St. Thomas, *Hb. Sonder.*   Danish West Indian Islands, *Hohenack*! (Meer-algen, No. 515).   [The Guadeloupe specimen is a very young one, and Hohenack's is mature.   Allowing for this, I am very doubtful of their being the same species.]

Z. PARVULA Ag.   Bermuda, *Kemp.*
    *Geogr. Distr.*   N. Atlantic, North Sea, Mediterranean.

HALYSERIS DELICATULA J. Ag.   Grenada, *Murray*!   Barbadoes, *Dickie*!   Guadeloupe, *Mazé*!   Cuba, *R. de la Sagra.*   "Shores of Mexico" (*fide J. G. Agardh*).
    *Geogr. Distr.*   Brazil.

H. JUSTII J. Ag.   Barbadoes, *Dickie*!   Guadeloupe, *Mazé*!   Martin-ique, *Duperrey.*   Jamaica, *Chitty*!   Cuba, *R. de la Sagra.*   Bermuda, '*Challenger*'!
    *Geogr. Distr.*   Brazil.

H. PLAGIOGRAMMA Mont.   Barbadoes, *Dickie*!   Guadeloupe, *Mazé*!   Jamaica, *Chitty*!   Cuba, *R. de la Sagra.*   Bermuda, *Rein.*
    *Geogr. Distr.*   Brazil [Sandwich Islands?].

H. POLYPODIOIDES Ag.   Bermuda, *Kemp, Rein.*
    *Geogr. Distr.*   Atlantic, North Sea, Mediterranean, Tasmania.

SORANTHERA LEATHESIÆFORMIS Crn.   Guadeloupe, *Mazé*!

## ECTOCARPACEÆ.

ECTOCARPUS SILICULOSUS Lyngb.   Guadeloupe, *Mazé*!
    *Geogr. Distr.*   Atlantic (from Faroe to Cape of Good Hope and Cape Horn), Mediterranean, Australia.

E. MITCHELLÆ Harv.?   Guadeloupe, *Mazé*!

E. DUCHASSAINGIANUS Grünow.   Guadeloupe, *Mazé*! (*Duchassaing*).

E. GUADELUPENSIS Crn.   Guadeloupe, *Mazé*!

E. MACROCARPUS Crn.   Guadeloupe, *Mazé*!

E. FENESTROIDES Crn.   Guadeloupe, *Mazé*!

E. SPONGODIOIDES Crn.   Guadeloupe, *Mazé*!

E. HAMATUS Crn.   Guadeloupe, *Mazé*!

E. HETEROCARPUS Crn.   Guadeloupe, *Mazé*!

E. OBTUSOCARPUS Crn.   Guadeloupe, *Mazé*!

E. DENUDATUS Crn.   Guadeloupe, *Mazé*!

E. PUSILLUS Griff.   Barbadoes, *Dickie*!
    *Geogr. Distr.*   Britain.

E. MINUTULUS Mont.   Cuba, *R. de la Sagra.*

## SPHACELARIACEÆ.

SPHACELARIA TRIBULOIDES Menegh.   Guadeloupe, *Mazé*!   Gulf of Mexico, *Liebman.*
    *Geogr. Distr.*   Mediterranean, Red Sea, Australia.

S. BRACHYGONIA Mont.   Martinique, *Hb. Le Jolis*!
    *Geogr. Distr.*   Brazil.

## CHORDARIEÆ.

MYRIOCLADIA CAPENSIS J. Ag.   Guadeloupe, *Mazé*!
    *Geogr. Distr.*   Cape of Good Hope.

M. GRACILIS Crn. (= MESOGLOIA GRACILIS Berk.). Guadeloupe, *Mazé*!
    *Geogr. Distr.* N. Atlantic.
M. MEDITERRANEA Crn. Guadeloupe, *Mazé*!
    *Geogr. Distr.* Mediterranean.
MESOGLOIA VERMICULATA Le Jol. Bermuda, *Kemp, Rein.*
    *Geogr. Distr.* Atlantic and North Sea.
EUDESME VIRESCENS J. Ag. (= MESOGLOIA VIRESCENS Carm.). Guade-
    loupe, *Mazé*! Florida, *Harvey.* Bermuda, *Kemp,* '*Challenger.*'
    *Geogr. Distr.* N. Atlantic and North Sea.
CASTAGNEA GRIFFITHSIANA J. Ag. Bermuda, *Kemp.*
    *Geogr. Distr.* N. Atlantic and North Sea.
CLADOSIPHON ZOSTERICOLA Harv. Guadeloupe, *Mazé*!
    *Geogr. Distr.* Australia.
LIEBMANNIA LEVEILLEI J. Ag. Vera Cruz, *Liebman.*
    *Geogr. Distr.* Atlantic (France) and Mediterranean.

### PUNCTARIACEÆ.

STRIARIA FRAGILIS J. Ag.? Guadeloupe, *Mazé*!
    *Geogr. Distr.* North Sea (and Baltic).

### ARTHROCLADIACEÆ.

CHNOOSPORA FASTIGIATA var. PACIFICA J. Ag. Guadeloupe, *Mazé*!
    *Geogr. Distr.* Pacific (Mexico).
C.? IMPLEXA J. Ag. Guadeloupe, *Mazé*!
    *Geogr. Distr.* Red Sea.

### SPOROCHNACEÆ.

SPOROCHNUS PEDUNCULATUS Ag. Bermuda, *Kemp.*
    *Geogr. Distr.* N. Atlantic, North Sea, Mediterranean.
STILOPHORA ANTILLARUM Crn. Guadeloupe, *Mazé*!
ASPEROCOCCUS CLATHRATUS J. Ag. Guadeloupe, *Mazé*! Danish West
    Indian Islands, *Hohenack*! Florida, *Harvey*! *Melvill.*
    *Geogr. Distr.* Warm Atlantic, Mediterranean, Red Sea, Indian
    Ocean, Australia, Pacific.
A. INTRICATUS J. Ag. Guadeloupe, *Mazé*! Vera Cruz, *Liebman.*
    Florida? *Melvill.*
A. ORIENTALIS J. Ag. Guadeloupe, *Mazé*!
    *Geogr. Distr.* Indian Ocean.
A. RAMOSISSIMUS Zanard. Guadeloupe, *Mazé*!
    *Geogr. Distr.* Mediterranean,
A. SCHRAMMII Crn. Guadeloupe, *Mazé*!
A. SINUOSUS Roth. Barbadoes, *Dickie*! Vera Cruz, *Liebman.*
    Florida, *Harvey, Melvill*! Bermuda, *Kemp, Rein,*'*Challenger.*'
    *Geogr. Distr.* Throughout warm oceans.

### RALFSIACEÆ.

RALFSIA EXPANSA J. Ag. Vera Cruz, *Liebman.*

(To be continued.)

## NOTES ON THE FLORA OF BEN LAIOGH, &c.

### By G. Claridge Druce.

The comfortable hotel at Tyndrum, placed though it be in a " howling wilderness," as I once heard it called by a London tourist, is conveniently situated for botanists who wish to investigate the treasures of Ben Laiogh, which is only six miles away.

This mountain is in two counties, which are separated by a line traced over the summit or watershed, the Argyll water running into Loch Awe, the Perthshire streams into Loch Tay. The neighbouring mountain, Ben Oss—rival we can scarcely call it—also is in two of the Watsonian vice-counties, those of West and Mid Perth ; the streams of the latter draining into the Tay, while the southern waters of West Perth belong to the Forth system.

Ben Laiogh is an interesting mountain, whether considered from a scenic, a geologic, or a botanical point of view. As one looks at it from an altitude of 900 ft., by the Coninish burn, close to the farm of that name near the mountain's base, it presents a very imposing appearance, as it rears itself up in steep and almost unbroken grassy slopes to the still steeper, slippery screes which are crowned by rocky cliffs of an even more acute angle, breaking into a handsome sky outline, which from the topmost summit bend outwards and downwards into wing-shaped ridges like those that encircle the Red Tarn on Helvellyn, except that the Stob Garb is not such a formidable rival to Ben Laiogh as Catchedicam is to the Lake District mountain. Here no mountain lake is enclosed in these rocky ridges, but a little stream issues from the corries, and bounding down through the grassy slopes increases the grandeur of the view, especially to one who has just come from the rounded summits of the granitic Cairngorms. This waterfall, which issues from this large Perthshire corrie, is principally fed by springs issuing from the base of the upper cliffs, and nurtures with its waters a rich growth of *Kobresia*, *Carex saxatilis*, *C. capillaris*, *Juncus castaneus*, *J. triglumis*, &c., as they rush down to join the Coninish burn.

At a height of about 1500 ft. above the sea-level may be seen, on each side of the waterfall, a low line of cliffs, not springing directly from its sides, but gradually appearing and increasing in size as they curve round to the south and north of the stream. These cliffs are of calcareous schist (the greater part of the mountain being made up of quartzose schist with masses of horneblende), which acts as a very important factor in the botany of this mountain ; generally they are much seamed and broken, but occasionally they have weathered into flat surfaces with very narrow ledges, on which a rich growth of treasures may be found. To the south they extend for some distance up the Laiogh Valley, but are not there particularly rich in typical plants. *Draba incana*, *Arabis hirsuta*, however, occur on them, as well as *Saussurea*. Northwards they increase in size, and turn sharply round to the west, 500 or 600 ft. above and parallel with the Coninish burn to the watershed, and continue in Argyleshire by the Eas Daimh burn for two or three miles, bending round in a south-western direction, when they again diminish in size, and eventually disappear under a slope covered up with moraine, which leads up into the great Argyll corrie on the west side of the mountain (corresponding to the similar Perth corrie on the eastern side). A short mile south-ward this line of calcareous cliffs again appears, but here not so rich in characteristic vegetation as the cliffs above the Coninish and Eas Daimh burns, to which I shall later on draw particular attention. From the burns to the cliffs are grassy slopes, and above the limestone cliffs again is a widish green slope, then a range of quartz-schist cliffs, then again calcareous cliffs, topped finally by a mass of quartz-schist. At about 2500 ft. on the north side is a small corrie with numerous springs at its base, principally draining into the Coninish burn; the western border of this small corrie is a ridge which forms the county boundary, the great Argyll corrie being situated on its south-western side, and is large and open, barren and stony, without a tarn, and in its upper portions free from springs. Lower down in the gap (on this side much wider), between the calcareous cliffs, great heaps of moraine, already alluded to, obscure the geologic features, but form a moor-land, boggy in places, whose peculiar plants are *Carex pauciflora*, *C. xanthocarpa*, *C. fulva*, *Drosera anglica*, *D. obovata*, and *Malaxis paludosa*.

The flora of the mountain is therefore separable into several zones : firstly, the moorland up to 1200 ft., whose typical features are the plants last noticed, in addition to the ordinary moorland vegetation of *Erica Tetralix*, *E. cinerea*, *Calluna*, *Myrica*, *Molinia*, *Carex binervis*, *C. flava*, *C. echinata*, *Narthecium*, *Drosera*, &c. Secondly, the vegetation of the calcareous cliffs (the occurrence of which, as I have pointed out, is such an important factor), and this is of a singularly rich and attractive character, consisting as it does of plants such as *Dryas*, *Draba incana*, *Pyrola rotundifolia*, *P. secunda*, *Saussurea*, *Bartsia alpina*, *Salix Arbuscula*, *S. Myrsinites*, *Carex atrata*, *C. vaginata*, *Juncus castaneus*, *Arabis petræa*, *Avena alpina*, *Saxifraga quinquefida*, *S. nivalis*, &c.

Then, thirdly, the higher grassy slopes which yield abundance of *Carex saxatilis*, *C. capillaris*, *C. echinata*, *Juncus castaneus*, *J. triglumis*, *Epilobium anagallidifolium*, *Veronica humifusa*.

And, fourthly, the higher cliffs and topmost corries, which have *Cerastium alpinum*, *Arabis petræa*, *Saxifraga nivalis*, *S. hypnoides*, *Poa alpina*, *Deschampsia alpina*, *D. montana*, *Arenaria sedoides*, &c.

I spent three days on the mountain ; the first, characterised by rain and mist, was occupied in working from Tyndrum over the moorland and ascending by the waterfall to the Perthshire corrie, climbing in dense mist the face of the steep central cliff to the summit and then crossing into the Argyll corrie, passing over into the small high Perth corrie, and working the higher limestone cliffs round the Stob Garb at about 2700 ft., into the large Perth corrie, whose southern side was again investigated, as well as the low line of cliffs stretching up Glen Laiogh.

The second day I drove about ten miles along the Dalmally Road and forded the River Lochy, worked the moorland on the Argyll side, and then the calcareous cliffs up the Eas Daimh stream to the Perth boundary; finding them so rich as to induce one to return by them at a slightly different level into the great Argyll corrie, whose cliffs, up to its head, were worked, and then its southern cliffs. This gave one of the best day's botanising of the season, and the day was clear and pleasant. A third and less agreeable day was occupied in working the Tyndrum moorland, the rocks, ravine and waterfall of Ben Chuirn, and the Perthshire cliffs of Ben Laiogh parallel with the Coninish burn up to the watershed. A fourth day was spent at Dalmally and the upper end of Loch Awe, and of course has nothing to do with Ben Laiogh. I have to thank Mr. Dakin, of the Geological Survey, and Mr. Franklin T. Richards, for kind assistance.

Below is a detailed list of interesting plants noticed. Many critical specimens are still under investigation. Mr. Arth. Bennett, Prof. Babington, Prof. Haussknecht, Dr. Buchanan White, and Mr. F. J. Hanbury have rendered help in plant discrimination. Supposed additions to Top. Bot., ed. ii., are marked.* Those species without personal authority in that work are marked †. Without giving precise localities, I may say that I gathered in Argyllshire *Cystopteris montana* Bernh., which occurred in great profusion and luxuriance.

†*Trollius europæus* L. Laiogh Cliffs, 98.

*Arabis petræa* Lamk. This is entered in Top. Bot. for Mid Perth on the authority of Lightfoot's Flora. In the Perthshire corrie and on the cliffs in both counties occurred a form of this species very distinct in appearance from the Cairngorm plant. It is well marked by its very much larger size, by its much larger, less deeply-cut leaves which are of a thinner texture and different colour. The flowers are larger and with more spreading petals. It appears to be worthy of varietal distinction and may be designated var. *grandifolia*. The leaves vary much as to leaf cutting, but are never so deeply cut as in the Cairngorm plants. They vary also from nearly glabrous to hairy.

*Arabis sagittata* DC., *A. hirsuta* Br., non Scop.? Calcareous cliffs.

*Draba incana* L., var. *contorta* (Ehrh.) with above, 88, *98.

*Drosera obovata* Mert. et Koch. Moorland by Coninish Farm, very rare; growing with *anglica* and *rotundifolia*, *88. Moorland near River Lochy, *98.

*Silene acaulis* Jacq. In flower in upper corries, 88, 98.

*Arenaria sedoides* Schultz. Descends to 1800 ft. on Argyll side, †98.

*Cerastium alpinum* L., var. *lanatum* (Lamk.). Upper part of Argyll corrie, rare, *98.—*C. triviale* Link, var. *alpinum* Koch. With above in both corries.

*Dryas octopetala* L. Very abundant and luxuriant on the lower line of calcareous cliffs, especially in †Argyll, extending into Perth but thinning out eastwards and apparently absent from the cliffs in the Laiogh Glen.

*Potentilla Sibbaldi* Haller fil.   In the upper portion of †Argyll and Perthshire corries.

*Rubus saxatilis* L.   On the lower cliffs, †Argyll, it occurred with smaller flowers than usual; they are sweetly scented like hawthorn.   In Perth, specimens occurred with the leaves quite bronzed.

*Alchemilla vulgaris* L., var. *minor* Huds., common, †98.

*Pyrus Aucuparia* Gaertn.   Ascends to 2800 ft. on southern cliffs of the Argyll side.

*Epilobium anagallidifolium* Lamk.   Upper grassy slopes, 88, †98, but less abundant on Laiogh than on the Grampians.

*Saxifraga nivalis* L.   Calcareous cliffs †Argyll and Perth, but not confined to them. — *S. quinquefida* Haworth.   Quartzose and calcareous cliffs *Argyll and *Mid Perth.—*S. hypnoides* L.   Upper part of †Argyll corrie, rare; and in the small corrie, Perth.

*Heracleum Sphondylium* L.   Ascends to 3300 ft., sometimes becoming very dwarfed.

*Galium sylvestre* Poll.   Lower calcareous cliffs, rather rare. *Argyll and Perth.

*Taraxacum officinale* Lamk., var. *palustre* DC., *98.

*Solidago Virgaurea* L., with its vars. *cambrica* and *angustifolia*, on Laiogh, 88 and *98.

*Loiseleuria procumbens* Desv.   †Ben Ohran, Mr. Dakin.

*Vaccinium Vitis-idaea* L.   Lower cliffs, †98.   The flowers are hawthorn-scented.—*V. uliginosum* L.   Fruited on the hill.

*Pyrola rotundifolia* L.   Splendid specimens on lower cliffs; they have a lilac-like odour.   It ascended to 2500 ft.—*P. secunda* L.   With above, but only on the lower cliffs in †Argyll only.

*Veronica humifusa* Dickson.   Upper slopes and corries, 88, *98.

*Utricularia intermedia* Hayne.   In a small plash on the watershed in the Coninish Valley at 1800 ft. elevation, principally draining into *Mid Perth.

*Armeria maritima* Willd.   A broad-leaved form, ? *planifolia* Syme.   On the grassy slopes, &c., descending in some cases to 1700 ft.

*Plantago maritima* L.   A luxuriant erect, broad-leaved plant on lower calcareous cliffs, 98.

*Betula alba* L., *B. verrucosa* Ehrh.   Lower cliffs and streamsides, *98.

*Habenaria viridis* Br.   †Argyll cliffs.

*Tofieldia palustris* Huds.   †Grassy slopes and cliffs, 98.

*Juncus castaneus* Sm.   Wet ledges, grassy slopes in several parts of the mountains, on the *Argyll but commoner on the Perthshire side.   Especially prefers a shallow stream-side and there spreading for a distance by the creeping roots.—*J. triglumis* L.   Common, †98, ascending to at least 3300 ft.

*Scirpus pauciflorus* Lightf.   Moorland, 88 and *98.

*Carex saxatilis* L.   Descends to 1500, ascends to 2300 ft.   A typical plant of the mountain, 88 and †98.   It varies considerably, but I saw nothing like *Grahami* or *dichroa*.   One form appeared to me to have a *flava* parentage, but Mr. Arthur Bennett does not coincide in this opinion.—*C. flava* L., var. *minor* Towns.   Common.

—*C. xanthocarpa* Deseg. Coninish moorland, Perth, *88, and by the Lochy. Argyll, 98, — *C. fulva* Good. Moorland, †98. — *C. vaginata* Tausch. Frequent and variable on †Argyll cliffs. The var. ? *borealis* was noticed in one place.—*C. flacca* Schreber, **1771**, var. *stictocarpa* D. Don, or near it. On Argyll cliffs. — *C. Goodenowii* Gay. Var. *juncella* Fries. Lochy side, 98. A variety occurred on the lower cliffs which is (Mr. Bennett tells me) analogous to the variety *prolixa* of *acuta*, but it is not the *prolixa* of *vulgaris* figured by Boott. It is a pretty and noticeable and apparently rare form.—*C. obtusangula* Ehrh., occurred as a small dark-coloured mountain form at about 1800 ft. in the Coninish Valley.

*Deschampsia alpina* Roem. et Schultz. Higher corries, 88 and †98.—*D. cæspitosa* Beauv., var. *alpina* Gaud. With above; also viviparous forms of *cæspitosa* by the Coninish burn, 88, and the so-called var. *pallida* Koch.—*D. flexuosa* Trin., var. *montana* (Huds). Higher cliffs, 88, 98.

*Agrostis vulgaris* With., var. *pumila* (L.). Higher corries, 88, 98.—*A. canina* L., f. *grandiflora*. Moorland, 98.

*Poa alpina* L. Higher cliffs, *Argyll, rare.

*Sieglingia decumbens* Bernh. Common, 88, 98.

*Phegopteris Dryopteris* Fée. †Argyll cliffs.

*Cryptogramma crispa* Br., †.

*Cystopteris fragilis* Bernh. Abundant in 88 and †98, with its var. *dentata*.—*C. montana* Bernh. *Argyll.

*Lastrea dilatata* Presl, var. *dumetorum* Newm. Corrie, 98.

*Athyrium alpestre* Wilde, †Argyll corrie.

*Lycopodium alpinum* L. Not uncommon, *98.—Var. *decipiens* Syme, under which it is safer to place the barren flattened forms suggestive of *complanatum*, 88, 98.—*L. Selago* L. †Argyll.

*Selago selaginoides* Gray. †Argyll.

*Equisetum palustre* L., var. *polystachyum*. By the Coninish lead mines, 88.

I spent a day at the head of Loch Awe, Dalmally, &c., where the following plants were noted :—

*Nymphæa lutea* L., *Barbarea vulgaris* Br., *Sisymbrium Thalianum* Gay, *Erophila vulgaris* DC., *Montia fontana* L., var *major* All., *Lychnis Flos Cuculi* L., *Cerastium glomeratum* Thuill., var. *apetalum* Dumort *Geranium molle* L., *Trifolium hybridum* L., *Potentilla palustris* Scop., *Fragaria vesca* L., *Rubus affinis* W. & N., *R. rhamnifolius* W. & N., *Rosa mollis* Sm., *var. *cærulea* (Woods), *R. canina* L., *var. *lutetiana* (Lem.) *var. *biserrata* (Merat), *var. *dumalis* (Bechst.), *var. *urica* (Lem.), var. *tomentella* (Lem.), *var. *decipiens* (Dum.), var. *Reuteri* (Godet), var. *platyphylla* Rau., *Alchemilla arvensis* Scop., *Epilobium montanum* L., f. *umbrosa* * *E. obscurum* × *palustre*, *Callitriche stagnalis* Scop., *C. hamulata* Kuetz., *Scleranthus annuus* L., †*Parnassia palustris* L., *Hedera Helix* L., *Meum Athamanticum* Jacq., plentiful in meadow near Loch Awe ; *Ægopodium Podagraria* L., *Galium palustre* L., *var. *Witheringii* (Sm.)., †*Hieracium Dewari* Bosw., *H. auratum* Fries, in the meadows near Kilchurn Castle ; *Arctium intermedium* Lange, *Cnicus lanceolatus* Willd., *C. palustris* Willd., *Centaurea Cyanus* L.,

*Chrysanthemum Parthenium* Pers., *Melampyrum pratense* L., var. *montanum* (Johns.), \**Teucrium Scorodonia* L., \**Stachys sylvatica* L., \**S. palustris* L., *Galeopsis Tetrahit* L., var. *bifida* (Boengh.) \**Myosotis cæspitosa* Schultz, \**M. repens* Don., *M. palustris* Relh., var. *strigulosa* Reichb., *Atriplex paluta* L., \**Polygonum Hydropiper* L., \**Rumex aquaticus* L., p.p. (*domesticus* Hartm.), *Corylus Avellana* L., \**Betula glutinosa* Fries., \**Salix rubra* Huds., \**S. ferruginea* G. And., \**S. rugosa* Leefe, the latter three probably introduced; \**S. nigricans*, Sm. *Carpinus Betulus* L., planted; *Pinus sylvestris* L., probably native; \**Alisma Plantago-aquatica* L., \**Juncus conglomeratus* L., *J. supinus* Mœnch, var. *uliginosus* Fries, \**Eleocharis multicaulis* Sm., †*Carex vesicaria* L., \**Alopecurus myosuroides* Huds., *Phragmites communis*. Fries., var. *nigricans* Gr. et Gods., *Agrostis alba* L., var. *stolonifera* (Linn.), \**Poa nemoralis* L., \**Festuca arundinacea* Schreb. †*Bromus commutatus* Schrad., *Agropyron repens* Beauv., var. *Leersianum* Gray, *Athyrium Filix-fœmina* Roth., var. *convexum*, †*Botrychium Lunaria* L., *Equisetum sylvaticum* L., var. *capillare* Hoffm., \**Nitella opaca* Agardh.

# ON BOTANICAL NOMENCLATURE.

## By Prof. C. C. Babington, F.R.S.

I think that we are going too far in enforcing the rule of priority in nomenclature as is now attempted. It is an admirable rule in itself, but one to which there must be some exceptions; as there are to all rules. I certainly should not go out of my way to adopt some almost unanimously neglected name to be found in an obscure essay, which has been put aside by all the great authorities such as DeCandolle and others. I think that we unnecessarily introduce confusion by doing so: certainly not remove it, as is to be desired. Why should we rake up some Inaugural Dissertation or obscure local Flora, for the purpose of finding an old name for some plant: a book quite forgotten in its own country and very difficult of access, not to be found even in the great libraries? I consider it to be riding to death a good rule when we do so. If the continental authorities have almost unanimously neglected these obscure names, why give unnecessary trouble by recalling them to notice?

But there are also cases in which names are found in the books of well-known authors, but have nevertheless been neglected by all modern botanists of note. I may take as an example the case of the *Nymphæaceæ* noticed in ' Journal of Botany ' for this year. It is a singular fact—but fact it is—that Salisbury's names have been universally neglected. Why should we revive them now, through some fancied idea of supporting his reputation? Certainly it seems to me that Smith showed good reason for applying *Nymphæa* and *Nuphar* as he did, when he thought he had shown that they are the old classical names of the plants (Eng. Fl. iii. 15), although that seems rather doubtful. I quite agree with Planchon (Ann. des Sc. Nat. Ser. 3, t. 19, p. 59) in thinking that it is not advisable now to change the names of these *Nymphæaceæ*. Salisbury's application of the names may have the priority, but that is somewhat doubtful; and as long since as the publication of Necker's 'Elementa' (1791), the *N. alba* was recognized as the type of the genus *Nymphæa*.

A singularly unfortunate alteration seems to have been made in 'London Catalogue' (ed. 8), when the original and only *Azalea* of Linnæus is displaced, probably because the gardeners chose to call some species of *Rhododendron* by that name. Surely the rule of priority, now so strictly enforced, is totally opposed to giving a new name to the Linnæan genus. This is a case in which a return to the old name is unquestionably right.

Again, I do not see why *Dabeocia* should be changed into *Daboecia*, because Don's printers made a not unnatural blunder, and he did not discover it when correcting the press. Palpable errors should certainly be corrected, especially in terms derived from the names of persons, such as this is. St. Dabeoc is a well-known person : who ever heard of Daboec? Such corrections have always been made by the best authorities, and I do not see any reason for differing from them. Let us try to accord with our neighbours; not to differ from them, even when we see that a strictly enforced rule is favourable to doing so. We do not advance science thereby but, in my view, confuse it. When it is doubtful if an author intended to give a name to one plant including others which we now separate from it, and especially in those cases when he can be shown to have confused various forms, or what we call species, under this name, I do not think it desirable to take the name and apply it as if its author meant one plant only by it. The name is a confused name, and ought to be put aside as undeterminable and liable to be applied by different people to different plants with equal justice. Of course if we can discover what modern species the author had especially in view, his name ought to appertain now to that species and be so retained.

Focke says (Consp. Rub. Germ. p. 58) :—" We have far too many botanical rag-collectors who, with their priority of thought and opinion, penetrate every turning-point, dragging matters again into the light of day which had better have been left in the shades of night." Darwin said ('Life,' i. 366):—" The names are adopted by Cuvier . . . and almost every well-known writer, but I find that all these were anticipated by a German : now I believe that if I were to follow the strict rule of priority, more harm would be done than good." These, I think, are sound views.

Since this paper was written I have seen the discussion upon nomenclature in the Aug. and Sept. numbers of this Journal. I simply wish to add my name to the list of those who in all possible cases retain the original specific name, being even inclined to select the supposed typical form to retain the old name, when it can be done without causing more difficulty than by dropping it as undeterminable. I think that this is only just to the original author.

But when a species is removed from its original genus and placed in another, I feel bound to follow the plan adopted by most botanists until very recently, of giving as the authority for the binomial name the author who placed it in its new and apparently more correct genus. I admit that this causes some trouble, and perhaps injustice, as the original author would probably be overlooked by many modern users of the name. But this trouble seems to me far less than that caused by the plan advocated by Mr. Britton and others. Personally, in the few cases in which I have had to place my own name after the new combination I have done it much against my will. I would far rather have seen that of the original describer of the species in that place; but I do not see how it could be done without causing unnecessary and most undesirable confusion in the nomenclature. I have always thought that botanists have acted wisely in not adopting the " Stricklandian " zoological plan. But I do not wish to prolong this discussion, and only hope that new ideas may not be introduced which will, I think, cause trouble without any adequate advantage.

---

## ON A NEW *ACROSTICHUM* FROM TRINIDAD.

### By J. G. Baker, F.R.S.

Amongst the ferns of the Trinidad herbarium lately sent home for comparison by Mr. Hart, is the following new species:—

105. **Acrostichum** (Gymnopteris) **Hartii** Baker, n. sp. — Rhizome wide-creeping; basal paleæ lanceolate, membranous, pale brown. Stipe of the barren frond above a foot long, naked, pale brown, deeply channelled down the face. Sterile lamina oblong-lanceolate, simply pinnate, $1\frac{1}{2}$–2 ft. or more long, moderately firm in texture, glabrous and entirely free from scales. Pinnæ sessile, lanceolate, 4–6 in. long, $1\frac{1}{4}$–2 in. broad, narrowed gradually to the tip, entire or shallowly lobed. Veins in pinnate groups, as in *Eunephrodium*, with 4–5 simple ascending veinlets on each side, the lower usually, but not always, anastomosing. Fertile frond bipinnate, distinct, or in one specimen there are several fertile pinnæ at the top of a sterile frond, as in *Photinopteris;* fertile final segments torulose, at most $\frac{1}{3}$–$\frac{1}{2}$ in. long.

Hab. Trinidad, *Government herbarium* 228 !

Allied to *A. suberectum* Baker in Hook. Ic. t. 1692; *A. juglandifolium*, tab. 1691; and *A. polybotryoides*, tab. 1690.

# BIOGRAPHICAL INDEX OF BRITISH AND IRISH BOTANISTS.

### By James Britten, F.L.S., and G. S. Boulger, F.L.S.

(Continued from p. 348).

**Fitz-Roberts, John** (fl. 1696). Of the Gill, Kendal. Correspondent of Petiver and Ray. R. Syn. ed. 2, 57, &c.; C. Nicholson, Annals of Kendal, 344.

**Fleming, John** (d. 1815): d. London, 10th May, 1815. M.D. F.R.S. F.L.S. 'Cat. Indian Medicinal Plants with Sanskrit names, Asiat. Researches, xi. 153 (1810); Pritz. 108; Jacks. 11. *Flemingia* Roxb.

**Fleming, John** (1785–1857) : b. Kirkroads, Bathgate, Linlithgow, 1785 ; d. Edinburgh, 18th Nov. 1857. Clerk. D.D., St. Andrews, 1813. F.R.S.E., 1814. Minister of Bressay, Shetland, 1808 ; Flisk, Fife, 1810–1834. Lect. Nat. Hist., Cork Instit., 1816. Prof. Nat. Phil., Aberdeen, 1834. Prof. Nat. Sci., New Coll., Edinb., 1845. 'Outline . . . . Fl. of Linlithgow,' Trans. Wern. Soc., Edinb., 1807. 'On a Law of Vegetable Life,' Journ. Agric. i. (1829). R. S. C. ii. 636 ; vii. 676. Mem., with portr., prefixed to 'Lithology of Edinburgh,' 1859.

**Foot, Frederick James** (1831 ?–1867): b. Ireland, 1831 ? ; drowned Lough Kay, near Boyle, 17th Jan. 1867. M.A., T.C.D. F.R.G.S.I. Corresp. Memb. Dublin Nat. Hist. Soc., 1857. Assist. Geologist, Irish Geol. Survey, from 1856. 'Botany and Marine Zoology of Clare,' 'Occurrence of *Hymenophyllum tunbridgense* in Co. Longford,' &c., in Proc. Dublin Nat. Hist. Soc. R. S. C. ii. 253 ; Geol. Mag. 1867, 95.

**Forbes, Arthur** (1819–1879): b. Douglas, I. Man, 25th Jan. 1819 ; d. Aldershot, 16th March, 1879. Ninth laird of Culloden. Had a herbarium. Trans. Bot. Soc. Edin. xiv. 20.

**Forbes, James** (1773–1861): b. Bridgend, Perthshire, May, 1773 ; d. Woburn Abbey, 6th July, 1861. Gardener, Woburn Abbey. A.L.S., 1832. 'Hortus Woburnensis,' 1838, &c. Pritz. 109 ; Jacks. 415 ; Proc. Linn. Soc. 1861–2, civ.

**Forbes, Edward** (1815–1854): b. Douglas, Isle of Man, 12th Feb. 1815 ; d. Wardie, Edinburgh, 18th Nov. 1854 ; bur. Dean Cemetery. F.L.S., 1843. F.R.S., 1845. Prof. Botany, King's Coll., 1842. Regius Prof. Nat. Hist. Edin., 1854. Pres. Geol. Soc., 1853. Visited Norway, 1833 ; Alps, 1835 ; Carniola, 1838 ; Aegean, 1841–2. Contributed Manx List in 1833 or 1834 to Top. Bot. Trans. Bot. Soc. Edin., 1836–1841 ; Mag. Zool. Bot. iii. (1839), 236 ; iv. (1840), 307 ; Ann. & Mag. vii. (1841), 157. 'Travels in Lycia' [with Spratt], 1846. Jacks. 230 ; R. S. C. ii. 654 ; Memoir by Wilson & Geikie, 1861 ; Proc. Linn. Soc. ii. 408 ; Trans. Bot. Soc. Edin. v. 23 ; Gard. Chron. 1854, 771 ; Proc. Geol. Soc. 1855, xxvii. Bust by Steel in College Museum, Edinb. ; one by J. C. Lough, Jermyn St. Museum. Portr. in Ipswich Mus. series ; one in Geikie's 'Life of Murchison.'

**Forbes, John** (1800 ?–1823) : d. Senna, E. Africa, Aug. 1823 ; tablet in Chiswick Churchyard. A.L.S., 1822. Collector for R. Hort. Soc. in E. and S. Africa, Madagascar, &c. Lasègue, 376 ; Trans. Hort. Soc. viii., iii. Plants in Mus. Brit. *Forbesia* Eckl. = *Curculigo* Gærtn.

**Forby, Joseph** (fl. 1801). Clerk. Brother of the following. Discovered *Salix Forbyana* Sm. Eng. Bot. 1344.

**Forby, Robert** (1759 ?–1825) : b. Stoke Ferry, Norfolk, 1759 ? ; d. Fincham, Norfolk, 20th Dec. 1825. Of Barton. Clerk. Brother of preceding. B.A., Camb., 1781. M.A., 1784. F.L.S., 1798. Rector of Fincham, Norfolk, 1799. Eng. Bot. 1344. Memoir by Dawson Turner, with portr. litho. from painting by M. Sharp, 1822, prefixed to 'Vocabulary of East Anglia,' 1830. Gent. Mag. vol. 96 (1826), i. 281. *Salix Forbyana* Sm.

**Ford, John** (fl. 1789). M.D. F.L.S., 1789. Linn. Letters, ii. 47 ; Eng. Bot. 78.

**Fordyce, George** (1736–1802) : b. Aberdeen, 18th Nov. 1736 ; d. Essex St., Strand, 25th May, 1802 ; bur. St. Ann's, Soho. Grandfather of George Bentham. M.A., Aberd., 1750. M.D., Edin., 1758. F.R.S., 1776. F.R.C.P., 1787. 'Elements of Agriculture and Vegetation,' 1765. Gent. Mag., June, 1802 ; Munk, ii. 373. Portr. by Phillips at St. Thomas's Hospital, engr. by Keating. Medal designed by Flaxman.

**Forster, Benjamin Meggot** (1764–1829) : b. Walbrook, 16th Jan., 1764 ; d. Scotts, Hale End, Walthamstow, 8th March, 1829 ; bur. at Walthamstow. Sent plants for Eng. Bot. Fungologist, electrician, and philanthropist. Brother of the two following. '*Peziza cuticulosa*,' 1786. 'Introduction to the Knowledge of Fungusses,' 1820. 'Epistolarium Forsterianum,' ii. pp. xiii.–xv. Pritz. 110 ; Jacks. 546 ; Nich. Illustr. viii. 553 ; Gent. Mag. xcix. (1829), 279.

**Forster, Edward** (1765–1849) : b. Wood Street, Walthamstow, 12th Oct. 1765 ; d. Ivy House, Woodford, Essex, 23rd Feb. 1849 ; bur. at Walthamstow. F.L.S., 1800 ; Treas., 1816 ; V.-P., 1828. F.R.S. Banker. Contributed to Gough's Camden, Phyt. i., ii., iii. ; E. B. S. i., ii., iv. (2790) ; Mag. Zool. Bot. ii. (1839), 95 ; Ann. & Mag. viii. (1842), 433 ; &c. Herbarium purchased by R. Brown, and presented to Mus. Brit. Pritz. 110 ; Jacks. 262 ; R. S. C. ii. 669 ; Proc. Linn. Soc. ii. 39 ; Gent. Mag. xxxii. (1849), 431 ; Nich. Illust. viii. 554 ; Gibson, Fl. Essex, 448 ; 'Recueil de ma vie,' T. I. M. Forster, 1837 ; 'Epistolarium Forsterianum,' ii. p. xv. Oil portr. by Eddis, Linn. Soc. Portr. in Ipswich Mus. series. *Luzula Forsteri* DC. *Sedum Forsterianum* Sm.

**Forster, Thomas Furley** (1761–1825) : b. Bond Court, Walbrook, 5th Sept. 1761 ; d. Walthamstow, 28th Oct. 1825 ; bur. Walthamstow. F.L.S., 1800. Joint-author with two preceding of the plant-lists in Gough's 'Camden.' Additions to Warner's 'Plantæ Woodfordiensis,' 1784. 'Flora Tonbrigensis,' 1816. Discovered *Asplenium lanceolatum* and *Viola lactea*. Eng. Bot. 240, 445, &c.; Pritz. 110 ; Jacks. 547 ; Gent. Mag. xxxii. (1849), 481 ; R. S. C. ii. 671 ; 'Recueil de ma vie,' T. I. M.

Forster, 1837; 'Epistolarium Forsterianum,' i. 33–41; Memoir in Fl. Tonbrig. ed. 2, 1842; Nich. Illustr. viii. 553.

**Forster, Thomas Ignatius Maria** (1789–1860): b. London, 9th Nov., 1789; d. Brussels, 2nd Feb, 1860. M.B., Camb., 1818. F.L.S., 1811. Son of preceding. 'Liber rerum naturalium,' 1805. 'Index Fungorum,' 1819. Edited Fl. Tonbrigensis,' ed. 2, 1842. Jacks. 14, 547; R. S. C. ii. 670; 'Recueil de ma vie,' 1837; 'Epistolarium Forsterianum,' 1845–50; Proc. Linn. Soc. 1860, xxiii.; Mag. Nat. Hist. i. 64; 'Annual Register,' vol. 102 (1860), p. 440.

**Forsyth, J. S.** (fl. 1827). 'First Lines of Botany,' 1827. Pritz. 110; Jacks. 547.

**Forsyth, William** (1737–1804): b. Old Meldrum, Aberdeen, 1737; d. Kensington Gardens, 25th July, 1804. Gardener, under Philip Miller, 1763; at Syon; at Chelsea, 1771–1784; then at St. James' and Kensington Palaces. Pritz. 110; Cott. Gard. iv. 233; Journ. Hort. xxi. (1876), with portr.; Semple, 112; Felton, 186. Portr. in 'Treatise on Fruit-trees,' 1802. *Forsythia* Walt. = *Decumaria*. *Forsythia* Vahl.

**Forsyth, William** (1772?–1835): b. Chelsea? 1772?; d. Nottingham Place, N.W., 28th July, 1835; bur. Chelsea. Son of preceding. 'Botanical Nomenclator,' 1794. Correspondence in Cott. Gardener, vii. 350, &c. Pritz. (ed. 1), 88; Jacks. 14; Loud. Gard. Mag. xi. (1835), 496.

**Fortune, Robert** (1813–1880): b. Kelloe, Edrom, Berwick, 16th Sept. 1813; d. Brompton, 13th April, 1880. Bot. Gard. Edin.; Chiswick, 1842. Collector to R. H. S. in China, Java, &c., 1843–1845 and 1848–1850. Curator, Chelsea, 1846–1848. 'Wanderings in China,' 1847. Introduced Tea into India. Pritz. 110; Jacks. 380; Journ. Bot. 1880, 160; Trans. Bot. Soc. Edin. xiv. 161; R. S. C. ii. 672; Gard. Chron. 1880, i. 487; Cott. Gard. xix. 192; Lasègue, 436. *Fortunæa* Ldl. = *Platycarya*.

**Fothergill, John** (1712–1780): b. Carr End, Wensleydale, York, 8th March, 1712; d. Harper St., Red Lion Square, 26th Dec. 1780; bur. Winchmore Hill. M.D., Edin., 1736. L.R.C.P., 1744. F.R.S., 1753. Travelled in Flanders before 1740. Practised in Lombard St. from 1740. Had bot. gard. at Upton, West Ham, from 1762. Loud., 'Arboretum,' 71, 82; Jacks. 415; Life, by Sir John Elliott, 1781; by Dr. Gilb. Thompson, 1782; by Lettsom, 1783; by J. Hack Tuke, 1880; Nich. Anecd. ix. 737, with portr.; Cott. Gard. vii. 327; Munk, ii. 154; Friends' Books, i. 629; 'Memorials of Bartram and Marshall,' by W. Darlington, 495–515. Full-length portr. by Hogarth at R. C. P. Mezzot. by V. Green, after G. Stuart, 1781. Copy at Kew. Wedgwood medallion and bust. *Fothergilla* L.

**Foulkes, Robert** (fl. 1727): b. Llanfrothen, Merioneth, 1702? B.A., Oxon, 1725. M.D. Of Llanbeder, Denbigh. Recorded *Cucubalus baccifer* and contradicted it. Linn. Letters, ii. 171; Rich. Corr. 132.

**Fox** (or **Foxe**), **John** (fl. 1695). Surgeon. Sent plants to Petiver from Cape and Bengal. Mus. Petiv. 44, 80.

**Fox, Joseph** (fl. 1779-1804). Of Norwich. Weaver. Raised *Lycopodium* from spores. Linn. Trans. ii. 315 ; vii. 297.

**Fox, W. Tilbury** (1836-1879): b. 1836; d. Paris, 7th June, 1879. M.D. 'Chignon Fungus,' Journ. Bot. 1867 ; 1879, 224. Jacks. 165.

**Francis, George William** (d. 1865): d. South Australia, Aug. (?) 1865. F.L.S., 1839. Superintendent, Adelaide Bot. Gard. 'Grammar of Botany,' 1840. 'Little English Flora,' 1840. 'Analysis of British Ferns,' 1850. 'Field Flowers,' 1853. Pritz. 111 ; Jacks. 547 ; R. S. C. ii. 696 ; Mag. Nat. Hist. viii. 1835, 221 ; Gard. Chron. 1865, 1226 ; 'South Australia Register,' 10th Aug. 1865.

**Francis, Robert Bransby** (1768 ?-1850): b. 1768 ? ; d. East Carleton, Norfolk, 27th April, 1850. Clerk. M.A., Cambridge, 1794. F.L.S., 1798. Studied *Jungermanniæ*. Eng. Bot. 605, 2569. Proc. Linn. Soc. ii. 132. *Jungermannia Francisci* Sm.

**Frankland, Sir Thomas, Bart.** (1750-1831): b. Westminster, 1750; d. 4th Jan. 1831. Of Thirkleby, York. M.P. for Thirsk, 1774. M.A., Oxon, 1771. F.L.S., 1796. Drew Algæ at Scarboro'. Had Hudson's marine plants. Sent plants to Smith for Eng. Bot. Smith Lett. i. 450; ii. 167, &c.; Eng. Bot. 2340; Linn. Trans. x. 157. *Franklandia* R. Br.

**Franklyn, George** (fl. 1700). Apothecary. Sent plants to Petiver from Charlestown, Carolina. Mus. Petiv. 80 and no. 744 ; Herb. Sloane, clix. 182, 3.

**Franqueville, John de** (fl. 1629). Merchant in London. Had a bot. gard. Introduced *Rosa sulphurea*, 1629. Lobel, 'Adversaria' (1605), 486, 487 ; Parkinson. *Franquevillia* Gray (Salisbury MS.) = *Cicendia* Adans.

**Fraser, Charles** (1852 ?): d. Sydney ? Colonial botanist, New South Wales. Bot. Misc. i. 221 ; Pritz. 112 ; R. S. C. ii. 702; Lasègue, 498 ; Comp. Bot. Mag. ii. 230 ; Lindley, 'Swan River,' ii. Plants at Mus. Brit. and Kew. *Hakea Fraseri* Br.

**Fraser, John** (1750-1811): b. Tomnacloich, Inverness, 1750 ; d. Sloane Square, Chelsea, 26th April, 1811 ; bur. Old Burial-ground, Chelsea. Hosier and collector. F.L.S., 1810. Published Walter's 'Flora Caroliniana.' To Newfoundland, 1780 ; Southern States, 1785 ; and seven times across Atlantic between 1780 and 1810. Established nursery at Sloane Square, 1795. Collector to the Czar, 1798. Herbarium presented to Linn. Soc., 1849. Comp. Bot. Mag. ii. 300 ; Pritz. 112 ; Jacks. 122, 145; Lasègue, 199 ; Cott. Gard. viii. 250; Loudon, 'Arboretum,' 119 ; Faulkner, 'Chelsea,' ii. 41. Litho. portr. in Comp. Bot. Mag. *Frasera* Walt.

**Fraser, John** (fl. 1799-1852). F.L.S., 1810. A.L.S., 1848. Son of preceding. Accompanied his father in his travels, and subsequently to N. America. Introduced *Dahlia*. Had nursery at the Hermitage, Ramsgate, 1817-1835. Comp. Bot. Mag. ii. 302 ; Pritz. 112.

**Freeman, Charlotte** (fl. 1797-1809). Botanical artist. 'Select Specimens of British Plants' [with Juliana S. Strickland-

Freeman and Dr. George Shaw], **1797–1809**.   Pritz. **112**; Jacks. 233.

**Freeman, Joseph** (fl. 1860).   Of Stratford, Essex.   ' Flora of Stratford,' 1860.   ' Hints on . . . . describing . . . . species,' Proc. Bot. Soc. Lond. p. 28.   ' Pl. from Stratford,' *ibid.* p. 48.

**Freeman, Juliana Sabina Strickland-** (fl. 1797–1809). Botanical artist.   (*See* FREEMAN, CHARLOTTE.)

**French, Alfred** (1839–1879) :  b. Banbury, 1839 ;  d. London, 22nd Oct. 1879.   Journeyman baker at Banbury.   From 1874, attendant in Bot. Dep. Mus. Brit.   ' *Salvia pratensis*,' Journ. Bot. 1875.   Pl. in Mus. Brit.   Journ. Bot. 1879, 352 ; Druce, ' Fl. Oxfordsh.' 396, and preface.

**Frost, John** (1803–1840) :  b. London, 1803 ;  d. Berlin, 17th March, 1840.   F.L.S., 1825.   Knt. of Imperial Brazilian Order of Southern Star.   Lecturer at Royal and London Institutions, St. Thomas' and St. George's Hospitals.   Founder of the Medico-Botanical Soc.   ' Mustard Tree,' 1827.   ' Science of Botany, 1827.   Edited Bingley's ' Introduction,' 1831.   Pritz. 114 ; Jacks. 19, 37 ; R. S. C. ii. 736.

**Frost, Philip** (1804–1887) ;  b. Moreton Hampstead, Devon, 10th July, 1804 ;  d. 10th May, 1887.   Foreman, Chelsea, 1829. Interested in British Botany.   Gard. Chron. i. 117, and 1887, i. 649, with portraits.

**Furber, Robert** (fl. 1724–1732).   Nurseryman, of Kensington. ' Catalogue of . . . . trees,' &c., 1724.   ' The Flower-garden displayed, 1732.   Pritz. 115.

(To be continued.)

---

## SHORT NOTES.

NEW COUNTY RECORDS. — *Bromus erectus* Huds. is not recorded for Carnarvon (Vice-county 49) in Top. Bot., ed. 2.   My son, Wm. West, found it in August last upon the Great Orme's Head.   The plant grew on the dry turf on the N.E. side, but was not plentiful. — *Potentilla Tormentilla* Scop.   The only county for which this is not recorded is 73 (Kirkcudbright).   Mr. James McAndrew, of New Galloway, states that it is common there, in his ' List of the Flowering Plants of Dumfriesshire and Kirkcudbrightshire,' published in 1882.—WM. WEST.

I have to record *Vinca minor* and *Erysimum cheiranthoides* from Tintern, Monmouth, the former in the woods towards Chepstow, the latter near the Abbey.   At Symond's Yat, West Gloucester, *Epipactis latifolia* and *Habenaria chlorantha* are both common ; all these I collected in July last.—H. W. MONINGTON.

ADDITIONS TO THE FLORA OF WILTS.—The following additions have been made during the past season by Mr. E. J. Tatum, of Salisbury ; the Rev. W. Moyle Rogers, of Bournemouth ; and F. A. Rogers, a pupil of Marlborough College.   (The numbers refer to the districts into which the county is now divided) :— *Sagina subulata* Presl., 5, Hamptworth, *Tatum*.   *Rhamnus Frangula* Linn., 5, Hamptworth, *Tatum*.   *Rosa involuta* var. f. *Robertsoni*

Baker, 7, Alton Barnes, *F. A. Rogers*. *R. tomentosa* var. b. *sub-globosa* Sm., 4, Marlborough, *F. A. Rogers;* and 9, Semley, *Rogers*. *R. canina* var. j. *dumetorum* Thuill., 7, Alton Barnes, *F. A. Rogers;* var. o. *Andegavensis* Bast., 6, Westbury, *Tatum;* var. p. *verticill-acantha* Mérat, 7, Martinsell, *F. A. Rogers*. *R. stylosa* var. *pseudo-rusticana?*, 11, Hagler's Hole, *F. A. Rogers*. *Taraxacum officinale* var. d. *udum* Jord., 9, Semley, *Rogers*. *Erythræa pulchella* Fr., 5, Bentley Wood, *Tatum*. *Myosotis arvensis* var. b. *umbrosa* Bab., 10, Harnham, *Tatum*. *Galeopsis Tetrahit* var. b. *bifida* Boenn., 5, Whiteparish, *Tatum*. *Juncus supinus* var. c. *subverticillatus* Wulf., 5, Hamptworth, *Tatum*. *Glyceria plicata* var. b. *pedicellata* Towns., 9, Semley, *Rogers;* and 11, Hagler's Hole, *Rogers*. I have also much pleasure in stating that Mr. J. Horsefield, of Heytesbury, has found that *Carduus tuberosus* still flourishes at "Great Ridge," but not where it is usually asserted to grow, quite on the open down; I prefer not to give the locality more exactly. *Rosa pseudo-rusticana* is apparently an addition to the British Flora; it was so named by M. Crépin, to whom Mr. Rogers sent specimens. I am also indebted to Mr. Rogers for pointing out that a specimen in the Herb. Brit. Mus. named by him *Carex fulva*, is really " inland *C. distans*." —T. A. PRESTON.

BOTANICAL NOMENCLATURE. — I observe that incidentally the method pursued, or attempted to be pursued, in the 'Flora of North-east Ireland' has been drawn into the recent discussion on nomenclature in this Journal. On one side it has been claimed that, in the Flora alluded to, the practice that prevails in most branches of Natural History has been followed, while on the other hand it has been shown that this has been done inconsistently, and with apparent hesitation. As I alone am accountable for the execution of this part of the work in question, allow me to confess at once that the charge of inconsistency is deserved, as I am well aware that the system adopted was not carried out as completely as it should have been, and that the failure was more probably due to incompetence than to hesitation, due to the difficulty experienced in a provincial city of having access to original sources of information. I hold strongly that once a name has been legally imposed on a plant, no succeeding author should be permitted to change either the generic or the specific term at his own arbitrary will. If it becomes necessary to institute new genera, let this be done subject to the approval of the botanical world, but no right exists (save with well-known exceptions) by which an author may replace existing specific terms by others of his own selection. The inconvenience now experienced in the restoration of names legally imposed whould have been avoided had the two preceding generations of botanists exhibited less laxity in the matter of nomenclature, and there seems to be no better method of preventing the creation of superfluous names than the refusal to accept them. That changeless specific names are a vast advantage need not be argued; lépidopterists know the two letters *Io* indicate the "peacock butterfly," and that *Cardamines* stands for the "orange-tip" requires no explanation. With regard to the authority for the name, botanists surely will not consent to ignore the first describer of the species in favour of him who merely revises the work. That such a practice would be a trifle more convenient than stating the whole truth goes for nothing. The inconvenience only arises when the species is referred to in a formal manner, and when the names become as inviolable as the *Io* and *Egeria* of the lepidopterist, will seldom be incurred.—S. A. STEWART.

ARUM ITALICUM Mill. AND A. MACULATUM Linn.—I have read with much interest Mr. J. Cosmo Melvill's remarks on *Arum italicum* in the last number of the 'Journal, as it is a plant that I have been observing much since I discovered it at this place (Fursdon, Egg Buckland) some years ago. It would seem that Mr. Melvill, on May 15th last, found both this plant and its near ally, *A. maculatum*, in flower near Folkestone; but here, at Fursdon, all, or nearly all, the flowers of *A. maculatum* pass out of bloom before those of *A. italicum* open, and, as a rule, no doubt this difference prevails between the two,—a difference not favourable to the production of hybrids. In the consideration of another question touched on by Mr. Melvill,—a possibility of the two plants being extremes of one variable species, several differences would have to be dealt with; not the least important of which, as opposed to an identity of species, would be the different periods of growth belonging to them. The earliest leaves of *A. italicum* spring up and expand about the beginning of November, or even before October is over, whereas those of *A. maculatum* do not appear until February. It is remarkable that the species producing its leaves so long before the other should be the later to flower. The leaves of *A. italicum* and *A. maculatum* have, notwithstanding their succulent character, a remarkable power of withstanding frost; in earliest spring, those of *A. maculatum* appear as the greenest of the vegetation on many a hedgebank near Plymouth. I have seen in a wood in this neighbourhood the spathes of this species torn in a noticeable manner, apparently by some small quadruped or bird, for the purpose of extracting the flowers. White, in his 'Natural History of Selborne,' tells us of the thrush feeding on its tubers. Reverting to *A. italicum*, I would add, that a good time to search for it in England is the winter season, as any green Arum leaves found between October and the commencement of February would, according to my experience, mark *A. italicum*. When first I found the plant here at Fursdon, I quite expected to meet with it elsewhere in the country around, but, having failed to do so, now suspect that it has become wild in the immediate vicinity of this house through early cultivation in proximity to an ancient dwelling, the meadow below here, close to which the plant occurs, still bearing the name of Undertown.—T. R. ARCHER BRIGGS.

THE TWO VALERIANS.—Since the publication of the last number of this Journal, Mr. Arthur Bennett has drawn attention to a valuable communication on this subject in the 'Botanical Gazette' for 1849. This consists of an editorial article comprising translations of papers by Prof. Schlechtendal (Bot. Zeitung, 1847); these papers recount the then recent investigations and experiments made by Schlechtendal, Wenderoth, and others, with reference to the various allied species of *Valeriana*. The general conclusion seems to have been that they are best regarded as distinct species, and I am glad to find that several of my observations which I had considered among the most important, are quite in accord with those made forty years ago, and in another country. Thus Schlechtendal says that the fruit of *V. sambucifolia*, "though but a little larger, is more attenuated upwards"; while Dr. Wenderoth remarks, "I always recognise *V. sambucifolia*, even at a distance, by the direction of the lower and middle stem-leaves. This is occasioned by the almost erect position of the petiole," &c. The latter also states that, after cultivating the various plants under all kinds of different artificial conditions, and altering the treatment of plants raised from seed, they still retain their distinctive characters as to suckers, stolons, &c.—W. H. BEEBY.

VALERIANA MIKANII.—With reference to Mr. Beeby's valuable paper in the 'Journal' for November, I may say that this plant grows abundantly in a decidedly *moist* situation, in and about Burgate Wood, near Mellis, E. Suffolk, associated with several species common on chalky soil. Though not recorded for the vice-county in 'Topographical Botany,' I gather from Dr. Hind that it is more frequent thereabouts than *sambucifolia*. — EDWARD S. MARSHALL.

RUBUS THYRSIGER (Bab.).—I have been asked to point out in print at once the fact that this name cannot stand. I published it on page 226 of vol. xxiv. of this Journal (1886). Focke published that same name (in 1877) as given by himself and Banning to a "form" closely allied to *R. scaber*, of which "a solitary shrub" was found by Banning at Volmardingsen, in Westphalia, in company with *R. scaber*. (See Focke, Syn. 341). I have now learned, through the kindness of Mr. N. E. Brown, that my *R. thyrsiger* may really claim the name of *R. rhenanus* (Müll.), which was published in 'Flora,' 1858, p. 184. I possess the specimen Wirtgen, Rub. Rhen. 58 and 59, as *R. thyrsiflorus*), upon which Müller founded his species. He does not inform us of the edition of Wirtg. R. R. to which he refers, but the dates show that it is the first. They are apparently the same as my *R. thyrsiger*, but not derived from so strong a plant. They are very nearly allied to *R. thyrsiflorus*, and seem to connect that plant with *R. Bloxamii*. Those same numbers in Wirtgen's 2nd edition do not refer to the same plants, and Focke refers one of them (58) to *R. candicans*, a very different species.—C. C. BABINGTON.

GOODYERA REPENS IN YORKSHIRE.—On Aug. 6th the members of the Yorkshire Naturalists' Union paid a visit to Market Weighton, and, during our ramble through Houghton Wood, I found several specimens of *Goodyera repens*. It was found growing in what is believed to be virgin soil, of a damp, peaty nature, beneath some old Scotch firs. The suggestion has been thrown out that it was conveyed there at the time of the introduction of these, which are supposed to be eighty years old.—J. J. MARSHALL.

## NOTICES OF BOOKS.

*The Flowering Plants of Wilts, with sketches of the Physical Geography and Climate of the County.* By the Rev. T. A. PRESTON, M.A. Published by the Wiltshire Archæological and Natural History Society. 1888. Pp. lxix. 436, with map.

THE author of this book, so well known for his contributions to the observation of phenological phenomena at Marlborough, claims his preface some consideration from readers and others. Having made a promise to edit the work, he has, after having left the county, been " forced to adhere " to that promise, though long before he could "elaborate materials," so that instead of using the title of Flora, he has " adopted without regret the unpretentious title which it now bears." This is conscientious, but may cause his book to be classed with the older lists, &c., by those not reading it. But the work is more than this, and is a fitting precursor to a complete Flora, which one may hope the author will be spared to bring out as a second edition.

The book begins with a map of the county, small certainly, but giving a fair space of the adjoining counties,—an excellent idea; too often the maps show the county only, and blank space around. The geological and botanical divisions of the county are shown on it.

Following a modest preface is a well-written account of the topography, by the Rev. J. Sowerby, made interesting even to those who may never have visited the county, by its good descriptive style. " Geological features," " Drainage, &c." by Prof. Boulger ; " Climate," with a rainfall chart of the river basins ; " Botanical districts," " Plan of the Flora," " Works and Herbaria quoted or referred to," " List of Authorities quoted in the Flora," followed by " The Flowering Plants of Wilts " occupying pp. 1 to 361, but ending with the Grasses, the higher Cryptogams not being given ; "Addenda," consisting of a " Table of Distribution in Wilts," " Notes to different species," " On the Batrachian Ranunculi," with a plate of the upper leaves ; " *Carduus tuberosus*," " On the Orcheston Long Grass," " Comparison with adjacent counties," " Additions and Index."

It will perhaps from this be considered that the author's regret of incompleteness is somewhat too modest. In one matter the book has pre-eminence: that is, in showing the first flowering, mean time, and latest date—the result of actual and continued observations in the field.

The tables given under Climate are some of them of much interest, and the given instances of great variation in limited areas, which joined to the time of flowering of the plants in the *same* places (these, however, are not collated, but must be sought in the body of the work), give some interesting results, though here again, local influences would have to be taken account of, such as protection by hills, woods, aspect, &c., before we are prepared to dogmatize too much on local climatic effects.

The list of plants of the county follows out the usual sequence : —through eleven districts under the river basins, the localities being fully given ; those of Mr. T. B. Flower's 'Flora of Wiltshire,' in the 'Wilts Archæological and Natural History Magazine' being given in brackets,—a wise plan.

The flora is mostly the usual one of a southern county, having large surfaces of chalky downs, about two-thirds of the county being of the Cretaceous formation, a very small portion being Eocene and Lias, the rest belonging to the Oolites. As usual, the Batrachian Ranunculi are a difficulty, but the notes on them at pp. 386—393 ought to be read by all British botanists; the admission of *R. Lenormandi* is stated in the Addenda to have been an error.

After many of the species some interesting notes are appended, either from old authors or by Mr. Preston himself. Here and there a species is admitted that might well be expunged: such as *Sisymbrium Irio* (p. 29). *Thlaspi perfoliatum* is traced into Wilts on seemingly safe grounds, and is an interesting addition to the county. *Hypericum montanum* is afterwards noted as an error; it will be strange if it is really an absentee.

Prof. Babington is thanked for valuable assistance in drawing up the list of Rubi, and this as it stands is probably " the only correct list of the county Rubi" since the alterations in nomenclature, &c., have made old records very doubtful as to names. Are the records under " *R. rudis* Weihe " really the true plant, or *R. echinatus* Lind. ?

At page 134 " *Œnanthe peucedanifolia* Poll. " is admitted on the authority of Prof. Babington's " Flora Bathoniensis, 1834, supp. 1839 "; this is probably what would now be named *Lachenalii*. No one would be more ready to admit than Prof. Babington himself, that these plants were then not properly separated in Britain, as the " old series " of the ' Phytologist' would show.

The note on the county rarity, *Carduus tuberosus*, is of much interest, the plant being confined on present knowledge in Britain to Wilts. But I must confess that I am surprised at *C. acaulis* having been mistaken for it. By the kindness of the Rev. W. M. Rogers, I have had growing for some years a root from Mr. Wheeler's garden (originally from Great Ridge Wood), and it never occurred to me that at any stage of its growth it resembled *acaulis*; still it may be different when wild. The theory of its being a hybrid is also discussed.

Under *Primula vulgaris* a long note is given of its forms or hybrids. Mr. Preston claims *Mentha pratensis* Sole, for Wilts, although it has been put in Hants ere this; there is no record of it since Sole's time (1798). Under *Carex humilis*, p. 333, the first record is given as " Flora, 1873 "; I think a much earlier one will be found in the ' Phytologist.' The Wilts Carex, *C. tomentosa*, seems to be extinct, from the partial drainage of the water-meadows, although specimens were sent to the Record Club in 1872, " to give confirmatory evidence of its recent occurrence." In another edition, the note under this interesting British plant might well be extended.

There are very few unlikely species introduced, and when given some doubt or caution is expressed; hence the book is a trustworthy one. Much of course still remains to be done in confirming or refuting (a difficult matter) many records. Throughout the book references are given to ' Topographical Botany,' Syme's ' English Botany,' Flower's ' Flora,' Babington's ' Manual,' Hooker's ' Students' Flora,' and the ' London Catalogue.' Some additions have been made to the county Flora since its publication, i. e.,

*Viola lactea, Cerastium pumilum, Sagina subulata, Rhamnus Frangula, Erythræa pulchella,* and *Juncus Gerardi.* The type is good and clear, but the body of the book is unnecessarily extended by too wide space and margins.

ARTHUR BENNETT.

*The Species of Ficus of the Indo-Malayan and Chinese Countries.* Part II. Synæcia, Sycidium, Covellia, Eusyce, and Neomorphe. By GEORGE KING, M.B., F.R.S., &c., Superintendent of the Royal Botanic Garden, Calcutta. London: L. Reeve & Co. 1888. Annals of the Royal Botanic Garden, Calcutta, vol. i., part, 2, pp. 67–185, tt. 87–125.

THE first part of this fully-illustrated monograph of the Asiatic Figs was noticed in some detail in this Journal (1887, pp. 218–220), and the basis of Dr. King's classification is given in the same volume, page 189. There is little to add now, except congratulations on the completion of so important a work, and on the excellence of the plates, executed by native artists, for botanical purposes. More exact delineations of specimens it is impossible to have. Throughout this part, too, the dissections are given on the same plate as the representation of the species. Altogether there are descriptions of 207 species, and a notice accompanies this part to the effect that a supplement will follow containing descriptions and figures of the new species recently collected by Mr. H. O. Forbes in New Guinea. This supplement will also contain a memoir, by Dr. D. D. Cunningham, on the fertilization of *Ficus Roxburghii* Wall. A photograph of a tree of this species in full fruit in the Calcutta Garden is given as a frontispiece to this volume. It is one of a number which produce their fruit in large clusters on the trunk, and in this the fruit lies on the ground, heaped up, as it were, around the base of the trunk. In some other species, such as *F. hypogæa, F. grocarpa,* and *F. conglobata,* the long ropes of fruit insinuate themselves more or less into the ground,—some become quite subterranean,— and the fruit ripens in this situation. The very common *Ficus hispida,* better known under the name of *F. oppositifolia,* sometimes bears receptacles both in the axils of the leaves and in clusters on the trunk.

To give some idea of the difficulties of the genus *Ficus,* it may be mentioned that Dr. King has seven pages devoted to doubtful and imperfectly-known species, and in these seven pages he deals with upwards of 130 names. This number includes some, indeed many, that are definitely reduced to accepted species, and there seems no reason why they should have been included here. On the other hand, the majority have proved indeterminable to Dr. King, who has seen nearly all the existing types, and most of them will in all probability always remain so.          W. B. H.

NEW BOOKS.—S. E. MAURIN, 'Formulaire de l'Herboristerie' (Paris, Meau; 12mo, pp. 575: 4 fr.)—.P. BOERY, 'Les Plantes oléagineuses' (Paris, Baillière, "1889": 8vo, pp. 160: 2 fr.).—R. SULZBERGER, 'La Rose: histoire, botanique, culture' (Namur Charlier: 8vo, pp. 148: 10 plates, 20 maps).—E. LABORIE, 'Recherches sur l'anatomie des axes floraux' (Toulouse, Durtend: 8vo, pp. 198).—A. F. W. Schimper, 'Die epiphytische Vegetation Amerikas' (Jena, Fischer: 8vo, 162: 6 plates).—P. Sorauer, 'Die Schäden der einheimischen Kulturpflanzen' (Berlin, Parey: 8vo, pp. vii. 250).—W. Rattan, 'A Popular Californian Flora and Analytical Key to West Coast Botany' (8th ed., San Francisco: 8vo, pp. xxviii. 118, 128).—H. Semler, 'Tropische und nord-amerikanische Waldwirtschaft und Holzkunde' (Berlin, Parey: 8vo, pp. xvi. 736).

### ARTICLES IN JOURNALS.

*Bot. Centralblatt.* (No. 45). — A. Tomaschek, 'Ueber *Bacillus muralis*.'—(Nos. 46-48). A. Prazmowski, 'Ueber die Wurzelknöll-chen der Leguminosen.'

*Bot. Gazette* (Oct.). — E. L. Gregory, 'Development of cork-wings on certain trees.' — G. Vasey, 'Characteristic Vegetation of N. American Desert.'--W. H. Evans, 'Stem of *Ephedra*' (1 plate). L. N. Johnson, 'A Tramp in N. Carolina Mountains.'

*Botaniska Notiser* (häft. 5). — C. G. Westerlund, 'Några bidrag till Blekings flora.'--A. A. Lundström, 'Bidrag till Södermanlands Växtgeografi.' — G. Lagerheim, 'Mykologiska Bidrag' (*Urocystis Junci*, n. sp.).—R. Jungner, 'Om *Rumex crispus* L. × *Hippolapathum* Fr. (= *R. similatus* Hausskn.).'--T. M. Fries, 'Några anmärkningar om slagtet *Pilophorus*.' — O. Juel, 'Morfologiska undersökningar öfver *Koenigia islandica* L.'--T. M. Fries, 'Om *Stenanthus curviflorus* Lönnr.' — B. Kaalaas, 'Nogle nye Skandinaviske moser.' — H. Kleban, '*Peridermium Strobi*.'

*Bot. Zeitung* (Nov. 9).—T. W. Engelmann, 'Die Purpurbacterien und ihre Beziehungen zum Lichte.'—(Nov. 16). M. W. Beyerinck, 'Die Bacterien der Papilionaceenknöllchen.'

*Bull. Torrey Bot. Club* (Nov.).—N. L. Britton, 'The genus *Hicoria* of Rafinesque.'—E. L. Greene, '*Unifolium*.'—F. Roth, 'On the opening of stomata.'

*Flora* (Oct. 1, 11).—P. Teitz, 'Ueber definitive Fixirung der Blatt-stellung durch die Torsionswirkung der Leitstrange' (1 plate).—J. Schrodt, 'Beiträge zur Oeffnungs-Mechanik der Cycadeen-Antheren.'

*Gardeners' Chronicle* (Nov. 3).— *Cypripedium Elliottianum*, n. sp. H. N. Ellacombe, 'Plant-names a thousand years ago.'—*Cæsalpinia japonica* (fig. 73). — (Nov. 10). *Dendrophylax Fawcetti* Rolfe, n. sp. — (Nov. 17). P. Newberry, 'The early history of Vine-culture in England.'—*Decaschista ficifolia* Mast., n. sp.—(Nov. 24). *Cynoches versicolor* Rchb. f., n. sp.

*Journal de Botanique* (Nov. 1).—P. van Tieghem, 'Sur la limite du cylindre central et de l'écorce dans les Cryptogames vasculaires.' —A. Franchet, '*Lefrovia*, genre nouveau de Mutisiacées' (*O. rha-ponticoides*, sp. unica). — G. Macgret, 'Le Tissu Sécréteur des Aloès.' — P. A. Dangeard, 'La Sexualité chez quelques Algues inférieures' (*Corbierea*, gen. nov.) — (Nov. 15). P. Maury, 'Cy-péracées de l'Ecuador et de la Nouvelle Grenade' (*Cyperus flexibilis*, *C. Andreanus, Dichronema fasciata*, spp. nn.). — C. Savageau, 'Sur un cas de protoplasme intercellulaire.' — P. van Tieghem, 'Sur le dédoublement de l'endoderme dans le Cryptogames vasculaires.'— *Nevrophyllum viride* Patouillard, n. sp.

*Midland Naturalist* (Nov.). — T. P. Blunt, 'Life-History of a

Myxomycete.' — W. Mathews, 'History of County Botany of Worcester' (contd.). — W. B. Grove & J. E. Bagnall, 'Fungi of Warwickshire' (contd.).

*Nuovo Giornale Bot. Italiano* (Oct. 22).—G. Massalongo, 'Sulla germogliazione delle sporule nelle *Sphæropsideæ*.' — A. N. Berlese, 'Sopra due parassiti della Vite per la prima volta trovati in Italia.' —G. Gasperini, 'Il Leghbi o vino di Palma.' — A. Borzi, '*Eremothecium Cymbalariæ*, n. sp.' — L. Micheletti, 'Raccomendazioni intese ad ottonere che l'Italia abbia la sua Lichenografia.' — A. Batelli, 'Escursione al M. Terminillo.'—G. Arcangeli, 'Sul germogliamento della *Euryale ferox*.' — L. Macchiati, 'Xantofillidrina.'— A. Borzi, 'Xerotropismo nelle Felci.'

*Oesterr. Bot. Zeitschrift.* (Nov.). — B. Blocki, '*Rumex Kerneri* (*R. conferta* × *obtusifolius*).' — K. Vandas, 'Beiträge zur Kenntniss der Flora von Sud-Hercegovina' (contd.).—A. F. Entleutner, 'Anlangen in Meran.'— L. Simonkai, *Tunica Haynaldiana.*— M. Kronfeld, 'Bemerkungen über volksthümliche Pflanzennamen.' — E. Formánek, 'Zur Flora von Bosnien' (contd.: *Stachys Zepcensis, Scutellaria hercegovinica*, spp. nn.).

*Trans. Bot. Soc. Edinburgh* (xvii., pt. 2: Nov.). — G. W. Traill, 'Marine Algæ of Elie.' — E. Janczewski, 'Fruits of *Anemone*.'— Arthur Bennett, 'Additions to the Scottish Flora, 1887.' — W. Craig, 'Excursion of Scottish Alpine Botanical Club to Hardanger district of Norway, 1887.' — A. Gray & L. W. Hinxman, 'Flora of West Sutherland.' — W. Couts, 'Visit to Glenure.' — R. Lindsay, 'Heterophylly in Veronicas' (1 plate). — R. Christison, 'Annual Increase in Girth of Trees.' — P. Sewell, 'Colouring Matter of Leaves and Flowers.' — W. E. Fothergill, 'Leaves of Climbing Plants.'—Obituaries of Sir Walter Elliot, Asa Gray, and A. de Bary.

---

*LINNEAN SOCIETY OF LONDON.*

*Nov.* 1, 1888.—W. Carruthers, F.R.S., President, in the chair. The following were elected Fellows of the Society: William Overend Priestley, M.D., F.R.C.P., John Way, M.D., and John Evans, Esq. —Prof. Bower exhibited and made remarks upon some adventitious buds on a leaf of *Gnetum Gnemon.*—The Rev. R. Baron read a paper 'On the Flora of Madagascar,' in which he gave an interesting account of his explorations and collections in that island. In a second paper, entitled 'Further Contributions to the Flora of Madagascar,' Mr. J. G. Baker, F.R.S., described the principal novelties brought home by Mr. Baron, and paid a well-deserved tribute to his energy and ability as a botanical explorer.

*Nov.* 15.—Mr. J. W. Stroud was elected a Fellow.— On behalf of Mr. H. Bolus, F.R.S., Mr. J. G. Baker exhibited a specimen of *Eriospermum folioliferum*, a plant showing a very remarkable type of leaf-structure. It was figured by Andrews in his 'Botanists' Repository' in 1807, and lost sight of until recently re-found by Mr. Bolus in Namaqua-land. — A paper was read by Mr. B. D. Jackson, on behalf of Mr. H. Chichester Hart, 'On the Mountain-range of Plants in Ireland,' and was criticised by Mr. J. G. Baker, who gave an interesting sketch of the characteristics of the Irish flora.

# INDEX.

*For classified articles, see—County Records; Journals, Articles in; Obituary; Reviews. New genera and species published in this volume are distinguished by an asterisk.*

Acrostichum Eggersii,* 34 ; exsculptum,* 326 ; Hartii,* 371
Adiantum Faberi,* 225 ; Hosei,* 324
Alchemilla vulgaris, 311
Algæ, W. Indian Marine, 193, 237, 303, 331, 358
Allen's 'Characeæ of America' (rev.), 125
Allium vineale, monstrosity of, 219
Alsophila dubia,* 1 (t. 279)
Anneslia, 10
Arum italicum, 348, 378
Aspidium perakense,* 4 ; xiphophyllum, 227 ; Wattii,* 234
Asplenium Mactieri,* 3 ; lastreoides,* 227 ; capillipes,* 228 ; caruifolium,* 228
Aulacodiscus, abnormal forms of (t. 281), 97

Babington, C. C., on Nomenclature, 369 ; Rubus thyrsiger, 379
Bagnall, J. E., Braithwaite's Moss Flora (rev.), 349 ; Tripp's British Mosses (rev.), 350
Bailey, L. H., Carex notes from Brit. Mus., 32
Baker, J. G., Synopsis of Tillandsiæ, 12, 39, 79, 104, 137, 167 ; Selaginella angustiramea,* 26 ; St. Domingo Ferns, 33 ; Memoir of J. T. Boswell, 82 ; of John Smith, 102 ; Ferns of W. China, 225 ; Salix fragilis, 249 ; Bucks Rubi, 248 ; 'Handbook of Amaryllideæ' (rev.) 253 ; Ferns of W. Borneo, 323 ; Acrostichum Hartii, 371
Beddome, R. H., Ferns of Perak, 1 ; of Manipur, 234
Beeby, W. H., on Nomenclature, 35 ; on Potentilla reptans, 78 ; Nomenclature of Sparganium, 115 ; Callitriche polymorpha in Britain, 233 ; the Two Valerians, 340, 378

Bloomfield, E. N., Mosses of Suffolk, 69
Boswell, Dr., Memoir of, 82 ; his herbarium, 157
Botanical appointments, 288, 352
Botanists, Biographical Index of British and Irish, 50, 85, 111, 145, 180, 213, 244, 273, 307, 345, 372
Botryocytinus, 127
Boulger, G. S., Endosperm, 37 ; Index of Botanists, 50, 85, 111, 145, 180, 213, 244, 273, 307, 345, 372
Bovista, Revision of (t. 282), 129 ; fulva,* 136 ; obovata,* 134 ; olivacea,* 133 ; radicata,* 133
Braithwaite's Moss Flora (rev.), 349
Brebner, G., Gymnosporangium Juniperi, 218
Briggs, T. R. A., Pyrus latifolia, 236 ; Arum italicum, 378
Brit. Mus., Bot. Dept., report of for 1887, 250
Britten, J., Flora of N.E. Ireland (rev.), 283 ; 'Index Generum' (rev.), 316 ; Nomenclature of Nymphæa, 6 ; Index of Botanists, 50, 85, 111, 145, 180, 213, 244, 273, 307, 345, 372 ; Flora of Herts (rev.), 58 ; Memoir of Asa Gray, 161 ; Nomenclature, 257, 296
Britton, E. G., Ulota phyllantha, 282
Britton, N. L., on Nomenclature, 292
Brown, Robert, 285
Bunbury's Herbarium, 69

Callitriche polymorpha in Britain, 233
Camillea, 95
Carex stricta, 154 ; rigida, 154 ; pelia, 154 ; lagopina, 156 ; ovata, 321 ; nova,* 322 ; affinis, 322 ; concolor, 322 ; podocarpa, 322 ; pedata, 322

Carruthers, W., Report of Bot.Dept. Brit. Mus., for 1887, 250; Sowerby's Models of Fungi, 268
Castalia, 7
Cathisinia, 256
Cheilanthes patula, 225
Chinese Ferns, 225
Chondriopsis cnicophylla* & leptacremon* (t. 284), 333
Clarke, C. B., Root-pressure, 201
Clarke, W. A., Cerastium pumilum in Wilts, 248
Closteridium, 256
Colgan, N., Flora of Grand Tournalin, 90
Conringia, the name, 90
Corbierea, 383
COUNTY RECORDS:—
  Berks, 156
  Bucks, 248
  Cambridge, 79, 277
  Carnarvon, 376
  Cornwall, 56
  Derby, 329
  Devon, 156, 236, 378
  Dorset, 156, 312
  Glamorgan, 57
  Gloucester, 376
  Herts, 58
  Hants, 156, 283, 311
  Huntingdon, 277
  I. Wight, 219
  Kent, 219, 311, 348, 349
  Lancashire, 79
  Monmouth, 376
  Somerset, 26, 313
  Suffolk, 69, 184
  Surrey, 28, 79, 249, 277, 342
  Sussex, 27, 79, 277, 312
  Westmoreland, 27
  Wilts, 248, 376, 380
  Worcester, 312
  Yorkshire, 219, 379

Dactylococcopsis, 352
Davallia Hosei,* 323; oligophlebia,* 323
Day's Ferns of Perak, 1
Daydonia,* 11
De Bary's Bacteria (rev.), 61; memoir of, 65
DeCandolle, A., on Nomenclature, 289
Desmids of Maine, 339
Druce, G. C., on Scottish plants, 17, 116, 364; Nomenclature of Sparganium, 115; E. Kent Plants, 349

Durand's 'Index Generum' (rev.), 316

Eagle's Herbarium, 69
Eggers' St. Domingo Ferns, 33
Elymus arenarius, 312
Endosperm, 37
Erythroxylon, 96

Ferns from W. China, 225; of St. Domingo, 33; of Manipur, 234; of Perak, 1; from W. Borneo, 323
Ficus, Species of (rev.), 382
Flower, T. B., Botany of Steep Holmes, 26; Mentha pratensis, 89
Forbes, H. O., Polypodium Annabellæ,* 33
Fry, D., Glamorganshire Plants, 57; Helianthemum polifolium, 313
Fryer, A., on Potamogeton, 57; P. fluitans, 273; P. flabellatus, 297

Geldart, H. D., Vicia hybrida, 219
Goodyera repens in Yorkshire, 379
Gray, Asa, memoir and portrait, 161; Elements of Botany (rev.),92
Greene, E. L., on Nomenclature, 327
Grove, W. B., Pimina,* 206
Groves, H. & J., Characeæ of America (rev.), 125
Gymnogramme acuminata,* 326; Dayi,* 5 (t. 279); valleculata,* 325; vitality of spores of, 185
Gymnosporangium Juniperi, 218

Hanbury, F. J., Hieracia new to Britain, 204
Hansgirgia, 352
Hart's 'Flora of Howth' (rev.), 121
Helianthemum polifolium, 313
Hemsley, W. B., 'King's 'Species of Ficus' (rev.), 382
Henslow's 'Origin of Floral Structures' (rev.), 312
Hetley's 'Flowers of N. Zealand' (rev.), 157
Hieracia new to Britain, 204
Hieracium Langwellense,* 206; pollinarium,* 206; scoticum,* 206; tridentatum, 312
Hieronyma alchornioides, 96
Hillebrand's Hawaiian Flora (rev.), 122
Hydrothrix, 30

Irish Plants, 56, 72, 121, 282, 283
Ito, T., Ranzania, 302

Jackson, B. D., The name Conringia, 90; Plates of 'Astrolabe', 269
JOURNALS, ARTICLES IN :—
American Naturalist, 30, 223, 255, 319
Annals of Botany, 30, 125, 255, 319
Ann. Sciences Nat., 223, 255, 287.
Bot. Centralblatt, 30, 62, 94, 126, 157, 191, 223, 255, 287, 319, 352, 383
Bot. Gazette, 30, 62, 94, 126, 158, 191, 255, 287, 319, 383
Bot. Notiser, 94, 158, 191, 383
Bot. Zeitung, 30, 62, 94, 126, 191, 223, 255, 320, 352, 383.
Bull. Soc.. Bot. Belg., 126
Bull. Soc. Bot. France, 30, 94, 158, 223, 288, 320
Bull. Torrey Club, 32, 62, 94, 126, 158, 191, 223, 255, 288, 320, 352, 383
Flora, 30, 126, 191, 256, 383
Gardener's Chronicle, 30, 62, 95, 126, 158, 192, 223, 256, 288, 352, 320, 352, 383
Icones Plantarum, 29, 125, 223
Journal de Botanique, 30, 63, 95, 127, 158, 223, 256, 288, 320, 383
Journ. Linn. Soc., 31, 127, 223, 288
Journ. R. Microscop. Soc., 159, 223
Magyar Novenytani Lapok. 95, 159, 352
Midland Naturalist, 127, 192, 223, 256, 383
Notarisia, 63, 159, 256, 352.
Nuovo Giorn. Bot. Ital., 95, 159, 256, 384
Oesterr. Bot. Zeitschrift, 31, 63, 95, 127, 159, 192, 223, 256, 288, 320, 352, 384
Pharmaceutical Journal, 63
Proc. Bristol Nat. Society, 94
Scottish Naturalist, 159, 256, 352
Trans. Bot. Soc. Edinburgh, 384
Trans. Linn. Soc., 159
Trans. Perthshire Soc. Nat. Science, 93

King, B., Hants Plants, 283
King's (G.) Species of Ficus (rev.), 382

Lange's 'Nomenclator Floræ Danicæ' (rev.), 254
Lee's 'Vegetable Lamb,' 93.
Lees, F. A., Botanical Record Club Report (rev.), 91; Flora of West Yorkshire (rev.), 219
Lefrovia, 383
Lemna trisulca, Photolysis in, 353 (t. 285).
Lindsayia indurata,* 324
Linnean Society, 31, 98, 127, 160, 192, 203, 382; index to Journal of, 286.
Linton, E. F., Carex trinervis in Ireland, 56
Linton, W. R., South Derbyshire Plants, 329
Lomaria deflexa,* 226
Lophopyxis, 29
Lygodium gracile,* 35

Madagascar Mosses, 263
Mahoæ, 123
Marshall, E. S., Westmoreland and Sussex Plants, 27; Cornish Plants, 56; Highland Plants, 149; Suffolk Plants, 184; Pulmonaria, 184; E. Kent Plants, 311; Valeriana Mikanii, 379
Marshall, J.J., Goodyera repens, 379
Massee, G., Revision of Bovista, (t. 282), 129
Masters, M. T., A heterodox Onion, 219
Matula, 159
Megaphyllæa, 29
Melvill, J. C., Species of Chondriopsis, 333; Arum italicum, 348
Mentha pratensis, 89
Microcoryne, 63
Monington, H. W., Alchemilla vulgaris in Kent, 311; New County Records, 376
Moore, S., Photolysis in Lemna trisulca (t. 285), 353
Mosses of Suffolk, 69; of Madagascar, 263
Mueller, Baron F., Selaginella angustiramea,* 26; 'Iconography of Acacia,' 29
Murray, G. R. M., Marine Algæ of West Indian Region, 193, 303, 331, 358; Memoir of De Bary, 65; on De Bary's Bacteria (rev.), 61
Murray, R. P., Botany of N. Portugal, 173
Myles, P. W., 'Origin of Floral Structures' (rev.), 313

Nephrodium melanorachis,* 325 ; myriolepis,* 34 : unifurcatum,* 228 ; pteropodina,* 325 ; simulans,* 325

New Books, 29, 190, 254, 382

New Phanerogams published in 1887, 115, 186

Nomenclature, 6, 35, 115, 157, 257, 289, 326, 369, 377

Nototrichium, 124

Nouelia, 127

Nymphæa, Nomenclature of, 6

OBITUARY :—
    J. T. I. Boswell, 82
    William Curnow, 128
    H. A. De Bary, 65
    Alexander Dickson, 63
    John Goldie, 299
    John Price, 32
    John Smith, 102

Orestia, 31

Perak and Penang Ferns, 1

Petrocosmea, 29

Photolysis in Lemna trisulca (t. 285), 353

Phycocelis, 63

Pimina* parasitica,* 206

Plants of the Expedition of the 'Astrolabe,' 269

Polydragma, 29

Polygala austriaca in Surrey, 249

Polygonum maritimum, 311

Polypodium alcicorne,* 229 ; Annabellæ,* 33 (t. 280) ; asterolepis,* 230 ; braineoides,* 229 ; deltoideum,* 230 ; gymnogrammoides,* 229 ; holophyllum,* 325 ; manipoorense,* 235 ; omeiense, 229 : stenopterum,* 229

Portugal, Botany of, 173

Potamogeton, Leaf-bearing stipules in, 57 ; fluitans, 274 ; flabellatus, 297

Potentilla reptans and allies, 78

Preston, T. A., Wilts Plants, 376 ; Flora of Wilts (rev.), 380

Prototremella, 288

Pryor's Flora of Herts (rev.), 58.

Pteris deltodon,* 227 ; furcans,* 324 ; Walkeri,* 374

Pterotropia, 123

Pyrus latifolia, 236

Ranzania* japonica,* 302

Rattray, J., Abnormal forms of Aulacodiscus (t. 281), 97

REVIEWS :—
    Manual of Orchidaceous Plants, 29, 222
    Flora of Hertfordshire. By R. A. Pryor, 58
    Lectures on Bacteria. By A. de Bary, 61
    Botanical Record Club Reports. By F. A. Lees, 91
    Elements of Botany. By Asa Gray, 92
    Flora of Howth. By H. C. Hart, 122
    Flora of the Hawaiian Islands. By W. Hillebrand, 122
    A School Flora. By W. M. Watts, 124
    Characeæ of America. B. T. F. Allen, 125
    Flowers of New Zealand. By Mrs. Hetley, 157.
    Flora of West Yorkshire. By F. A. Lees, 219
    Handbook of Amaryllideæ. By J. G. Baker, 253
    Nomenclator Floræ Danicæ, By J. Lange, 254
    Flora of North-east of Ireland. By S. A. Stewart, 283
    Index to Linnean Journal, &c. By B. D. Jackson, 286
    Origin of Floral Structures. By G. Henslow, 313
    Index Generum Phanerogamarum. By T. Durand, 316
    British Moss-Flora. By R. Braithwaite, 349
    British Mosses. By F. E. Tripp, 350
    Flowering Plants of Wilts. By T. A. Preston, 380
    Species of Ficus of Indo-Malayan and Chinese Countries. By G. King, 382

Rogers, W. M., Rubi records, 156 ; Polygonum maritimum, 311 ; Elymus arenarius, 312

Rolfe, R. A., 'Manual of Orchidaceous Plants' (rev.), 28, 222

Root-pressure, 201

Roper, F. C. S., Rumex maritimus, 312

Rosa mollis var. glabrata, 67 ; coriifolia var. Lintoni, 68

Rubi Records, 156, 248

Rubus lusitanicus,* 178 ; thyrsiger, 374

Rumex maritimus, 312

Salix fragilis, 196, 249; S. Russelliana and S. viridis, 196
Scheutz, N. J., de duabus Rosis britannicis, 67
Schizostege, 124
Scortechinia, 29
Scottish Plants, 17, 68, 69, 79, 116, 149, 364, 376
Scully, R., Kerry Plants, 71
Selaginella angustiramea,* 26
Sharland, A., Vitality of Spores of Gymnogramma, 185
Sindechites, 223
Smith, W. G., Sowerby's Models of British Fungi, 231, 268
Sparganium, Nomenclature of, 115, 157
Sphyranthera, 29
Stewart's ' Flora of N. E. Ireland' (rev.), 283; on Botanical Nomenclature, 377
Stricklandia, 253
Suffolk Mosses, 69

Thamniastrum, 256

Tillandsieæ, Synopsis of, 12, 39, 79, 104, 137, 167; index of species, 169
Towndrow, R. F., Hieracium tridentatum, 312
Tripp's British Mosses (rev.), 350

Ulota phyllantha at Killarney, 282
Utricularia, crackling sound of, 28

Valeriana Mikanii and V. sambucifolia, 340, 379
Vicia hybrida, 219

Watts's School Flora (rev.), 124
West, W., Notice of W. Curnow, 128; Desmids of Maine, 339; New County Records, 376
West Indian Marine Algæ, 193
White, F. B., Salix fragilis, S. Russelliana, and S. viridis, 196
Whitwell, W., Polygala austriaca in Surrey, 249
Wright, C. H., Madagascar Mosses, 263

ERRATA.

Page 69, line 9 from top, *for* " Charles Moss's," *read* " Charles's Moss.'

85,      6 and 7, delete sentence beginning " Found *Trichomanes.*"

85,      29, delete sentence beginning " Second edition."

121,      5 from bottom, *for* " Hunt," *read* " Hart."

244, omit lines 23 and 24 from bottom.

245, line 2 from bottom. *for* " Warrington," *read* " Liverpool."

309, delete entry of " Donovan, Edward."

WEST, NEWMAN AND CO., PRINTERS, HATTON GARDEN, LONDON, E.C.

*Title-page & Index for Vol. XXV. are issued with this number.*
**SUBSCRIPTIONS ARE NOW DUE.**

No. 301.      JANUARY, 1888.      Vol. XXVI.

# THE
# JOURNAL OF BOTANY

## BRITISH AND FOREIGN.

EDITED BY

## JAMES BRITTEN, F.L.S.,

BRITISH MUSEUM (NATURAL HISTORY), SOUTH KENSINGTON.

## CONTENTS.

PAGE

Ferns collected in Perak and Penang by Mr. J. Day. By Col. R. H. BEDDOME, F.L.S. (Plate 279)..   1

The Nomenclature of *Nymphæa*, &c. By JAMES BRITTEN, F.L.S.   ..   6

A Synopsis of *Tillandsieæ*. By J. G. BAKER, F.R.S., F.L.S. (Continued)   ..   ..   ..   ..   12

Notes on the Flora of Easterness, Elgin, Banff, and West Ross. By G. CLARIDGE DRUCE ..   ..   17

On a New *Selaginella* from New Guinea. By Baron VON MUELLER and J. G. BAKER   ..   ..   26

SHORT NOTES.—Botany of the Steep Holmes. — *Hieracium Gibsoni*

PAGE

Backh. and *Carex irrigua* Hoppe in Westmoreland. — *Equisetum sylvaticum* L., var. *capillare* Hoffm. in W. Sussex. — Crackling sound of *Utricularia*   ..   26

NOTICES OF BOOKS :—

A Manual of Orchidaceous Plants cultivated under Glass in Great Britain   ..   ..   ..   ..   28

Short Notices   ..   ..   ..   ..   29

New Books ..   ..   ..   ..   ..   29

Articles in Journals   ..   ..   ..   30

Linnean Society of London   ..   ..   31

Obituary ..   ..   ..   ..   ..   32

LONDON:

WEST, NEWMAN & CO., 54, HATTON GARDEN, E.C.

DULAU & CO., SOHO SQUARE.

*Price One Shilling and Threepence.*

# NOTICE.

The JOURNAL OF BOTANY is printed and published by WEST, NEWMAN & Co., 54, Hatton Garden, London, E.C., to whom Subscriptions for 1887 (in advance, Twelve Shillings; if not paid in advance, chargeable at the rate of 1s. 3d. per number) should be paid. Post-Office Orders may be drawn on the Hatton Garden Office.

The Volume for 1887 (price 16s. 6d., bound in cloth) is nearly ready; also covers for the Volume (price 1s. 2d. post free).

The Volumes for 1884, 1885, and 1886 can still be had.

For Volumes and back numbers for 1872—82 application should be made to DULAU & Co., Soho Square, W.

The Editor will be glad to send the JOURNAL OF BOTANY in exchange for other Journals of a similar character. Such Journals, Books for review, and Communications intended for publication, to be addressed to JAMES BRITTEN, Esq., 18, West Square, Southwark, S.E.

He will be greatly obliged to the Secretaries of Local Natural History Societies if they will forward him copies of their Transactions, so that any paper of botanical interest may be recorded in this Journal.

## TERMS FOR ADVERTISEMENTS.

| | | | |
|---|---|---|---|
| Whole page | £2 0 0 | Quarter page | 0 15 0 |
| Half page | 1 5 0 | Six lines and under | 0 5 0 |

Every additional line, 9d.

A liberal reduction is made for a series.

To be sent to WEST, NEWMAN & Co., 54, Hatton Garden, not later than the 24th of each month.

## AUTHOR'S SEPARATE COPIES.

Authors who require copies of their articles are requested to order separate copies from the Publishers, and to notify this and state the number required at head of their MS.; otherwise the type may be distributed before the order is received. By the buying of additional copies of the Journal sets are destroyed. The charges for separate copies are as under:—

| 2 pages | 25 copies | 4s. | 4 pages | 25 copies | 5s. | 8 pages | 25 copies | 8s. 0d. |
|---|---|---|---|---|---|---|---|---|
| ,, | 50 ,, | 5s. | ,, | 50 ,, | 6s. | ,, | 50 ,, | 9s. 0d. |
| ,, | 100 ,, | 7s. | ,, | 100 ,, | 8s. | ,, | 100 ,, | 10s. 6d. |

A greater number of pages to be charged in equal proportion. Separate Titles, Wrappers, &c., extra.

Annual Subscription, paid in advance, Twelve Shillings, post free within the Postal Union. Single Numbrs, 1s. 3d.

No. 302. FEBRUARY, 1888. Vol. XXVI.

# THE
# JOURNAL OF BOTANY

## BRITISH AND FOREIGN.

EDITED BY

## JAMES BRITTEN, F.L.S.,

BRITISH MUSEUM (NATURAL HISTORY), SOUTH KENSINGTON.

## CONTENTS.

PAGE

A new Fern from New Guinea. By H. O. FORBES, F.R.G.S., A.L.S. (Plate 280) .. .. .. .. .. 33

On a Collection of Ferns made by Baron Eggers in St. Domingo. By J. G. BAKER, F.R.S. .. .. 33

On Nomenclature. By W. H. BEEBY, A.L.S. .. .. .. .. 35

"Endosperm." By G. S. BOULGER, F.L.S. .. .. .. .. .. .. 37

A Synopsis of *Tillandsieæ*. By J. G. BAKER, F.R.S., F.L.S. (Continued) .. .. .. .. .. 39

Biographical Index of British and Irish Botanists. By JAMES

PAGE

BRITTEN, F.L.S., and G. S. BOULGER, F.L.S. .. .. .. .. 50

SHORT NOTES. — West Cornish Plants. — *Carex trinervis* Degl. in Ireland. — Glamorganshire Plants. — On Leaf-bearing Stipules in *Potamogeton* .. .. 56

NOTICES OF BOOKS :—

A Flora of Hertfordshire. By the late ALFRED REGINALD PRYOR, B.A., F.L.S. .. .. .. .. 58

Lectures on Bacteria. By A. DE BARY .. .. .. .. .. .. 61

Articles in Journals .. .. . 62

Obituary .. .. .. .. .. .. 63

LONDON:

WEST, NEWMAN & CO., 54, HATTON GARDEN, E.C.

DULAU & CO., SOHO SQUARE.

*Price One Shilling and Threepence.*

# NOTICE.

The JOURNAL OF BOTANY is printed and published by WEST, NEWMAN & Co., 54, Hatton Garden, London, E.C., to whom Subscriptions for 1888 (in advance, Twelve Shillings; if not paid in advance, chargeable at the rate of 1s. 3d. per number) should be paid. Post-Office Orders may be drawn on the Hatton Garden Office.

The Volume for 1887 (price 16s. 6d., bound in cloth) is now ready; also covers for the Volume (price 1s. 2d. post free).

The Volumes for 1884, 1885, and 1886 can still be had.

For Volumes and back numbers for 1872—82 application should be made to DULAU & Co., Soho Square, W.

The Editor will be glad to send the JOURNAL OF BOTANY in exchange for other Journals of a similar character. Such Journals, Books for review, and Communications intended for publication, to be addressed to JAMES BRITTEN, Esq., 18, West Square, Southwark, S.E.

He will be greatly obliged to the Secretaries of Local Natural History Societies if they will forward him copies of their Transactions, so that any paper of botanical interest may be recorded in this Journal.

## TERMS FOR ADVERTISEMENTS.

| | | | | |
|---|---|---|---|---|
| Whole page | £2 0 0 | | Quarter page | 0 15 0 |
| Half page | 1 5 0 | | Six lines and under | 0 5 0 |

Every additional line, 9d.

A liberal reduction is made for a series.

To be sent to WEST, NEWMAN & Co., 54, Hatton Garden, not later than the 24th of each month.

## AUTHOR'S SEPARATE COPIES.

Authors who require copies of their articles are requested to order separate copies from the Publishers, and to notify this and state the number required at head of their MS.; otherwise the type may be distributed before the order is received. By the buying of additional copies of the Journal sets are destroyed. The charges for separate copies are as under :—

| 2 pages | 25 copies | 4s. | 4 pages | 25 copies | 5s. | 8 pages | 25 copies | 8s. 0d. |
|---|---|---|---|---|---|---|---|---|
| ,, | 50 ,, | 5s. | ,, | 50 ,, | 6s. | ,, | 50 ,, | 9s. 0d. |
| ,, | 100 ,, | 7s. | ,, | 100 ,, | 8s. | ,, | 100 ,, | 10s. 6d. |

A greater number of pages to be charged in equal proportion. Separate Titles, Wrappers, &c., extra.

Annual Subscription, paid in advance, Twelve Shillings, post free
within the Postal Union.   Single Numbrs, 1s. 3d.

No. 303.  MARCH, 1888.  Vol. XXVI.

# THE
# JOURNAL OF BOTANY

## BRITISH AND FOREIGN.

EDITED BY

## JAMES BRITTEN, F. L. S.,

BRITISH MUSEUM (NATURAL HISTORY), SOUTH KENSINGTON.

## CONTENTS.

PAGE

Heinrich Anton De Bary.  By GEO. MURRAY  .. .. .. .. .. 65

De Duabus Rosis Britannicis scripsit N. J. SCHEUTZ  .. .. .. 67

The Moss Flora of Suffolk.  By the Rev. E. N. BLOOMFIELD, M.A. ... 69

Notes on some Kerry Plants.  By REGINALD SCULLY  .. .. .. 71

On *Potentilla reptans* and its Allies. By W. H. BEEBY, A.L.S. .. .. 78

A Synopsis of *Tillandsieæ*.  By J. G. BAKER, F.R.S., F.L.S.  (Continued)  .. .. .. .. .. 79

The late Dr. Boswell.  By J. G. BAKER, F.R.S., F.L.S.  .. .. 82

Biographical Index of British and Irish Botanists.  By JAMES

PAGE

BRITTEN, F.L.S., and G. S. BOULGER, F.L.S.  (Continued)  .. 85

SHORT NOTES. — Note on *Mentha pratensis* Sole. — The Summit Flora of the Grand Tournalin. —The Name *Conringia* .. .. 89

NOTICES OF BOOKS :—

The Botanical Record Club : Phanerogamic and Cryptogamic. Report for the years 1884, 1885, 1886, by the Editor, F. ARNOLD LEES, M.R.C.S. .. .. 91

The Elements of Botany for beginners and for schools.  By ASA GRAY .. .. .. .. .. 92

Short Notices  .. .. .. .. 93

Articles in Journals .. .. .. 94

Linnean Society of London .. .. 95

LONDON:

WEST, NEWMAN & CO., 54, HATTON GARDEN, E.C.
DULAU & CO., SOHO SQUARE.

*Price One Shilling and Threepence.*

# NOTICE.

The JOURNAL OF BOTANY is printed and published by West, Newman & Co., 54, Hatton Garden, London, E.C., to whom Subscriptions for 1888 (in advance, Twelve Shillings; if not paid in advance, chargeable at the rate of 1s. 3d. per number) should be paid. Post-Office Orders may be drawn on the Hatton Garden Office.

The Volume for 1887 (price 16s. 6d., bound in cloth) is now ready; also covers for the Volume (price 1s. 2d. post free).

The Volumes for 1884, 1885, and 1886 can still be had.

For Volumes and back numbers for 1872—82 application should be made to Dulau & Co., Soho Square, W.

The Editor will be glad to send the Journal of Botany in exchange for other Journals of a similar character. Such Journals, Books for review, and Communications intended for publication, to be addressed to James Britten, Esq., 18, West Square, Southwark, S.E.

He will be greatly obliged to the Secretaries of Local Natural History Societies if they will forward him copies of their Transactions, so that any paper of botanical interest may be recorded in this Journal.

## TERMS FOR ADVERTISEMENTS.

## AUTHOR'S SEPARATE COPIES.

Authors who require copies of their articles are requested to order separate copies from the Publishers, and to notify this and state the number required at head of their MS.; otherwise the type may be distributed before the order is received. By the buying of additional copies of the Journal sets are destroyed. The charges for separate copies are as under :—

| 2 pages | 25 copies | 4s. | 4 pages | 25 copies | 5s. | 8 pages | 25 copies | 8s. 0d. |
|---|---|---|---|---|---|---|---|---|
| ,, | 50 ,, | 5s. | ,, | 50 ,, | 6s. | ,, | 50 ,, | 9s. 0d. |
| ,, | 100 ,, | 7s. | ,, | 100 ,, | 8s. | ,, | 100 ,, | 10s. 6d. |

A greater number of pages to be charged in equal proportion. Separate Titles, Wrappers, &c., extra.

---

LONDON : WEST NEWMAN AND CO., 54, HATTON GARDEN, E.C

Annual Subscription, paid in advance, Twelve Shillings, post free within the Postal Union. Single Numbers, 1s. 3d.

No. 304.   APRIL, 1888.   Vol. XXVI.

# THE
# JOURNAL OF BOTANY

## BRITISH AND FOREIGN.

EDITED BY

## JAMES BRITTEN, F.L.S.,

BRITISH MUSEUM (NATURAL HISTORY), SOUTH KENSINGTON.

## CONTENTS.

PAGE

Notes on some abnormal forms of *Aulacodiscus* Ehrb. By JOHN RATTRAY, M.A., B.Sc., F.R.S.E. (Plate 281) .. .. .. .. .. 97

The late John Smith, A.L.S. By J. G. BAKER, F.R.S., F.L.S. .. 102

A Synopsis of *Tillandsieæ*. By J. G. BAKER, F.R.S., F.L.S. (Continued) .. .. .. .. .. 104

Biographical Index of British and Irish Botanists. By JAMES BRITTEN, F.L.S., and G. S. BOULGER, F.L.S. (Continued) .. 111

SHORT NOTES.—The Nomenclature of *Sparganium*. — Notes on the Flora of Easterness, Banff, Elgin, and West Ross .. .. .. .. 115

New Phanerogams published in Britain in 1887 .. .. .. .. 116

NOTICES OF BOOKS :—
The Flora of Howth. With map and an introduction on the

PAGE

Geology and other features of the promontory. By H. C. HUNT, B.A., F.L.S. .. .. .. 121

Flora of the Hawaiian Islands : a description of their Phanerogams and Vascular Cryptogams. By WILLIAM HILLEBRAND, M.D. Annotated and published after the author's death by W. F. HILLEBRAND .. .. .. .. 122

A School Flora for the use of Elementary Botanical Classes. By W. MARSHALL WATTS, D.Sc. 124

The *Characeæ* of America. Part I. By Dr. T. F. ALLEN .. .. 125

Short Notice .. .. .. .. 125

Articles in Journals .. .. .. 125

Linnean Society of London .. .. 127

Obituary .. .. .. .. .. 128

LONDON :

WEST, NEWMAN & CO., 54, HATTON GARDEN, E.C.
DULAU & CO., SOHO SQUARE.

*Price One Shilling and Threepence.*

# NOTICE.

The JOURNAL OF BOTANY is printed and published by WEST, NEWMAN & Co., 54, Hatton Garden, London, E.C., to whom Subscriptions for 1888 (in advance, Twelve Shillings; if not paid in advance, chargeable at the rate of 1s. 3d. per number) should be paid. Post-Office Orders may be drawn on the Hatton Garden Office.

The Volume for 1887 (price 16s. 6d., bound in cloth) is now ready; also covers for the Volume (price 1s. 2d. post free).

The Volumes for 1884, 1885, and 1886 can still be had.

For Volumes and back numbers for 1872—82 application should be made to DULAU & Co., Soho Square, W.

The Editor will be glad to send the JOURNAL OF BOTANY in exchange for other Journals of a similar character. Such Journals, Books for review, and Communications intended for publication, to be addressed to JAMES BRITTEN, Esq., 18, West Square, Southwark, S.E.

He will be greatly obliged to the Secretaries of Local Natural History Societies if they will forward him copies of their Transactions, so that any paper of botanical interest may be recorded in this Journal.

## BOTANICAL DRYING PAPER

### FOR DRYING FLOWERING PLANTS, FERNS, AND SEA-WEEDS.

Preserves form and colour in the best possible manner. Used by the Naturalists on board the Arctic ships, and also on the cruise of H.M.S. 'Challenger.'

| 16 in. by 10 | when folded, | 15s. per ream, | 1s. 1d. per quire. |
| 18 „ 11 | „ | 19s. „ | 1s. 4d. „ |
| 20 „ 12 | „ | 23s. „ | 1s. 9d. „ |
| 20 „ 16 | „ | 30s. „ | 2s. 2d. „ |

### ☞ EXTRA THICK DRYING PAPER.

*Made in response to a demand for a paper such as is used in the American Herbaria. Of the same quality as the ordinary paper, but more than four times the thickness; only one sheet is required between specimens. Extremely durable.*

**Price** (only one size), 18 by 22 in., flat, **£4** per ream; **5s.** per quire.

☞ This Paper is too heavy to send by Post.

WEST, NEWMAN & CO., 54, HATTON GARDEN, LONDON.

Annual Subscription, paid in advance, Twelve Shillings, post within the Postal Union. Single Numbers, 1s. 3d.

No. 305.    MAY, 1

THE

# JOURNAL OF BOTAN

## BRITISH AND FOREIGN.

EDITED BY

## JAMES BRITTEN, F.L.S.,

BRITISH MUSEUM (NATURAL HISTORY), SOUTH KENSINGTON.

## CONTENTS.

| | PAGE |
|---|---|
| A Revision of the Genus *Bovista* (Dill.) Fr. By GEORGE MASSEE. (Plate 282) .. .. .. .. .. | 129 |
| A Synopsis of *Tillandsieæ*. By J. G. BAKER, F.R.S., F.L.S. (Continued) .. .. .. .. .. | 137 |
| Biographical Index of British and Irish Botanists. By JAMES BRITTEN, F.L.S., and G. S. BOULGER, F.L.S. (Continued) .. | 145 |
| Notes on Highland Plants. By the Rev. E. S. MARSHALL, M.A., F.L.S.. .. .. .. .. .. | 149 |

| | PA |
|---|---|
| SHORT NOTES. — Some new Rubi Records for 1887. — *Carex lagopina* Wahlenberg.—The Nomenclature of *Sparganium.* — Dr. Boswell's Herbarium .. .. | 1 |
| NOTICES OF BOOKS:— | |
| The Native Flowers of New Zealand illustrated in Colours. By Mrs. CHARLES HETLEY. Part I. .. .. .. .. .. | 1? |
| Articles in Journals .. .. .. | 1 |
| Linnean Society of London .. .. | 1? |

LONDON:

WEST, NEWMAN & CO., 54, HATTON GARDEN, E.C.

DULAU & CO., SOHO SQUARE.

*Price One Shilling and Threepence.*

# NOTICE.

The JOURNAL OF BOTANY is printed and published by WEST, NEWMAN & Co., 54, Hatton Garden, London, E.C., to whom Subscriptions for 1888 (in advance, Twelve Shillings; if not paid in advance, chargeable at the rate of 1s. 3d. per number) should be paid. Post-Office Orders may be drawn on the Hatton Garden Office.

The Volume for 1887 (price 16s. 6d., bound in cloth) is now ready; also covers for the Volume (price 1s. 2d. post free).

The Volumes for 1884, 1885, and 1886 can still be had.

For Volumes and back numbers for 1872—82 application should be made to DULAU & Co., Soho Square, W.

The Editor will be glad to send the JOURNAL OF BOTANY in exchange for other Journals of a similar character. Such Journals, Books for review, and Communications intended for publication, to be addressed to JAMES BRITTEN, Esq., 18, West Square, Southwark, S.E.

He will be greatly obliged to the Secretaries of Local Natural History Societies if they will forward him copies of their Transactions, so that any paper of botanical interest may be recorded in this Journal.

## TERMS FOR ADVERTISEMENTS.

| | | | | |
|---|---|---|---|---|
| Whole page ........ | £2 0 0 | | Quarter page ...... | 0 15 0 |
| Half page .......... | 1 5 0 | | Six lines and under | 0 5 0 |

Every additional line, 9d.

A liberal reduction is made for a series.

To be sent to WEST, NEWMAN & Co., 54, Hatton Garden, not later than the 24th of each month.

## AUTHOR'S SEPARATE COPIES.

Authors who require copies of their articles are requested to order separate copies from the Publishers, and to notify this and state the number required at head of their MS.; otherwise the type may be distributed before the order is received. By the buying of additional copies of the Journal sets are destroyed. The charges for separate copies are as under :—

| 2 pages | 25 copies | 4s. | 4 pages | 25 copies | 5s. | 8 pages | 25 copies | 8s. 0d. |
|---|---|---|---|---|---|---|---|---|
| „ | 50 „ | 5s. | „ | 50 „ | 6s. | „ | 50 „ | 9s. 0d. |
| „ | 100 „ | 7s. | „ | 100 „ | 8s. | „ | 100 „ | 10s. 6d. |

A greater number of pages to be charged in equal proportion. Separate Titles, Wrappers, &c., extra.

Annual Subscription, paid in advance, Twelve Shillings, post free
within the Postal Union.   Single Numbers, 1s. 3d.

No. 306.                JUNE, 1888.                Vol. XXVI.

# THE

# JOURNAL OF BOTANY

## BRITISH AND FOREIGN.

EDITED BY

## JAMES BRITTEN, F.L.S.,

BRITISH MUSEUM (NATURAL HISTORY), SOUTH KENSINGTON.

## CONTENTS.

                                        PAGE
Asa Gray. (With Portrait).   By
  JAMES BRITTEN, F.L.S.  ..  .. 161
A Synopsis of *Tillandsieæ*. By J. G.
  BAKER, F.R.S., F.L.S. (Con-
  cluded)  ..  ..  ..  ..  .. 167
Notes on the Botany of Northern
  Portugal.   By the Rev. R. P.
  MURRAY, M.A., F.L.S.  ..  .. 173
Biographical Index of British and
  Irish Botanists.   By JAMES
  BRITTEN, F.L.S., and G. S. BOUL-
  GER, F.L.S. (Continued)  .. 180

                                        PAGE
SHORT NOTES. — Suffolk Plants. —
  *Pulmonaria officinalis* L. as a
  native of Britain. — Vitality of
  Spores of *Gymnogramma lepto-
  phylla* ..  ..  ..  ..  ..  .. 184
New Phanerogams published in
  Britain in 1887 (Concluded.).. 186
NOTICES OF BOOKS:—
  New Books ..  ..  ..  ..  .. 190
  Articles in Journals  ..  .. 191
  Linnean Society of London ..  .. 192

## LONDON:

WEST, NEWMAN & CO., 54, HATTON GARDEN, E.C.

DULAU & CO., SOHO SQUARE.

*Price One Shilling and Threepence.*

Annual Subscription, paid in advance, Twelve Shillings, post free
within the Postal Union. Single Numbers, 1s. 3d.

No. 307.    JULY, 1888.    Vol. XXVI.

# THE
# JOURNAL OF BOTANY

## BRITISH AND FOREIGN.

EDITED BY

## JAMES BRITTEN, F.L.S.,

BRITISH MUSEUM (NATURAL HISTORY), SOUTH KENSINGTON.

## CONTENTS.

PAGE

Catalogue of the Marine Algæ of the
West Indian Region. By GEO.
MURRAY, F.L.S. .. .. .. 193

Salix fragilis, S. Russ lliana, and
S. viridis. By F. BUCHANAN
WHITE, M.D., F.L.S. .. .. 196

Root-pressure. By C. B. CLARKE,
M.A., F.R.S. .. .. .. .. 201

Notes on some Hieracia new to
Britain. By FREDERICK J. HAN-
BURY, F.L.S. .. .. .. .. 204

Pimina, Novum Hyphomycetum
Genus. Descripsit W. B. GROVE,
B.A. .. .. .. .. .. .. 206

Centenary of the Linnean Society
of London .. .. .. .. 207

Biographical Index of British and
Irish Botanists. By JAMES

PAGE

BRITTEN, F.L.S., and G. S. BOUL-
GER, F.L.S. (Continued) .. 213

SHORT NOTES.—Experiments with
Gymnosporangium Juniperi. —
A Heterodox Onion.—Vicia hy-
brida L. .. .. .. .. .. 218

NOTICES OF BOOKS:—

The Flora of West Yorkshire, with
a Sketch of the Climatology and
Lithology in connection therewith. By FREDERICK ARNOLD
LEES, M.R.C.S., L.R.C.P. .. 219

A Manual of Orchidaceous Plants.
Part III. JAS. VEITCH & SONS. 222

Short Notice .. .. .. .. 223

Articles in Journals .. .. .. 223

Linnean Society of London .. .. 224

LONDON:

WEST, NEWMAN & CO., 54, HATTON GARDEN, E.C.

DULAU & CO., SOHO SQUARE.

*Price One Shilling and Threepence.*

## AUTHOR'S SEPARATE COPIES.

Authors who require copies of their articles are requested to order separate copies from the Publishers, and to notify this and state the number required at head of their MS.; otherwise the type may be distributed before the order is received. By the buying of additional copies of the Journal sets are destroyed. The charges for separate copies are as under :—

| 2 pages | 25 copies | 4s. | 4 pages | 25 copies | 5s. | 8 pages | 25 copies | 8s. 0d. |
|---|---|---|---|---|---|---|---|---|
| ,, | 50 ,, | 5s. | ,, | 50 ,, | 6s. | ,, | 50 ,, | 9s. 0d. |
| ,, | 100 ,, | 7s. | ,, | 100 ,, | 8s. | ,, | 100 ,, | 10s. 6d. |

A greater number of pages to be charged in equal proportion. Separate Titles, Wra ers. &c.. extra.

# SOWERBY'S ENGLISH BOTANY.

Containing a Description and Life-sized Coloured Drawing of every British Plant. Edited and brought up to the Present Standard of Scientific Knowledge by the late J. T. BOSWELL, LL.D., F.L.S., &c. The Figures by J. E. SOWERBY, J. W. SALTER, A.L.S., and N. E. BROWN. Third Edition, in 12 vols., super-royal 8vo, £24 3s. in cloth, £26 11s. in half-morocco, and £30 9s. whole morocco. Also 89 parts, 5s. each, except the Index Part, 7s. 6d. *₊* The Work has for nearly a century been the chief authority on English Botany, and in its present form may fairly be taken as the most complete representative of the present state of knowledge of British Plants.

## CONTENTS:

**Vol. I.** Ranunculaceæ, Berberidaceæ, Nymphæaceæ, Papaveraceæ, and Cruciferæ. Cloth, £1 18s.; half-morocco, £2 2s.: morocco, £2 8s. 6d

**Vol. II.** Resedaceæ, Cistaceæ, Violaceæ, Droseraceæ, Polygalaceæ, Frankeniaceæ, Caryophyllaceæ, Portulacaceæ, Tamariscaceæ, Elatineæ, Hypericaceæ, Malvaceæ, Tiliaceæ, Linaceæ, Geraniaceæ, Ilicineæ, Celastraceæ, Rhamnaceæ, Sapindaceæ. Cloth, £1 18s.; half-morocco, £2 2s.; morocco, £2 8s. 6d.

**Vol. III.** Leguminiferæ and Rosaceæ. Cloth, £2 3s.; half-morocco, £2 7s.; morocco, £2 13s. 6d.

**Vol. IV.** Lythraceæ, Onagraceæ, Cucurbitaceæ, Grossulariaceæ, Crassulaceæ, Saxifragaceæ, Umbelliferæ, Araliaceæ, Cornaceæ, Loranthaceæ, Caprifoliaceæ, Rubiaceæ, Valerianaceæ, and Dipsaceæ. Cloth, £2 8s.; half-morocco, £2 12s.; morocco, £2 18s. 6d.

**Vol. V.** Compositæ. Cloth, £2 3s.; half-morocco, £2 7s.; morocco, £2 13s. 6d.

**Vol. VI.** Campanulaceæ, Ericaceæ, Jasminaceæ, Apocynaceæ, Gentianaceæ, Polemoniaceæ Convolvulaceæ, Solanaceæ, Scrophulariaceæ, Orobanchaceæ, and Verbenaceæ. Cloth, £1 18s.; half-morocco, £2 2s.; morocco, £2 8s. 6d.

**Vol. VII.** Labiatæ, Boraginaceæ, Lentibulariaceæ, Primulaceæ, Plumbaginaceæ, Plantaginaceæ, Paronychiaceæ, and Amarantaceæ. Cloth, £1 18s.; half-morocco, £2 2s.; morocco, £2 8s. 6d.

**Vol. VIII.** Chenopodiaceæ, Polygonaceæ, Eleagnaceæ, Thymelaceæ, Santalaceæ, Aristolochiaceæ, Empetraceæ, Euphorbiaceæ, Callitrichaceæ, Ceratophyllaceæ, Urticaceæ, Amentiferæ, and Coniferæ. Cloth, £2 13s.; half-morocco, £2 17s.; morocco, £3 3s. 6d.

**Vol. IX.** Typhaceæ, Araceæ, Lemnaceæ, Naiadaceæ, Alismaceæ, Hydrocharidaceæ, Orchidaceæ, Iridaceæ, Amaryllidaceæ, Dioscoreaceæ, and Liliaceæ. Cloth, £1 18s.; half-morocco, £2 2s.; morocco, £2 8s. 6d.

**Vol. X.** Juncaceæ and Cyperaceæ. Cloth, £1 18s.; half-morocco, £2 2s.; morocco, £2 8s. 6d.

**Vol. XI.** Graminaceæ. Cloth, £1 13s.; half-morocco, £1 17s.; morocco, £2 3s. 6d.

**Vol. XII.** Marsiliaceæ, Isoetaceæ, Selaginellaceæ, Lycopodiaceæ, Ophioglossaceæ, Filices, Equisetaceæ, and Characeæ; and an Index to the whole Work. Cloth, £1 15s.; half-morocco, £1 19s.; morocco, £2 5s. 6d.

## By J. G. BAKER, F.R.S., F.L.S.,
First-Assistant in the Herbarium of the Royal Gardens, Kew.

HANDBOOK OF THE AMARYLLIDEÆ, including the ALSTRŒMERIEÆ and AGAVEÆ. Demy 8vo. 5s. *Just Published.*

HANDBOOK OF THE FERN ALLIES. A Synopsis of the Genera and Species of the Natural Orders, Equisetaceæ, Lycopodiaceæ, Selaginellaceæ, Rhizocarpeæ. Demy 8vo. 5s.

"Mr. Baker's intimate acquaintance with the class of plants of which he treats is sufficient guarantee of the excellence of the work. It fills a gap which has long been felt.' —*Academy.*

A FLORA OF THE ENGLISH LAKE DISTRICT. Demy 8vo. 7s. 6d. "A model of what a local flora should be."—*Guardian.*

THE BOTANIST'S POCKET BOOK. By W. R. HAYWARD. Containing, arranged in a tabulated form, the chief characteristics of British Plants. Fcap. 8vo, flexible binding for the pocket, Fifth Edition Revised, with new Appendix, 4s. 6d.

"The diagnoses seem framed with considerable care and judgment, the characteristics having been well selected and contrasted."—*Journal of Botany.*

Annual Subscription, paid in advance, Twelve Shillings, post free within the Postal Union. Single Numbers, 1s. 3d.

No. 308.      AUGUST, 1888.      Vol. XXVI.

# THE
# JOURNAL OF BOTANY
## BRITISH AND FOREIGN.

EDITED BY

## JAMES BRITTEN, F.L.S.,

BRITISH MUSEUM (NATURAL HISTORY), SOUTH KENSINGTON.

## CONTENTS.

PAGE

On two recent Collections of Ferns from Western China. By J. G. BAKER, F.R.S. .. .. .. .. 225

Sowerby's Models of British Fungi. By WORTHINGTON G. SMITH, F.L.S. .. .. .. .. .. .. 231

On *Callitriche polymorpha* Lönnroth as a British Plant. By W. H. BEEBY, A.L.S. .. .. .. .. 233

New Manipur Ferns collected by Dr. Watt. By Col. R. H. BEDDOME, F.L.S. .. .. .. .. 234

Remarks on *Pyrus latifolia* Syme. By T. R. ARCHER BRIGGS, F.L.S. 236

Catalogue of the Marine Algæ of the West Indian Region. By GEO. MURRAY, F.L.S. (Continued).. 237

Biographical Index of British and Irish Botanists. By JAMES BRITTEN, F.L.S., and G. S. BOULGER, F.L.S. (Continued) .. 244

PAGE

SHORT NOTES.—*Cerastium pumilum* in Wilts. — Note on Buckinghamshire Rubi.—Note on *Salix fragilis.* — *Polygala austriaca* Crantz, in Surrey .. .. .. 248

Report of Department of Botany, British Museum, for 1887. By W. CARRUTHERS, F.R.S. .. .. 250

NOTICES OF BOOKS :—

Handbook of the Amaryllideæ, including the Alstrœmerieæ and Agaveæ. By J. G. BAKER, F.R.S. .. .. .. .. .. 253

Nomenclator 'Floræ Danicæ' sive index systematicus et alphabeticus operis, quod 'Icones Floræ Danicæ' inscribitur, cum enumeratione tabularum ordinem temporum habente, adjectis notis criticis. Auctore JOH. LANGE .. .. .. .. 254

New Books .. .. .. .. .. 254

Articles in Journals .. .. . 255

LONDON:

WEST, NEWMAN & CO., 54, HATTON GARDEN, E.C.
DULAU & CO., SOHO SQUARE.

*Price One Shilling and Threepence.*

Annual Subscription, paid in advance, Twelve Shillings, post free
within the Postal Union.   Single Numbers, 1s. 3d.

No. 309.                 SEPTEMBER, 1888.                 Vol. XXVI.

# THE
# JOURNAL OF BOTANY
## BRITISH AND FOREIGN.

EDITED BY

## JAMES  BRITTEN,  F.L.S.,

BRITISH MUSEUM (NATURAL HISTORY), SOUTH KENSINGTON.

## CONTENTS.

PAGE

Recent Tendencies in American
Botanical Nomenclature.   By
the EDITOR.. .. .. .. .. 257

Mosses of Madagascar.   By C. H.
WRIGHT  .. .. .. .. .. 263

Note on Sowerby's Models of British
Fungi.  By WM. CARRUTHERS,
F.R.S. .. .. .. .. .. .. 268

Note on the Botanical Plates of the
Expedition of the 'Astrolabe'
and the 'Zélée.   By B. DAYDON
JACKSON, Sec. L.S.  .. .. .. 269

Notes on Pondweeds.   By ALFRED
FRYER .. .. .. .. .. ..

Biographical Index of British and
Irish Botanists.   By JAMES
BRITTEN, F.L.S., and G. S. BOUL-
GER, F.L.S.  (Continued)  .. 273

PAGE

SHORT NOTES.—Ulota phyllantha in
Fruit from Killarney. — Hants
Plants .. .. .. .. .. .. 282

NOTICES OF BOOKS:—

A Flora of the North-east of Ireland, including the Phanerogamia, the Cryptogamia, Vascularia, and the Muscineæ.  By
SAMUEL ALEXANDER STEWART
and the late THOMAS HUGHES
CORRY   .. .. .. .. .. 283

General Index to the First Twenty
volumes of the Journal (Botany), and the Botanical portion of the Proceedings, Nov.
1838 to June, 1886, of the
Linnean Society.   Edited by
B. DAYDON JACKSON, Sec. L.S. 286

Articles in Journals  ..      . 287

LONDON:

WEST, NEWMAN & CO., 54, HATTON GARDEN, E.C.
DULAU & CO., SOHO SQUARE.

*Price One Shilling and Threepence.*

# NOTICE.

The JOURNAL OF BOTANY is printed and published by WEST, NEWMAN & Co., 54, Hatton Garden, London, E.C., to whom Subscriptions for 1888 (in advance, Twelve Shillings; if not paid in advance, chargeable at the rate of 1s. 3d. per number) should be paid. Post-Office Orders may be drawn on the Hatton Garden Office.

The Volume for 1887 (price 16s. 6d., bound in cloth) is now ready; also covers for the Volume (price 1s. 2d. post free).

The Volumes for 1884, 1885, and 1886 can still be had.

For Volumes and back numbers for 1872—82 application should be made to DULAU & Co., Soho Square, W.

The Editor will be glad to send the JOURNAL OF BOTANY in exchange for other Journals of a similar character. Such Journals, Books for review, and Communications intended for publication, to be addressed to JAMES BRITTEN, Esq., 18, West Square, Southwark, S.E.

He will be greatly obliged to the Secretaries of Local Natural History Societies if they will forward him copies of their Transactions, so that any paper of botanical interest may be recorded in this Journal.

## AUTHOR'S SEPARATE COPIES.

Authors who require copies of their articles are requested to order separate copies from the Publishers, and to notify this and state the number required at head of their MS.; otherwise the type may be distributed before the order is received. By the buying of additional copies of the Journal sets are destroyed. The charges for separate copies are as under :—

| 2 pages | 25 copies | 4s. | 4 pages | 25 copies | 5s. | 8 pages | 25 copies | 8s. 0d. |
|---|---|---|---|---|---|---|---|---|
| ,, | 50 ,, | 5s. | ,, | 50 ,, | 6s. | ,, | 50 ,, | 9s. 0d. |
| ,, | 100 ,, | 7s. | ,, | 100 ,, | 8s. | ,, | 100 ,, | 10s. 6d. |

A greater number of pages to be charged in equal proportion. Separate Titles, Wrappers, &c., extra.

Annual Subscription, paid in advance, Twelve Shillings, post free
within the Postal Union. Single Numbers, 1s. 3d.

No. 310.        OCTOBER, 1888.        .Vol. XXVI.

# THE

# JOURNAL OF BOTANY

## BRITISH AND FOREIGN.

EDITED BY

# JAMES BRITTEN, F.L.S.,

BRITISH MUSEUM (NATURAL HISTORY), SOUTH KENSINGTON.

## CONTENTS.

PAGE

Botanical Nomenclature. By ALPH.
DECANDOLLE, N. L. BRITTON,
and JAS. BRITTEN .. .. .. 289

Notes on Pondweeds. By ALFRED
FRYER .. .. .. .. .. .. 297

John Goldie .. .. .. .. .. 299

*Ranzania:* a new Genus of Berberidaceæ. By TOKUTARO ITO,
F.L.S. .. .. .. .. .. .. 302

Catalogue of the Marine Algæ of the
West Indian Region. By GEO.
MURRAY, F.L.S. (Continued).. 303

Biographical Index of British and
Irish Botanists. By JAMES
BRITTEN, F.L.S., and G. S. BOUL-
GER, F.L.S. (Continued) .. 307.

PAGE

SHORT NOTES.—*Alchemilla vulgaris*
L. in Kent. — *Polygonum maritimum* still in S. Hants. — East
Kent Plants —*Elymus arenarius*
L. in Dorset.—*Rumex maritimus*
and *R. palustris* in East Sussex.
*Hieracium tridentatum* in Worcestershire.—*Helianthemum polifolium* Pers. in N. Somerset .. 311

NOTICES OF BOOKS :—
The Origin of Floral Structures
through Insect and other Agencies. By the Rev. GEORGE
HENSLOW .. .. .. .. .. 313

Index Generum Phanerogamarum. By TH. DURAND .. .. 316

Articles in Journals .. .. .. 319

LONDON:

WEST, NEWMAN & CO., 54, HATTON GARDEN, E.C.

DULAU & CO., SOHO SQUARE.

*Price One Shilling and Threepence.*

# NOTICE.

The JOURNAL OF BOTANY is printed and published by WEST, NEWMAN & Co., 54, Hatton Garden, London, E.C., to whom Subscriptions for 1888 (in advance, Twelve Shillings; if not paid in advance, chargeable at the rate of 1s. 3d. per number) should be paid. Post-Office Orders may be drawn on the Hatton Garden Office.

The Volume for 1887 (price 16s. 6d., bound in cloth) is now ready; also covers for the Volume (price 1s. 2d. post free).

The Volumes for 1884, 1885, and 1886 can still be had.

For Volumes and back numbers for 1872—82 application should be made to DULAU & Co., Soho Square, W.

The Editor will be glad to send the JOURNAL OF BOTANY in exchange for other Journals of a similar character. Such Journals, Books for review, and Communications intended for publication, to be addressed to JAMES BRITTEN, Esq., 18, West Square, Southwark, S.E.

He will be greatly obliged to the Secretaries of Local Natural History Societies if they will forward him copies of their Transactions, so that any paper of botanical interest may be recorded in this Journal.

## TERMS FOR ADVERTISEMENTS.

| | | |
|---|---|---|
| Whole page ........ £2 0 0 | Quarter page ...... 0 15 0 |
| Half page .......... 1 5 0 | Six lines and under 0 5 0 |

Every additional line, 9d.

A liberal reduction is made for a series.

To be sent to WEST, NEWMAN & Co., 54, Hatton Garden, not later than the 24th of each month.

## AUTHOR'S SEPARATE COPIES.

Authors who require copies of their articles are requested to order separate copies from the Publishers, and to notify this and state the number required at head of their MS.; otherwise the type may be distributed before the order is received. By the buying of additional copies of the Journal sets are destroyed. The charges for separate copies are as under:—

| 2 pages | 25 copies | 4s. | 4 pages | 25 copies | 5s. | 8 pages | 25 copies | 8s. 0d. |
|---|---|---|---|---|---|---|---|---|
| ., | 50 ,, | 5s. | ,, | 50 ,, | 6s. | ,, | 50 ,, | 9s. 0d. |
| ,, | 100 ,, | 7s. | ,, | 100 ,, | 8s. | ,, | 100 ,, | 10s. 6d. |

A greater number of pages to be charged in equal proportion. Separate Titles, Wrappers, &c., extra.

Annual Subscription, paid in advance, Twelve Shillings, post free within the Postal Union.  Single Numbers, 1s. 3d.

No. 311.  NOVEMBER, 1888.  Vol. XXVI.

# THE
# JOURNAL OF BOTANY

## BRITISH AND FOREIGN.

EDITED BY

## JAMES BRITTEN, F.L.S.,

BRITISH MUSEUM (NATURAL HISTORY), SOUTH KENSINGTON.

## CONTENTS.

PAGE

*Carex* Notes from the British Museum.  By L. H. BAILEY .. .. 321

On a Third Collection of Ferns made in West Borneo by the Bishop of Singapore and Sarawak.  By J. G. BAKER, F.R.S. .. 323

Botanical Nomenclature in North America.  By EDWD. L. GREENE 326

South Derbyshire Plants.  By the Rev. W. R. LINTON, M.A. .. .. 329

Catalogue of the Marine Algæ of the West Indian Region.  By GEO. MURRAY, F.L.S.  (Continued.) (Plate 284) .. .. .. .. .. 331

The Desmids of Maine.  By WM. WEST, F.L.S. .. .. .. .. 339

PAGE

On the Two Valerians.  By W. H. BEEBY, A.L.S. .. .. .. .. 340

Biographical Index of British and Irish Botanists.  By JAMES BRITTEN, F.L.S., and G. S. BOULGER, F.L.S.  (Continued) .. 345

SHORT NOTES. — *Arum italicum* (Mill.). — East Kent Plants .. 348

NOTICES OF BOOKS :—

The British Moss-Flora.  By R. BRAITHWAITE, M.D., F.L.S. .. 349

British Mosses: their Homes, Aspects, Structure, and Uses. By F. E. TRIPP .. .. .. 350

Articles in Journals .. .. .. 352

LONDON:

WEST, NEWMAN & CO., 54, HATTON GARDEN, E.C.

DULAU & CO., SOHO SQUARE.

*Price One Shilling and Threepence.*

# NOTICE.

The JOURNAL OF BOTANY is printed and published by WEST, NEWMAN & Co., 54, Hatton Garden, London, E.C., to whom Subscriptions for 1888 (in advance, Twelve Shillings; if not paid in advance, chargeable at the rate of 1s. 3d. per number) should be paid. Post-Office Orders may be drawn on the Hatton Garden Office.

The Volume for 1887 (price 16s. 6d., bound in cloth) is now ready; also covers for the Volume (price 1s. 2d. post free).

The Volumes for 1884, 1885, and 1886 can still be had.

For Volumes and back numbers for 1872—82 application should be made to DULAU & Co., Soho Square, W.

The Editor will be glad to send the JOURNAL OF BOTANY in exchange for other Journals of a similar character. Such Journals, Books for review, and Communications intended for publication, to be addressed to JAMES BRITTEN, Esq., 18, West Square, Southwark, S.E.

He will be greatly obliged to the Secretaries of Local Natural History Societies if they will forward him copies of their Transactions, so that any paper of botanical interest may be recorded in this Journal.

## TERMS FOR ADVERTISEMENTS.

| | | | |
|---|---|---|---|
| Whole page ........ £2 0 0 | | Quarter page ...... | 0 15 0 |
| Half page .......... 1 5 0 | | Six lines and under | 0 5 0 |

Every additional line, 9d.

A liberal reduction is made for a series.

To be sent to WEST, NEWMAN & Co., 54, Hatton Garden, not later than the 24th of each month.

## AUTHOR'S SEPARATE COPIES.

Authors who require copies of their articles are requested to order separate copies from the Publishers, and to notify this and state the number required at head of their MS.; otherwise the type may be distributed before the order is received. By the buying of additional copies of the Journal sets are destroyed. The charges for separate copies are as under :—

| 2 pages | 25 copies | 4s. | 4 pages | 25 copies | 5s. | 8 pages | 25 copies | 8s. 0d. |
|---|---|---|---|---|---|---|---|---|
| ,, | 50 ,, | 5s. | ,, | 50 ,, | 6s. | ,, | 50 ,, | 9s. 0d. |
| ,, | 100 ,, | 7s. | ,, | 100 ,, | 8s. | ,, | 100 ,, | 10s. 6d. |

A greater number of pages to be charged in equal proportion. Separate Titles, Wrappers, &c., extra.

SUBSCRIPTIONS HAVE EXPIRED. Those for 1889, 12s., may
be sent to Messrs. West, Newman & Co. Single Numbers, 1s. 3d.

No. 312. DECEMBER, 1888. Vol. XXVI.

# THE
# JOURNAL OF BOTANY

## BRITISH AND FOREIGN.

EDITED BY

## JAMES BRITTEN, F.L.S.,

BRITISH MUSEUM (NATURAL HISTORY), SOUTH KENSINGTON.

## CONTENTS.

PAGE

Photolysis in *Lemna trisulca*. By
SPENCER LE M. MOORE, F.L.S.
(Plate 285) .. .. .. .. .. 353
Catalogue of the Marine Algæ of the
West Indian Region. By GEO.
MURRAY, F.L.S. (Continued.)
(Plate 284) .. .. .. .. .. 358
Notes on the Flora of Ben Laiogh,
&c. By G. CLARIDGE DRUCE .. 364
On Botanical Nomenclature. By
Prof. C. C. BABINGTON, F.R.S... 369
On a new *Acrostichum* from Trini-
dad. By J. G. BAKER, F.R.S... 371
Biographical Index of British and
Irish Botanists. By JAMES
BRITTEN, F.L.S., and G. S. BOUL-
GER, F.L.S. (Continued) .. 372
SHORT NOTES. — New County Re-
cords. — Additions to the Flora

PAGE

of Wilts. — Botanical Nomen-
clature. — *Arum italicum* Mill.
and *A. maculatum* Linn. — The
Two Valeriaus.— *Valeriana Mi-
kanii.*— *Rubus thyrsiger* Bab.—
*Goodyera repens* in Yorkshire.. 376
NOTICES OF BOOKS :—
The Flowering Plants of Wilts,
with sketches of the Physical
Geography and Climate of the
County. By the Rev. T. A.
PRESTON, M.A. .. .. .. .. 380
The Species of *Ficus* of the Indo-
Malayan & Chinese Countries.
By GEORGE KING, M.B., F.R.S. 382
New Books .. .. .. .. .. 382
Articles in Journals .. . . 383
Linnean Society of London .. .. 384

LONDON:

WEST, NEWMAN & CO., 54, HATTON GARDEN, E.C.
DULAU & CO., SOHO SQUARE.

*Price One Shilling and Threepence.*

Lightning Source UK Ltd.
Milton Keynes UK
UKHW020719201118
332647UK00010B/784/P